P9-BTN-822

The
URBANIZATION
of the
SUBURBS

Volume 7, URBAN AFFAIRS ANNUAL REVIEWS

INTERNATIONAL EDITORIAL ADVISORY BOARD

HOWARD S. BECKER
Northwestern University

ASA BRIGGS
University of Sussex

JOHN W. DYCKMAN
University of California (Berkeley)

H. J. DYOS
University of Leicester

MARILYN GITTELL
City University of New York

JEAN GOTTMANN
Oxford University

SCOTT GREER
Northwestern University

ROBERT J. HAVIGHURST
University of Chicago

WERNER Z. HIRSCH
University of California (Los Angeles)

EIICHI ISOMURA
Tokyo Metropolitan University

WILLIAM C. LORING
U.S. Public Health Service

MARTIN MEYERSON
University of Pennsylvania

EDUARDO NEIRA
Interamerican Development Bank

HARVEY S. PERLOFF
University of California (Los Angeles)

P. J. O. SELF
The London School of Political Science and Economics

KENZO TANGE
Tokyo National University

The URBANIZATION of the SUBURBS

Edited by
LOUIS H. MASOTTI
and
JEFFREY K. HADDEN

Volume 7, URBAN AFFAIRS ANNUAL REVIEWS

LIBRARY, ST. LAWRENCE UNIVERSITY
CANTON, NEW YORK 13617

 SAGE PUBLICATIONS / BEVERLY HILLS / LONDON

HT
108
U7
v.7

Copyright © 1973 by Sage Publications, Inc.

All rights reserved. No part of this book may be reproduced or utilized in any form or by any means, electronic or mechanical, including photocopying, record- ing, or by any information storage and retrieval system, without permission in writing from the publisher.

For information address:

SAGE PUBLICATIONS, INC.
275 South Beverly Drive
Beverly Hills, California 90212

SAGE PUBLICATIONS LTD
St George's House / 44 Hatton Garden
London EC1N 8ER

Printed in the United States of America

International Standard Book Number 0-8039-0198-4

Library of Congress Catalog Card No. 72-98038

FIRST PRINTING

75-3591

JUL 1 9 78

In memory of

HARLAN PAUL DOUGLASS

Pioneer Suburbanologist

PREFACE

Suburbia, the subject of considerable ridicule and defamation in the fifties and largely ignored in the sixties, is rapidly being revived as a central focus of consideration and concern. The marked decrease in the number and intensity of urban disorders at the end of the sixties and the startling evidence of suburban development contained in the 1970 Census has combined to redirect the attention of the media, public officials and urban scholars to the rim of the city where change is unceasing and increasingly dramatic. The growth and development of the "urban fringe" in recent years has been too impressive and the implications too significant to ignore.

This is not to say that the serious problems of the central cities have been solved or forgotten; they have not. It is only to suggest, as noted suburbanologist Robert C. Wood has said recently, that "the battle of the cities will be fought on the suburban front." The consequences of what is now occurring outside the central cities will influence not only the future of suburbia but may well determine the viability and perhaps even the survival of the central cities as we know them. The suburbs are no longer merely "a footnote to urban civilization" (Douglass, 1925). The nature and rapidity of suburbanization is in the process of transforming the "outer city" into a new form of urban civilization.

We felt that this was an urban development sufficient to warrant scholarly exploration in the Urban Affairs Annual Review series. The purpose in this seventh volume in that series was to stimulate a group of prominent urban scholars to direct their attention to various aspects of contemporary suburbanization with a serious hope on the part of the editors that such an exercise would begin to review what we know, conceptualize the problems and define the issues that need to be researched and analyzed.

At the time the project was conceived, there had been very few new published thoughts on the subject of suburbia and suburbanization[1] since the age of the "suburban myth." In the interim, our initial conviction of the topic's significance has been confirmed by a sudden and welcome spate of new books on the topic. The first new reader on suburbs since Dobriner's 1958 classic was assembled by John Kramer (*North American Suburbs,* 1972). This was followed by another, edited by Charles Haar (*The End of Innocence,* 1972), who had served as chairman of President Johnson's Task Force on Suburban Problems.[2] A third reader, *Suburbia in Transition,* edited by Masotti and Hadden, is in press. In 1972, Frederick M. Wirt and his colleagues published a volume entitled *On the City's Rim* which focuses, albeit not exclusively, on the electoral and national policy implications of suburbanization. Leo Schnore issued a small data-filled book called *Class and Race in Cities and Suburbs* (1972); and David Birch produced a monograph for the Committee for Economic Development, *The Economic Future of City and Suburb* (1970). Some other recent publications of note are Dennis P. Sobin, *The Future of the American Suburbs* (1971), Bryan T. Downes (ed.) *Cities and Suburbs* (1971), and two case studies by George Sternlieb and his associates: *The Affluent Suburb* (1971) and *The Zone of Emergence* (1972).[3] In addition, at least two national newsweekly magazines (*Time* and *Newsweek*) have done recent cover stories on the suburban phenomenon and *City* magazine has devoted an entire issue to it. In 1971, the New York *Times* did two extensive series on the topic. Other metropolitan newspapers have begun in-depth coverage of suburban development; some, e.g., the Chicago *Tribune,* have suburban inserts for suburban editions.

Obviously, all this new attention by journalists and scholars to the subject of suburbs is welcome. Most of the published work has been or will prove to be valuable as the nation begins to explore seriously the challenge which the "new suburbanization" presents. It is because the challenge is so great, the topic so vast and the "handles" on it so few that we offer still another title to the growing suburban bibliography.

The editors had considerable assistance in preparing this volume in addition to the work represented in the book by the contributing authors. I would like in particular to express appreciation to Deborah E. Dennis, Gaye K. Haverkos, Mary Anne Richardson, and Deborah E. Holmes of the staff at the Center for Urban Affairs, Northwestern University. I am extremely grateful for the many stimulating conversations about suburbia held with John L. McKnight, Calvin Bradford, Darel Grothaus, and Leonard Rubinowitz over the past eighteen months, which resulted in numerous new insights now incorporated into the Prologue and Epilogue, even though they are not always specifically noted. They are, of course, absolved of any and all injustices I may have done to their ideas. Finally, we would like to thank Connie Greaser of Sage Publications, Inc., for her perseverance and good humor in the face of obstinacy and procrastination.

−LHM

Evanston, Illinois
February 1973

NOTES

1. Some obvious exceptions are Gans (1968, 1959), Whyte (1968), Clark (1966), and Donaldson (1969).
2. The six-volume study by that Task Force has never been released, although Haar's book contains several selections from the report. See also Part V, "National Policy Problems," in Wirt et al. (1972); it was written by Francine Rabinovitz, who worked with Haar and had access to the report.
3. The lengthy bibliography at the end of this volume contains the most relevant items from a much more extensive bibliography compiled by the editors in 1971 (*Suburbs, Suburbia and Suburbanization*, Northwestern University Center for Urban Affairs). That bibliography is presently being revised and updated.

REFERENCES

CLARK, S. D. (1966) The Suburban Society. Toronto: Univ. of Toronto Press.
DONALDSON, S. (1969) The Suburban Myth. New York: Columbia Univ. Press.
DOUGLASS, H. P. (1925) The Suburban Trend. New York and London: Century.
GANS, H. J. (1968) People and Plans. New York: Basic Books.
――― (1959) The Levittowners. New York: Pantheon.
WHYTE, W. H., Jr. (1968) The Last Landscape. Garden City, N.Y.: Doubleday.
WIRT, F. M., B. WALTER, F. F. RABINOVITZ, and D. R. HENSLER (1972) On the City's Rim: Politics and Policy in Suburbia. Lexington, Mass.: D. C. Heath.

CONTENTS

Prologue: Suburbia Reconsidered
—Myth and Counter-Myth

LOUIS H. MASOTTI

☐ SUBURBIA IS UNDERGOING a significant transition from its traditional role as dependent "urban fringe" to independent "neo-city." It is a process which began after World War II, when a major housing shortage, aided by an increasing, automobilized population and FHA mortgages, resulted in the "suburban boom" of the early fifties. This flight from the cities created a suburban residential existence for much of middle-class America and stimulated innumerable shallow novels, normative journalistic criticism, and "pop" sociology, which distorted much of what was happening, and why. The suburban myth created by this literature may have warped our perception of the suburban phenomenon, but it did little to slow the exodus from the central city in the ensuing years. The "suburban trend" which Harlan Paul Douglass identified in 1925 has continued unabated, but until recently it was relatively subtle and took place in the shadow of the much more visible and dramatic urban crisis. The significance of suburbia was not put in proper perspective until the fires of the urban riots began to fade in the late sixties, and 1970 census figures confirmed what many had suspected about the outward movement of urban populations—that America had become suburbanized on such a scale and in such a comprehensive way as to threaten the viability of the historic city. An explanation of the dynamics of that process and its consequences is what this book is about.

Myths die hard. Despite the numerous retrospective critiques of the mythological literature of the fifties, the conventional wisdom still holds tenaciously to an image of suburbia which is either incorrect or factually outdated and, in either case, dangerously misleading for the present situation. To be sure, suburbia still houses organizational men in "split-level traps" with "cracks in the picture windows" and crabgrass in the lawn. But the pervasive consequences that the mythologists assumed were *caused* by the suburbanization of the middle class—homogenization, standardization, conformity, isolation, superficiality, other-directedness, sexual promiscuity, and so on—were either not demonstrated or based on highly selective case studies.

Even granting the accuracy of the charges at the time they were made, much has happened since to suggest strongly that the suburbia of today is a significantly different phenomenon, which requires new research and new analysis unencumbered by the assumptions of the past or even the conventional wisdom of today. The suburbia of the seventies is a new ball game.

TOWARD AN URBAN SUBURBIA

At least since suburbia lost its distinction as the encampment of the rich (or, as Lewis Mumford put it, "green ghettos dedicated to the affluent"), suburbia has been assumed to consist of politically independent residential communities outside the corporate limits of a large central city, but within reasonable commuting distance of it, and culturally and economically dependent upon it.

Despite considerable pressure to metropolitanize or annex suburban governments, they have effectively resisted, and remain politically distinct entities which make every effort to preserve the power to determine their distinctive "character." But commuting patterns have changed to reflect the marked decentralization of manufacturing, service industry, commercial activities and retail trade and the jobs that go with them to the suburbs. While many of the older, established, and affluent suburbs are able to maintain their "residential only" character by paying the increased residential property tax bill, some of the older, and all of the new "frontier" suburbs have tried to provide for industrial parks, office complexes, major retail (shopping) centers, or some combination of the three. Thus, fewer suburbanites enter the central city to work, although intersuburban commuting is extensive, and more urbanites, especially skilled and semi-skilled workers, travel beyond the city line to earn a living.

The most significant change in the definition of contemporary suburbia may be its growing economic independence of the central city. It is becoming increasingly less *sub*urban and more urban. More and more of what have historically been central-city functions are now being shared by suburbia. The net effect is to urbanize suburbia and undermine the viability of the central city.

Note that we speak of an urbanized *suburbia* and not urbanized *suburbs*. The distinction is an important one. The individual suburb is a functional part of suburbia, or perhaps of a suburban subregion in much the same way that an urban neighborhood is a functional part of the city. The major difference is that cities have a single governmental system with the real potential for authoritative control and regulation of the functions within its jurisdiction. Suburbia lacks a comparable control system and operates, willy-nilly, on the basis of intersuburban competition for scarce resources like industrial, commercial, retail, and residential developments with low public service costs and high tax production.

Individual suburbs may not be urbanized in the same sense that we are using the term here—i.e., they may not have a full range of urban functions and a diverse population. But suburbia, we submit, is rapidly becoming a highly competitive and increasingly attractive, although more loosely organized, alternative to the historic city for economic activities and population groups. In major metropolitan areas, there is now scarcely an urban function, activity, or group which cannot also be found in suburbia.

In some cases, the sharing of economic functions between center city and suburb is facilitated by a process of branching. This is particularly true of major retail stores (department stores, super-markets, and so on) who hedge their sales bets by locating an outlet in every regional shopping center; other less likely activities, such as night clubs, quality restaurants and specialty shops, are following suit. State law in some cases prohibits this practice for selected activities—e.g., banking. This *physical decentralization or relocation of economic activities* throughout the metropolitan area is one of the major dimensions of the process which we are calling the urbanization of the suburbs.

Another dimension of this process, which is in symbiotic relation to the first, is *migration of a larger and more heterogeneous population* to suburbia. Although there is a minor problem with the Census Bureau's definition of what is suburban, there is no question about the rapidly increasing size of the suburban population, both absolutely and relative to the urban population. During the 1960-1970 decade, the suburban population increase was 28% (16 million) while the central cities grew only 6% (3.1 million). Well over half (57%) of the total metropolitan population of the United States is now suburban, and almost 38% of the nation's population are suburbanites, compared to 31% each for city and farm areas. Even allowing for some suburban overestimation due to Census Bureau definitional ambiguities, the figures are impressive. Like it or not, suburbia *is* indeed becoming the "new America."

Before World War II, suburbia was the haven of the rich plus a relatively small number of blue-collar workers and white-collar service employees. The postwar boom enticed large numbers of middle-class families. Subsequent housing construction, government subsidies, and employment opportunities outside the city, and increasing discontent with the quality of life inside, have now made suburbia a desirable residential option for a wide spectrum of the urban population. Perhaps the only segment of the urban population is the underclass of marginal men who continue to rely for subsistance and protection on the opportunities provided by the urban density and the physical deterioration of the city.

Most suburbanites are not fully aware of the increasing heterogenity of the outer city because individual suburbs tend toward homogeneity in much the same way that urban neighborhoods do. The probability of an urban resident encountering a person significantly different from himself is still quite high because urban neighborhoods are more likely to be juxtaposed than are suburban communities, which tend to phase more gradually from one type to another.

Nonetheless, population diversity does exist in the suburbs along all dimensions—class, race, ethnicity, religion, age, marital status, and life-cycle stage. The first four characteristics have been represented in the suburban population for some time, although some groups, notably the lower class, Blacks, Latins, and Jews, have experienced rather severe discrimination until relatively recently. Civil rights legislation has forced a strategic shift in exclusionary practices to economic discrimination with stringent ("snob") zoning and building codes as the main tactical weapon. The last three population groups represent a relatively new phenomenon for suburbia; they are related to a third dimension of the suburban urbanization process—*the adoption of urban life styles.*

If the causal relationship between a suburbanizing population and decentralizing economy is unclear, the confusion is only compounded by introducing the concept of life styles into the discussion. For some time now suburbia has been synonymous with familism, children, and single-family, owner-occupied homes. While all the data are not in, there is at least enough evidence to suggest that suburban life styles are becoming as varied as the changing population.

New suburban housing continues to provide single-family homes for those who want and can afford them, and, of course, they are available in the older communities. But increased land and construction costs, plus the demand for apartments (both condominium and rental), have forced much of the new construction into multiunit dwellings. For the first time in 1972, there were as many multiunit as there were private home starts. The availability of apartments (including town houses, condominiums, and "amenities" complexes) attracts childless couples, singles, and the elderly, who historically have found it difficult to find suitable housing outside the city. It also makes it unnecessary for "empty-nesters" (parents whose children have left the home) to return to the city. Tax-conscious suburban governments are pleased to have these childless households, because children require expensive public services like schools and recreational facilities.

The suburban apartment boom seems to be a logical concomitant of urban functional decentralization. The increasing availability of multifamily units provides a fuller range of residential options for both urbanites and suburbanites as their employment opportunities and family status change.

As the proportion of childless residents without house maintenance responsibilities becomes significant, other indicators of an increasing suburban life-style mix emerge—singles bars, nightclubs, more and better restaurants, first-run movies, legitimate theatres, and adult recreational facilities. Suburbia is clearly no longer just a "family place." Regardless of marital status, age, or stage in life cycle, there is, or soon will be, a residential option with a reasonably good sociocultural support structure. Another indicator of "urbs in suburbs."

The decentralization of the urban economy and its most productive citizens has obvious and serious consequences for the historic
central city. The tax base is depleted and the class and race
disparities between city and suburb, between the haves and have-
nots, are exacerbated. Suburban progress is largely at the expense of
the cities, which are increasingly becoming "reservations" for those
who would like to get out if they could, but cannot.

But the urbanization of the fringe has costs as well as benefits
outside the city. These costs constitute the fourth dimension of an
intensified suburbanization process. The "outer city" has been
considered by many as a refuge from the problems and tribulations
of the central city. There one could escape the negative aspects of
the urban condition—crime, grime, congestion, poverty, high taxes
and bad services, inadequate schools, pollution, deterioration, and so
on. But migrants to suburbia have begun to realize that crossing the
city line does not guarantee immunity from the problems they
thought they had left behind; and long-time residents are beginning
to associate the emergent problems of suburbia with new migrants,
especially the lower class and minority groups.

Despite a vast difference in scale and some important differences
in kind, the problems of city and suburb tend toward isomorphism.
Suburban crime increases at an accelerating rate; taxes are soaring,
but services in some suburbs, particularly the older ones, are
deteriorating, as is the physical plant; the poverty level in suburbia is
two-thirds that of the central cities; pollution and drug trafficking
refuse to stop at the city line; and automobile congestion is growing
in the absence of mass transit. And at least two of the physical
symbols of the city—high-rise buildings and apartments—are invading
the urban fringe. All things considered, the situation is much better
in suburbia than in the city, and in some suburbs it is vastly better
than others. But leaving the city is clearly less a solution to the urban
condition than it is a temporary respite; the handwriting is on the
suburban wall.

TAKING SUBURBIA INTO ACCOUNT: RETHINKING
URBAN RESEARCH PRIORITIES

As recently as 1967, students of suburbia were able to argue, with
some justification, that the subject was not very fashionable (Wrong,
1967: 358). Indeed, recent serious research on the subject has been

rare. In the fifties, when mass suburbanization caught America's attention, the analysis tended to be superficial and normative. Suburbia was viewed as a way of life which, as Max Lerner (1957: 177) put it, "defied all the traditional claptrap about American individualism." Suburbia was blamed for creating an undesirable life style when it seems it was really only the "new home for old values" held by a resettled middle class (Ktsanes and Reissman, 1959-1960). Much of the research and analysis of the sixties, especially the work of Herbert Gans (1968, 1959) and Bennett Berger (1961, 1960) can be interpreted as counter-myth; it served to untangle the web of mythology spun by the social critics of the postwar period. Despite an impressive suburban bibliography, we are not much further along toward an understanding of the suburban phenomenon than we were when the boom began more than a quarter of a century ago.

The relative paucity of suburban research may be attributable to the subject's ambiguous and amorphous nature and to an understandable commitment to the cities as "civilization's Great Adventure" (Douglass, 1925: 304). But there seems to be some agreement now that suburbia has undergone a significant transformation and has become the "new city"—the cutting edge of an emerging urban form. As the New York *Times* concluded recently, "The suburbs are the city now."

It is time for suburban research to become fashionable again. The subject begs for serious scholarly attention, and it is beginning to get it. Much research must be done before suburbia is adequately understood and its consequences for the urban condition properly evaluated. In the meantime, there is an urgent need for an urban research agenda which takes suburbia into account as an influence in state and national affairs, as an integral part of the metropolitan community, as a significant factor in the changing nature of the central city and as a dynamic new development in and of itself.

It was intended that the essays solicited for this volume would begin to give shape to that agenda by providing new data, insights, and analyses and by posing significant questions for future suburban research. The authors represent a variety of perspectives on the subject—sociological, demographic, political, economic, legal, historical, and practical (Babcock is a zoning lawyer; Rubinowitz was a HUD official until recently; Weissbourd is a developer).

We have organized their contributions into six sections. The first offers three historical perspectives on the topic. Part II deals with social structure, life style, ethnicity, familty, kinship, and race, all major sociological dimensions of suburbanization. In the third section, four political scientists discuss governmental structures, models of metropolitan politics, intersuburban relationships, and suburban influence in state and national politics. Part IV assesses four significant "points of contention" in suburbia—exclusionary zoning, low- and moderate-income housing, education, and police services—and provides a framework for analyzing them. The economics of suburbia and the metropolitan area are the focus of Part V. The last section discusses the suburban future. It consists of a major proposal for planned suburban decentralization and a closing commentary on the factors which will play a major role in determining the alternative futures of suburban America.

REFERENCES

BERGER, B. (1961) "The myth of suburbia." J. of Social Issues 17 (November): 38-49.
——— (1960) Working-Class Suburb: A Study of Auto Workers in Suburbia. Berkeley: Univ. of California Press.
DOUGLASS, H. P. (1925) The Suburban Trend. New York and London: Century.
GANS, H. (1968) People and Plans. New York: Basic Books.
——— (1959) The Levittowners. New York: Pantheon.
KTSANES, T. and L. REISSMAN (1959-1960) "Suburbia—new homes for old values." Social Problems 7 (Winter): 187-195.
LERNER, M. (1957) "The suburban revolution," pp. 172-182 in America as a Civilization. New York: Simon & Schuster.
WRONG, D. H. (1967) "Suburbs and the myth of suburbia," pp. 358-364 in D. H. Wrong and H. L. Gracey (eds.) Readings in Introductory Sociology. New York: Macmillan.

HISTORICAL PERSPECTIVES

ON SUBURBANIZATION

HISTORICAL PERSPECTIVES

ON SUBURBANIZATION

Before launching into an assessment of contemporary suburban-ization, we thought it appropriate to glance back at the historical development of suburbia. Thus, this first section is devoted to three perspectives on the past. The first is social history, the second demographic development, and the last ideological.

Singleton reviews three trends in American social history—peripheral settlements, residential segregation patterns, and middle-class consciousness—which make the modern suburb understandable not as a "social mutation" but as a result of historical processes.

Glenn, in reviewing and interpreting the demographic history of the urban fringe since 1945, cautions about overinterpreting the data. Much suburbanization is simply a consequence of the "natural city" outgrowing its political boundaries. The familistic life style identified with suburbia is more a function of selective migration and not a causal effect of suburban residence; suburban familism will become less obvious as suburbs attract a more diversified population at various stages in their life-family cycle. His review of the socioeconomic differences between cities and suburbia suggests a finding which will surprise many people: the SES polarization was substantially less pronounced in 1960-1970 than it had been in 1950-1960 and it may not continue.

The third essay, by Hadden and Barton, reviews the history of man's images of cities and suburbs and argues that the negative image of the city, the result of an anti-urban ideology, reinforces the rejection of the "unnatural" city in favor of the suburbs where "pastoral rapport" between man and nature can be recreated.

1

The Genesis of Suburbia:
A Complex of Historical Trends

GREGORY H. SINGLETON

☐ IN THE BEGINNING, GOD CREATED AN AMERICA with
wilderness, farms, plantations, villages, and a few seaport towns.
Villages grew into towns, and some of the towns became cities. Then
man created suburbs.

Most urbanists would dismiss the theological elements of this
interpretation, but many have articulated the teleological implications. The development of cities in American history is regarded as
natural—perhaps inevitable—and the process has been subjected to
close scrutiny, with research often extending back into the seventeenth century. The suburb, by contrast, is seen as a social mutation:
a barking cat or a foaling cow. There appears to be nothing in the
genesis of our society, nor in the eighteenth- and nineteenth-century
developments, to suggest the emergence of the peripheral community
in metropolitan America. It seems to have appeared overnight as a
parasitic growth on the urban body politic. The metaphors are
mixed, but they may be taken as adequate expression of the
scholarly confusion concerning the origin and development of the
suburb. When its history is written, our research often extends back
no further than World War II and is almost always confined to the
twentieth century. At the very earliest, urbanists cite 1880 as the

beginning of suburbanization. In that year, the U.S. Bureau of the Census first used the word "suburb," in connection with the term "metropolitan." The sibling concepts then applied only to New York City and were not used as general statistical categories until 1910. Census data indicate that the most rapidly growing areas until 1920 were central cities, and since that time the suburbs have been in the ascendancy up to the present decade, when they became the modal residence for Americans. The concentration on the recent past of the American suburb, therefore, seems justified.[1]

It seems justified, that is, with important exceptions. The works of Leo Schnore (1959), Sam Bass Warner, Jr. (1962), and Kenneth T. Jackson (1970) have already raised serious questions about the viability of suburban theories which do not take the realities of the spatial dimensions of urban residential patterns during the nineteenth century into account. In addition to the literature on deconcentration, the nature of community interaction throughout American history tends to diminish the justification for treating suburbanization as a totally recent development. To what extent, for example, were peripheral settlements such as Roxbury and Cambridge "suburbs" of Boston in the seventeenth century? We do know that the pastor of the Roxbury congregation, John Eliot, and many of his parishioners made frequent trips to Boston for public and private business, and the proximity of Boston and Cambridge facilitated a very high rate of pedestrian traffic between the two towns (see Zuckerman, 1970: 10-45; Warden, 1970: 15-33; Winslow, 1968: throughout). A survey of the residents of Llewelyn Park, New Jersey, in 1860 and Lake Forest, Illinois, in 1875 indicates that a majority of the citizens of these two residential communities held business addresses in New York City and Chicago.[2] Aside from settlements which were to some extent deliberately founded as satellite communities well before the turn of the century, to what extent did comparatively independent villages and towns, such as Evanston, Illinois, and Pasadena, California, provide some of the functions of suburbs prior to the metropolitanization of their regions? This is an area that is sorely underresearched, but recent investigations indicate that no matter how strong the desire for autonomy and independence, small towns founded near growing cities in the nineteenth century almost immediately provided minimal suburban functions.[3]

Clearly, these exceptions are only isolated examples, but they represent many more cases which could be built into a history of peripheral settlements from 1631 to the present. To do this would not be very useful, and it would invite endless semantic gymnastics over the definition of "suburb." Obviously, most commentators on this form of communal life do not define suburb simply as a settlement on the periphery of a city, usually in a supportive relationship to the central city and often providing residential facilities for a portion of the urban economic community. It has been suggested by a number of scholars that, by this definition, the suburb, far from being a phenomenon of the twentieth century, is at least as old as the Roman Empire (e.g., Mumford, 1961; Carcopino, 1940: 6 ff.; Arnold, 1971: xi). The silent modifier in suburban scholarship is "modern." The *modern* suburb is said to be a recent development. Ignoring the very serious question of just how useful that information is, it must be admitted that the kinds of pre-twentieth-century peripheral settlements mentioned above bear relatively little resemblance to the present suburb. In the first place, the early settlements had specialized functions: education in Cambridge, and elegant, isolated residences for the elite of Chicago in Lake Forest. Although there are still suburbs that provide these functions, we have become accustomed to suburban diversification, at least during the last decade.

Of greater importance, there are a number of elements of modern suburban life, some of which have achieved mythic proportions (see Berger, 1961), which had no easily discernible counterparts in peripheral settlements prior to our era. The predominance of single-family dwellings, the ubiquitous barbecue, car pools, day camps, Little League baseball, the Rolling Hills Presbygational Church, St. Swithin's in the Swamp Episcopal Church, encounter groups meeting on patios, and the golden arches of McDonalds do indicate a comparatively new system of social existence in American history. Other essays in this volume will follow the development of that system since World War II and will offer analyses of the components of the suburban way of life. This essay does not provide a history of the modern suburb as a holistic entity. Such an essay would be either ridiculously narrow in time span or a rather banal exercise in the antequarian relating of the chronology of towns founded near cities. This essay will attempt, rather, to suggest various trends in American social history which make the modern suburb understandable as a result of historical processes rather than as an aberration. The attempt to come to terms with the origins of suburbia in the context of a holistic social institution has led many gifted urbanists into an ahistoric dead end. The historical background of the modern suburb is not linear. It is multivariate. Like Frankenstein's monster, the modern suburb has no past as an entity, but its cultural components do have a past and can be subjected to historical inquiry. The three components chosen for analysis here have been suggested by the corpus of literature on the modern suburb that has emerged since the 1950s and by recent developments in the analytical study of the history of American communities. These are: peripheral settlements, residential patterns, and middle-class consciousness. These are by no means all of the historical components of the modern suburb, but this list contains most of the vital organs. It is in the development of these factors, rather than in "case studies" of individual suburbs, that the historical background of the American suburb is to be found.

European commentators have often remarked on the peculiarly juridical definition of the American suburb. Unlike the English and Continental use of the term, which can include residential sections within the city, the American suburb is a settlement on the periphery of an urban area that is politically independent and distinct from the city. From his English perspective, Peter Hall (1966: 196-197) has argued that the process of "suburbanization" in a European sense is rather old in American society, but that the peculiarly American definition of the term "tends ... to describe development in the recent past—since 1945."

Hall's point is well taken, but he is incorrect (along with a number of urban sociologists and geographers in this country) in his assumption that the peripheral settlement is a rather recent development in American society. It has already been mentioned that some of the towns founded in seventeenth-century New England shared a great deal of interaction with Boston. This is not to argue that they were indeed founded as "suburbs." Boston was hardly a metropolis at this time, and Roxbury and Cambridge were by no means bedroom communities. The settlements were more interdependent with Boston than dependent upon it as a "central city." What is important about the relationship between these towns (and other clusters of colonial towns such as those around New Haven early in the eighteenth century) is the tradition of social and economic interaction between politically distinct settlements in close proximity. This was a part of the social experience of a large portion of town dwellers in American history. Certainly, the later development of dominant and subservient relationships between central cities and dependent peripheral towns should be seen as the development and alteration of older social patterns rather than the beginnings of a unique form of communal life. Indeed, by the late eighteenth century, Boston, New York, and Philadelphia were all surrounded by dependent peripheral suburbs which had, at the beginning of the century, shared an interdependent relationship with the town that was to become their central city (Bridenbaugh, 1938).

Some of the early American towns were quite small—less than the 2,500 population required for classification as an urban place by the U.S. Bureau of the Census—and it is difficult to retrieve exact information on peripheral settlements. Using the census data as a rough guide, however, it is quite clear that proximal urban settlements—towns within one mile of another town or city— continued to be a significant part of the American social experience throughout the nineteenth and into the twentieth centuries (see Table 1). The proportion of urban settlements which were proximal between 1790 and 1840 diminished somewhat, perhaps as a result of the opening of the West. After a slight rise in the 1840s and 1850s, the proportion of proximal urban places declined until 1900 when the proportion rose steadily and sharply. Table 2 indicates that the rate of growth of the urban population was greater than that of

TABLE 1

NUMBER OF TOTAL AND PROXIMAL URBAN PLACES IN THE UNITED STATES 1790-1940

	Number of Urban Places	Number of Proximal Urban Places	Proximal Urban Places as % of Total	% of New Urban Areas Which are Proximal
1790	24	4	16.6	—
1800	33	4	12.2	0.0
1810	46	6	13.0	15.4
1820	61	9	14.4	20.0
1830	90	13	14.4	13.8
1840	131	23	17.6	24.4
1850	236	49	20.7	24.7
1860	392	66	16.8	10.9
1870	663	80	12.0	5.2
1880	939	113	12.1	11.9
1890	1,348	198	14.6	20.7
1900	1,737	307	17.8	28.1
1910	2,262	583	25.7	52.4
1920	2,722	781	28.6	43.0
1930	3,165	1,179	37.0	89.9
1940	3,464	1,396	40.1	72.8

SOURCE: U.S. Census Data (Printed), 1790-1940, compared with land survey maps.
NOTE: "Proximal" is defined as within one mile of another urban place.

TABLE 2
DIFFERENTIAL RATES OF GROWTH U.S. URBAN POPULATION, URBAN PLACES, AND PROXIMAL URBAN PLACES: 1790-1940
(in percentages)

	Urban Population	Urban Places	Proximal Urban Places
1790-1800	59.2	37.5	0.0
1800-1810	63.1	39.4	50.0
1810-1820	32.0	32.6	50.0
1820-1830	62.3	47.5	44.4
1830-1840	63.5	45.6	76.9
1840-1850	91.7	80.2	113.0
1850-1860	75.8	66.1	34.6
1860-1870	59.1	69.1	21.2
1870-1880	42.6	41.6	41.3
1880-1890	56.2	43.5	75.2
1890-1900	36.4	28.6	55.1
1900-1910	39.3	30.2	89.9
1910-1920	28.9	20.3	33.8
1920-1930	27.2	16.2	50.9
1930-1940	7.9	9.4	19.4

SOURCE: Derived from data in Table 1 on U.S. Bureau of the Census (1965), series A 181-195.

NOTE: The relative dependence of the rate of growth in the number of urban places in general and the urban population can be further demonstrated by the low average variation (5.03 percentage points) from the mean differentiation between the two variables, with the greatest variation being 13.1 percentage points from the mean differentiation. The relative independence of the growth in the number of proximal urban places can be further demonstrated by the high average variations from the mean between this variable and both the number of total urban places (15.52 percentage points) and the urban population (15.44 percentage points) with high variations 35.4 and 36.8, respectively.

urban places, but that the configuration of these two trends was conformable. The rate of growth of proximal urban places, however, has been relatively independent of the other two factors and has most often been higher than the growth rate of either the urban population or of urban places in general.

The early slump in the percentage and growth of proximal urban areas can, in part, be explained by the westward movement, although the founding of widespread and isolated towns did not diminish the statistical significance of the basis for town interaction. The late nineteenth-century slump is somewhat misleading. Adna Weber (1898), in his studies of the statistical implications of urban growth, noted the tendency of large urban areas to increase through annexation and consolidation. More recently, Kenneth Jackson (1970; 1972: 445-446) has subjected the process of peripheral accretion to greater scrutiny and has demonstrated that, without the incorporation of proximal settlements and towns in the nineteenth century, only New York City would today have a population of one million. Other recent scholarship tends to indicate that the act of consolidation did not significantly alter the social pattern of the former proximal town, nor did it change the relationship, except in a political sense, between the new city neighborhood and the city core (see Sennett, 1970).[4] Therefore, the comparatively low proportion of proximal towns between 1860 and 1900 indicates a substantial increase in the number of social areas with similar relationships to other towns and central cities. In spite of the octopus-like growth of nineteenth-century cities, the peripheral communities continued to be an important part of the urban ecology of large cities throughout the century. Indeed, Leo Schnore (1959) has presented some intriguing data which suggest that New York City experienced deconcentration as early as mid-century, and that Cincinnati, San Francisco, New Haven, Boston, Albany, Baltimore, St. Louis, Scranton, and Duluth began to show signs of this process before 1900. Undoubtedly, the process has greatly increased in the twentieth century, but scholars who have seen it as a phenomenon of our era have overlooked the rather deep historical roots (see Hawley, 1956; Thompson, 1947). In his pioneering study, Arthur Schlesinger, Sr. (1933: 108) found impressionistic data to suggest that, by the late nineteenth century, urban Americans were well aware of the metropolitan area and suburbia as categories of social reality, not only in New York and Boston, but also in Pittsburgh, Providence, and Cincinnati.

These data need not diminish the importance of modern sub-urbanization as a phenomenon of the twentieth century; and the establishment of metropolitan districts in 1910 and the Standard Metropolitan Area in 1950 are indicative of an important change in social perception (see Glaab, 1968: 399-402). Cities ceased to expand through peripheral accretion, and the peripheral settlements —many of them residential communities—grew rapidly (McKenzie, 1933: 183). In terms of social relationships, however, rapid deconcentration is less a new trend than an alteration of an established pattern.

The data presented in Tables 1 and 2 are by no means an exact statement on peripheral settlements in urban America. Two settlements of 2,500 people within one mile of each other hardly constitute a suburban experience. The data are suggestive, however. First, they indicate a social experience prior to suburbanization which included some sort of intercommunity contact as an important and constant factor. Second, settlements on the periphery of urban centers, which emerged first in the growing cities in the Northeast, rapidly became a feature of cities established in the West. By the 1820s, residential communities began to appear around the newly founded "River Cities" (Wade, 1959: 305). At the time of the Gold Rush, San Francisco already had an established peripheral ring of settlements (Lotchin, 1972).

The data also do not give any indication of the diversity or changing function of peripheral settlements. There are no easy generalizations that can be made about these important factors, but most of the communities that have been studied intensively give evidence of a similar pattern up to the mid-nineteenth century. Charleston, Roxbury, and Cambridge, for example, were not "bedroom" communities for Boston, nor were they subservient and supportive in their economic relations to the growing city. Rather, they shared an interdependence in regional growth. Although there was a great deal of interaction between the peripheral towns and the city, all three had "walking cities" and minor manufacturing (Warner, 1962: 18-19). Peripheral settlements around other large cities were similar. In the second half of the nineteenth century, however, residential suburbs gained increasing attention. As early as 1851, a loyal Cincinnati citizen complained about local businessmen who lived outside the city, presumably to escape taxes (Miller, 1968b: 4). Fashionable residential communities were built in proximity to all the growing urban areas in the late nineteenth century.

It would, however, be a mistake to create rigid typologies of pre-1850 residential suburbs. Some communities, such as Pullman, Illinois, were established as industrial centers and rapidly became satellite cities (see Buder, 1967: 118-130). In the early twentieth century, both Graham Taylor (1915) and Harlan Paul Douglass (1925) found many examples of commercial and industrial suburbs. Older communities which had existed as small towns with relatively little contact with the city were brought into the metropolitan region and continued to practice more traditional forms of life, often in defiance (see Dobriner, 1960). Although the amount of human and economic traffic between the core of the central city and the peripheral settlements increased by the turn of the century, commercial decentralization and the rapid diversification of land use in suburban areas were noticeable trends at least as early as the 1920s (Glaab, 1968: 428-429). All these qualifications aside, the dormitory function of the suburb increased greatly between 1850 and 1940 and became almost the defining characteristic of the peripheral community in the popular mind.

Transportation technology facilitated the increased contact between the core city and the peripheral settlement. The streetcar and the automobile in particular have been cited as primary agents in the suburbanization of America (Rae, 1965: 219-235; Warner, 1962). While it would be ridiculous to discount the significance of spatial mobility in an urban area, the importance of transportation technology was primarily in the facilitation of changing residential patterns. The beginnings of intraurban traffic on the early nineteenth-century omnibus (Taylor, 1951: 390-391) was roughly coincident with one of the most important social and spatial changes in American urban history: residential segregation. Mass interurban traffic on the railway and private mobility in the automobile were simply extensions and technological sophistications of a pattern that was well established by the late nineteenth century.

From the time of the founding of the seaport towns to the early nineteenth century, the "walking city" was a residential as well as a commercial concept. The center of the city contained public buildings and most of the major commercial concerns. Near the center of the city were the residences of the community elite. Surrounding this core were residential neighborhoods where merchants, clerks, "professionals,"[5] artisans, and laborers lived and practiced their trades in close proximity. Just outside the city limits were the less desirable elements of urban life, such as shanties, butcher-houses, and brothels (Jackson, 1970: 24; Bridenbaugh, 1938: 238-239). Thus, within the city, the majority of the citizenry lived in relatively heterogeneous neighborhoods. Various classes and interest groups shared daily contact, simply by direct visual sensation on the street, or in more intimate conversation in taverns and coffee houses (Warner, 1968: 14-21; Taylor, 1966: 39; Sjoberg, 1960: 323). Furthermore, from one neighborhood to another or to the core of the city was easy walking distance.

This pattern of community life was maintained fairly intact until the third or fourth decade of the nineteenth century. There were, however, early signs of changing patterns. In the 1750s, Benjamin Franklin noted the growth of residential neighborhoods, composed of wealthier merchants, beyond the southern boundary of Philadelphia (Bridenbaugh and Bridenbaugh, 1942: 12). Elite housing on the periphery of the city (in those areas without the above-mentioned drawbacks) was not uncommon in the eighteenth century, but it was not the norm. In any event, the presence or absence of a portion of the elite near the core had very little effect on the social dimensions of town life in most neighborhoods.

Industrialization and immigration began to change the residential pattern of American city life in the early nineteenth century. During the period of industrialization, the factory replaced the residential workshop as the locus of manufacturing. The owners of the means of production were no longer required to enter into direct daily contact with their workers, especially after the decline of crafts and the rise of industrial labor (Baltzell, 1958: 130-141; Sullivan, 1955: 29-83; Ware, 1924: 18-70; Warner, 1968: 49-98). The timing was different for each city, but the inevitable result was abandonment of the previously heterogeneous neighborhood by the wealthier residents. The movement of the middle class, able to afford more comfortable housing in homogeneous neighborhoods, was an easily discernible phenomenon in cities as diverse as New York, Boston, Philadelphia, Cincinnati, and San Francisco by the middle of the nineteenth century (Jackson, 1970; Warner, 1962, 1968; Miller, 1968a; Lotchin, 1972). Areas which experienced urbanization and industrialization later, realized the same trend, such as Los Angeles in the early twentieth century (Fogelson, 1967: 186 ff.). It would seem that one of the features of American urbanization—what Michael Frisch (1972: 249-250) has characterized as the development from a community that is experienced directly and informally to one that is perceived as a formal abstraction—is residential segregation along economic lines.

Cities have also been segregated by ethnic groups. Of course, the immigrant population was composed primarily of laborers who, both native and immigrant, tended to stay in the central city near the factories until the advent of more efficient urban and interurban transportation. As new groups entered the urban labor market, older groups which had acquired industrial skills or had saved enough of their earnings moved into areas that were predominantly residential (Ward, 1971: throughout). As a result of industrialization and the growth of ethnic neighborhoods, the American city as a whole became more heterogeneous. Ironically, this increase in diversity was contemporaneous with the development of more homogeneous neighborhoods.

The specifics of the development of residential segregation were different for each city. In Philadelphia, the segregation of neighborhoods by income and ethnicity was perceptible, but mild, between

1830 and 1860 and became pronounced by 1930 (Warner, 1968: 169-171). In Los Angeles, residential segregation by income was slight until the 1920s; until the 1940s, residential segregation by national origin among people of European background was unknown. The Mexican population, however, was highly segregated.[6] The migration of southern Blacks to industrial cities between 1890 and World War I did create distinctively Black neighborhoods, but it was not until the great increase in the migration during and after the war that rigid and hostile segregation became the norm. Indeed, fear of and hostility toward the Black population were factors in white suburbanization in the early twentieth century (see Green, 1967: 235).

Throughout the nineteenth century and into the twentieth century, therefore, American cities have increasingly become the sites of a variety of distinctive neighborhoods. The modern suburb is a direct result of residential segregation. Before Lake Forest, Illinois, existed, there were exclusive residential districts in the city of Chicago. There were many fashionable (and expensive) homes along Adams Boulevard in Los Angeles prior to the creation of Beverly Hills. The suburb as an exclusive upper- or middle-class residential area is simply an extension of a rather old pattern of residential segregation in American cities. That they have become established suburbs rather than districts within cities can be attributed to the decline of peripheral accretion as a method of urban growth. The growth of Chicago in the late nineteenth century, for example, can be seen as the incorporation of one residential suburb after another. As a result of similar developments in other cities, the central core of most metropolitan areas by the 1920s contained residential neighborhoods that were socially indistinguishable from many of the surrounding suburbs.[7]

Of course, not all of the suburbs that have developed in the past century have been or are middle-class or elite. By the end of the nineteenth century, immigrant and labor-class suburbs were in evidence near all major cities. With the introduction of commercial and industrial function into the suburb from the turn of the century on, many suburbs began to experience residential segregation within their boundaries. The same pattern that had occurred in cities in the early nineteenth century—development of homogeneous neighborhoods in the face of increasing heterogeneity for the community as a whole—was recapitulated in some suburbs. By 1900, the peripheral towns of the Boston area had developed class-specific neighborhoods

(Warner, 1962: 64). By 1920, Pasadena contained neighborhoods that were segregated by income and ethnicity. This is not to argue that suburbs are simply cities writ small. Not all suburbs became diversified, and those that did still had characteristics that distinguished them from the central cities. It is simply to argue that, in terms of social and spatial arrangements, the modern suburb can and should be understood in the context of a rather long historical development.

Enough has been written to disprove the older assumptions about the suburb as the exclusive domain of the American middle class to create some fear about including this factor, but there are three good reasons to do so. First, although one must recognize that there have existed and do exist working-class suburbs, the majority of suburbanites in the twentieth century have been middle-class.[8] Second, it has been the middle class that has been most conscious of being "suburban."[9] Finally, so much of the literature on suburbia has not been on the sociology of the peripheral communities, but on the middle-class commuter. An early twentieth-century study of peripheral settlements which were predominantly working-class ethnic neighborhoods differentiated the older WASP population by saying that "the rear guard of the city's American population is to be found on the *suburban* side of the zone" (Woods and Kennedy, 1962: 36). To address the class-consciousness factor in the development of middle-class suburbs is to speak of a limited area of the suburban experience, but it is to speak of an important part of that experience.

Middle-class consciousness developed in the late nineteenth century. There had always been people in American cities who had known that they were richer than some and poorer than others and in that sense belonged to a "middle class." Until the onset of industrialization and modernization, however, this class identification was less important than heterogeneous neighborhoods. Direct daily contact across what we would call class lines tended to diminish whatever class identification existed, if any. By the end of the century, however, the American middle class was as self-conscious as the Parisian bourgeois (see Daumard, 1970).

To a certain extent, the rise of middle-class consciousness was facilitated by residential segregation, and the resultant homogeneous neighborhoods, which were extremely important in the development of industrial-class consciousness in Europe (see J. Scott, 1969). Its origins have at least as much to do with changing economic and social function, however. Robert Wiebe (1967: 111-132) has suggested that a "new" urban middle class emerged in the late nineteenth century as a result of increased urban migration, immigration, corporatization, and professionalism. One may question how "new" this class was in its constituency, but Wiebe has correctly specified the factors that led the urban middle class to be aware of themselves as a distinctive social group. The older identification with diverse and multifunctional neighborhoods disappeared with those neighborhoods in the years following the onset of industrialization, and was replaced by new means of identification.

It is in the development of the American WASP that the emerging middle-class consciousness can be seen most clearly. As the working class became increasingly identified with the immigrant, Catholic, and esoteric Protestant populations, spokesmen for the more traditional American Protestant denominations warned that their churches were in danger of becoming "mere middle-class institutions" (Carter, 1971: 141). Their churches had always been predominantly "middle-class," but the immigrant, Catholic, and Jewish challenge to the hegemony of social and moral authority in American cities made American Protestants quite aware of the class constituency of their denominations (see Singleton, 1973).[10] Prior to the end of the nineteenth century, the urban Protestant middle class had not seen their leadership role in terms of class interest, but had assumed that they were acting in behalf of the entire community (Rosenberg, 1971; Huggins, 1971: 15-56). The appearance of ethnic benevolence organizations, labor unions, and ethnic and labor-based political "machines" challenged this assumption and contributed to the creation of a class consciousness.

Equally important to the development of the new middle class was the separation of work function, along with residential segregation, between the middle class and labor. Not only were these groups separated by residence, but the amount of direct contact through the work day decreased. As corporations increased in complexity, so did the spatial separation of the work experience. The new managerial and clerical functions occurred in partitioned offices, and at times in different buildings, often in different parts of the city (see Brody, 1960; Wright, 1883). It was no longer necessary to live near the place of work, and it was indeed desirable not to do so.

Not all of the new middle class took up residence in the suburbs, but a survey of managerial and professional types in three major urban areas in 1890 indicates that most of them lived on the periphery of the central city.[11] It was during the late nineteenth century that suburbs became conscious of themselves both as suburbs and as middle-class communities. The Chicago *Sunday Times* in 1873 celebrated the glories of the North Shore residential communities. Among their chief benefits were pleasant and quiet homes for the city's business community (Glaab, 1963: 229-233). In the 1880s, Pasadena real estate promoters proudly proclaimed their town a "beautiful suburb" combining the advantages of a peaceful residential village and proximity to a growing city (Farnsworth, 1883). The term "suburb" became part of the urban lexicon, and the middle-class commuter became its symbolic citizen. At the turn of the century, *Harper's Bazaar* (1900) mildly complained that "the suburban husband and father is almost entirely a Sunday institution" (see Schmitt, 1969).

In more fashionable suburbs, various forms of community identification began to emerge in the late nineteenth century. For the wealthy, country clubs became available, and for those of more modest income, there were various social and civic organizations specifically organized for suburbanities (Baltzell, 1964: 121 ff.; Schmitt, 1969: throughout). Beyond formal organizations, some suburbs adopted various "covenants" to restrict housing to single-family dwellings, which indicated a middle-class bias, and to exclude factories and saloons, the former to prevent the introduction of working-class neighborhoods and the latter as an extension of the predominantly middle-class Temperance Movement (Warner, 1962: 122).

The development of middle-class consciousness had definite political implications. Samuel Hays (1964) has suggested that the municipal reform movement in the late nineteenth century should be understood in the context of an urban middle class which had developed an obsession with efficiency and nationalism and which saw the political machine as inefficient and parochial. Richard Wade (1968) has added an interesting dimension to this interpretation by suggesting that the class division had spatial dimensions. It was, he argues, a case of the peripheral middle class against the immigrant and labor center of the city. Research on urban progressivism in at least one city tends to confirm this hypothesis (Miller, 1968a). All the factors involved in the development of middle-class consciousness suggest that the cessation of urban growth through peripheral accretion at the turn of the century was no accident. In at least one suburb—Evanston, Illinois—the middle-class citizenry consciously articulated a desire to remain socially and legally separate from the core city.[12] By the turn of the century, middle-class consciousness became the most important factor in molding the image and reality of suburbia. A full social analysis of early twentieth-century suburbs would be necessary to test this proposition, but a few words can be said here about two of the most obvious examples: the middle-class market for community planning and the perpetuation of small-town government in suburbia.

A frequently expressed middle-class suburban goal was to avoid the seeming chaos and disorder of the central city (see Schmitt, 1969: throughout; Tunnard, 1947), and many exurbanites in the twentieth century have been willing to pay the price of moving to planned communities. The development of planned middle-class housing around the turn of the century marked a significant change in the planners' clientele. Early practitioners, such as Andrew Jackson Downing, Frederick Law Olmstead, and Calvert Vaux assumed that planned residential communities could be supported only by the wealthy (Reps, 1965: 341; Burchard and Bush-Brown, 1961: throughout). The latter two planners, in their plan for Riverside, Illinois, clearly stated the elitist nature of the future community (Olmstead and Vaux, 1868), and, indeed, early planned communities such as Riverside, Lake Forest, and Llewelyn Park, New Jersey, were far beyond the means of most middle-class Americans.

By the turn of the century, the increase of middle-class suburbs captured the attention of planners (M. Scott, 1969: 1-109), and the writings of the British planner, Ebenezer Howard (1902), advocating planned communities for a greater spectrum of the population, gained increasing acceptance. Community planning lacked direction and coordination until 1923, when the Regional Planning Association of America was founded. The RPAA was an ambitious organization composed of architects, planners, and social critics, such as Lewis Mumford, Clarence Stein, Henry Wright, and Frederick L. Ackerman. As a group, they were opposed to both metropolitan centralization and suburban diffusion in the major urban areas (Lubove, 1963: 1-31). In spite of the excellent attempts of the RPAA to bring about a systematic ordering of the growing metropolis, planners continued to build almost exclusively for the urban middle-class, who so desired new communities on the periphery, without regard for the total metropolitan ecology. The association was somewhat vindicated when, shortly after October 1929, it became apparent that planners had overestimated the economic growth potential of the American middle class and were left with half-finished suburban communities (Glaab, 1968: 408-409).

The New Deal administration also attempted to bring some equitable order to suburbanization, through the Suburban Resettlement Division of the Resettlement Administration. The "Greenbelt Town" program built suburbs near Washington, D.C., Cincinnati, and Milwaukee. These turned out to be publicly supported housing programs primarily for the middle class, however. In Washington, D.C., for example, the Greenbelt suburb was used primarily as an escape route for the city's white middle-class residents fleeing the growing Black population (see Conkin, 1959: 305-325; Arnold, 1971). It was not until after World War II that community planning was applied to nonelite and non-middle-class populations. Within the limits of this essay, community planning was important in the context of the development of middle-class suburbs since the turn of the century. The same is true of the peculiarities of the government of suburbia.

Robert Wood (1959: 11-12), in his classic study of suburbia, raised an interesting paradox in considering the social and political dimensions of the suburban man. In his business relations, the suburbanite lives in a social order that has been built on centralization, efficiency, and largeness. In his suburban government, however, the same individual not only tolerates, but actually supports, a small, ineffective, and decentralized government. This is an intriguing problem, and one that cannot be easily dismissed, but Wood's formulation is presentistic, and he does not consider the historical dimensions. The oldest political tradition that has been directly experienced by Americans throughout our history has been small-town government. While the New England town meeting of the seventeenth and eighteenth centuries was hardly a forum of democracy (see Lockridge, 1970; Zuckerman, 1970), it was a model of politics as community interaction that was recapitulated throughout the nineteenth century, and one that many suburbs attempted to retain into the twentieth century.

Cities did not adopt the concept of professional and centralized government in the nineteenth century as a simple result of population growth. Direct and personal contact between elected officials and their constituents continued to be an important feature of New York City politics until the middle of the century (Rosenberg, 1971: 15-43, 163-185). Springfield, Massachusetts, experienced rapid population growth between 1840 and 1850, but retained a small-town government through town meetings, citizens committees, and direct contact between officials and residents (Frisch, 1972: 44-47). Big-city government came as a result of diversity. Both the political machine and urban reform were attempts to deal with cities that were ethnically, religiously, and economically diverse, which resulted in what Sebastian de Grazia (1948: 74-75) has termed a general separation of rulers from the community.

For the Protestant middle class in American cities, this late nineteenth-century anomie was not simply a separation from political leadership, but a loss of leadership status by their class. If the growth of middle-class suburbs in the late nineteenth century and the subsequent resistance to annexation can be understood as a reaction to the implications of diversity and loss of status in the central city, then we should not be surprised to find these new suburbanites, in spite of their participation in the development of a corporate economy, seeking to recapture a form of political community under which they, as a group, had enjoyed stature and no small measure of social control. Furthermore, it would have been somewhat surprising to see them adopt the governmental forms of the cities they had abandoned. Studies of middle-class suburbs in the 1920s indicate that the personalistic nature of town government continued to be important to suburbanites (see Douglass, 1925; McKenzie, 1933; Cosse, 1928; Merino, 1968). This was the result of both a historical tradition and a reaction to what the city came to represent to them.

The modern American suburb is indeed a new phenomenon which does not have a clearly delineated historical tradition. It is the result of a number of important developments in American history, however, a few of which have been briefly surveyed here. Peripheral settlement is the oldest of these developments. With increased efficiency in transportation technology and the beginning of residential segregation in American cities, these peripheral areas began to serve new functions. Residential segregation, industrialization, and the challenge of ethnic and labor groups to the middle-class control of the city created a class consciousness which was expressed in many ways, including the "flight" to the suburbs. By the end of the nineteenth century, the middle-class suburb, in spite of the fact that other kinds of suburbs existed, became a social category. The development of planned communities during the first four decades of this century and the persistence of an antiquated small-town government can be seen as having their origins in the middle-class reaction to the modern American city. This does not tell the whole story. No brief essay could, but it is a significant part of the story.

This sketch of the complex of historical trends cannot end without saying that suburbanites have not been able to escape some of the more serious social and psychological problems that have plagued urban dwellers. This is a commonplace statement today, and we are all familiar with suburban slums, racial tensions, and even signs of "urban decay" in some suburbs. It was true almost fifty years ago. Although it was not yet clear that what was true of the city could become true of the suburb, Harlan Paul Douglass (1925: 12-13) reported that the most serious problem facing the suburbanite was anomie and the lack of community identity. Having escaped the city, he could not escape the implications of modern corporate America. Indeed, his new community was an extension of patterns of American society resulting from the collapse of traditional culture. In creating the suburb, he merely gave a spatial formulation to the paradox that has faced most Americans since the simultaneous rise of social doctrines of egalitarianism and industrial class stratification—a desire to proclaim and deny democracy in one breath. The American suburb, in its middle-class form, through the homogeneity of social composition and the maintaining of the form of small-town government, became an expression of equality *within* a community, which was qualitatively different from other communities. The suburb, both as a particular kind of community and as the sum of important social processes in our history, was, by the mid-twentieth century, an excellent symbol of the uniquely American reaction to a rapidly changing social structure. Suburbia attempted to deny these changes. Ironically, it is also the evidence of these changes.

NOTES

1. The suburb is one of a number of "modern" American phenomena that need to be understood in terms of historical trends rather than in terms of the histories of the specific phenomena themselves. For examples, see Ernest Sandeen (1970) on "fundamentalism" and Paul Koistinen (1967) on the "military-industrial complex."

2. This is based on a survey (in progress) by the author of social and economic indices relating to the residents of Chicago and New York and their suburbs. At this point, the data are indicative, but not precise.

3. I am indebted to John E. Drodow of Northwestern University for sharing information and insights from his dissertation to be entitled, "From Village to Suburb: Evanston, Illinois, 1880-1930," now in progress.

4. I am indebted to Barbara M. Posadas of Northwestern University for bringing this factor of neighborhood analysis to my attention and for sharing information and insights from her dissertation to be entitled, "Community Structures of Chicago's Northwest Side: Patterns of Ethnic Neighborhoods, 1880-1934," now in progress.

5. The term "professional" is used in a loose sense here and refers primarily to lawyers and physicians. The development of professions in a modern sense occurred at roughly the same time as the rise of "middle-class consciousness," and is related to it.

6. This is based on research summarized in my own dissertation, entitled "Religion in the City of the Angels: Urbanization and American Protestant Culture, Los Angeles, 1850-1930," and worked on for University of California (Los Angeles) 1973.

7. The source for this is the unpublished research of Barbara M. Posadas, mentioned above.

8. This is based on the survey mentioned in note 2.

9. This point is difficult to prove, but it is indicated by a significant body of impressionistic literature from the late nineteenth and early twentieth centuries. For an interesting survey and interpretation of this literature, see Schmitt (1969).

10. American Protestantism (primarily the Presbyterian, Protestant Episcopal, Methodist, Baptist, and Congregational denominations) as a means of social identification since the rise of industrialism needs to be more fully explored in the context of the developing middle-class consciousness, but some suggestive insights can be found in Huggins (1971) and Carter (1971).

11. This is based on the survey mentioned in note 2.

12. The source for this is the unpublished research of John E. Drodow, mentioned above.

REFERENCES

ARNOLD, J. L. (1971) The New Deal in the Suburbs: A History of the Greenbelt Town Program, 1935-1954. Columbus: Ohio State Univ. Press.

BALTZELL, E. D. (1964) The Protestant Establishment: Aristocracy and Caste in America. New York: Alfred A. Knopf.

––– (1958) Philadelphia Gentlemen: The Making of a National Upper Class. New York: Free Press.

BERGER, B. M. (1961) "The myth of suburbia." J. of Social Issues 17 (November): 38-49.

BRIDENBAUGH, C. (1938) Cities in the Wilderness: Urban Life in America, 1625-1742. New York: Ronald Press.

––– and J. BRIDENBAUGH (1942) Rebels and Gentlemen: Philadelphia in the Age of Franklin. New York: Reynal & Hitchcock.

BRODY, D. (1960) Steelworkers in America: The Nonunion Era. Cambridge, Mass.: Harvard Univ. Press.

BUDER, S. (1967) Pullman: An Experiment in Industrial Order and Community Planning, 1880-1930. New York: Oxford Univ. Press.

BURCHARD, J. and A. BUSH-BROWN (1961) The Architecture of America: a Social and Cultural History. Boston: Little, Brown.

CARCOPINO, J. (1940) Daily Life in Ancient Rome: The People and the City at the Height of the Empire. New Haven, Conn.: Yale Univ. Press.

CARTER, P. A. (1971) The Spiritual Crisis of the Guilded Age. De Kalb: Northern Illinois Univ. Press.

CONKIN, P. K. (1959) Tomorrow a New World: The New Deal Community Program. Ithaca: Cornell Univ. Press.

COSSE, M. V. (1928) The Suburban Weekly. New York: Columbia Univ. Press.

DAUMARD, A. (1970) Les bourgeois de Paris au XIXe siècle. Paris: Flammarion.

DE GRAZIA, S. (1948) The Political Community: A Study in Anomie. Chicago: Univ. of Chicago Press.

DOBRINER, W. M. (1960) "The natural history of a reluctant suburb." Yale Rev. 49 (Spring): 398-412.

DOUGLASS, H. P. (1925) The Suburban Trend. New York: Century.

FARNSWORTH, R.W.C. (1883) A Southern California Paradise, in the Suburbs of Los Angeles, Being a Historic and Descriptive Account of Pasadena, San Gabriel, Sierra Madre, and La Cañada; With Important Reference to Los Angeles and All Southern California. Pasadena: Pacific.

FOGELSON, R. M. (1967) The Fragmented Metropolis: Los Angeles, 1850-1930. Cambridge, Mass.: Harvard Univ. Press.

FRISCH, M. H. (1972) Town Into City: Springfield, Massachusetts, and the Meaning of Community, 1840-1880. Cambridge, Mass.: Harvard Univ. Press.

GLAAB, C. N. (1968) "Metropolis and suburb: the changing American City," in J. Braeman et al. (eds.) Change and Continuity in Twentieth-Century America: the 1920's. Columbus: Ohio State Univ. Press.

––– (1963) The American City: A Documentary History. Homewood, Ill.: Dorsey.

GREEN, C. M. (1967) The Secret City: A History of Race Relations in the Nation's Capital. Princeton: Princeton Univ. Press.

HALL, P. (1966) The World Cities. New York: McGraw-Hill.

Harper's Bazaar (1900) "Rapid transit and home life." Volume 33 (December): 2001-2005.

HAWLEY, A. (1956) The Changing Shape of Metropolitan America: Deconcentration Since 1920. New York: Free Press.

HAYS, S. (1964) "The politics of reform in municipal government in the progressive era." Pacific Northwest Q. 55 (October: 157-169).

HOWARD, E. (1902) Garden Cities of Tomorrow. London: Swan Sonnenschein. (Originally published in 1898 as Tomorrow.)

HUGGINS, N. I. (1971) Protestants Against Poverty: Boston's Charities, 1870-1900. Westport, Conn.: Greenwood.

JACKSON, K. T. (1972) "Metropolitan government versus political autonomy: politics on the crabgrass frontier," in K. T. Jackson and S. K. Schultz (eds.) Cities in American History. New York: Alfred A. Knopf.

––– (1970) "Urban deconcentration and suburbanization in the nineteenth century." Presented at the Conference on the New Urban History, University of Wisconsin.

KOISTINEN, P.A.C. (1967) "The 'industrial-military complex' in historical perspective: World War I." Business History Rev. 41 (Winter): 819-839.

LOCKRIDGE, K. A. (1970) A New England Town the First Hundred Years: Dedham, Massachusetts, 1636-1736. New York: W. W. Norton.

LOTCHIN, R. (1972) "San Francisco: the patterns of chaos and growth," in K. T. Jackson and S. K. Schultz (eds.) Cities in American History. New York: Alfred A. Knopf.

LUBOVE, R. (1963) Community Planning in the 1920's: The Contribution of the Regional Planning Association of America. Pittsburgh: Univ. of Pittsburgh Press.

McKENZIE, R. D. (1933) The Metropolitan Community. New York: McGraw-Hill.

MERINO, J. A. (1968) "A great city and its suburbs: an attempt to integrate metropolitan Boston, 1865-1920." Ph.D. dissertation. University of Texas.

MILLER, Z. L. (1968a) "Boss Cox's Cincinnati: a study in urbanization and politics, 1880-1914." J. of Amer. History 54 (March); 823-838.

––– (1968b) Boss Cox's Cincinnati: Urban Politics in the Progressive Era. New York: Oxford Univ. Press.

MUMFORD, L. (1961) The City in History: Its Origins, Its Transformations, and Its Prospects. New York: Harcourt, Brace & World.

OLMSTEAD, F. L. and C. VAUX (1868) Preliminary Report Upon the Proposed Suburban Village at Riverside, Near Chicago. New York: Sutton Browne.

RAE, J. B. (1965) The American Automobile: A Brief History. Chicago: Univ. of Chicago Press.

REPS, J. W. (1965) The Making of Urban America. Princeton: Princeton Univ. Press.

ROSENBERG, C. S. (1971) Religion and the Rise of the American City: The New York City Mission Movement, 1812-1870. Ithaca: Cornell Univ. Press.

SANDEEN, E. R. (1970) The Roots of Fundamentalism: British and American Millenarianism, 1800-1930. Chicago: Univ. of Chicago Press.

SCHLESINGER, A. M., Sr. (1933) The Rise of the City, 1878-1898. New York: Macmillan.

SCHMITT, P. J. (1969) Back to Nature: The Arcadian Myth in Urban America. New York: Oxford Univ. Press.

SCHNORE, L. F. (1959) "The timing of metropolitan decentralization: a contribution to the debate." J. of Amer. Institute of Planners 25 (November): 200-206.

SCOTT, J. (1969) "The glassworkers of Carmaux, 1850-1900," in S. Thernstrom and R. Sennett (eds.) Nineteenth Century Cities: Essays in the New Urban History. New Haven, Conn.: Yale Univ. Press.

SCOTT, M. (1969) American City Planning Since 1890. Berkeley: Univ. of California Press.

SENNETT, R. (1970) Families Against the City: Middle Class Homes of Industrial Chicago, 1872-1890. Cambridge, Mass.: Harvard University Press.

SINGLETON, G. H. (1973) " 'Mere middle-class institutions': urban protestantism in nineteenth-century America." J. of Social History (June).

SJOBERG, G. (1960) The Preindustrial City: Past and Present. New York: Free Press.

SULLIVAN, W. A. (1955) The Industrial Worker in Pennsylvania, 1800-1840. Harrisburg: Pennsylvania State Univ. Press.

——— (1951) The Transportation Revolution, 1815-1860. New York: Holt, Rinehart & Winston.

TAYLOR, G. G. (1966) "The beginnings of mass transportation in urban America." Smithsonian J. of History 1 (Summer): 35-50.

THOMPSON, W. D. (1947) The Growth of Metropolitan Districts in the United States: 1900-1940. Washington, D.C.: Government Printing Office.

TUNNARD, C. (1947) "The romantic suburb in America." Magazine of Art 40 (May): 184-187.

U.S. Bureau of the Census (1965) Statistical Abstract of the United States: Colonial Times to the Present. Washington, D.C.: Government Printing Office.

WADE, R. (1968) "Urbanization," in C. V. Woodward (ed.) The Comparative Approach to American History. New York: Basic Books.

——— (1959) The Urban Frontier: The Rise of Western Cities 1790-1830. Cambridge, Mass.: Harvard Univ. Press.

WARD, D. (1971) Cities and Immigrants: A Geography of Change in Nineteenth Century America. New York: Oxford Univ. Press.

WARDEN, G. B. (1970) Boston, 1689-1776. Boston: Little, Brown.

WARE, N. (1924) The Industrial Worker, 1840-1860. Boston: Houghton Mifflin.

WARNER, S. B., Jr. (1968) The Private City: Philadelphia in Three Periods of Its Growth. Philadelphia: Univ. of Pennsylvania Press.

——— (1962) Streetcar Suburbs: The Process of Growth in Boston, 1870-1900. Cambridge, Mass.: Harvard Univ. Press.

WEBER, A. F. (1898) "Suburban annexations." North Amer. Rev. 166 (May): 612-617.

WIEBE, R. H. (1967) The Search for Order, 1877-1920. New York: Hill & Wang.

WINSLOW, O. E. (1968) John Eliot, "Apostle to the Indians." Boston: Houghton Mifflin.

WOOD, R. C. (1959) Suburbia, Its People and Their Politics. Boston: Houghton Mifflin.

WOODS, R. A. and A. J. KENNEDY (1962) The Zone of Emergence: Observations of the Lower Middle and Upper Working Class Communities of Boston, 1905-1914. Cambridge, Mass.: MIT Press.

WRIGHT, C. D. (1883) "The factory system of the United States," in U.S. Census Office, Report on the Manufactures of the United States at the Tenth Census, June 1, 1880. Washington, D.C.: Government Printing Office.

ZUCKERMAN, M. (1970) Peaceable Kingdoms: New England Towns in the Eighteenth Century. New York: Alfred A. Knopf.

2

Suburbanization in the United States Since World War II

NORVAL D. GLENN

☐ THERE IS WIDESPREAD AGREEMENT AMONG STUDENTS of American society that suburbanization has been one of the most important trends since World War II, but there is not agreement on precisely what suburbanization entails. A traditional and perhaps still the most widely used definition of a suburb is that it is a settlement of urban or semi-urban density which is near, and perhaps adjacent to, a larger urban settlement.[1] Although a suburb is not within the legally defined geographic limits of the larger city, it may be surrounded by the "central" city. If a settlement is separated from the larger city and its contiguous suburbs by a band of open countryside or by water, its being considered a suburb may depend more on the degree of economic and social interdependence between it and the city than on the distance separating them.

The crucial aspect of this distinction between central city and suburb is that it is based on political and jurisdictional boundaries. The suburb may become part of the central city through the legal act of annexation. Therefore, this definition of the suburb may be called the "legal" or "political" definition.

The definition of suburbanization which corresponds to the legal definition of suburb is simply the growth of suburbs and their populations. The process is not restricted to population movement from a central city to its suburbs or even to movement from all central cities, considered collectively, to all suburbs. Rather, suburban population growth may come from migration from rural areas and small independent cities, immigration, and natural increase. Furthermore, stable residents of once independent communities are added to the suburban population when those communities are engulfed by the expansion of the suburban zone of larger cities.

The legal distinction between central cities and suburbs is almost always used in discussions based on census data. Since 1950, the Bureau of the Census has designated two areal units which are often used in studies of suburbs, the Urbanized Area and the Standard Metropolitan Statistical Area (SMSA). The latter was called a Standard Metropolitan Area (SMA) in 1950 and was defined somewhat differently from the SMSA in 1960 and 1970; the definition of an Urbanized Area has changed slightly since 1950.[2] Nevertheless, these areal units are roughly comparable from 1950 to 1960 to 1970, and the small differences in definitions have a negligible effect on most comparisons among these dates.

Urbanized Areas and SMSAs have in common the fact that each contains at least one legal city with a population of at least 50,000, or else contiguous "twin cities" each having at least 15,000 residents and together having 50,000 or more. In addition to the central city or cities, the Urbanized Area includes the "Urban Fringe," or the suburban zone, consisting of all closely settled territory (1,000 or more people per square mile) either adjacent to the central city or cities or separated from the main body of the Urbanized Area by no more than a mile and one-half.[3] The Urban Fringe usually includes a number of incorporated places (suburbs) and often includes in addition some unincorporated, densely settled territory. If the corporate limits of either a central city or a suburb extend into the open countryside, the rural portion of the incorporated territory is excluded from the Urbanized Area. The SMSA, in contrast, consists of an entire county or of two or more contiguous counties, except in New England, where SMSAs consist of towns and cities instead of counties. Counties (or towns in New England) which do not contain the central city, or one of the central cities, are included in the SMSA if they meet certain criteria of economic and social interdependence with the central city or cities. Therefore, an SMSA usually includes some rural population and people in small urban settlements which are not suburbs of the central city. Also, some of the closely settled territory adjacent to the central city or its contiguous suburbs often extends outside the SMSA. Nevertheless, the population of SMSAs outside the central cities is sometimes designated the suburban population of the United States, especially by journalists apparently insensitive to the artificial nature of SMSAs, which grows out of their boundaries coinciding with county lines or the boundaries of New England towns and cities. Social scientists also sometimes use SMSAs for central city-suburban comparisons, but usually only because the Bureau of the Census tabulates some data for SMSAs which it does not tabulate for Urbanized Areas.

The Urbanized Area is very useful for many kinds of urban research, because it roughly corresponds to what James Quinn (1955: 12-14) has called a "natural city" and what Europeans call a "conurbation." That is, it is an urban population concentration separated from other such concentrations by broad bands of water or of sparsely populated or unpopulated land not devoted primarily to such urban uses as commerce and industry. Even though "natural cities" in the United States are increasingly joined by narrow strips of urban or semi-urban land usage along major highways, an aerial view reveals an unmistakable separation into distinct population concentrations.

Although the boundary of an Urbanized Area is not essentially artificial or arbitrary, the same cannot be said for the boundary between the central city and the suburbs within an Urbanized Area. To be sure, the boundaries of the municipalities in one conurbation are important in many ways, especially with regard to the financing and control of local governments and the many public services local governments provide. However, local government is just one of many aspects of social organization, and it is relatively unimportant with regard to many of the social and cultural phenomena with which social scientists are concerned. Therefore, some social scientists believe that the central city-suburban distinction should be made on the basis of demographic and ecological characteristics of neighborhoods rather than on the basis of political boundaries. For instance, a suburban neighborhood may be conceived of as having lower population density than a central city or urban neighborhood, as having a larger proportion of its dwelling units in one-unit structures, and as having residents who travel a greater mean distance to work. According to this definition, many neighborhoods in legal central cities are suburban, and many neighborhoods in legal suburbs are urban rather than suburban.[4]

Related to this "demographic" definition of suburban neighborhoods is the definition of suburbanization as decentralization of population and activities within the natural city. Whereas suburbanization in the legal sense can be retarded, offset, or reversed by annexation, suburbanization in the demographic sense may occur even if the legal boundaries of the central city are extended to enclose all or most of the expanding natural city. In some states, such as Texas, state law makes annexation relatively easy, and some cities have extended their corporate limits well into the surrounding countryside in anticipation of future settlement and commercial development. However, the tract developments, shopping centers, and general "suburban sprawl" around the periphery of these cities are essentially the same as those around the periphery of natural cities undergoing rapid suburbanization in the legal sense. Is it realistic to exclude these cities from discussions of suburbanization and related social, demographic, and economic changes? Generally, the answer is no, although these cities do not suffer from the problems of political fragmentation which result from suburbanization in the legal sense.

A third alternative is to define suburbs and suburbanization in terms of a way of life (for an at least implicit, early definition of this kind, see Fava, 1956). Much of the early literature on the postwar suburbs dwelt on an alleged suburban way of life, including such characteristics as family-centeredness, neighboring, conformity, a nonpartisan approach to local politics, and a nonsectarian approach to religion (see, e.g., Whyte, 1956; Wood, 1958; Riesman, 1957). However, the accuracy of these characterizations of suburbia has been challenged, and empirical evidence has given no more than qualified support to these stereotypic views of suburban life in the absence of conclusive evidence on the existence and nature of a suburban way of life (see especially Berger, 1960; Gans, 1967), it is better left a topic for empirical investigation rather than being incorporated into the definitions of suburb and suburbanization. Even if a satisfactory definition of suburbia as a way of life could be devised, it could not be used to delineate suburbs on the basis of census data or any other readily available evidence.

In fact, the requirement for readily available evidence compels me to adopt the legal definition for my statistical treatment of suburbanization since World War II, even though I believe that suburbanization in the demographic sense may be at least as important and I critically assess the importance of the data on legal suburbs. Suburbanization in the demographic sense is amenable to empirical investigation, of course, through special census tabulations and data gathered specifically to study the phenomenon, and urban ecologists and geographers have intensively studied decentralization within some natural cities during certain periods of time. However, neither the published studies of decentralization nor published census data provide material for easily assembled indicators of the process in the nation as a whole from World War II to the recent past, and to assemble the needed evidence would be an enormous task beyond the resources available for preparing this paper. Therefore, I limit my empirical treatment of suburbanization in the demographic sense to some evidence that it has not always closely corresponded with suburbanization in the legal sense.

THE GROWTH OF URBAN FRINGES FROM
1950 TO 1970

The slowing down of residential construction during the Great Depression of the 1930s, the near cessation of such construction during World War II, and the increase in marriages and fertility during the war produced a huge unmet need for housing by the time the war ended in 1945. The accumulation of savings during the war and the GI Bill home-loan program, which provided low-cost loans to veterans with little or no down payment, assured that much of the need for housing would be met by construction of one-unit structures for family ownership. Widespread ownership of auto-mobiles had already made possible a great dispersion of population around the cities, and the search for relatively inexpensive land for the new residential development provided the impetus for peripheral expansion. The young families of the postwar era may also have preferred to live away from the center of the cities, but more important was the fact that their opportunities for home ownership with a low initial outlay and affordable monthly payments were mainly in the suburban and other peripheral neighborhoods. Therefore, the suburban housing boom was under way almost as soon as the nation disarmed and the economy was retooled for civilian production.

There was considerable postwar suburban growth before April of 1950, when the decennial census of population was conducted, but the lack of comparable data for 1940 and 1950 makes estimation of this growth difficult. At any rate, the suburban population was still a small fraction of the total in 1950, only 13.9% of the people in the nation being in the Urban Fringes of Urbanized Areas (see Table 1). There were almost two and a half times as many people in the central cities and more than two and a half times as many in rural areas. Even the people in small urban places outside Urbanized Areas outnumbered the Urban Fringe residents.

The data in Table 1 reveal the growth of the population of the Urban Fringes from 1950 to 1970. The Fringe population grew 161%, an increase of more than 33.5 million people, and it doubled as a percentage of the total population. In contrast, the population of central cities increased only 32% and declined slightly as a percentage of the total population. Both the rural population and the population of small urban places outside Urbanized Areas declined as a percentage of the total population, although small cities gained more than 3.5 million people and rural areas lost a third of a million.

If the residents of the Urban Fringes are considered the suburban population, suburbanites were more than a quarter of the nation in 1970 and exceeded the rural population by a half-million. Although they were not yet as numerous as the residents of central cities, suburban residents defined in this manner were a substantial and rapidly growing segment of the total population.

In fact, considering only the residents of the Urban Fringes of Urbanized Areas to be suburban leads to an understatement of the size of the suburban population. Cities with fewer than 50,000 people which are not in Urbanized Areas often have suburbs, and the residents of these suburbs are listed either as "other urban" or

TABLE 1

Distribution of the United States Population by Kind of Community, 1950, 1960, and 1970

Kind of Community	1950[a]		1960		1970	
	n	%	n	%	n	%
Urbanized areas						
central cities	48,377,240	32.1	59,975,132	32.3	63,921,684	31.4
Urban Fringes	20,871,908	13.9	37,873,355	21.1	54,524,882	26.8
Other urban	27,218,538	18.1	29,420,263	16.4	30,878,364	15.1
Rural	54,229,675	36.0	54,054,425	30.1	53,886,996	26.5
Total	150,697,361	100.0	179,323,175	100.0	203,211,926	100.0

a. Conterminous United States (excluding Alaska and Hawaii).

"rural" in Table 1. I know of no good estimate of the number of people who live in the suburbs of the smaller cities, and there is no practical way to derive an estimate from the census reports or census tapes, but these suburbanites may number in the millions. To be sure, the suburbs of small cities probably differ appreciably from the suburbs of larger cities in their social, economic, and demographic characteristics, but they should not differ markedly from the Urban Fringes of the smaller Urbanized Areas. In any event, a truly comprehensive study of suburbanization would treat this uncounted and generally ignored segment of the suburban population.

In addition, there are a number of commuter settlements near larger cities which are separated from the central cities and their contiguous suburbs by more than a mile and one-half and thus are not included in Urbanized Areas. These spatially separated but dependent settlements fit the definition of suburb in the first paragraph of this paper and are usually very similar in most respects to the contiguous suburbs, especially if they were never independent communities of any appreciable size. Their residents should, therefore, be included in any complete enumeration of the suburban population. Their populations are generally included if the suburban population is defined as all persons in SMSAs outside the central cities, since most of the detached suburbs are largely or entirely in SMSAs. Unfortunately, however, there were also more than 16 million rural people in SMSAs in 1970, including some who lived great distances from the central cities, and these people should not be included in the suburban population. There was also a substantial but unknown number of people in small independent urban places in the SMSAs who were not suburban in any realistic sense.[5] Therefore, one can only guess at the number of suburbanites outside the Urban Fringes of Urbanized Areas, but the residents of the detached suburbs of the larger cities and the people in the suburbs of small cities probably raise the suburban population to almost a third of the total population in the United States.

The rate of suburban growth has varied greatly among the Urbanized Areas, partly because the total Urbanized Areas have grown at different rates, because varying amounts of land have been available for residential construction in the central cities, and because some central cities have been able to annex territory in which peripheral growth has occurred and other cities have not been able to do so. In order to illustrate the diversity of suburban growth patterns, I selected twenty Urbanized Areas of varying sizes, of varying rates of growth, and in different regions.[6] The percentage of the total population of each Urbanized Area which was in the Urban Fringe in 1950, 1960, and 1970 is shown in Table 2. The percentage increased in 1950-1960 in all the Urbanized Areas except Phoenix and Columbus, although the increase was quite small for Atlanta. The large decline for Phoenix came about because the central city annexed a very large proportion of its former suburban zone during the decade. The percentage increased in 1960-1970 in all Urbanized Areas except Columbus and Memphis, although the increase was hardly more than negligible for Shreveport and Albuquerque. The increase from 1950 to 1970 was quite substantial for the largest Urbanized Areas (Detroit, Boston, Cleveland, St. Louis, Seattle, Atlanta, Kansas City, and New Orleans) and for all those in the Northeast and East North Central Region. In 1950, only Phoenix and Boston had more than half their population in the Urban Fringe, whereas by 1960 ten of the twenty Urbanized Areas had more

TABLE 2

Percentages of the Population of Each of Twenty Urbanized Areas in the Urban Fringe, 1950, 1960, and 1970

Urbanized Area	1950	1960	1970
Kansas City	34.6	48.3	54.0
St. Louis	49.0	55.0	67.0
Albuquerque[a]	—	16.6	18.1
Omaha	19.1	22.6	29.4
Lansing	31.3	36.3	42.7
Detroit	38.7	52.8	61.9
Columbus, Ga.	32.8	26.3	26.1
Atlanta	34.8	36.5	57.6
Chattanooga	21.9	36.6	46.7
New Orleans	13.5	25.8	38.3
Shreveport	15.3	21.2	22.4
Boston	64.1	71.1	75.8
Cleveland	38.9	50.9	61.7
St. Petersburg	15.6	44.2	56.3
Seattle[b]	24.8	35.5	57.1
Spokane	8.1	20.0	25.7
Memphis	2.5	8.6	6.1
Phoenix	50.6	20.4	32.6
New Haven	32.8	45.5	60.5
Hartford	41.0	57.5	66.0
Mean[c]	30.0	37.6	46.7

a. Not designated an Urbanized Area in 1950.
b. In 1970, Everett was designated a central city and the Urbanized Area was titled Seattle-Everett. However, for reasons of comparability, the population of Everett is included in the Urban Fringe at all three dates in this table.
c. Excluding Albuquerque.

people in the Urban Fringe than in the central city, and six had more than sixty percent outside the central city.

Of course, suburbanization in one sense occurs if the Urban Fringe population increases in numbers but not as a percentage of the population of the Urbanized Area. The growth rates (in percentages) of the central cities and the Urban Fringes of the twenty Urbanized Areas from 1950 to 1960 and from 1960 to 1970 are shown in Table 3. The only case of an Urban Fringe population which failed to grow during a decade was Memphis in 1960-1970, and this one decrease was undoubtedly effected by annexation. Some of the Urban Fringe populations grew in spite of large losses to the central cities through annexation, Phoenix in 1950-1960 being the most notable case.

One might infer from the data in Table 3 that most central cities inevitably reach a stage at which their populations stabilize and then begin to decline, although some of the cities were already losing

TABLE 3

Percentage Change in the Population of the Central City and Urban Fringe of Each of Twenty Urbanized Areas, 1950 to 1960 and 1960 to 1970

Urbanized Area	1950 to 1960		1960 to 1970	
	Central City	Urban Fringe	Central City	Urban Fringe
Kansas City	4.4	84.3	6.4	33.5
St. Louis	−12.5	11.4	−17.0	37.3
Albuquerque[a]	−−	−−	21.2	34.2
Omaha	20.1	49.2	15.2	63.3
Lansing	17.0	46.7	22.0	59.3
Detroit	−9.7	60.2	−9.5	31.5
Columbus, Ga.	46.7	7.0	32.0	30.9
Atlanta	47.1	59.0	2.0	140.8
Chattanooga	−0.8	104.6	−8.4	39.1
New Orleans	10.0	143.7	−5.4	69.1
Shreveport	29.2	92.2	10.8	18.7
Boston	−12.9	19.8	−8.1	17.2
Cleveland	−4.2	93.9	−14.3	33.0
St. Petersburg	87.4	703.8	19.3	94.3
Seattle[b]	19.1	99.5	−4.7	130.4
Spokane	12.3	217.4	−6.1	30.4
Memphis	37.5	368.2	14.5	−13.9
Phoenix	311.1	3.3	32.4	149.7
New Haven	−7.5	57.7	−9.4	66.2
Hartford	−8.6	77.8	−2.6	39.9
Mean[c]	30.8	121.0	3.6	56.4

a. Not designated an Urbanized Area in 1950.
b. In 1970, Everett was designated a central city and the Urbanized Area was titled Seattle-Everett. However, for reasons of comparability, the population of Everett is included in the Urban Fringe at all three dates in this table.
c. Excluding Albuquerque.

population in 1950-1960, and some were still gaining in 1960-1970. The apparent evolutionary tendency from population growth to decline is illustrated by the central cities of New Orleans and Seattle, which went from an increase in 1950-1960 to a decline in 1960-1970. The central city of Atlanta experienced rapid growth in 1950-1960 but had a virtually stable population in 1960-1970 as its Fringe population burgeoned, and Chattanooga went from a stable to a declining population. Reaching the stages of suburbanization at which the central-city population stabilizes and declines obviously is not simply a function of the size of the Urbanized Area, since the central city of Kansas City, in an Urbanized Area of over a million people, continued to grow in 1960-1970, whereas Chattanooga and Spokane, each in an Urbanized Area of less than a quarter of a million people, lost population. However, all the central cities of the largest Urbanized Areas (900,000 or more population in 1970) except Kansas City and Atlanta lost population in 1960-1970, and all except Kansas City, Atlanta, and Seattle lost in 1950-1960. Of the medium-sized Urbanized Areas (400,000 to 900,000 in 1970), the central cities of Omaha, St. Petersburg, Memphis, and Phoenix continued to grow in 1960-1970, and only Hartford declined. The central cities of the smallest Urbanized Areas were almost equally divided into gainers (Albuquerque, Lansing, Columbus, and Shreveport) and losers (Chattanooga, Spokane, and New Haven). The variation among cities in Urbanized Areas of similar size probably reflects such factors as age of the cities, recent rates of growth of the Urbanized Areas, state laws regarding annexation, and a multitude of local historical events and developments.

Population figures for the central cities and Urban Fringes (not shown in tabular form) illustrate the extent of suburbanization perhaps even more dramatically than the data on percentage change. The Urban Fringes of the eight largest Urbanized Areas in Tables 2 and 3 increased by 4,732,878 people from 1950 to 1970, while their central cities lost 594,534. The Detroit Urban Fringe alone grew by 1,292,473, while its central city declined by 338,086. The Fringe of Cleveland gained 740,186, while its central city lost 163,905; the Fringe of Boston gained 579,500, while its central city lost 160,373; the Fringe of St. Louis gained 436,223, while its central city lost 234,560.

SOME COMMON MISINTERPRETATIONS OF THE DATA ON SUBURBANIZATION

Data such as those which I present and discuss above are often reported in newspapers and news magazines, usually with the suggestion or implication that the large central cities are being abandoned, that there is a mass migration out of decaying central cities into suburbs. Even many social scientists tend to view the data on suburbanization in those terms. For reasons which I discuss below, the almost universal growth of suburban zones and the decline of the populations of many central cities *are* important trends which have created some serious and intractable problems, but the data showing these changes are often overinterpreted. In many cases, the process of suburbanization, in the legal sense, is little more than the growth of the natural city, while the corporate limits of the central city remain fixed, or almost so. If the corporate limits of the central city are not extended as the natural city expands in area and population size, the central city will almost inevitably contain a progressively smaller proportion of the population of the natural city, and, beyond a certain point, it will almost inevitably suffer an absolute decline in population. The central city contains the central business district, which typically has a large daytime population but a small residential population, and if the central business district expands, it takes land from residential usage. Some central business districts have not grown much since World War II, due to a decentralization of commercial activities or a slow rate of growth of their natural cities, but others have continued to grow, and a few previously stagnant ones have resumed growth in the past few years. Whatever happens to the central business district, as the natural city grows, more space near its core is likely to be devoted to civic buildings, freeways, warehouses, and similar nonresidential uses. Therefore, if the territory of the central city is not expanded and if the population density in its remaining residential areas is not substantially increased, its population is bound to decline after the growth of the natural city reaches a certain point. In some instances, then, the decline of the central-city population may be simply a function of the growth of the natural city as a whole.

The decline of a central-city population is often more than that, of course, since there are blocks of vacant residential and commercial buildings in some central cities (and in some suburbs), but the census data on population decline in central cities are not in themselves evidence of abandonment and decay. In large measure, the people leaving the central cities are not fleeing the cities but are being pushed out by essentially the same processes of succession which have characterized growing American cities since the nineteenth century and which antedate postwar suburbanization.

Another common misinterpretation of data on the growth of legal suburbs is that this growth indicates an increase in, or an influence toward, certain life styles which are believed to characterize suburbia. Although I point out above that the evidence for a distinctive suburban way of life is rather tenuous, there are certain life styles and values which are more prevalent in suburbs than in central cities. For instance, a larger percentage of suburban adults are married, and suburban families are more fertile on the average than central-city families,[7] which is rather conclusive evidence for greater family-centeredness in the suburbs. Furthermore, a larger percentage of suburban people live in one-unit residential structures—an important difference in life style which implies other differences as well.

However, it is important that the central city-suburban differences in the prevalence of certain life styles are apparently small to moderate; all life styles prevalent in the urban population as a whole are fairly common in both central cities and suburbs. And only if all suburbanites had the same life style and were categorically different from all central-city residents would growth of the suburban population necessarily imply any change in life styles in the country as a whole. Furthermore, differences in behavior between central-city and suburban residents are probably very largely (I suspect almost entirely) the result of selective movement to the suburbs rather than the result of influences from living in suburbia. For instance, peripheral suburban neighborhoods have tended to attract family-centered individuals or individuals during a family-centered stage of their lives. New residents of suburbia who are not inclined to be family-centered are unlikely to move into the more family-centered neighborhoods, and, if they do, their values and behavior may be relatively unaffected. It is apparent that the considerable suburbanization during the 1960s did not increase fertility in the nation as a whole, since fertility declined during the decade, and it is doubtful that suburbanization was even an important pro-fertility influence.

Living in a detached one-unit structure, in a house surrounded by a yard, is perhaps the aspect of life style most often associated with suburbia. Since postwar suburbanization has consisted to a large extent of construction of one-unit structures in tract developments, it seems reasonable to assume that living in such houses has increased, or tended to increase, with suburbanization. In fact, I point out above that suburbanization in a demographic sense consists of, among other things, an increase in neighborhoods of one-unit residential structures.

The U.S. summary data from the 1970 census of housing were not yet available as this was written, but there was little increase in the percentage of housing units in one-unit structures between the census of housing conducted in April of 1950 and the census conducted in December of 1959. In 1950, 69.4% of all dwelling units were in one-unit structures and semi-detached two-unit structures, whereas in 1959, 70.6% were in one-unit structures.[8] Since semi-detached two-unit structures are rare, the increase in percentage of dwelling units in one-unit structures could not have been great. This does not mean, however, that suburbanization was not an important influence for one-unit structures. Rather, the decline of the rural population as a percentage of the total was an influence away from one-unit structures which largely offset any effects of suburbanization.

Significantly, the percentage of occupied dwelling units which were owner-occupied increased from 54.9 in 1950 to 62.2 in 1959, and almost nine-tenths of the owner-occupied units were in one-unit structures at both dates. However, the increased occupancy of one-unit structures resulting from increased homeownership was offset by an increased tendency for renters to live in multiple-unit structures. Whereas 44.7% of the renter-occupied dwelling units were in one-unit structures or semi-detached two-unit structures in 1950, only 39.7% were in one-unit structures in 1959. The percentage of renter-occupied units in structures with five or more units increased from 24.0 to 26.3. Therefore, there was a tendency for renters either to become homeowners (no doubt often in the suburbs) or to move into multiple-unit structures. Probably these offsetting influences on the percentage of housing units in one-unit structures were largely aspects of the dual processes of urbanization and suburbanization.

The 1970 data can hardly show an increase in the percentage of housing units in one-unit structures, since there was a marked shift in residential construction away from detached one-unit structures to apartments, condominiums, and town houses during the 1960s. Whereas 80.5% of all the new housing units put in place during 1959 were in one-unit structures, only 55.6% were in one-unit structures in 1970. This change came about largely in two spurts of apartment-building, in 1960-1963 and in 1967-1969 (U.S. Savings and Loan

League, 1971: 24). Most predictions are that this trend will continue and perhaps accelerate as land in and near the major population concentrations becomes more scarce and expensive and as one-unit structures become too expensive for middle-income and even upper-middle-income families (for one of the many such predictions, see Sternlieb, 1972). If these predictions are correct, and if residential construction continues to be predominantly in the suburbs, a house of one's own with a lawn and a back yard for barbecuing and for the children and pets to romp in will become less and less characteristic of suburbia. Already, many people are leaving one-unit structures in central cities to live in apartments (sometimes high-rise apartments) in the suburbs. If suburbanization ever entailed a shift away from an essentially urban way of life, it is a mistake to assume that it will continue to do so.

The extent to which some suburbs were already "urbanized" in 1970 is revealed by data on the percentage of housing units in one-unit structures in some of the suburbs in the Urbanized Areas listed in Tables 2 and 3. Although the percentage was generally lower for the central city than for the total of its suburbs with 2,500 or more people, most of the Urbanized Areas had at least one or two suburbs with a lower percentage than the central city. For instance, the percentage was 49.3 for Atlanta and 46.1 for two of its suburbs, Astell and Chamblee. The percentage was 68.7 for Omaha but was only 17.0 and 27.2 respectively for two Omaha suburbs, Offutt East and Offutt West. One suburb of Seattle (Tukwila) had a percentage of housing units in one-unit structures of only 36.4, compared with 59.8% for Seattle.

So far, it may seem that the process of suburbanization in the legal sense, as indicated by the data in Tables 1, 2, and 3, is not very important, and, indeed, I maintain that it is not as important as some people have believed. However, the next section should make it clear that the growth of legal suburbs since World War II has had some important consequences and created a number of problems.

SOCIAL AND ECONOMIC POLARIZATION
OF CENTRAL CITIES AND SUBURBS

Even if people in the suburbs generally had the same charac-teristics as people in central cities, suburbanization would have some important effects on the financing of local government, especially in the central city. In some respects, the municipal government of the central city provides services for people who live in all parts of the natural city, including especially the suburban residents who work in the central city. Many suburbanites use the streets, parks, zoos, museums, libraries, and other facilities of the central city, and, when they work, shop, or seek entertainment there, they require police protection and other services from the city. And yet, since central cities have tended to rely on property taxes for their revenues, suburbanites have largely avoided direct taxation by the central city. Some central cities do directly tax suburban residents through income and sales taxes, and suburbanites pay various taxes indirectly to the central city if they buy goods and services there, but there is a widespread belief that suburbanites do not bear their fair share of the costs of local government for the natural city.

The consequences of suburbanization on the finances of the central city are exacerbated by the fact that, in most natural cities, the suburbanites tend to be more prosperous and of higher social status than the residents of the central city. Duncan and Reiss (1956) demonstrated that this difference already existed in 1950, and, from 1950 to 1960, socioeconomic differences between cities and suburbs generally increased (for evidence of this increase, see Farley, 1964; Fine et al., 1971, among others). The movement of more prosperous people to the suburbs has, of course, cut into the tax base of the central city, and the growing slum and ghetto areas in the central city have placed heavier demands on the welfare services and law enforcement agencies there. And as a larger percentage of the people in a natural city come to live in the suburbs, the effects of a given degree of central city-suburban economic differentiation become greater.

One important aspect of the socioeconomic polarization is the concentration of blacks in central cities, as the black population has become highly urbanized, and the more limited movement of blacks into suburbs. The greater average affluence of suburbanites has resulted largely from the fact that most persons moving into suburbia have had to be able to afford the new or relatively new housing there, and the movement of blacks into new housing has been retarded by discrimination as well as by economic factors. Therefore, the percentage of the population of the central cities of Urbanized Areas which was nonwhite went from 13.1 in 1950 to 22.5 in 1970 (see Table 4) and the number of nonwhites in central cities increased

TABLE 4

Representation of Nonwhites in Different Kinds of Communities,
1950, 1960, and 1970

Kind of Community	Number of Nonwhites			% of the Population Nonwhite		
	1950[a]	1960	1970	1950[a]	1960	1970
Urbanized areas						
central cities	6,335,272	10,347,900	14,375,113	13.1	17.8	22.5
Urban Fringes	988,840	1,730,652	3,119,951	4.7	4.6	5.7
Other urban	2,387,139	2,761,866	3,056,626	8.8	9.4	9.9
Rural	6,044,082	5,651,025	4,911,261	11.1	10.5	9.1
Total	15,755,333	20,491,443	25,462,951	10.4	11.4	12.5

a. Conterminous United States (excluding Alaska and Hawaii).

by more than eight million. In contrast, nonwhites in Urban Fringes increased by just over two million, and nonwhites decreased slightly as a percentage of the Urban Fringe population from 1950 to 1960 and increased by only one percentage point, to 5.7% by 1970. The difference in the percentages of nonwhites in the central city and in the Urban Fringe increased from 8.4 percentage points in 1950, to 13.2 in 1960, to 16.8 in 1970.

I need not discuss all of the several important implications of racial polarization of central cities and suburbs since most of them are treated elsewhere in this volume. I need only point out that racial differences contribute substantially to the general social and economic differences I discuss below, and they mean that the suburbs are havens for the relatively privileged segments of the population to an even greater extent than the data presented below would otherwise suggest.

The 1970 aggregated data on occupation, education, and income in the central cities and Urban Fringes of all Urbanized Areas in the United States had not been published when this paper was written, so I rely here on the socioeconomic data for each of the twenty Urbanized Areas listed in Tables 2 and 3. In many respects, data on individual Urbanized Areas are more valuable than the aggregated data, since the latter are disproportionately affected by the characteristics of a few of the larger Urbanized Areas, and since trends in the aggregated data are affected by the addition of Urbanized Areas from one census year to the next as well as by changes within individual Urbanized Areas. Furthermore, data on individual Urbanized Areas illustrate the variation in central-city suburban differences and the variation in changes in those differences.

Summary measures of the differences between the twenty central cities and their Urban Fringes in occupation, education, and family income in 1950, 1960, and 1970 are presented in Tables 5, 6, and 7. The sample of Urbanized Areas is not precisely representative of Urbanized Areas as a whole, but comparison with data on all Urbanized Areas in 1950 and 1960 reveals that the sample is not highly unrepresentative for those dates. Furthermore, Urbanized Areas in the sample changed in the same direction, on the average, and to a similar degree as all the cases in 1950-1960, and they probably did so in 1960-1970.

The data show a sharp increase in central city-suburban socioeconomic polarization from 1950 to 1960 and a small increase from 1960 to 1970. On the average, the central cities and Fringes did not differ much in 1950, but in 1970 most of the Fringes had a moderate to marked advantage over their central cities on each of the three

TABLE 5

Difference Between Central City and Urban Fringe in Percentage of Employed Workers in Nonmanual Occupations, in Each of Twenty Urbanized Areas, 1950, 1960, and 1970 (minus indicates smaller percentage in urban fringe)

Urbanized Area	1950	1960	1970
Kansas City	−7.0	3.5	5.2
St. Louis	7.3	14.0	16.0
Albuquerque[a]	−−	−24.4	−25.6
Omaha	−8.4	−1.0	−2.8
Lansing	3.9	12.1	14.6
Detroit	−1.0	6.4	10.9
Columbus, Ga.	−10.5	−6.0	−16.9
Atlanta	13.9	17.3	16.3
Chattanooga	11.6	13.5	13.2
New Orleans	−15.4	5.8	4.7
Shreveport	−12.7	0.9	−3.1
Boston	4.1	7.8	5.3
Cleveland	23.7	22.9	22.0
St. Petersburg	5.1	−2.1	0.6
Seattle[b]	−7.5	−1.4	−0.5
Spokane	−10.5	−4.1	−3.1
Memphis	−17.8	1.4	11.7
Phoenix	−19.6	1.0	4.6
New Haven	6.2	6.4	6.6
Hartford	5.2	15.6	17.8
Mean[c]	−1.5	6.0	6.5

a. Not designated an Urbanized Area in 1950.
b. In 1970, Everett was designated a central city and the Urbanized Area was titled Seattle-Everett. However, for reasons of comparability, the population of Everett is included in the Urban Fringe at all three dates in this table.
c. Excluding Albuquerque.

variables. In many cases, the change was rather pronounced and was monotonic from 1950 to 1960 to 1970. St. Louis, Lansing, Detroit, Memphis, Phoenix, and Hartford changed monotonically on all three variables, and the Fringe experienced a net gain in relative status from 1950 to 1970 in seventeen of the nineteen cases in occupation, in eighteen cases in education, and in sixteen cases in family income. By 1970, there were still six of the twenty Urbanized Areas in which the percentage of employed workers in nonmanual occupations was greater in the central city than in the Fringe, but the difference was marked only with Albuquerque and Columbus. Two of the central cities still exceeded their Fringes in percentage of high school graduates, the difference in the case of Albuquerque being pronounced, and median family income was higher in three of the central cities and was virtually the same in one central city and its Fringe.

TABLE 6

Difference Between Central City and Urban Fringe in Percentage of Persons Age 25 and Older who had Completed 12 or More Years of School, in Each of Twenty Urbanized Areas, 1950, 1960, and 1970
(minus indicates smaller percentage in urban fringe)

Urbanized Area	1950	1960	1970
Kansas City	−4.5	6.6	9.0
St. Louis	8.1	17.0	22.8
Albuquerque[a]	−	−21.4	−24.4
Omaha	−9.7	−0.3	4.3
Lansing	14.9	18.2	22.6
Detroit	3.6	13.1	16.9
Columbus, Ga.	−5.0	−9.9	−1.4
Atlanta	12.6	17.5	15.4
Chattanooga	11.7	16.4	16.3
New Orleans	0.3	11.3	9.9
Shreveport	1.9	10.8	3.5
Boston	7.7	11.2	13.1
Cleveland	24.5	25.4	23.2
St. Petersburg	1.3	0.9	3.1
Seattle[b]	1.5	7.4	7.0
Spokane	−2.4	5.6	4.8
Memphis	−15.3	4.2	17.0
Phoenix	−6.9	7.6	7.8
New Haven	10.0	9.1	11.7
Hartford	14.1	20.1	24.7
Mean[c]	3.6	10.1	12.2

a. Not designated an Urbanized Area in 1950.
b. In 1970, Everett was designated a central city and the Urbanized Area was titled Seattle-Everett. However, for reasons of comparability, the population of Everett is included in the Urban Fringe at all three dates in this table.
c. Excluding Albuquerque.

TABLE 7
Ratio of Central-City to Urban Fringe Median Family Income, in
Each of Twenty Urbanized Areas, 1949, 1959, and 1969

Urbanized Area	1950	1960	1970
Kansas City	1.00	.87	.88
St. Louis	.86	.76	.70
Albuquerque[a]	--	1.35	1.38
Omaha	1.02	.99	1.00
Lansing	1.09	.94	.86
Detroit	.90	.81	.76
Columbus, Ga.	1.03	.94	1.08
Atlanta	.71	.72	.70
Chattanooga	.69	.72	.70
New Orleans	.90	.78	.73
Shreveport	1.09	.98	1.05
Boston	.88	.83	.77
Cleveland	.74	.74	.76
St. Petersburg	1.02	.95	.94
Seattle[b]	1.02	.93	.89
Spokane	1.03	.89	.87
Memphis	1.17	.85	.74
Phoenix	1.00	.98	.96
New Haven	.89	.82	.76
Hartford	.80	.74	.69
Mean[c]	.94	.85	.83

a. Not designated an Urbanized Area in 1950.
b. In 1970, Everett was designated a central city and the Urbanized Area was titled Seattle-Everett. However, for reasons of comparability, the population of Everett is included in the Urban Fringe at all three dates in this table.
c. Excluding Albuquerque.

Even in these exceptional cases, the trend was usually toward the more typical pattern, Albuquerque being the most notable exception. However, the exceptions should not be overlooked, for they belie the universal validity of the popular view that the affluent and well-educated predominate in suburbs while the poor and under-privileged are concentrated in central cities. Furthermore, since the trend to polarization apparently slowed down in the 1960s, it is less than certain that the usual pattern will become universal. Significantly, of the nineteen Urbanized Areas in my sample for which data are available for all three dates, seventeen experienced a shift toward higher relative occupational status in their Fringes from 1950 to 1960, but only twelve did so from 1960 to 1970. Similarly, the shifts toward higher relative educational status in the Fringes fell from sixteen in 1950-1960 to twelve in 1960-1970, and the comparable changes with family income were sixteen and fourteen, respectively, during the two decades.

Schnore (1963) examined 1960 data on all Urbanized Areas and found that only the older ones (age being defined as the time elapsed since the central city first attained a size of 50,000 people) rather consistently had Fringe populations which ranked above the central-city populations in occupation, education, and income. On the basis of this finding, he hypothesized that as natural cities ("urban agglomerations," in his terminology) grow older, they tend to evolve toward the well-known Burgess concentric zonal pattern, whereby mean socioeconomic status varies positively and monotonically with distance from the center of the city.

The data in Tables 5, 6, and 7 generally support Schnore's hypothesis, although the tendency for the relative status of the Fringe residents to increase was not universal in 1950-1960 and was even less nearly universal in 1960-1970. If the older Urbanized Areas have evolved in the manner hypothesized by Schnore, it is less than certain, as Schnore is careful to point out, that the younger ones will follow the same course, and the deceleration of the trend to higher status in the Fringes in 1960-1970 suggests that the younger natural cities may well not follow the traditional evolutionary course. Furthermore, higher mean status in the Urban Fringe than in the central city is not in itself adequate evidence for close approximation of the Burgess pattern, and a trend toward higher relative status in the Fringe does not necessarily mean that an Urbanized Area is evolving toward the Burgess pattern. For instance, at least a moderate increase in the relative status of the Fringe population could result from a greater ability of the more affluent suburbs, many of which are now enclaves surrounded by the central city, to resist annexation. And if the mean status of the residents should vary from the center to the periphery of the Urbanized Area in a nonlinear fashion, growth of the natural city while the limits of the central city remain stationary could by itself effect a change in the relative status of the central-city and Fringe populations.

The summary measures in Tables 5, 6, and 7, of course, do not reveal many aspects of the socioeconomic differences between central cities and suburbs, and, in some cases, they tend to understate the differences. For instance, in those cases in which a larger percentage of the employed workers were in nonmanual occupations in the Fringe, the difference in the percentages usually understates the difference in occupational status because a larger percentage of the nonmanual workers were in the higher-status, nonmanual occupations in the Fringe. Furthermore, the summary measures are not good indicators of the consequences of central city-suburban socioeconomic polarization because those consequences depend on the distribution of the population between the central city and the Fringe as well as on the degree of polarization.

The data in Table 8 reveal some details on central city suburban family income differences in eight rather diverse Urbanized Areas and illustrate variation in the probable consequences of similar degrees of income polarization in different kinds of Urbanized Areas. At each of fifteen income levels (1969), I show the percentage of the total families who lived in the central city in 1970. Probable effects of central city-suburban income differences in three distinctly different kinds of Urbanized Areas can be inferred from the data on Detroit, Boston, Memphis, and St. Petersburg in the first four columns of the table.

The data for Detroit dramatically illustrate how great the adverse consequences for the central city are likely to be when there is high-income polarization combined with a high percentage of the population outside the central city. The central city had just over a third of the total families but almost two-thirds of the very poor families and less than a sixth of the very wealthy ones. People at all income levels below $10,000 were more than proportionately represented in the central city, and people at all higher-income levels were less than proportionately represented. The effects of income polarization were probably quite similar for Boston, which had less income polarization than Detroit but a smaller percentage of the population in the central city. As in Detroit, the central city of Boston had almost twice its proportionate share of very poor families and well under half its proportionate share of very wealthy families.

In contrast to Detroit and Boston, a relatively high degree of income polarization between the central city and Fringe of Memphis (about the same degree as in Detroit, according to the summary measures in Table 7) could not have had a very great adverse effect on the central city, since only 6.1% of the families were in the Fringe. At only one income level ($15,000-49,999) did the percentage of families in the central city fall slightly below 90% and the range in the percentages among the income levels was less than ten points.

St. Petersburg illustrates still another combination of characteristics—namely, low central city-suburban income polarization and a large percentage of the population in the Fringe. As I point out above, the latter of these characteristics alone may place a

TABLE 8

Percentage of the Families at Each Income Level in 1969 who were in the Central City in 1970 in each of Eight Urbanized Areas

Family Income	Detroit	Boston	Memphis	St. Petersburg	Phoenix	Albuquerque	Omaha	Spokane
Less than 1,000	63.7	39.0	97.9	51.0	68.8	71.2	70.9	78.3
1,000-1,999	61.5	35.0	96.5	48.3	72.9	74.7	76.2	83.4
2,000-2,999	62.4	37.8	98.1	47.4	69.4	78.0	77.6	83.0
3,000-3,999	58.4	38.0	97.5	43.0	66.0	77.1	78.8	79.4
4,000-4,999	56.1	34.8	97.5	41.1	65.4	76.3	73.4	83.7
5,000-5,999	55.3	32.6	97.2	41.6	67.0	75.0	70.5	79.7
6,000-6,999	53.2	30.7	95.6	41.4	67.7	78.4	66.3	77.9
7,000-7,999	50.0	27.5	96.5	40.7	68.2	77.6	67.3	77.0
8,000-8,999	46.5	25.4	94.2	42.4	69.4	81.5	69.9	72.8
9,000-9,999	40.8	23.8	93.1	37.4	69.5	82.7	69.7	72.9
10,000-11,999	36.4	20.3	91.7	39.1	68.7	86.3	69.4	70.5
12,000-14,999	31.6	18.1	90.0	39.0	65.8	89.2	69.1	68.5
15,000-24,999	27.3	15.0	89.9	38.7	64.4	92.6	71.2	69.8
25,000-49,999	21.1	11.2	93.3	41.4	60.9	95.8	78.6	77.1
50,000 and up	14.9	7.5	92.6	47.7	65.6	90.4	76.0	93.1
Total	37.7	22.2	93.9	41.4	66.9	83.2	70.8	74.4

disproportionate financial burden on the central city, but in the case of St. Petersburg, the plight of the central city was not exacerbated by its having an enormously disproportionate share of the poor families and a much smaller than proportionate share of the affluent families. Although poor families were somewhat overrepresented in the central city, so were very wealthy families. Upper-middle-income families ($9,000-24,999) were underrepresented in the central city, but not by a substantial margin.

The remaining Urbanized Areas in Table 8 represent other variations in central city-suburban income differences. For instance, Phoenix, Omaha, and Spokane (as well as Memphis) illustrate a quite common phenomenon—namely, overrepresentation of low-income families in the central city but a higher representation of very wealthy than of upper-middle-income families in the central city. In both Omaha and Spokane, the proportion in the central city was greater for the very wealthy than for all families, the difference being very substantial in Spokane. Albuquerque is the only case for which I present data of underrepresentation of families at all the lower-income levels and overrepresentation of families at all the higher levels ($10,000 and up) in the central city. Albuquerque also differed from many other Urbanized Areas by having its very wealthy families less concentrated in the central city than are its upper-middle-income families. Omaha illustrates that there can be important income differences between the central city and the Fringe even when the median incomes are virtually identical; families both in the highest and in four of the lower-income levels were distinctly overrepresented in the central city.

In addition to creating financial problems for the central city, the socioeconomic polarization between central cities and suburbs which has evolved in the larger and older natural cities in the United States may have a variety of other important consequences, although there is no conclusive evidence for most other possible effects. For instance, the quality of education in the public schools is affected not only by level of financial support but by the typical backgrounds of the pupils (Coleman et al., 1966), and the movement of middle-class people to the suburbs may increase segregation in the schools by social level (as well as by race). Furthermore, increased socioeconomic polarization between central cities and suburbs may limit contact and communication outside the schools among people of different social levels, and, if so, the development of empathy and understanding may be impaired.

However, there is no conclusive evidence that suburbanization since World War II and the increased socioeconomic polarization as shown by the data on central cities and Urban Fringes has had any great effect on degree of contact and communication among people at different social levels. Residential segregation by social level has been considerable for many decades, and, at least in eight Urbanized Areas, there was no increase in measured residential segregation by occupational level from 1950 to 1960, and the segregation was somewhat greater in the central cities than in the Fringes at both dates (Fine et al., 1971).[9] Even before there was substantial suburbanization, the larger central cities had different public schools which served the poorer and the more affluent sections of the cities, so suburbanization may have had little effect on segregation by social level in the schools.

It is widely believed that suburbs tend to contain people within a narrow range of income, and critics of suburbia have attributed a number of undesirable effects, especially on children, to living in homogeneous, "one-class" suburbs (for a more or less typical allegation of this nature, see Lee, 1963). However, 1970 census data (for 1969) reveal that at least the larger suburbs did not typically have a great deal less income inequality among families than did the central cities, and a few suburbs had very high income inequality. For all cities with 50,000 or more people, the census reports give an "index of income concentration" for families. The value of the index can vary theoretically from zero (all families having the same income) to 1.0 (one family having all the income). The differences in the index between Boston and its suburbs with 50,000 or more people are rather typical of the differences between the larger central cities and their larger suburbs. The index was .350 for Boston and ranged from .288 to .457 among its eight larger suburbs, the mean for the suburbs being .334. One of the larger central city-suburban differences was the case of Detroit, which had an index of .348, while the mean for its thirteen larger suburbs was .274. Many of the central city-suburban differences were smaller than the typical difference between cities in the South, which have relatively high income inequality, and cities in other regions. Therefore, if there are many suburbs whose residents generally have incomes within a narrow range, most of those suburbs must be relatively small and probably are near other suburbs or portions of the central city with quite different economic characteristics.

SUMMARY AND CONCLUSIONS

Suburbanization in the United States since World War II has been a dramatic and substantial change which has had many important consequences for American society, many of which are treated in detail in other chapters of this volume. However, most discussions of suburbanization in the popular press, and not a few of the academic treatments, have exaggerated its importance and consequences. Suburbanization in the usual sense of the term (which I call the "legal" or "political" sense) has been in large measure simply a consequence of "natural cities" or "conurbations" expanding more rapidly than the political boundaries of the central cities. It has been augmented by a decentralization of population and activities within natural cities, but suburbanization in the legal sense would have been substantial even in the absence of decentralization.

Although certain life styles are more prevalent in suburbs than in central cities, there is no convincing evidence that suburbanization has played an important causal role in changes in life styles in the country as a whole since World War II. Suburbanites have been more family-centered and fertile on the average than residents of central cities, but perhaps only because family-centered individuals—and individuals during a family-centered stage in their lives—have been attracted to the newer neighborhoods on or near the periphery of natural cities. As more jobs are moved to suburbs, many of the new residents of suburbia will have different motives for moving there, and they may not typically exhibit the life styles, values, and behavior which have tended to characterize suburbanites during the past two and one-half decades.

In 1950, most of the older and larger central cities had suburbs whose populations ranked higher on the average in occupation, education, and income than the people in the central cities; by 1970, many newer and smaller cities and their suburbs had evolved those same kinds of differences, and the differences in the older and larger cities had generally become greater. However, the trend to central city-suburban socioeconomic polarization was substantially less pronounced in 1960-1970 than it was in 1950-1960, and it may not continue. A good many natural cities still had no substantial central city-suburban status differences in 1970, and a few had higher mean status in their central cities. Furthermore, all central cities and at least most of the larger suburbs each retained a high degree of socioeconomic diversity among its residents, and there is no convincing evidence that suburbanization has substantially lessened contact and communication among the social levels.

Suburbanization is sometimes conceived of in a demographic sense, in which case it is considered to consist of decentralization of people and activities within natural cities or growth of low-density neighborhoods of detached one-unit housing structures. I present no systematic evidence in this paper of suburbanization in this sense, but undoubtedly considerable decentralization has occurred in most natural cities in the United States since World War II. However, any trend toward one-unit housing structures, which has never been pronounced since World War II, has apparently been reversed. The stereotypical suburban neighborhoods, consisting of detached one-family houses surrounded by yards, are apparently destined to become less characteristic of the American way of life.

NOTES

1. Many definitions of suburbs are similar to this one but are more elaborate. For instance, Boskoff (1970: 109) defines suburbs as "those urbanized nuclei located outside (but within accessible range) of central cities that are politically independent but economically and psychologically linked with services provided by the metropolis." He defines "urbanized nuclei" in turn as "those areas outside of the central city that have relatively substantial population densities, a preponderance of nonrural occupations, and distinctively urban forms of recreation, family life, and education."

2. Complete definitions of Urbanized Areas, SMAs, and SMSAs are in almost any census publication which reports data for those units.

3. The census term "Urban Fringe" is capitalized in this paper to distinguish it from the terms "urban fringe," "fringe," and "rural-urban fringe," which are often used in urban sociology to refer to territory of less than urban population density but near cities and socially and economically integrated with the cities. The Urban Fringe of Urbanized Areas contains little or none of the "urban fringe" (see Martin, 1953).

4. Although this definition is often implicit in the writings of social scientists and journalists, it is rarely stated explicitly. For instance, when a recent article on suburbs in *U.S. News and World Report* (on August 7, 1972) refers to suburbs of Houston "both inside and outside the city's 450-square-mile limits," the legal definition of suburbs obviously is not being used, but no formal definition is given.

5. Since most Urbanized Areas were largely within SMSAs and most SMSAs contained at least most of one Urbanized Area, subtracting the rural SMSA population and the population of all Urbanized Areas from the population of all SMSAs provides a rough estimate of the urban SMSA population outside Urbanized Areas. This estimate for 1970 is 4,560,705, or 2.2% of the total U.S. population. There is no practical way to arrive at a good estimate of the division of this population between detached suburbs and independent communities, but the independent communities are likely to contain half of it or more.

6. This is a purposive sample which may not be as representative of the universe as a random sample usually would be. The selection of cases was dictated in part by the availability of data, since the 1970 census reports which give most of the relevant data on Urbanized Areas (the state reports of *General Social and Economic Characteristics*) were not all published when this paper was written. For instance, data for New York and California were not available, and data for Texas became available only as this paper was nearing completion.

7. In 1970, the percentages of females age 14 and older who were married were 56.1 and 63.8, respectively, for central cities and Urban Fringes of Urbanized Areas, and the corresponding percentages for males were 62.5 and 68.3. The number of children under five years of age per 1,000 women ages 15 through 49 (the fertility ratio) was 336 for central cities and 342 for Urban Fringes. Significantly, the central city-Fringe difference in the fertility ratio was much greater in 1960, the ratio being 442 for central cities and 498 for Urban Fringes. Whereas the Fringe ratio was only about two percent higher than the central-city ratio in 1970, it was about thirteen percent higher in 1960. In this as well as several other respects, the suburbs seem to be losing their distinctiveness, perhaps because they are coming to contain many people who do not live in neighborhoods on or near the periphery of natural cities.

8. However, 81% of the new units (built since the last census of housing) were in one-unit structures in 1959, and new units were more than a fourth of the total units. Therefore, there would have been a more marked trend toward one-unit structures during the 1950s if many one-unit structures (perhaps largely in rural areas and small towns) had not been razed or abandoned.

9. Although the measures of segregation for entire Urbanized Areas in 1950 and 1960 were computed in the study by Fine, Glann, and Monts, they are not reported in the cited article.

REFERENCES

BERGER, B. (1960) Working-Class Suburb. Berkeley: Univ. of California Press.

BOSKOFF, A. (1970) The Sociology of Urban Regions. New York: Appleton-Century-Crofts.

COLEMAN, J. S. et al. (1966) Equality of Educational Opportunity. Washington, D.C.: Government Printing Office.

DUNCAN, O. D. and A. J. REISS (1956) Social Characteristics of Urban and Rural Communities, 1950. New York: John Wiley.

FAVA, S. F. (1956) "Suburbanism as a way of life." Amer. Soc. Rev. 21 (February): 34-37.

FARLEY, R. (1964) "Suburban persistence." Amer. Soc. Rev. 29 (February): 38-47.

FINE, J., N. D. GLENN, and J. K. MONTS (1971) "The residential segregation of occupational groups in central cities and suburbs." Demography 8 (February): 91-101.

GANS, H. (1967) The Levittowners: Ways of Life and Politics in a New Suburban Community. New York: Pantheon.

LEE, D. (1963) "Suburbia reconsidered: diversity and the creative life," in E. Green (ed.) Man in the Modern City. Pittsburgh: Univ. of Pittsburgh Press.

MARTIN, W. T. (1953) The Rural-Urban Fringe: A Study in Adjustment to Urban Location. Eugene: Univ. of Oregon Press.

QUINN, J. A. (1955) Urban Sociology. New York: American Book.

RIESMAN, D. (1957) "The suburban dislocation." Annals of Amer. Academy of Pol. and Social Sci. (November): 123-142.

SCHNORE, L. (1963) "The socioeconomic status of cities and suburbs." Amer. Soc. Rev. 28 (February): 76-85.

STERNLIEB, G. (1972) "Death of the American dream house." Society 9 (February): 39-42.

U.S. Savings and Loan League (1971) 1971 Yearbook. Chicago.

WHYTE, W. H., Jr. (1956) The Organization Man. New York: Simon & Schuster.

WOOD, R. (1950) Suburbia—Its People and Their Politics. Boston: Houghton Mifflin.

An Image that Will Not Die: Thoughts on the History of Anti-Urban Ideology

JEFFREY K. HADDEN
JOSEF J. BARTON

□ IT IS THE PURPOSE OF THIS ESSAY to trace the history of man's thought about cities and suburbs. We take as axiomatic that man, in his yet short experience, still struggles to accommodate himself to the reality of the urban revolution. Without denying the significance of the political, economic, technological, and organizational dimensions of urban problems, we argue that the negative *image* man holds of the city looms as a major obstacle to the maturation of urban civilization. In substantial measure, the discontents of urban civilization are a function of man's deep-seated and fundamental rejection of the city as an *idea*.[1]

The suburb is, in part, merely a consequence of rapid population expansion along the lines of least resistance toward cheaper land at the fringes of cities, facilitated by expanding technology's ability to transport men greater and greater distances between place of residence and place of work. But the suburb is also an attempt to compromise the imperatives of the industrial revolution, to re-create the pastoral rapport between man and nature, at least on weekends and holidays.

We have known that some—indeed, many—men of history disliked the city, but scholars have only recently begun to explore the implications this has. It is now rather thoroughly documented that many of the key figures in the framing of this nation held city life and all that cities represent in utter disdain (Rourke, 1964). Similarly, research has pointed to the like disaffection of many intellectual giants of the American past (White and White, 1962). The historical roots of anti-urbanism remain, however, inadequately plumbed. Nor can such a sweeping topic be comprehensively explored here. Our task, thus, is exploratory, a seeking to establish only the plausibility of a line of reasoning, and thereby hopefully stimulate others to join in more thorough research.

Our analysis proceeds in three distinct directions. First, we attempt to establish plausibility for the proposition that anti-urban sentiment is as old as cities themselves. Second, we examine the objective conditions and consequences of the early industrial-urban revolution. And third, we explore in some detail the mosaic of several complex ideological threads in American history which have interacted to preserve and give peculiar shape to a negative image of cities and the consequent suburbanization of our own culture. Underlying each of these tasks is a theoretical assumption that man's basic value orientations—i.e., his taken-for-granted assumptions about the nature of reality—significantly mold the physical environment he creates.

THEORETICAL ORIENTATION

The theoretical heritage of the sociology of knowledge informs our inquiry, particularly as this perspective has been elaborated by Berger and Luckmann (1966). Reality is socially constructed. Man the creator is in turn molded and continually reshaped both by the objective structures of his creation and by the meanings he internalizes about social reality. Man created the city and gave meaning to his creation. But this product is in turn an objective reality which, on the one hand, imposes restraints and limitations and, on the other hand, presents new vistas and opportunities. In wrestling with the possibilities and meanings of the city with its sui generis realities, man is ever internalizing new meaning.

Thus, this task of world-building (city-building) is an ongoing process. Still, meanings once internalized in the collective human consciousness may transcend the generations. Meanings internalized by men constitute the essential social bonds of human societies—that is, assumed truths about the nature of social reality. But that which is internalized as truth and the objective imperatives of social structures are not necessarily congruent. Man's capacity to create (invent) new artifacts results in a process of continually changing social structures. Meanings, thus, constantly shift to accommodate new realities. But, again, there is neither inherent necessity nor guarantee that meaning and structure will move toward harmony. Indeed, meaning and structure may drift apart to the point of ultimate collapse of a human group. But, more important for our purposes, meaning and structure can be unbalanced for long periods of time by mutually reinforcing the imbalance.[2]

To apply this to our task, we believe the meaning or image of the city developed in human history to be antithetical to the structural reality of the city. From the onset, man viewed the city as contrary to the "natural" relationship between man and environment. As man altered his nomadic life style for more permanent or quasi-permanent settlements, he came to define the wilderness as "unnatural." But to his subsequent transition from this pastoral-agrarian environment to an urban environment, he did not bring a parallel redefinition of the meaning of "natural" environment. Momentarily we will attempt to show some compelling objective reasons for this failure to reconstruct reality. For the present, however, we wish simply to postulate the initial imbalance between meaning and structure and to further suggest the reciprocal nature of the incongruity. Early cities were not only a departure from the natural as man had come to define it, but also the very structure of urban life lent objective reason for man's dislike and distrust. This internalized meaning, in turn, affected the ongoing process of city-building by structuring and restructuring the city in a manner to reinforce rather than reduce estrangement. City builders have been men *in* but ideologically not *of* the cities—that is, not attuned to the proposition (or internalized meaning) of cities as man's "natural" habitat. Hence, efforts to utilize man's resources to create an ideal urban milieu have hardly even been attempted. Stated otherwise, man has approached the city as an unnatural and unpleasant environment and has thus contributed to a self-fulfilling prophecy. Moreover, the circuit's effects are felt *both* in terms of the physical structure of the city and in terms of man's feelings toward this environment.

Let us hasten to add that the history of man's thought about the city has not been monolithic. Greek philosophers loved their cities, as have some other men of history. Many have viewed the city as the locus of nearly all man's intellectual, ideological, cultural, and technological achievements (Spengler, 1928: 90). Indeed, the words city and civilization have a common origin in the Latin *civis*. But to acknowledge that some men have liked the city and that many, if not most, of man's achievements have occurred in the city is not to disqualify the proposition that anti-urban sentiment has undergirded and dominated man's thinking about the city.

INEQUALITY AND
THE EMERGENCE OF CITIES

Answers to the questions of where, when, and how cities emerged can never be known with certainty. But the space-time/order-of-magnitude aspects of these questions now seem in reasonable focus (Hammond, 1972; Mumford, 1961; Sjoberg, 1960). While the fairly immediate ancestors to homo sapiens may have been around for millions of years, physically modern man seems to date back some forty to fifty thousand years. Cities did not appear until roughly five or six thousand years ago. The process of transition from permanent farming villages to "real" cities was in one sense very gradual. Still, in a relative sense, it happened very quickly. The transition from food-gathering to food production, implying the first permanent settlement, did not occur until the Mesolithic period, roughly ten to twelve thousand years ago. Evidence of the embryonic foreshadowings of cities are apparent by 4000-3500 B.C. and by 2400 B.C. there is evidence of well-developed urban culture (Hammond, 1972).

It is neither central nor essential to our task to elaborate the necessary and sufficient conditions for the existence of cities (Childe, 1950). Rather, we will discuss only those conditions having a bearing on man's image of the city.

In our view, the development of organizational principles radically different from those previously existing stands as *the* critical imperative for the emergence and development of cities. Central to this process was the development of principles of hierarchy not based on kinship. In embryonic form, some cities may well have involved organization along the lines of elaborate kinship systems. But leaders of extended kin networks were necessarily caught in a role conflict between their fellow kinsmen and the imperatives of urban development as perceived by those at the pinnacle of the hierarchy. Only as the influence of kinship declined was it possible to develop the elaborate division of labor which distinguishes a city from a large village.

The second characteristic of urban organization was simply the accentuation of the principle of hierarchy. Cities generated not only a greater division of labor than existed in farming settlements, but the stratifying of people by function also produced a more elaborate hierarchy. In short, the emergence of cities engendered greater inequities than existed in villages organized along lines of kinship. Moreover, leaders demanded new bonds of allegiance, sans kinship ties. The creation of new gods and other ideological principles no doubt served to legitimize emerging hierarchies, but old gods and old traditions died slowly. The memories of other social arrangements lingered, especially among those who gained the least (or suffered the most) from the new principles of organization.

It is reasonably clear that more than natural increase among inhabitants accounts for city growth. Promise of opportunity lured some, but a substantial proportion of the growth occurred as men were brought to the city in bondage. Other probabilities for providing new residents are territorial disputes and nomadic bands. In some instances, nomadic bands conquered cities and wandered on leaving them leveled. In other instances, they remained as conquerers, imposing their superior military organization upon the social organization already present. It is possible that the first cities emerged as powerful nomadic bands settled down to reap the benefits of the conquest of agricultural villages.

Theories on the emergence of cities abound. However it happened, the very nature and growth of cities almost inevitably implies an intensification of the principle of hierarchy and, hence, marked inequality (Childe, 1954). Those at the bottom of the totem pole, and there is reason to believe they constituted a sizeable proportion, must necessarily have viewed the city as an inferior environment.

The emergence of cities also demanded a stable working relationship with an agricultural hinterland. This, again, raised a problem of new forms of social organization. How such relationships were initially established leaves much room for conjecture. Perhaps contractual agreements between free men were sufficient at first, but the survival of a growing urban population required more than an assumption of continuing good will between farmers and artisans. Without reasonable guarantee of a continual flow of food from the hinterland, the settlement of large aggregates of people is inconceivable. It seems, thus, reasonable to speculate that such assurances were achieved by superior strength—that is, the ability to impose sanctions for failure to comply with the needs and expectations of the city. But this guarantee of food is minimum. As cities grew in size and strength, many of them must have demanded more from their hinterland and for diminishing compensation. Moreover, the conditions which gave rise to a city must have resulted in several cities within the same area. As cities grew, they required larger hinterlands and struggles to claim the same territory must have ensued, with the agrarian peasants being caught in the cross-fire.

In short, the emergence of cities not only resulted in heightened inequalities for those within them, but also imposed radically different social relationships among those within the range of a city's hinterland. It is not inconceivable that some farmers profited from cities, but, on balance, the cities imposed disturbing restraints on herders, fishermen, and agriculturists. Hence, another possible source of anti-urban sentiment emerges from those beyond the city gates but nevertheless deeply affected by the city.

An elaborate accounting of life in early cities and of the relationships between cities and their hinterlands is unnecessary. Our major point is that the emergence of cities demanded a more marked system of stratification than was previously the case. This argument is, of course, speculative, as we have no known written records of the earliest cities. The earliest written records available, however, indicate rigid stratification with a sizeable proportion of the population in bondage. In the Greek city-state, the concept of citizen applied to only a small minority. It is our position, thus, that the emergence and growth of cities depended on the development of new principles of organization involving significant intensification of hierarchical principles. While many within and outside the cities profited from the new social order, a very large proportion did not. And from those victimized rather than benefitted, there emerged dislike, distrust, and fear of the city.

Important as this proposition may be in tracing sources of anti-urbanism, we cannot ignore the possibility that many who experienced status enhancement in the city may also have found urban life repugnant, or at best a mixed blessing. Cities were dirty and noisy; disease was more easily contracted. At least a small minority found the city an escape from the sanctions of informal social control in the village or clan—i.e., they were more likely to cheat or steal. In short, in miniature scale, all of today's urban shortcomings existed. Personal success in the city did not, then, necessarily reflect personal satisfaction with the environment.

BIBLICAL EVIDENCE OF ANTI-URBANISM

Though cities may have provided fertile turf for the spawning of anti-urbanism, people's early feelings about cities remain yet an empirical question. That is, is there any evidence attesting to a negative image of the city from its beginning? Public opinion surveys were nonexistent, very little of detailed development records have been uncovered, diaries and memoirs are unavailable, and, further, it is unlikely that the early philosophers whose work remains extant reflected or represented the populace.

Probably the nearest we come to possessing a social history of the masses are the books of the Old Testament. While recognizing its limitations, we can yet defend the Bible as an embodiment of popular opinion by virtue of its prestige as the most influential document in Western Civilization. It emerged from and carried on a folk wisdom. We may view religious values as the most basic of human values, orienting one's total *weltanschauung.* Thus, the Judeo-Christian tradition, including its reflections on the city, has affected Western man's thinking.

The scholarly literature on cities contains a few passing references to anti-urbanism in the Bible (e.g., Ericksen, 1954: 68), but not until recently has anyone explored it in depth. Jacques Ellul, French philosopher and author of the controversial and provocative book, *The Technological Society* (1964), has recently undertaken this task in a volume entitled *The Meaning of the City* (1970).

The city, as viewed by Ellul (1970: 8, 58) is a hiding place man created to escape the judgment of God, and as such remains forever cursed:

> [The] city was, from the day of its creation, incapable, because of the motives behind its construction, of any other destiny than that

of killing the country, where God put man to enable him to live his life as best he could. . . . The City was built as protection for man. It turns out that she is nothing other than his ruin: the word of God pierces her ramparts and she takes all her inhabitants down with her in her destruction, precisely because man hoped that the city would shelter him from destruction. Condemned herself, the city brings on man's destruction. He is not directly condemned but is implicated in her crime by having sought her protection.

There is a remarkable parallelism between Ellul's earlier treatment of *la technique* and his analysis of the city. Man is incapable of controlling either technology or the city. Now, in *The Meaning of the City,* Ellul lays bare his own theological presuppositions in explanation: God's judgment allows both to rage out of control until, at the time of the second coming, His grace will restore order.

Such a view may reassure those professing benign neglect as an appropriate policy for dealing with urban problems. Similarly, were they inclined toward reading, fundamentalist preachers would delight in learning of Ellul's "text-proof" of God's condemnation of the wicked city. But can any serious scholar view this work as anything other than a diatribe by an angry and disillusioned man whose mind and soul belong to another era?

Harvey Cox (1965), one of America's most prominent theologians, says yes. "Jacques Ellul," writes Cox (1971: 357), "is neither a purblind Luddite nor a quaint religious fanatic. Though his theology has some serious thin spots and even some holes here and there, his instincts are usually dependable. He cannot be ignored." Cox quarrels mainly with Ellul's preoccupation with human evil to the neglect of grace and redemption promised in the New Testament. But of Ellul's tedious explication of anti-urbanism in the Bible, Cox feels the argument is neither strained nor out of context.

Few social scientists will accept Ellul's interpretation of the *meaning* of this anti-urbanism—i.e., an indicator of God's condemnation of the city and of the futility of man's attempts to change this. But his personal presuppositions notwithstanding, *The Meaning of the City* stands as a bench mark in illuminating the historical depths of anti-urbanism. Moving back and forth from Ellul to his scripture citations, one suspects that Ellul's exuberance has occasionally led him to quote passages out of context. But the strength of Ellul's scholarship lies not in particulars but rather in demonstrating the consistency with which the Bible regards cities.

Ellul's argument begins with the story of creation. Adam and Eve bore two sons, Cain and Abel. After Cain slew his brother, God sent him forth, a fugitive, wandering the face of the earth. God promised, however, to protect him from revenge seekers. But Cain refused this protection. He went, rather, to the land of Nod, where he built and named the first city after his son Enoch. Thus, the first murderer is also the first city builder. The city is man's attempt to control his own destiny in defiance of God. Symbolically, the city stands as a fortress where man attempts to hide from God's judgment.

Nimrod, who appears in the tenth chapter of Genesis, was the first mighty man of earth and builder of many cities. Descended from Ham, he bore a curse for the latter's improprieties. Again, the censured man builds the city which in turn shares his condemnation. It was in the most famous of Nimrod's cities where the Babylonians attempted to build a tower to the heavens to become as gods. And, of course, the Lord's response was to confuse their tongues, to cause babbling and misunderstanding. The city, hence, became a place where men lived in discord and hostility.

The Israelites themselves were a nomadic people who knew nothing of cities until as captives in Egypt they were forced to build them. From that day forth,

> when Israel built their own cities, it was always for them the sign of a curse, and the proclamation of slavery renewed. . . . Israel bound herself to slavery, and, even more, to the land of sorrow and sin; by the cities she built, cities that were always the imitation of what she had learned in Egypt, before the deliverance. For the greatest significance is in the fact that Israel was initiated in this art by a king other than God. The people whom God chose for himself obeyed a worldly king, a king whose power was the most impressive of his time, the king of the land of Mizraim. And the power of this king forced the chosen people into ways not meant for a people of God [Ellul, 1970: 25].

The plight of the Israelites echoes our earlier argument that those brought to the city in bondage have an objective basis for their dislike of the urban environment. Yet the grasp of the city loomed mightier than the mere locale of captivity. Delivered from Egypt, the Israelites built their own cities and bound themselves to repeated violation of God's covenant.

Jericho is probably the most familiar instance. The Israelites, having broken God's covenant, found their beloved Jericho besieged and miraculously destroyed by a small band following Joshua, who then promised a curse on he who would rebuild the city. "At the cost of his first-born shall he lay its foundation/ And at the cost of his youngest son shall he set up its gates" (Josh. 6 : 26). And later Hiel of Bethel did indeed rebuild Jericho and paid the price prophesied by Joshua (I Kings 16 : 34).

Again and again the Old Testament recounts the breaking of God's covenant in the city and the judgment of God which follows. The city appears as an environment of disobedience to God, presumably because the city is an "unnatural" setting created by man in defiance of God's will.

The New Testament contains fewer negative references to the city. Yet, if not preoccupied with cities, the New Testament magnifies the Old Testament lessons. The eleventh chapter of Matthew contains Jesus Christ's strongest reproachment of cities, but nonetheless the Gospels reflect a subtle—and sometimes not so subtle—rejection of urban life throughout. At no point does Jesus speak favorably about cities. He often retreats to the countryside to teach and pray. Frequently he admonishes typically urban behavior and life styles. And, in the end, a big-city boss releases him to an unruly urban mob to be crucified.

Ellul (1970: 123) summarizes his interpretation of the scriptural message about cities as follows: "If man keeps up Cain's reaction, if man continues to take the city as his port and as his security, then Jesus' work is in vain—or, rather, man will forever be ignorant of his true port and his true security." Such a conclusion is altogether unacceptable as social scientific scholarship. But a careful reading of the scores of texts Ellul cites can leave little doubt as to the Bible's pervasive negative sentiment toward cities. From Genesis to Revelations, the Bible denounces the city. Such consistency cannot be dismissed as incidental or insignificant.

The Scriptures lend credibility and plausibility, though certainly not proof, to our thesis that the very nature of social organization in early cities contained seeds for anti-urban sentiment. We will now postulate a further logical link between contemporary anti-urbanism and the earliest cities. The validity of our thesis, however, does not depend on absolute and specific conclusions here. To repeat our earlier stated goal, we are trying simply to establish plausibility for the thesis that anti-urbanism stems from roots deeper than the late industrial revolution; its origins are far older—perhaps, indeed, as old as cities themselves.

We predicate our argument on the assumption of persistent anti-urban ideology over the millennia. To trace the pattern in broad strokes, we know that Greek and Roman philosophers were urbanists by residence and sentiment. Yet in their writings we find repeated cautions against the dangers of urban life and affirmations of the virtues of farmers. Plato, for example, models his Republic in an agrarian setting. His utopian city would have contained only 5,040 people. As Mumford (1961: 180-181) notes, "No city could have shrunk into the form Plato desired without ceasing to be a city." Though the urban philosophers do govern, the agriculturists garner much prestige while trade, money-lending, commerce and other urban occupations do not even rank in this vision. An aristocratic distrust of the urban masses underlies his ideal state. Aristotle, too, rated farmers as superior to the urban riff-raff. Excluding them from the governing class reflected his belief that governing required special skills, not his sentiments on farmers' worth. Clearly the health of the city-state depended on protecting and fostering the vitality of the farming class. Like Plato, Aristotle distrusted merchants and artisans, viewing their life style as "ignoble and inimical to virtue." Seneca similarly felt the vitality of the city-state depended on the well-being of the agrarian population. In his *Epilogue,* he notes that farmers make superior soldiers and chastises urban dwellers for laziness.

The quest for salvation of the soul dominated men's thought in Medieval times, and no scholar was more significant than Saint Augustine. Of his more than one hundred titles, none carried influence equal to *The City of God.* He began by refuting a popular notion of his time which traced the fall of the Roman Empire to the establishment of Christianity as the state religion and the concomitant disobedience of the "pagan" gods of Rome. He then posited that only the Christian God offered immortality, and followed this with the development of a radical dichotomy between the city of God and the cities of man. Cain, the first city builder, and Romulus, the founder of Rome, were both murderers. The cities of man, thus, are founded in sin. They manifest demonic forces and deserve the epithet "city of the Devil." Immortality, however, proceeds from renouncement of materialism, worldly power, and the pleasures of the flesh. In other words, the achievement of immortality and a place in the city of God depends upon one's disciplined rejection of the cities of the world.

Harvey Wheeler (1971: 10-11) refers to Augustine's absolute separation of heaven and earth as "the Christian heresy": "one effect of the Augustinian maneuver was to secularize nature, giving the things of this world a monopoly of evil and visiting the heavenly city with a monopoly of virtue." Having so defined reality, man could waste and ravage this evil environment with impunity. "The consummation came," Wheeler (1971: 11) argues, "with the industrializing Protestants, who interpreted their Christianity as giving them free license to indulge themselves in the merciless exploitation of man and nature alike." And, we would argue, this merciless exploitation masked a total disregard and disdain for attempts to create inhabitable cities.

The Judeo-Christian tradition can claim no monopoly on anti-urban underpinnings. One of history's most intriguing statements about cities comes from Ibn Khaldun, a fourteenth-century Muslim. A great thinker, Khaldun was a city dweller who won favor with many political regimes and thereby shared the riches and pleasures of urban civilization. Yet he saw the city as fundamentally corrupting of mind and body. In his three-volume classic *The Muqaddimah,* he devotes more than a hundred pages to analyzing the differences between nomads and sedentary urban people.[3]

Urban civilization constitutes the pinnacle of human development, but, once attained, the populace becomes corrupted and degenerate, incapable, even, of caring for their own necessities. They eat and drink to excess. They indulge in sexual perversions which result in the inability to replenish population (homosexuality) and in loss of community (because promiscuity makes it impossible to identify kinsmen). For Khaldun, the maintenance of a strong sense of community equates with the highest human virtues. Nomadic people, in contrast to urbanites, must work hard for the necessities of life. They are physically stronger, braver, more loyal to their kinsmen and community, and more faithful to their moral and religious heritage. Cities, however, must continually ingest nomad populations, who, in turn, assimilate the corrupting influences of urban life. Though a highly original thinker, Khaldun's writing probably reflects accurately the *zeitgeist* of the broader Arab world. The veil of anti-urbanism stretches beyond the Judeo-Christian world.[4]

HUMAN CONSEQUENCES OF
THE URBAN-INDUSTRIAL REVOLUTION

With the breakup of the Roman Empire, urban life in the West gradually disintegrated. By the ninth century, cities as large political, administrative, commercial, and religious centers had largely vanished (Pirenne, 1956: 17). Besieged by repeated invasions, cities became essentially fortresses for protection. Though the twelfth and thirteenth centuries witnessed a resurgence of trade centers struggling for autonomy from feudal lords, the fourteenth century brought political anarchy, and most of Europe slipped again into urban and economic coma until the sixteenth century.

The Industrial Revolution kindled the first serious and extensive city-building. The forces were a long time in mustering sufficient cohesiveness and direction, but by the late eighteenth century the components had gathered together and thrust the Western world into a phenomenal surge in urban growth.

People were pushed as well as pulled into the city. The guild system had largely broken down. Craftsmen working for merchants under a putting-out or domestic system experienced increasing exploitation. Expanding competition intensified the plight of the artisans to the point of hopeless indebtedness to merchants who required more production for less compensation. The emerging factory system had already made many crafts obsolete and thus forced village workers into cities. Agrarian peasants suffered similar difficulties, and tens of thousands chose to cast off rural impoverishment for the dream of urban prosperity.

Concomitant with burgeoning urbanization and industrialization came unprecedented population growth. Demographers have traditionally attributed rapid population increase to improved food supplies and health conditions which lower mortality rates. William Petersen (1960) has argued persuasively, with historical data from the Netherlands, that the population expansion more likely reflects increased fertility resulting from the breakup of the joint family household, which had previously functioned to restrict marriage and fertility. Urbanization meant a larger proportion of the population marrying, possibly at a younger age. In the Netherlands, a significant decline in mortality does not occur until the last decade of the nineteenth century (see also Eversley, 1965).

While cities eventually produced a sizeable middle class, they initially offered mainly misery to the large majority of their inhabitants. Living conditions were deplorable. The tiny flats which helped destroy extended families soon overflowed in unrestrained fertility. Cities could not cope with the pace of migration and fertility, and the crowded results were hothouses for communicable diseases. Petersen (1960: 340) reports that, in some urban locales, infant mortality ran as high as one-third to one-half.

Early factories were dens of inhumane work and wage conditions. To escape starvation, a large proportion of the population sent their children into the mills at a very young age, often by six or eight. These children, working twelve to sixteen hours a day, six days a week, could expect severe beatings for tardiness and a strap across the back for drowsiness or dreaminess over machines they could barely manage. In England, an investigation of the working conditions of children conducted by Michael Sadler (1832) helped enact the Factory Act of 1833, which regulated child labor practices.[5] While somewhat effective in reducing the long hours and mistreatment, it failed to remove many children from the labor force. By the end of the century, a large proportion of children still worked.

In 1889, Charles Booth classified 35% of the population of London as poor.[6] His category of "regular standard earnings—above the line of poverty" consisted of 42% of the population, but these families typically maintained this level by sending their children into the labor force (Fried and Elman, 1968: 18). In addition to the impoverished, a small proportion of the population was absolutely dispossessed and without any means of livelihood. Booth estimates this group as only about one-and-one-quarter percent of the population, but hastens to clarify this as a "very rough estimate, as these people are beyond enumeration" (Fried and Elman, 1968: 11). Given our own difficulties in calculating certain groups within the urban poor, we can reasonably assume Booth's tally to underestimate actual percentages. Clearly, such a population roaming the streets, begging, and stealing would generate fear and more suffering.

Poor housing, despicable working conditions, poverty or marginal subsistence, high rates of disease and mortality, rampant crime, and the like made cities hell incarnate. During the nineteenth century, millions migrated to the United States seeking more than their factory or peasant existence in Europe. But, for many, the American city offered little solace, only more factories and tenements.

Eventually, in Europe and the United States, the impoverished achieved middle class and escaped the slums, leaving them for others to overcome. But scars remained. And the image of the past lingered as a memory of dirty, crowded, crime-ridden streets, cold, rat-infested walk-up tenements, and noisy, inhumane mills.

Millions came to the cities by need rather than choice. Often the realities they then faced magnified their deepest apprehensions. Objective conditions notwithstanding, they constructed a nostalgic image of a village or pastoral environment where life was better and simpler. The centrifugal growth of cities was as much an ideological movement to escape the city as it was a natural pursuit of cheaper land. The city alienated her majority long before she could offer them the joys of the culture, variety, and excitement of urban living. But the roots of anti-urbanism in America run deeper than the objective conditions suffered by millions of immigrants. America, the new world, had been shaping her identity since the seventeenth century, and threads of her ideals had gradually formed a complex pattern of philosophic thought which now reacted to the reality of industrial cities.

IMAGES OF
CITY AND COUNTRY IN AMERICA

When J. Hector St. John de Crevecoeur ringingly asked in 1782, "What then is the American, this new man?" he partially responded with a vision of the American appearing in a middle landscape, a place midway between the primitive and the civilized. He saw this new man following a path to a village, "where, far removed from the accursed neighborhood of Europeans, its inhabitants live with more ease, decency, and peace, than you imagine; where, though governed by no laws, yet find, in uncontaminated simple manners all that laws can afford." This system of life, this pastoral setting, encompassed enough "to answer all the primary wants of man, and to constitute him a social being, such as he ought to be in the great forest of nature" (Crevecoeur, 1912: 211).

This image of the self-sufficient American in the plenitude of the garden blossomed into one of the dominant symbols of nineteenth-century American culture, a collective representation defining the promise of American life. "The master symbol of the garden," as Henry Nash Smith (1950) reminds us, "embraced a cluster of metaphors expressing fecundity, growth, increase, and blissful labor in the earth." The image of an agricultural paradise, dominated by the heroic figure of the frontier farmer and embodying memories of an earlier, simpler, and happier state of society, long forged a strong bond between memory and desire. "So powerful and vivid was the image," Smith (1950: 123) adds, "that down to the very end of the nineteenth century it continued to remain a representation in Whitman's words, of the core of the nation, 'the real genuine America.' " Never oblivious to life across the ocean, America sought to define herself, and perhaps thus actually mold herself, differently.

This powerful image early became associated with an American animus toward the city. But this cluster of memories and desires represented more than a simple distrust of the city and its ways (White and White, 1962). The pastoral representation of the American landscape also reflected the dissonance created by urban growth and constituted an effort to resolve the conflict between urban and rural values. The mythic figure of the farmer and the poetic map of the middle ground between the wilderness and the city merged into American ideology because they resolved a profound conflict between the values of the "real genuine America" and the attraction of the city and its cluster of new technologies (Marx, 1964: 22-23, 29-32; Empson, 1950: 11-19). As the city threatened to break down the deeply held valuation of the countryside, the pastoral ideal reconciled Americans to the emergence of a new world.

Jefferson's (1955) *Notes on the State of Virginia,* written in 1785, furnishes the most famous use of the yeoman farmer as the vehicle of truly American values. In an answer to a query about the present state of manufactures in Virginia, Jefferson wonders whether man should improve the vast countryside, or whether half should devote their energies to manufacture and crafts. His answer, which has fascinated Americans as much as Crevecoeur's question, resolves the problem forthrightly:

> Those who labor in the earth are the chosen people of God, if ever He had a chosen people, whose breasts He has made His peculiar deposit for substantial and genuine virtue. It is the focus in which He keeps alive that sacred fire, which otherwise might escape from the face of the earth. Corruption of morals in the mass of cultivators is a phenomenon of which no age or nation has furnished an example. It

is the mark set on those, who, not looking up to heaven, to their own soil and industry, as does the husbandman, for their subsistence, depend for it on casualties and caprice of customers. Dependence begets subservience and venality, suffocates the germ of virtue, and prepares fit tools for the designs of ambition [Jefferson, 1955: 164-165].

The farmer was safe from the gross vices of the city dweller, for the agriculturist lived with his hands, rather than with his head. To work with one's hands meant independence; to live by artifice signaled moral degeneracy. Let the workshops remain in Europe, wrote Jefferson (1955: 165), for "the mobs of great cities add just so much to the support of pure government, as sores do to the strength of the human body" (see also Boorstin, 1948: 147). Jefferson nevertheless recognized the restless striving of Americans and, despite misgivings, later resigned himself to the necessity of industry and cities. But his image of "those who labor in the earth" magnified to continental size the cardinal image of American aspirations (Marx, 1964: 124-144).

The Jeffersonian equation of virtue and agrarianism continued to fascinate Americans during the nineteenth century. For the Jacksonians, though they lived in the midst of an industrial revolution, "the real people" were farmers and planters, the mechanics and laborers—in short, the productive classes, in the peculiar Jacksonian meaning of the term. This "bone and sinew of the country" was simple and stable, self-reliant and independent, honest and plain-dealing. Andrew Jackson's America, as described in his first inaugural address, was a happy countryside of "flocks and herds and cultivated farms, worked in seasonal rhythm and linked in republican community" (Meyers, 1957: 24). Look toward the West if you seek examples to follow, urged a Fourth of July orator in 1831: "There we can behold how the young American grapples with the wilderness and thence we can return and imagine how our fathers lived" (Ward, 1955: 39). The Jacksonians, then, never reconciled themselves to economic change and urbanization, but instead seized the enchanting Jeffersonian version of an agrarian society just as it threatened to slip away. And as they attempted to restore the world they had lost, they breathed new life into the pastoral ideal (Peterson, 1960: 84-85).

The political potential of Jeffersonian agrarianism slowly petered out in the late nineteenth century. A brief revival of the radical side of the tradition accompanied the rise of Populism in the 1890s. But the exhaustion of creative potential in Jeffersonianism is nowhere more apparent than in the last few years of Tom Watson, an agrarian rebel whose only rest seemed to come when he was near the earth. The world of the twentieth century, however, would not yield to agrarian solutions, and Watson spent his last few years in his Georgia household, spying everywhere the shadow of the Pope and the conspiracy of the international Jew. This biographer of Jefferson ended his days by releasing much of the rural malice which found its vehicle in the Ku Klux Klan (Peterson, 1960: 256-258; Woodward, 1938: 346, 416-450). The vexations of a rural world in flux found an ethnic, urban focus in anti-Semitism; but as rural America was increasingly integrated into the national culture, even this last perversion of Jeffersonianism disappeared (Higham, 1966: 249-250). However important these political varieties of Jeffersonian agrarianism may have been, the more important version of pastoralism has been the domesticated rural ideal of the urban middle class. This domestic pastoral had its roots in the Romantic evaluation of rural life as the point of equilibrium between the wilderness and urban civilization. This revolution in understanding found its intellectual expression in Henry Thoreau's *Walden* (1854), its popular form in the landscapes of Thomas Cole, and its occasional realization in summers in the Adirondacks (Nash, 1967: 61-62, 78-83, 90-95). Comfortable urban dwellers hung the landscapes of Cole, Asher Durand, and Thomas Doughty over their fireplaces because they sought a tamed wilderness, scenes in which man and nature recovered a lost goodness (Miller, 1967: 180-181). And so the American wilderness entered the urban home just as urban homes invaded the rural landscape around Boston, New York, and Philadelphia.

We can see this ideological current at work by tracing the rise of a new domestic architecture and its use in early American suburbs. Andrew Jackson Downing (1815-1852) provides, in his architectural career, one way to lay bare the ideal image of American society implicit in these early suburban developments. Downing valued above all the tranquilizing effect of landscape, for "under its enchanting influence, the too great hustle and excitement of our commercial cities will be happily counter-balanced by the more elegant and quiet enjoyments of country life." What was needed was a domestic architecture of equipoise in order to assuage the effects of American urban life, an environment to serve "for a counterpoise to the great tendency towards constant change and the restless spirit of emigration." Unable to put down roots anywhere, the American needed, in Downing's estimation, "the security and repose of society, the love of home" so that his unsettling energies might be kept within bounds. Downing thus translated a widespread concern about the boundless energies of American character into a domestic architecture (Ward, 1969: 270-281). And it was this domestic architecture, scattered on the peripheries of growing American cities, which would define the new suburban dimension (Scully, 1953; 1955: 1-18, 71-90).

The garden suburb of the second half of the nineteenth century was the logical outcome of this domestication of nature. In Llewellyn Park, New Jersey (1853), in Lake Forest, Illinois (1857), in Riverside, Illinois (1969), in Shaker Heights, Ohio (1910), the mid-twentieth-century suburb had its origins (Reps, 1965: 339-348; Gowans, 1964: 309-315). What was there in these rural and open suburbs to provide a sense of life in man's world, an association of the feeling of art and community? "There wants to be something in a city that produces a sense of its being a world in itself," wrote the American theologian Horace Bushnell (1864: 313) as he wondered about the growth of cities at mid-century. Frederick Law Olmsted provided a distinctively American answer to the theologian's disturbed questioning, a response which signals the emergence of a characteristic urban community. The suburb was to exhibit, Olmsted wrote in his 1868 plan for Riverside, "in regard to those special features whereby the town is distinguished from the country . . . the greatest possible contrast which is compatible with the convenient communication and pleasant abode of a community." The essential feature of a suburb was domesticity; all other aspects of community life were subordinate to the "domestic indoor and outdoor private life" of the family (Sutton, 1971: 299-300, 303).

American images of the city, then, were rarely simply anti-urban. Rather they were a bridge, a middle ground between the wilderness and urban civilization. Only occasionally did this ideology animate a pure hatred of the city, as in the novels of Thomas Dixon, Jr. Author of *The Clansman* (1905), the novel which furnished the script for D. W. Griffith's "The Birth of a Nation" (1915), Dixon spent ten frustrating years as a Baptist minister in New York. "In the roar of this modern Babylon," cried this Jeremiah, "is religion increasing its hold on man?" No. So he fled, first to an "airy apartment," then to a suburb, on to a five-acre place on Staten Island, finally to a stately colonial manor 200 years old, from which he poured forth his wrath in popular racist novels (Dixon, 1896: 13, 1905: 1-8). The intractable urban world vexed his Protestant moral sensibilities, as it also irked Tom Watson, so he sought shelter in a rural bastion. But the more general outcome was the suburb, a new Arcadia, where, as a promoter observed in 1915, urban dwellers hoped "to obtain attractive homes at a modest cost, to get into the genuine unspoiled country, to take their own social life with them, to restore to the land its elemental charm" (Schmitt, 1969: 18). The dream had its roots in the ancient promise of the garden, where a fecund earth produced under the hand of an honest toiler. The persistent attempt to redeem this promise of plenitude mediated the American response to the city.

PROMISE OF PLENTY

This people of plenty held to a second promise, equally as important as their search for the garden. The self-made man, the mobile American, stood beside the yeoman farmer, a hero of iron will who reconciled Americans to the city's production of social inequalities. At first glance, this mythical representative of equality should have been an urban figure, and so he was in his first appearance. But the apostles of the self-made man made him also a vehicle for rural values, and therein lies his significance for our account of suburbanization. For as a repository of pastoral virtue, the American entrepreneur broadcast a profound distrust of the city. As this ideology trickled down into the ranks of businessmen and clerks after 1850, the American middle classes acquired a distinctly anti-urban attitude.

The roots of the nineteenth-century myth of the self-made man lie in the career of Benjamin Franklin. A provincial in colonial Philadelphia, Franklin was nevertheless practiced in the arts of urbanity. Here Franklin found, like James Boswell, another famous eighteenth-century provincial in the big city, "the satisfaction of pursuing whatever plan is most agreeable, without being known or looked at" (Boswell, 1950: 69). This was a world of voluntary association, in which the main problem was to prepare men "to enter the World, zealously unite, and make all the Interest there can be made to establish them" (Franklin, 1959: vol. 3, 400). When Franklin wrote this in 1749, the problematical shape of American character was already evident: the individual was not born to an identity, but had to create one for himself. Hence the various associations of Philadelphia were Franklin's life enlarged and extended; this extension and enlargement furnished the material for myth (Van Doren, 1938: 77-78). In a pamphlet published in 1751, "Idea of the English School," Franklin projected his own experience to continental size. After having described a thoroughly innovative curriculum for an urban school, Franklin concluded with a remarkable judgment on the probable effects of this education: "Thus instructed, Youth will come out of this school fitted for learning any Business, Calling or Profession" (Franklin, 1959: vol. 4, 108). Change in status and occupation, not continuity, seemed to Franklin the common experience. This emerging attitude reflects, as Bernard Bailyn (1960: 36) has noted, "the beginnings of a permanent motion within American society by which the continuity between generations was to be repeatedly broken."

The rise of the city and the spread of the factory, as we have already pointed out, were dissonant elements in this hopeful creed. To accommodate themselves to the troublesome new inequalities of industrial cities and to the threat of disorder, Americans supplied themselves with a conservative version of Franklin's promise of mobility. The exponents of self-improvement in the age of Jackson and Lincoln tried to balance traditional values and innovation. The status of rich or poor was impermanent, these apostles insisted; a mobile, freely competitive society would give every man an opportunity to rise to his merited status. But success also grew out of a man's moral character. The secrets of success lay in the practices of industry and economy and in following a simple maxim: spend less than you earn. The morally meritorious always succeeded, yet there were weaknesses in the creed. Luck sometimes played an embarrassingly large part in success, as in the novels of Horatio Alger. And even the most hidebound exponent of self-help admitted that environment and institutions helped men to succeed. The big emphasis, however, falls on the conservative function of the myth of the self-made man. The heroes achieve, but they achieve through fidelity to their employers and by acquisition of a middle-class competence (Cawelti, 1965: 46-55, 101-123; Thernstrom, 1964: 56, 66, 72-79).

The tensions of urban life, however, produced pressures on the middle class which led to a transformation of the self-help gospel. For one thing, as the editor of the *Nation* pointed out in 1865, the myth of the self-made man left no room for decorum and for the deference which successful men expected (Godkin, 1896: 42-43). The very honorific nature of status, the constant threat of loss as well as gain, induced the proponents of success to seek some means of consolidating their positions. But as the businessmen of the late nineteenth century flaunted symbols of success, they also brought down criticism for introducing the very emblems of the inequality which the mobility ideology was to efface (Kirkland, 1956: 29-49; Wyllie, 1954).

More importantly, the persistent tension between the ideals of frugality and abstinence, and the virtues of conspicuous consumption, both necessary to status in the city, led to a modification of the mobility ideology. A. D. Mayo captured this tension well in 1859:

> All the dangers of the town may be summed up in this: that here, withdrawn from the blessed influence of Nature, and set face to face against humanity, man loses his own nature and becomes a new and artificial creature—an inhuman cog in a social machinery that works like a fate, and cheats him of his true culture as a soul [Wohl, 1969: 106-107].

The tradition of rural virtue lay ready at hand to reclaim the urban dweller from his fate, and to accommodate him to life in the city. For the exponents of success, urban life appeared too complex, too difficult, too hazardous for Americans. So they adapted a rural tradition which would define an accommodation between the promise of success and the desire for order (Wohl, 1969: 110, 139, 155). Urban society was a neutral arena within which each individual was free to pursue the main chance, then, but it was also fraught with the tensions of individualism. Hence the society trumpeted rich and conspicuous success, but continued to yearn for a lost rural society in which the country boy made his way through a simple world (Williams, 1958: 325; Wohl, 1953: 394; Cawelti, 1965: 125-164).

Urbanization, mobility, migration—all these social developments left Americans with a deep sense of loss. There was no longer in New England, Charles Eliot Norton lamented in 1889, a "common stock of things taken for granted" (Solomon, 1956: 21). Rather the competition for prestige and success, a struggle symbolized in the self-made man, had eroded social order. The status rivalries characteristic of late nineteenth-century society found a convenient vehicle in anti-immigrant and anti-urban ideologies. Hence as the Irish and Jews began to crowd the social ladder—to claim the dream of success as their own—they suffered anti-Catholic and anti-Semitic outbreaks. Americans expected immigrants to become Americans, but not too fast (Higham, 1958: 151-152, 156, 1966: 243-246). The loss of confidence which these status rivalries provoked, and the consequent unleashing of anti-immigrant and anti-urban hysteria, reveal the persistent tensions which American society could not overcome. The vexations of an assertive society found an urban and ethnic focus in a series of campaigns against the immigrants, from the American Protective Association in the 1880s to the Immigration Restriction League of the 1910s (Kinzer, 1964: 58, 80-84, 91-94, 180; Higham, 1963: 52-87, 158-186). One of the major sources of anti-urban ideologies, then, has been the conflict induced by the rapid assimilation of immigrant groups.

The clerks and stockboys of American cities dreamed of success, but when they woke they began to flee the site of their success. The urban origins of the self-made man wore off as the hidden costs of mobility became obvious in the late nineteenth century. Success became more and more closely linked to the preservation of an older, simpler world, and many of the urban reform movements of the 1890s looked to the restoration of an agrarian society. We see again, then, how the myth of rural virtue accommodated Americans to the strange new world they were busily building in the city.

FAMILY LIFE AND SUBURBS

These images of the good life performed their conciliatory function in the setting of middle-class households. Within the circle of bourgeois family life, as Andrew Jackson Downing noticed, the love of country was linked to the love of home. Whatever led a man to assemble the comforts of rural life in his household also strengthened the family. Here again we find the theme of restoration: "And as the first man was shut out from the *garden*, in the cultivation of which no alloy was mixed with his happiness, the desire to return to it seems to be implanted by nature, more or less strongly, in every heart" (Downing, 1844: ix). The battle of life is carried on in the cities, where mobile men incessantly struggle to achieve the promises of success. But the temper of life, Downing wrote, "is, for the most part, fixed amid those communings with nature and the family, where individuality takes its most natural and strongest development." The family hearth must symbolize "the dearest affections and enjoyments of social life," where "that feverish unrest and want of balance between the desire and fulfillment of life, is calmed and adjusted by the pursuit of tastes which result in making a little world of the family home, where truthfulness, beauty and order have the largest dominion" (Downing, 1850: v-vi, 23). The suburban home was a refuge where the ambitious American could "keep alive his love for nature, till the time shall come when he shall have wrung out of the nervous hand of commerce enough means to enable him to realize his ideal of the 'retired life' of an American landed proprietor" (Downing, 1853: 111, 212).

"A little world of the family home," for the "retired life." These hopes would provoke contemptuous criticism a hundred years later, but, in the mid-nineteenth century, they expressed the triumph of the middle class. A confidence in the civilizing mission of the bourgeois home, the achievement of equipoise between the anti-institutionalism of American life and the widening compass of middle-class decorum signaled an important reorientation of American culture after 1850 (Elkins, 1959: 27-37; Higham, 1969; Fredrickson, 1965). And it was in the suburbs that this middle class pursued its need for a stable and organized environment.

The major problem, then, is to grasp the relationship between an emerging middle-class family and the cumulation of individual decisions which created the characteristic American suburb. In order to do this, we shall have to make a frankly speculative foray into the history of the American family. Since the American literature is thin, some of our material comes from the abundant writing on the English and French middle-class family. Nonetheless, this effort in hypothesis provides a crucial connection between the ideologies of pastoralism and romantic capitalism and their realization in the suburb.

The origins of the link between middle-class aspirations and the residential development of American families lies in the changing relationship between family and community. The boundaries between family and community in the eighteenth century were ill-defined, so there was no distinct sphere of family life. We can see this most clearly in the variety of communal sanctions against early marriage. A recent study of Andover, Massachusetts, furnished an extraordinary insight into the nature of communal controls on the life of the family (Greven, 1970: compare Demos, 1970; Laslett, 1965: 53-106). Greven shows that the first generation of settlers created a stable social order and established a patriarchal family system. The population of the town as a whole grew very rapidly during the first three generations. Husbands and wives lived together longer than they had in England and shared the responsibilities of parenthood well into old age. The survival of large numbers of their children into adulthood meant that they must have land and houses to establish households. The fathers accomplished this by delaying their granting of land to the sons until they were mature and well-established. Thus more than 25% of both the second and third generations married after thirty years of age. The stability of this kind of control over fertility depends, however, on the persistence of communal institutions. The fourth generation of Andover's population effectively subverted the whole system. The sons reached maturity sooner, married younger, established their independence more effectively and earlier in life, and migrated from the community much more frequently than earlier generations. Thus the small farmers of the 1750s had set into motion the same constant break in communal and generational links which Benjamin Franklin celebrated in Philadelphia. In the loosening of the ties binding men to their parents and to their families of origin, the fourth generation of Americans also broke communal controls over family life.

Before the industrial revolution, then, the community exercised sanctions over the family, while, during the nineteenth century, such social controls disappeared, and family limitation—a matter of personal decision—became much more common. The dramatic fall in the birth rate during the nineteenth century illustrated the effectiveness of family limitation. During the sixty years from 1830 to 1890, a very sharp reduction occurred in American fertility. The crude birth rate dropped from 50 to about 28 per thousand, and the death rate fell from about 25 to 15 per thousand. Thus, the annual rate of reproductive increase fell from 25 per thousand in 1830 to 13 per thousand in 1890. The rate of reproductive increase was, in short, cut in half. Thus Americans reduced their fertility by half in the space of two generations, and without the expensive appliance of modern birth control (Potter, 1965: 631-688).

The decisive fact in this remarkable fall in fertility was the emergence of the modern, middle-class family. In 1790, the mean size of American households was about 6.5 members; discounting slaves reduces the figure to 5.8. Not until the early years of the twentieth century does the size dip to 4.75 members, which was the average size of English households during the nineteenth century. Within the upper-income levels, however, the small modern family was making its appearance early in the nineteenth century. Rather than live in the large households which characterized American society before 1850, these affluent families were cutting themselves off from the world around them and gathering into isolated groups of parents and children. All the effort of these families was directed toward helping the children to rise in the world, individually and without any collective ambition for the household; the children, rather than the whole family, became the center of the household. At first, this type of family was limited to the merchant aristocracy, the middle class, and the richer artisans. In the mid-nineteenth century, a large section of the population, the poorest and biggest part, still lived in extended households under the old dependencies. But the fact remains that, by the end of the eighteenth century, the modern family had appeared, and it has changed very little since. The important change in the nineteenth century was the gradual extension of this modern form of household life to other social strata. Late marriage, the precariousness of employment, the difficulty of finding shelter, the mobility of laborers, and the continuation of the tradition of apprenticeship—all these were obstacles which the evolution of the family would eventually remove. So family life finally embraced nearly the whole of society, to such an extent that historians and sociologists have forgotten its middle-class origins (Ariès, 1962: 365-406; Carey, 1830: 456; Jaher, 1968: 205-208; Warner, 1968: 63-78).

But how does this transformation of family life affect the whole of society? A tentative answers lies in the outcome of the remarkable population debate of the nineteenth century. The population controversy in the nineteenth-century Atlantic community—first, the argument between Thomas Malthus and Richard Godwin, then the reopening of the argument by Francis Place in the 1820s—revealed an important practice of middle-class families, the delay of marriage until self-sufficiency was achieved. The parties to the population controversy were engaged chiefly in setting up the prudential behavior of the middle class as an ideal for the rest of society. In doing so, the exponents of bourgeois morality made into an explicit ethos what had been merely a practice. If a man's ambition is awakened, wrote an American moralist in the 1850s, "he will become prudent not only in his expenditures, but in contracting any relations which may become a burden to him . . . In a normal state, then, the inclination of people to marry is controlled by the opinion of the effect which marriage will have upon their position in life" (Bowen, 1850: 155-156). This calculation was directly linked to the structure of middle-class careers. Unlike the workingman, whose maximum earnings were reached in early adulthood, the middle-class father could anticipate a steady series of income increases at least until middle age. Postponement of marriage was necessary for four or five years, until the clerk or businessman was established, but then children came at a fairly rapid pace. Families continued to be fairly large, then, until the 1880s, when contraception began to be widely practiced (Himes, 1938: 209-285; Calhoun, 1965: 1-19, 178-197).

What provoked a search for a new setting for this emergent family life was the tension following the economic and social crises of the 1870s and 1880s. The burgeoning growth of American cities, the threatening expansion of ethnic groups into middle-class neighborhoods, and the sense of economic restriction furnished powerful motives for the attempt to assimilate the whole of society into the economy of bourgeois family life (Banks, 1954; Sennett, 1970a; Kennedy, 1971: 36-107). Middle-class family life became the model for a whole generation of urban reformers in their attempt to bring order into the formless cities of the nineteenth century. Temperance campaigns, charity organization, the standardization and integration of school systems and police forces all sprang from the attempt to uplift the "dangerous classes" into middle-class family life (Gusfield, 1963: 79, 85; Bremner, 1956: 31-66; Katz, 1968: 39-40. 91; Lane, 1967: 118-141). This steady enlargement of the area of family life created the characteristic domestic suburb of the late nineteenth and early twentieth century.

ANTI-URBAN IDEOLOGIES AND SUBURBANIZATION: A TRIAL BALANCE

The rural ideal and the pleasures of private family life—these were the two images which encouraged the middle class to build the modern suburb. Before the streetcar and the automobile, two houses, one in the city and one in the country, were necessary to realize the "retired life." But the streetcar and automobile suburbs opened up a whole continent to settlement by middle-class homesteads. New transportation technology enabled families to move beyond the old city into the countryside. The result of thousands of family decisions was a weave of small patterns which created the social geography of the modern city. The activities of middle-class householders and small lenders and builders shaped the metropolis into a discontinuous series of residential neighborhoods segregated by income and ethnic identity. The poor, the immigrants and the blacks inherited the inner city; the affluent and white dispersed throughout the residential suburbs (Warner, 1962; Schnore, 1965: 222-241, 255-272; Taeuber and Taeuber, 1965: 43-62). The logical outcome was a new creation, a city like Los Angeles, a metropolis which was, as a city planner gushed in 1930, "a federation of communities coordinated into a metropolis of sunlight and air." It was toward this ideal that American cities would move until, as another Angeleno hoped in 1910, the metropolis spread to meet the country, "until beautiful forms of urban life blend[ed] almost imperceptibly into beautiful forms of rural life" (Fogelson, 1967: 163).

What gave shape and meaning to this protean growth of the American city was the juncture of the powerful ideal of rural virtue and the growing vexation with the assertiveness of urban society. The rural ideal promised relief from the "rasp and graze of splintered/ normality," from the "clamors of collision." Though it was a dream more than a little false, the rural ideal recovered the link between pastoral and family life whose loss Americans had begun to mourn in the 1830s. Thus the movement outward of the middle class was not simply an escape from the city; it was more importantly an attempt to find a pleasing context in which to enjoy the newly discovered pleasures of family life. These pleasurable haunts of family life were continually disturbed, however, by another myth, that of the self-made, mobile American. A creature of nature and custom, so the parable went, found the rural world restrictive and left at the first chance for the city. But while he found success and approval, he was haunted by dreams of peace and wholeness which must, he imagined, have been realized somewhere in the past of the village. So the self-made American returned to the countryside in the suburb, where he could find solace in the delights of family and in the rhythms of nature.

When we discover what a tenuous balance Americans have struck between pastoral dreams and urban realities, it should come as no surprise to find a welter of anti-urban ideologies in American culture. Americans have associated virtue with nature, and freedom with open space. Their ideals of family life and individualism have seemed incompatible with the impersonal character of the new urban world. Hence, when urban problems have impinged too much upon their retreats, they have lashed out at the dangers and corruptions of the city in hundreds of reform campaigns. But these campaigns never approached the city as a functioning community, but rather as a collection of evils. These reform efforts, typically dominated by the older middle class, sought to forcibly uplift working-class urban dwellers to the level of 100% Americans (Hofstadter, 1955; Huthmacher, 1962). The older Americans, still enchanted by pastoral dreams, felt no need to accept the city and make it livable. Rather, they thought to make it over into the rural village of their memories.

While the charter Americans remained deeply distrustful of the city, a new group of Americans, especially from eastern and southern Europe, were coming to terms with their urban world. "We *had* to live in it," a second-generation immigrant wrote in 1930, "and learn what it chose to teach us" (Gold, 1930: 19). This the immigrants and their children were prepared to do, for their own cultures had a long tradition of viewing cities as the intellectual and cultural heart of their society, while simultaneously placing far less emphasis on rural life and individualism than did the Anglo-American community. Thus, these immigrants, in shaping a livable urban habitat, took the lead in organizing trade unions and in creating a welfare state. While accommodating themselves to the city in these ways, the immigrants, and especially their children, were also forging a new urban mass culture to meet the needs which their broken communal life could no longer fulfill (Higham, 1968; Holli, 1969: 157-181; Huthmacher, 1968; Juergens, 1966). In some measure, then, as John Higham (1968: 103) has recently written, these immigrants taught the older Americans how to live with the city. But before the new immigrants could fully overcome the handicaps of high-density living and develop its potentials, the centrifugal growth of the city was already assured (Warner, 1962).

Throughout the course of the twentieth century, numerous forces have interacted to push and pull immigrants from the communal life of their ethnic neighborhoods. In some measure, the desire of second and third generations of immigrants to assimilate into the mainstream of American culture resulted in their conscious loss of accent, a change of name, and residence outside their ethnic ghettos. More important to the demise of ethnic communities was the systematic development of public policy which encouraged suburban development while reaping havoc on inner-city neighborhoods. For example, Cleveland's rapid transit system, built in 1917 to shuttle the city's elite between downtown and the elegant suburban Shaker Heights, destroyed Slovak and Polish neighborhoods.

Urban renewal, beginning with the Housing Act of 1937, no doubt razed much slum housing that was in a condition of hopeless disrepair. But as the program gathered steam, after World War II, it bulldozed indiscriminately through neighborhoods which could easily have been rehabilitated, and even leveled areas with little substandard housing (Rossi and Dentler, 1961). And, more tragically, the urban renewal agency uprooted thousands of urban poor whose fondest possession was a sense of identity and belonging in their neighborhood (Gans, 1962). Later, the nation's multi-billion-dollar superhighway program gobbled up thousands of acres of inner-city land, shattering hundreds of neighborhoods.

Urban renewal and expressways are hardly the only public policies which have functioned to encourage suburbanization while destroying inner-city neighborhoods. It is more difficult to obtain a full FHA or VA mortgage in aging residential areas than for new suburban housing. Since the larger proportion of the full mortgage is interest, and, hence, tax deductible, millions of Americans have found it economically possible to buy in the suburbs where they could not afford comparable housing in the central city. Moreover, rehabilitation of inner-city housing is certain to bring an appraiser and result in higher taxes. While many suburbs have higher mean property taxes than the mean for their parent city, the balance of tax laws favors the choice of suburban housing rather than the purchase and rehabilitation of an older, central-city house.

To this must be added the large influx of blacks into the central cities. Prejudice and fear have unmistakably played a role in encouraging the exodus of whites from the central city, but, equally important, the decline of municipal services and the conscious neglect of deteriorating, absentee-owned rental property have taken a heavy toll. Those who chose to stay and protect their neighborhoods were fighting high odds against the encroachment of malignant urban decay.

In short, the complex maze of public policies, the influx of another wave of poor immigrants, and thousands of individual decisions militated against the maturation and preservation of community life in the inner cities. Those individuals and groups who sought to preserve their neighborhoods gradually learned it was easier to switch than fight.

Finally, we must remember that the dominant cultural ideology professed the virtue of the suburb as an ideal urban life style. This ideology not only encouraged the policies which resulted in centrifugal growth, but its final crushing blow was to enter the consciousness of immigrants who for one brief shining moment believed they were building their Camelot in the heart of the metropolis.

CONCLUSION

Perhaps the experience of the past two hundred years is sufficient to account for man's dislike of cities. We have attempted in this paper to sketch an argument for the proposition that dislike and distrust of urban life is as old as cities themselves. Whether further research will refute or reinforce this proposition, there can be little question of the persistence of anti-urbanism in the American experience. We have tried to trace some of the unique features of American culture which have reinforced this sentiment. Recently, the Gallup organization asked the American public "If you could live anywhere in the United States that you wanted to, would you prefer a city, suburban area, small town or farm?" (Gallup, 1972). Only thirteen percent responded that their ideal place to live would be the city. Of those who live in a city of a half-million or more, four out of five would prefer not to live in the city.

We have tried to trace some of the unique features of American culture which have contributed to this anti-urban sentiment. While there are objective reasons why large proportions of our population should not like cities, these are criss-crossed and reinforced by an underpinning of agrarian philosophy which, as it were, never gave the cities a chance. It is our view that cities have been the birthplace of nearly everything we value in human society, but if the city has been the liberator of man, this has not occurred without a heavy price.

And that price, reinforced by agrarian philosophy, lingers closer to human consciousness than the realization that cities have made men free.

Only a tiny proportion of elites have ever realized the potential of high-density living. But, for the most part, they developed that potential only for themselves. In some measure, late-arriving immigrants made their peace with the city and found it a good environment, but their romance with the city was short-lived, for they were powerless to fight the tide of exodus and the encroachment of decay. By the time planners and other visionaries realized the potential of a high-density core, their efforts were too little and too late to prevent the inevitable flow of population to the periphery of the metropolis. Little by little, the great old neighborhoods have been engulfed by the malignancy of urban decay. And every year more and more ideologically urbane give up their dream of a city where art, culture, and culinary delight are at their doorstep and join the army of commuters who seek the promised land in the suburbs.

Those who imagine the city as a high-density repository of culture, maximum opportunity, and the fulfillment of man's highest aspirations will not find their city in this century—at least not in the United States. We are becoming an urban civilization without cities. Our commitment to sprawl is irreversible in this century.

This sprawl is not so bad as some have tried to portray. It is not easy to find the ideal-typical suburb so vociferously portrayed by journalists and academics alike during the 1950s and 1960s. Suburbia, while highly segregated along marginally differentiated cost lines, is not one endless row of homogeneous, split-level traps inhabited by a conforming middle class that is hyperorganized and hypersociable. On the whole, they are less organized and less sociable than the high-density ethnic ghettos of the late nineteenth and early twentieth centuries. If there is excessive drinking, adultery, divorce, and mental illness in the suburbs, it is not easy to draw a causal link to suburbanization. The suburbs may not be the Buelah Lands many expected, but there can be no question that the overwhelming majority of suburbanites prefer being where they are to high-density city life. The expectations of many planners that large numbers of affluent suburbanites beyond the child-rearing age would return to luxury apartments in the central cities has simply not materialized. We are reshaping urban America in the form of suburban sprawl because this is what Americans want.

But what of the future? Suburbs, too, will grow old and decay. The tiny shopping centers conveniently scattered throughout the suburbs will become increasingly hard-pressed to stay alive in the face of growing competition from giant shopping malls. As this happens, we will be forced to spend more and more time on increasingly congested concrete arteries which are obsolete before completion. And, in the long run, it is questionable whether the decentralization of commerce and industry will reduce the time/distance ratio of the journey to work. And what will be the long-range consequences of isolation in relatively homogeneous neighborhoods? Will people, as Richard Sennett (1970b) suggests, become less capable of coping with diversity and dissent in a pluralistic culture? Or is this just another shibboleth of a disheartened urbanist who grieves the passing of the city?

Perhaps there is no compelling reason for urban civilization to have high-density cores. We may succeed in suburbanizing fine restaurants, theatres, museums, orchestras, and professional athletics without destroying their vitality. And, in time, we may tear down those horribly decaying centers and make suburbs of them also (Banfield, 1970).

But if it is aimless nonsense to prolong nostalgia for the kind of city we have only imagined as we climbed Telegraph Hill, or strolled carelessly through the Vieux Carre, or whiled away a Sunday afternoon in Washington Square, it is also dangerous to fail to be hard-nosed about the long range implications of the new city we are creating. As individuals and as a society, we are captives of our images. Except for a small minority, our image of the high-density city has been negative. And now we are working at a frantic pace to create a new city on the fringe of the old. But we should recognize that the new creation is no more than a facsimile of what we had envisioned. What it will ultimately be is as problematic as the visionaries' image of a vertical city, which was never realized. But our new city most assuredly will come into full fruition. Suburbia is the new American frontier. We are too deeply committed to reverse the trend. But while we are completing the task, it is perhaps appropriate for some to ponder the question what happens next if we discover that the environment we have created has the advantages neither of urban nor of rural living, and the disadvantages of both?

NOTES

1. The concept "discontents of urban civilization" is taken from Irving Kristol (1970a). In an article, Kristol argues that it is the absence of human values which are appropriate and applicable to urban civilization that has created the "urban crisis." "The challenge to urban democracy," he writes, "is to evolve a set of values and a conception of democracy that can function as the equivalent of the 'republican morality' of yesteryear" (Kristol, 1970b: 47). In short, the urban crisis is essentially "a moral-philosophical one, and . . . it cannot be dealt with simply by a 'practical' pragmatic, matter-of-fact approach" (Kristol, 1970b). This is implicitly the argument of this paper.

2. This discussion is also informed by the writings of Kenneth Boulding. We view Boulding's (1956) proposal for a science of *eiconics* as substantially in the same genre of theoretical thought as developed by Berger and Luckmann (1966). The concept's meaning and image as used in this paper are essentially interchangeable.

3. The longest single passage on urban and nomadic life is in Volume 2 (1958: 233-307), but passing references to the character of men of the city and the country are scattered throughout the three volumes. See also Volume 1 (1958: 250-267, 308-310).

4. The literature on the emergence of cities in the Oriental world is sparse in comparison with that available for the Occidental world. We hesitate, at this point, to apply our argument to cities of the Far East, although there is evidence which tends to support both our assumptions and conclusions. For example, cities of the Shang Dynasty during the second millennium B.C. were politically organized as theocracies with a political order replacing kinship organization (Wheatley, 1971). Similarly, the evidence points to rigidly stratified social organization (Wright, 1967). But it is also clear that cities of the Orient were of a significantly different character from those which emerged in the West (Mote, 1970; Balazs, 1964). To conclude, either the presence or absence of anti-urbanism is premature.

5. Excerpts from this remarkable report are reprinted in Beatty and Johnson (1958: 534-543).

6. Charles Booth's first volume, which was thrice revised, on the social and economic conditions of London was published in 1889. The third edition of *Life and Labour in London* was 17 volumes (1902-1903). The materials cited here are from an edited volume of Booth's work (Fried and Elman, 1968).

REFERENCES

ARIES, P. (1962) Centuries of Childhood. New York: Alfred A. Knopf.

BAILYN, B. (1960) Education in the Forming of American Society. Chapel Hill: Univ. of North Carolina Press.

BALAZS, E. (1964) Chinese Civilization and Bureaucracy. New Haven, Conn.: Yale Univ. Press.

BANFIELD, E. C. (1970) The Unheavenly City. Boston: Little, Brown.

BANKS, J. A. (1954) Prosperity and Parenthood: A Study of Family Planning among the Victorian Middle Classes. London: Routledge & Kegan Paul.

BEATTY, J. L. and O. A. JOHNSON [eds.] (1958) Heritage of Western Civilization. Englewood Cliffs, N.J.: Prentice-Hall.

BERGER, P. L. and T. LUCKMANN (1966) The Social Construction of Reality. Garden City, N.Y.: Doubleday.

BOORSTIN, D. J. (1948) The Lost World of Thomas Jefferson. New York: Holt, Rinehart & Winston.

BOOTH, C. (1902-1903) Life and Labour of the People of London. London and New York: Macmillan.

——— (1889) Life and Labour of the People. London and Edinburgh: Williams & Norgate.

BOSWELL, J. (1950) London Journal, 1762-1763. New York: McGraw-Hill.

BOULDING, K. E. (1956) The Image. Ann Arbor: Univ. of Michigan Press.

BOWEN, F. (1856) The Principles of Political Economy. Boston: Little, Brown.

BREMNER, R. H. (1956) From the Depths: The Discovery of Poverty in the United States. New York: New York Univ. Press.

BUSHNELL, H. (1864) Work and Play; or Literary Varieties. New York: Harper.

CALHOUN, D. H. (1965) Professional Lives in America: Structure and Aspiration, 1750-1850. Cambridge, Mass.: Harvard Univ. Press.

CAREY, M. (1830) Miscellaneous Essays. Philadelphia: Carey & Hart.

CAWELTI, J. G. (1965) Apostles of the Self-Made Man. Chicago: Univ. of Chicago Press.

CHILDE, V. G. (1954) What Happened in History. Baltimore: Penguin.

——— (1950) "The urban revolution." Town Planning Rev. 21: 3-17.

COX, H. (1971) "The ungodly city: a theological response to Jacques Ellul." Commonweal (July 9): 351-357.

——— (1965) The Secular City. New York: Macmillan.

CREVECOEUR, J. H. J. (1912) Letters from an American Farmer. London: J. M. Dent.

DEMOS, J. (1970) A Little Commonwealth: Family Life in Plymouth Colony. New York: Oxford Univ. Press.

DIXON, T., Jr. (1905) The Life Worth Living: A Personal Experience. New York: Doubleday, Page.

——— (1896) The Failure of Protestantism in New York and its Causes. New York: Strauss & Rehn.

DOWNING, A. J. (1853) Rural Essays. New York: Leavitt & Allen.

——— (1850) The Architecture of Country Houses. New York: D. Appleton.

——— (1844) A Treatise on the Theory and Practice of Landscape Gardening. New York: Wiley & Putnam.

ELKINS, S. (1959) Slavery: A Problem in American Institutional and Intellectual Life. Chicago: Univ. of Chicago Press.

ELLUL, J. (1970) The Meaning of the City. Grand Rapids, Michigan: Eerdmans.

——— (1964) The Technological Society. New York: Alfred A. Knopf. (Originally published in French in 1954 as La Technique ou l'enjeu du siècle by Librairie Armand Colin.)

EMPSON, W. (1950) Some Versions of the Pastoral. London: Chatto & Windus.

ERICKSEN, E. G. (1954) Urban Behavior. New York: Macmillan.

EVERSLEY, D.E.C. (1965) "Population, economy and society," in D. V. Glass and D.E.C. Eversley (eds.) Population in History.

FOGELSON, R. M. (1967) The Fragmented Metropolis: Los Angeles 1850-1930. Cambridge, Mass.: Harvard Univ. Press.

FRANKLIN, B. (1959) Papers. New Haven, Conn.: Yale Univ. Press.

FREDRICKSON, G. (1965) The Inner Civil War: Northern Intellectuals and the Crisis of the Union. New York: Harper & Row.

FRIED, A. and R. M. ELMAN [eds.] (1968) Charles Booth's London. New York: Pantheon.

Gallup, G. and Associates (1972) "The Gallup poll." Washington Post (December 12).

GANS, H. J. (1962) The Urban Villagers. New York: Free Press.

GODKIN, E. L. (1896) Problems of Modern Democracy. New York: Charles Scribner's.

GOLD, M. (1930) Jews Without Money. London: N. Douglas.

GOWANS, A. (1964) Images of American Living: Four Centuries of Architecture and Furniture as Cultural Expression. Philadelphia: J. B. Lippincott.

GREVEN, P. J., Jr. (1970) Four Generations: Population, Land and Family in Colonial Andover, Massachusetts. Ithaca: Cornell Univ. Press.

GUSFIELD, J. R. (1963) Symbolic Crusade: Status Politics and the American Temperance Movement. Urbana: Univ. of Illinois Press.

HAMMOND, M. (1972) The City in the Ancient World. Cambridge, Mass.: Harvard Univ. Press.

HIGHAM, J. (1969) From Boundlessness to Consolidation: The Transformation of American Culture, 1848-1860. Ann Arbor: William L. Clements Library.

——— (1968) "Immigration," in C. V. Woodward (ed.) The Comparative Approach to American History. New York: Basic Books.

——— (1966) "American Anti-Semitism historically reconsidered," in M. H. Stember (ed.) Jews in the Mind of America. New York: Basic Books.

——— (1963) Strangers in the Land: Patterns of American Nativism, 1860-1925. New York: Atheneum.

——— (1958) "Another look at nativism." Catholic Historical Rev. 44: 147-158.

HIMES, N. E. (1938) Medical History of Contraception. Baltimore: Williams & Wilkins.

HOFSTADTER, R. (1955) The Age of Reform. New York: Alfred A. Knopf.

HOLLI, M. G. (1969) Reform in Detroit: Hazen S. Pingree and Urban Politics. New York: Oxford Univ. Press.

HUTHMACHER, J. J. (1968) Senator Robert F. Wagner and the Rise of Urban Liberalism. New York: Atheneum.

——— (1962) "Urban liberalism and the age of reform." Mississippi Valley Historical Rev. 49: 231-241.

JAHER, F. C. (1968) "The Boston Brahmins in the age of industrial capitalism," in The Age of Industrialism in America. New York: Free Press.

JEFFERSON, T. (1955) Notes on the State of Virginia. Chapel Hill: Univ. of North Carolina Press.

JUERGENS, G. (1966) Joseph Pulitzer and the New York World. Princeton: Princeton Univ. Press.

KATZ, M. (1968) The Irony of Early School Reform: Educational Innovation in Mid-Nineteenth Century Massachusetts. Cambridge, Mass.: Harvard Univ. Press.

KENNEDY, D. M. (1971) Birth Control in America: The Career of Margaret Sanger. New Haven, Conn.: Yale Univ. Press.

KHALDUN, I. (1969) The Maquaddimah. Princeton: Princeton Univ. Press.

KINZER, D. L. (1964) An Episode in Anti-Catholicism: The American Protective Association. Seattle: Univ. of Washington Press.

KIRKLAND, E. C. (1956) Dream and Thought in the Business Community, 1860-1900. Ithaca: Cornell Univ. Press.

KRISTOL, I. (1970a) "Urban civilization and its discontents." Commentary (July): 29-35.

——— (1970b) "Is the urban crisis real?" Commentary (November): 44-47.

LANE, R. (1967) Policing the City: Boston, 1821-1885. Cambridge, Mass.: Harvard Univ. Press.

LASLETT, P. (1965) The World We Have Lost. New York: Charles Scribner's.

MARX, L. (1964) The Machine in the Garden: Technology and the Pastoral Ideal in America. New York: Oxford Univ. Press.

MEYERS, M. (1957) The Jacksonian Persuasion: Politics and Belief. Stanford: Stanford Univ. Press.

MILLER, P. (1967) Nature's Nation. Cambridge, Mass.: Harvard Univ. Press.

MOTE, F. W. (1970) "The city in traditional Chinese civilization," in J.T.C. Liu and W. Tu (eds.) Traditional China. Englewood Cliffs, N.J.: Prentice-Hall.

MUMFORD, L. (1961) The City in History. New York: Harcourt, Brace & World.

NASH, R. (1967) Wilderness and the American Mind. New Haven, Conn.: Yale Univ. Press.

PETERSEN, W. (1960) "The demographic transition in the Netherlands." Amer. Soc. Rev. 25: 334-347.

PETERSON, M. D. (1960) The Jefferson Image in the American Mind. New York: Oxford Univ. Press.

PIRENNE, H. (1956) Medieval Cities. Garden City, N.Y.: Doubleday.

POTTER, J. (1965) "The growth of population in America, 1700-1860," in D. V. Glass and D.E.C. Eversley (eds.) Population in History. London: Arnold.

REPS, J. W. (1965) The Making of Urban America: A History of City Planning in the United States. Princeton: Princeton Univ. Press.

ROSSI, P. H. and R. A. DENTLER (1961) The Politics of Urban Renewal. New York: Free Press.

ROURKE, F. E. (1964) "Urbanism and American democracy." Ethics 74: 255-268.

SADLER, M. (1832) The Sadler Report, Report from the Committee on the Bill to Regulate the Labour of the Children in the Mills and Factories of the United Kingdom. London: House of Commons.

SCHMITT, P. J. (1969) Back to Nature: The Arcadian Myth in Urban America. New York: Oxford Univ. Press.

SCHNORE, L. F. (1965) The Urban Scene: Human Ecology and Demography. New York: Free Press.

SCULLY, V. J., Jr. (1955) The Shingle Style: Architectural Theory and Design from Richardson to the Origins of Wright. New Haven, Conn.: Yale Univ. Press.

——— (1953) "Romantic retionalism and the expression of structure in wood: Downing, Wheeler, Gardner, and the 'stick style'." Art Bull. 35: 121-142.

SENNETT, R. (1970a) Families Against the City: Middle Class Homes of Industrial Chicago, 1872-1890. Cambridge, Mass.: Harvard Univ. Press.

——— (1970b) The Uses of Disorder: Personal Identity and City Life. New York: Alfred A. Knopf.

SJOBERG, G. (1960) The Preindustrial City, Past and Present. New York: Free Press.

SMITH, H. N. (1950) Virgin Land: The American West as Symbol and Myth. Cambridge, Mass.: Harvard Univ. Press.

SOLOMON, B. M. (1956) Ancestors and Immigrants: A Changing New England Tradition. Cambridge, Mass.: Harvard Univ. Press.

SPENGLER, O. (1928) The Decline of the West. Volume 2 of Perspectives of World-History. New York: Alfred A. Knopf.

SUTTON, S. B. [ed.] (1971) Civilizing American Cities: A Selection of Frederick Law Olmsted's Writings on City Landscapes. Cambridge, Mass.: MIT Press.

TAEUBER, K. E. and A. F. TAEUBER (1965) Negroes in Cities: Residential Segregation and Neighborhood Change. Chicago: Aldine.

THERNSTROM, S. (1964) Poverty and Progress: Social Mobility in a Nineteenth-Century City. Cambridge, Mass.: Harvard Univ. Press.

VAN DOREN, C. (1938) Benjamin Franklin. New York: Viking.

WARD, J. W. (1969) Red, White and Blue: Men, Books and Ideas in American Culture. New York: Oxford Univ. Press.

——— (1955) Andrew Jackson: Symbol for an Age. New York: Oxford Univ. Press.

WARNER, S. B., Jr. (1968) The Private City: Philadelphia in Three Periods of its Growth. Philadelphia: Univ. of Pennsylvania Press.

——— (1962) Streetcar Suburbs: The Process of Growth in Boston, 1870-1900. Cambridge, Mass.: Harvard Univ. Press.

WHEATLEY, P. (1971) The Pivot of the Four Quarters: A Preliminary Inquiry into the Origins and Character of the Ancient Chinese City. Chicago: Aldine.

WHEELER, H. (1971) "The phenomenon of God." Center Magazine (March/April): 7-12.

WHITE, M. and L. WHITE (1962) The Intellectual Versus the City. Cambridge, Mass.: Harvard Univ. and MIT Presses.

WILLIAMS, R. (1958) Culture and Society, 1750-1950. New York: Columbia Univ. Press.

WOHL, R. R. (1969) "The 'country boy' myth and its place in American urban culture: the nineteenth-century contribution." Perspectives in Amer. History 3: 77-156.

――― (1953) "The rags to riches story: an episode of secular idealism," in R. Bendix and S. M. Lipset (eds.) Class, Status and Power. New York: Free Press.

WOODWARD, C. V. (1938) Tom Watson: Agrarian Rebel. New York: Macmillan.

WRIGHT, A. F. (1967) "Changan," in A. Toynbee (ed.) Cities of Destiny. New York: McGraw-Hill.

WYLLIE, I. G. (1954) The Self-Made Man in America: The Myth of Rags to Riches. New Brunswick, N.J.: Rutgers Univ. Press.

Part II

SOCIAL STRUCTURE

AND SOCIAL PROCESS

SOCIAL STRUCTURE AND

SOCIAL PROCESS

The last time suburbia was "fashionable" was in the fifties, when most of the attention it received focused on middle-class families and their life styles. It is fitting, therefore, to begin the substantive part of the book with two essays on the same topic before introducing two new developments in the sociology of suburbia, ethnic groups, and Blacks.

In the opening chapter, Marshall reviews the issues in the sociological debate over the cause of suburban life style, suburban residence versus class and life-cycle stage. His critique of life-style literature leads him to conclude that the urban-suburban life-style differences are not as sharp as commonly assumed, and tend to concentrate more on suburbanites' involvement with family, neighbors, and community. But the sources of these minor differences are not clear because of methodological deficiencies in the studies. No research to date has sorted out the independent, joint, and interactive effects of structure, social-psychological selection, class, and family composition. Further, the important question may not be the nature and intensity of city-suburban life-style differences, but rather the identity of the variables which influence life style regardless of place of residence.

Greer assesses the effects of suburbanization on ethnicity, kinship patterns, and family relationships in the second selection of Part II. He suggests that ethnicity ("tribalism") in the United States, which has been maintained through spatial concentration, ethnic endogamy, the functional role of the extended kin structure and the pressures of the community, is threatened by mobility in social rank and scatteration in space. Suburbs, homogeneous in class, rank, and life style, represent heterogeneous ethnic lineages which result in intermarriage and assimilation and in turn reduce support for the extended kin system and mitigate community pressure.

The paper by Harlan Hahn focuses on white ethnic families in the suburbs. Such families display characteristics that diverge from the dominant social values of suburbia. They do not fit the accepted nuclear family model because of a disproportionate number of large or extended families and single or unattached migrants. This difference between the family attributes of some ethnic groups and the suburban norm creates a potential for intense community conflict as the proportion of white ethnics in the suburbs increases. Hahn concludes that government agencies need to review and perhaps revise their policies and priorities in order to facilitate the accommodation of new suburban migrants.

The last essay in this section attempts to evaluate the phenomenon of Black suburbanization, which, like that of ethnic groups, is relatively recent. Using a selected sample of eleven metropolitan areas, Pendleton examines the data in an effort to answer two questions: Do Black and white suburbanization have the same consequences, and does Black suburbanization reduce racial differences? He concluded that Black suburbanization is similar to white on demographic dimensions (e.g., youth and aged dependency, fertility ratios), but not on social dimensions (e.g., value of owner-occupied units, contract rent, housing quality). This is explained in part by the fact that much Black suburbanization is a result of the growth of traditional Black urban fringe communities like East St. Louis.

4

Suburban Life Styles:
A Contribution to the Debate

HARVEY MARSHALL

INTRODUCTION

☐ THE RAPID GROWTH OF SUBURBS during recent decades has stimulated a massive amount of comment and research. Perhaps no aspect of this phenomenon has aroused as much interest as questions about the kinds of lives being lived in these communities. Various descriptions and speculations have appeared in both professional journals and the mass media, leading to widely held beliefs about the characteristics and activities of suburbanites. Many have seen suburbia as the home of frantic socializing, conformity, drab homogeneity, social climbing, and so on. More careful empirical work has shown this picture to be so distorted and overdrawn that it may appropriately be called a myth. Nonetheless, sociologists remain divided on the question: on one side are those who maintain that suburbanites are different from their urban counterparts in the way

AUTHOR'S NOTE: *I would like to express my appreciation to Russell Hamby and Robert Perrucci for their extremely helpful comments on earlier drafts of this paper. Of course, I alone am responsible for any remaining ambiguities and errors.*

they live and think; on the other side are those who contend that whatever variance exists between urban and suburban life styles is essentially a function of class and life-cycle stage and has nothing to do with the mere fact of suburban residence.

It is this polemic which is of concern here: To what extent do suburbanites differ from urbanites in their life styles? What are the causes of these differences, if they exist at all? These questions and related issues are the topic of this paper. First, however, it is necessary to systematically examine two of the central concepts about which the debate is centered: "style of life" and "suburb."

STYLE OF LIFE

While style of life is used in a variety of ways in the literature, there seem to be two general definitional strategies as it is applied to comparisons of cities and suburbs. The first, which might be labelled "specific," attempts to specify one or more "sociologically relevant" aspects of behavior, and urban and suburban residents are then compared on each. For example, Tallman and Morgner (1970: 337) define a number of activities and orientations, "each of which requires a distinctive investment of the individuals' time, energy or money." No attempt is made to show how these are interrelated or whether or not one activity or element is dominant.

The second usage is more inclusive and suggests that certain activities are dominant in the sense that a variety of others are arranged with relation to them. Also implicit in this usage is the idea that a particular set of activities form an interrelated constellation —given knowledge of one element, it is possible to predict others (compare Feldman and Theilbar, 1972: 1-3). For example, Bell (1958) seems to imply that "familism" is a cluster of interrelated activities and attitudes, with emphasis upon children being so dominant that virtually all other aspects of life are arranged with reference to it.

SUBURBS

"Suburb" is another term which is used in various ways, although there may be more agreement as to its meaning than is the case with "life style." From an ecological point of view, suburbs usually are regarded as one part of the territorial division of labor within metropolitan areas. These units are more or less specialized in terms of such functions as production, residence, or recreation. Within these categories, further specialization is possible; for example, there are blue- and white-collar residential areas, suburbs with relative concentrations of light manufacturing, and so on. Also found within the metropolitan area are agricultural activities, usually specialized and oriented toward production for the local urban population.

This areal specialization implies interdependence among physically separated units, with a "central city" providing coordination through its concentration of administrative, financial, wholesale, and specialized retail and entertainment activities (compare Bogue, 1949; Gras, 1922, 1926; Kish, 1954; McKenzie, 1933; and Schnore, 1957b). Although not strictly relevant to an ecological definition, the various specialized units are often separate political entities, each with its own police and fire departments, sewage and water services, and the like. In the concluding section of this chapter, the question of whether or not suburbs are selected for study on the basis of this, or any other theoretical definition is considered.

OVERVIEW

Life styles between cities and suburbs logically can become differentiated in three ways. First, structural and demographic characteristics of the suburb may *produce* a distinct style of life. Suburban settlement patterns are presumed to differ from urban patterns and these may be regarded as independent variables vis-à-vis life style. For example, it is often argued that the greater amount of commuting implied by suburban living tends to isolate the breadwinner from participation in the local community and limits contact with his family. Similarly, the institutional structure is presumably much simpler in suburbs, since not all aspects of the city are equally free to decentralize, thereby restricting the range of activities which suburbanites may be involved in. This is essentially a social-psychological argument—elements of structure are assumed to affect individual behavior (compare Schnore, 1965: 29-43).

Second, some authors argue that suburban life styles become differentiated through selective migration. There may be a tendency for families with certain attitudes and values to choose the suburbs, while those with different orientations remain in the city. For example, suburbanites may prefer gardening, a sense of effective participation in small governments, and knowing most of the people in town, isolation from crime, and so on. If people who choose to remain in the city do not have these tastes, then different life styles characterize the two areas. An explicit assumption is that the "suburban" life styles are imported.

Of course, there is the possibility that both selective migration and structural variables together operate to differentiate city and suburban life styles. In fact, writers who argue that suburbia is being selectively populated often emphasize the importance of structure as well. It may also be that these variables interact to produce an affect that neither would produce in isolation.

Opposing both the structural and selective migration explanations are a sizable number of sociologists who contend that the basic determinants of life styles are social class and life-cycle stage. The proponents of this third position suggest that, since that segment of the middle class in the childbearing stage of the family life cycle is heavily concentrated in the suburbs, gross comparisons between "cities" and "suburbs" may well show differences in life styles. But, according to this argument, such differences are spurious—class and family composition produce both suburbanization *and* life style. Consequently, a more appropriate comparison is between suburbanites of a particular social class and life-cycle stage with city dwellers with the same characteristics. If class and life-cycle stage is held constant in this way, it is argued that there are no life-style differences between the two groups.

Note that this argument is quite similar to the selective migration hypothesis, since it stresses the outward movement of certain population types rather than of others. In other words, if the upper-working and middle classes in the childbearing stage of the family life cycle are being concentrated in the suburbs, and other types in the city, then a different style of life probably characterizes the two areas. This is consistent with both the selective migration and the social class-life cycle explanations.

However, there is a key assumption which sharply differentiates the two. Advocates of the selective migration hypothesis argue that selection is *not* solely on the basis of class and life-cycle stage. Rather, they seem to be contending that *segments* of the middle and working classes in particular age categories value such things as family life and social involvement to a much higher degree than their class and life-cycle counterparts in the city. Put somewhat differently, some families may choose to remain in the city, even though they have children and are middle class, in order to maximize access to various activities which can only be found there; they are either not interested in suburban advantages or are willing to give them up in order to maintain others associated with access to the downtown areas of larger cities.

This suggests that an implicit assumption of the class-life cycle hypothesis is that suburbanites are a random sample of the various social classes living in cities, if family life-cycle stage is controlled. Or, at least, that factors underlying the move are unrelated to life style. Without such an assumption, the question of characteristics underlying selection—with the possibility of an associated life style—becomes an issue.

The next section of this chapter is devoted to discussion—first, of structural and demographic variables as determinants of life style; second, of selective migration and life styles; and third, of the social class-life cycle hypothesis. In the final section, these positions are evaluated, some conclusions about the debate drawn, and suggestions for future research outlined.

STRUCTURAL AND DEMOGRAPHIC VARIABLES

As suggested above, it is possible to regard suburban structure as the independent variable in determination of life styles. Structural characteristics, such as demographic homogeneity and spatial arrangements, are seen as *causes* of particular activities, and questions of why people initially move to the suburb tend to be ignored.

The first such structural variable to be considered is commuting. Martin (1956) argues that high rates of commuting to a central city, which he regards as an essential defining characteristic of "suburbs," has implications for neighborhood interaction patterns. He studied a village where more than half the labor force commuted to a nearby larger city, and his data indicated that the commuters were significantly less involved in the local community than was the case for the noncommuters. Scaff's (1952) analysis of the effects of commuting upon participation in local community organizations is consistent with Martin's conclusions. Scaff studied a Los Angeles suburb and found that not only did commuters participate less, distance of the job from the suburb was also related to participation. This finding is especially significant given that commuters tended to be better educated and have higher-status jobs than noncommuters —both of which are positively related to participation.

Another presumed effect of commuting is upon family life, although the data are somewhat contradictory. Lundberg et al. (1934), in an early study, found that, even if income is held constant, the husbands and wives of commuters spend *more* of their evenings together than do couples in families where the husband does not commute; he also found that the children of commuting families spend more time at home. The authors argue that these data support the interpretation that commuters tend to value home life more highly than noncommuters, as well as the hypothesis that commuting has an integrative effect. Of course, there are serious problems with inferring values from behavior, as Lundberg and his associates have done; for example, it may be that suburbanites have fewer places to go at night, and hence have little choice other than spending their evenings at home.

However, Gans' (1967: 222-223) study of residents of a new suburb leads him to conclude that commuting does not have much impact upon family life one way or the other; there seemed to be little relation between length of commute and time spent with either children or wives. It is clear that, given these somewhat contradictory data, firm conclusions cannot be reached about the effects of commuting either upon social participation or family life.

Another alleged characteristic of suburbs which may affect life style is homogeneity: while suburbs differ among themselves, individual communities are presumably populated by persons of approximately the same age, at the same point in their family cycle, of the same class, and with broadly similar backgrounds and interests. (The validity of this assumption is examined in the concluding section of this chapter.)

Although this homogeneity results from selective migration, it is hypothesized that homogeneity itself increases neighborhood inter-action. In fact, Martin (1956) maintains that it is lack of homo-geneity which is largely responsible for the small amount of neighborhood interaction in urban residential districts.

Fava (1956) also attributes the relatively high rates of informal interaction in suburbs to homogeneity. She emphasizes the preva-lence of middle-class married couples with their children as the underlying homogenizing factors. However, she also points to such physical characteristics as low population density and open spaces as contributing to the relatively higher volume of suburban, compared with urban, interaction.

There are some data which indicate that interaction is more extensive in homogeneous neighborhoods. For example, Caplow and Foreman (1950) compared respondents from a married student residential area on the University of Minnesota campus with a sample drawn from a large number of Minneapolis-St. Paul neighborhoods. They found that the amount of interaction in the student area was extremely high when compared with the city neighborhoods, despite the fact that the average length of residence in the latter was ten times as long as that in the student area. The causes of this variation in level of interaction between the two types of neighborhood appeared to be absence of status distinctions and the presence of similar interests and background of residents in the student area, which in turn are linked to ease of communication.

However, there are alternate views on the effects of homogeneity. For example, Hawley (1971: 194) argues that "the greater their similarity, the less people have to say to one-another." He contends that the increased mobility of urban populations means that they may select their associates from a wide geographical area. Even among working-class groups, the members of which are presumably less mobile and hence tied more closely to the local area, contacts with neighbors other than kin do not appear to be extensive (compare Shuval, 1956).

A third variable which has been linked to suburbs and neighboring patterns is density. For example, Dobriner (1963: 9-11) points to the possible effects of a "visibility pattern." Because of the openness and lack of congestion, suburbanites are able to follow the activities of their neighbors rather easily. Furthermore, when engaged in activities of their own outside the house, they are likely to encounter neighbors similarly involved, forming the basis for more or less casual exchanges. While there are few data bearing on this hypothesis, it is likely that visibility contributes to the volume of informal inter-action, which may form the basis for friendships.

However, there are other views. Gans (1967: 280-281) is skeptical about the effect of physical arrangement of dwelling sites. He says that while proximity and layout may lead to acquaintance, com-patibility is the basis for friendship—and the latter is obviously not guaranteed by proximity. Moreover, there is sufficient distance between houses that it is possible to ignore all but immediate neighbors if residents choose to do so. Thus, as with commuting, there are no definitive data which permit a conclusion one way or the other.

Another frequently cited factor which may affect life style is "ecological position," referring to location beyond city limits (compare Martin, 1956). Not only may this affect commuting rates, there is also the possibility that different "urban" institutions are not equally likely to decentralize. That is, those things that "are" the city are not equally amenable to suburbanization. As Douglas (1925: 34) observed, "It is by studying these 'cans' and 'cannots' that one comes to understand the suburb as a social fact in the accompanying mind of its people." The implication of this for life styles is that suburban residents are relatively isolated from the specialized activities of the large center, and this presumably affects their leisure behavior as well as the pattern of friendships and association memberships.

In this connection, Riesman (1958) contends that suburban leisure patterns are relatively unspecialized compared with city behavior. For example, city music lovers are found in sufficiently large numbers that they can form clubs oriented to particular musicians, such as Bach, Mendelssohn, or Beethoven. In the suburbs, if such clubs are to be formed, tastes must be generalized so that a local "classical" music club is formed. It is consequently argued that suburban leisure is becoming geared to mass tastes. Dobriner (1963: 56-57) also suggests that many urban cultural and recreational opportunities are not available in the suburbs; at least, if the suburbanite is to participate in them, he often must travel relatively long distances. These include museums, the theatre and opera, and, often, good restaurants.

Looking at other urban-suburban leisure patterns which may be influenced by isolation from an urban center, Gans (1967) found that, while the vast majority of respondents in a new suburban community reported changes in their leisure activities, compared with their behavior in the city, these reflected little alteration in their basic style. Among the men, the major change (not surprisingly) involved increased gardening and yard work, while the women tended to increase their participation in local organizations and game activities, as well as gardening. There were also fewer trips to parks and beaches, and movie attendance declined.

That there is no single "suburban" leisure pattern was shown by Meyersohn and Jackson (1958), who compared two suburbs. Both expectations about gardening and amount of time spent in this activity varied sharply between the two communities (these differences were probably attributable to the fact that one suburb was considerably more homogeneous in terms of both age and family composition than the other, as well as having been developed at a different time).

Ennis (1958) attempted to determine whether or not the suburban move was associated with greater "creativity" in leisure. He compared budget expenditures for different kinds of leisure activities, and his data suggest that, among low income workers—both white- and blue-collar—urban and suburban patterns were virtually identical. However, there was some indication that the leisure patterns of the relatively high-income white-collar workers were more creative in the suburbs than in the city. Among the relatively affluent blue-collar workers, there was no difference associated with settlement pattern, suggesting that income per se is not a crucial variable. Obviously, much work needs to be done before we can infer much about urban-suburban differences in leisure activities.

Regarding participation in organizations, a study by Martin (1952) of a fringe area suggests that the impact of suburban location upon participation in urban activities is at least partly a function of whether or not the husband works there. Specifically, he found that commuters were more likely to belong to formal organizations located in the urban center than was the case for noncommuters. This finding casts some doubt upon the hypothesis that suburban isolation necessarily limits participation in urban activities.

However, there are some data which suggest that the impact of suburban isolation may vary according to sex and class level. Tallman (1969), on the basis of interviews with a sample of working-class couples in an urban and a suburban neighborhood, found that the suburban women have fewer friends, experience a greater sense of isolation, and report more marital conflict than do their urban counterparts. Tallman interprets these data as reflecting the heavy reliance of working-class women upon childhood friends and kin for emotional support. The move to the suburbs disrupts this source of support and consequently affects them much more severely than is the case for working-class men, who still maintain contacts at work. It may be, although Tallman's data do not bear specifically on this question, that middle-class women adapt more readily to the suburban situation because they are presumably less involved with their kin (compare Gans, 1967: 226). Berger's (1960) data on a working-class San Jose suburb also suggest that women are more affected by suburban living than are men; he found that the women in his sample were more likely than their husbands to engage in informal visiting, identify themselves as middle class, discuss child-rearing, and belong to informal organizations.

Finally, some researchers have argued that aspects of the suburban environment affect the level and type of religious participation. Greer and Kube (1959) found a positive relation between frequency of church attendance and "suburbanization." Similarly, Tallman and Morgner (1970) found that, in a sample of working class families, the suburban group attended church with considerably more frequency than the urban group.

Other data suggest that church attendance is lower in suburban than in urban areas. For example, Zimmer and Hawley (1959) concluded that church attendance was lower in fringe areas even when distance from a church was controlled. Similarly, Berger (1960: 40-53) found that his working-class respondents were about as likely to decrease as to increase attendance.

However, despite these contradictory data, the weight of evidence suggests that church attendance is higher in suburban areas, and it may have a different meaning. A possible interpretation of this apparent tendency has been suggested by Carlos (1970), who contends that urban and suburban settlements differ in the degree to which they emphasize shared interests rooted in a particular territory. While both are, in Gans' (1962a) terms, characterized by "quasi-community" life styles, urban residents tend to base com-munication upon such categories as occupation and social status, suburban residents upon a territorially based community. In the latter communities, church attendance may reflect a need for public expression of solidarity and hence is an integrative device.

Gans' (1967: 264-266) study of a new Levittown community is consistent with this interpretation. Specifically, he found that, while a large proportion of his respondents reported an increase in the frequency of church attendance, there was considerable variation by affiliation—with Jews reporting the greatest increase, Catholics the least. (However, Catholic males did report an increase in the frequency of Mass attendance.) The causes of these increases appeared to differ when Jews were compared with Protestants. Among the latter, the increase may reflect a desire to be part of the community, resulting in turn from a shift from apartment or high-density city living to a low-density area and homeownership; this interpretation was supported by the fact that the greatest increases were among former city and apartment dwellers, as well as by the correlation between increases in "neighboring" and increases in church attendance. In the case of the Jews, the principal stimulus appeared to be a desire to maintain a Jewish identity in the heterogeneous suburban context. These data, along with Carlos', suggest that identification with a small community area and homeownership may produce greater expressions of public solidarity, one dimension of which is church attendance (compare Mueller, 1966; Whyte, 1957; Winter, 1961).

SELECTIVE MIGRATION

The second thesis, that families move to suburbia because there they can maximize certain values, has a long history. As early as 1925, Douglas (1925: 34) argued that "the people of the residential suburbs . . . live where they do by reason of natural selection based on a peculiar psychology and motivation." Specifically, he felt that the suburban move was associated with a desire for family privacy and independence: "they [suburbanites] are a chosen people separated from their fellow city men." Similarly, Lundberg et al. (1934), in another early analysis of suburban life styles, maintained that, since city dwellers often have the same demographic charac-teristics as their suburban counterparts, the latter must be selected on the basis of psychological characteristics.

In this tradition, Fava (1956) has argued that social-psychological selectivity is an important source of the distinctive life styles found in the suburbs. She particularly suggests that people oriented toward "neighboring" are likely to migrate to the suburbs. To test this hypothesis, she undertook a pilot study comparing central-city dwellers, persons living in an outer city area, and those residing in suburbs (Fava, 1958). The hypothesis that informal interaction is more extensive as distance from the center of the city increases was substantiated; this difference persisted within subsamples matched on the basis of age, sex, marital status, length of residence, nativity, education, and size of childhood community. She concludes that: "A complex set of factors selects a segment of the urban population for the suburban migration" (Fava, 1958: 128).

Numerous other studies are consistent with Fava's. For example, Tomeh (1964), using pooled data from three Detroit Area Studies, matched a sample of persons from three zones in the Detroit metropolitan area—inner city, outer city, and suburbs. Respondents were matched on age, marital and migrant status, sex, race, education, and religious affiliation. A summary score was computed for each respondent based upon the total amount of informal participation with neighbors, coworkers, relatives, and friends. There was a clear difference between suburban and city respondents, with the former having the highest scores. However, there was no difference between the inner- and outer-city samples. On the basis of these data, she concludes that "zone of residence has a small but independent effect upon such activity" (Tomeh, 1964: 34). Gans (1967: 261) also found an increase in the amount of neighboring and participation associated with the suburban move, but does not attribute this change to any unique "suburban" influence; rather, he argues that it results primarily from the shift from apartment to owner-occupied single dwelling units where compatible neighbors are more frequently found.

Tallman and Morgner (1970), in their study of working-class samples drawn from a central-city residential district and a suburb, also concluded that the amount of contact with neighbors was higher in the suburb, for both men and women. Furthermore, while gross comparisons indicated no differences between the two samples in participation in service and recreational organizations, suburban women with previous suburban residential experience did report greater levels of participation in both types of organization. Suburban men with previous suburban residential experience were significantly different from the urban sample only in group recreation activities. Although only working-class respondents were compared and roughly matched on the basis of family size, there were rather sharp differences in proportions of employed wives, homeownership, education, and age; consequently, it is difficult to determine from these data whether selective migration (or differences in settlement patterns) is responsible for the observed differences.

Bell (1958) has explicated the selective migration hypothesis somewhat differently than has Fava. He posits the existence of three different, and to some degree mutually exclusive, life styles open to residents of industrialized societies. Each involves emphasis on a particular pattern of interrelated values, to the exclusion of alternatives. These orientations center around family, career, or consumer values. Bell's basic argument is that the demographic differentiation of suburbs reflects emphasis on the familism complex.

To test this hypothesis Bell (1958) conducted a pilot study of Chicago and two of its suburbs. A high-status and a low-status census tract were located in the city and in the suburbs. Analysis of responses indicated that families in the low- and high-income suburbs were more likely than those in the matching city neighborhood to be classified as familistic.

Bell (1958) also reports data from a study of two additional Chicago suburbs in which respondents were asked to give their *reasons* for moving to a suburb; the vast majority gave responses which could be classified as familistic, and an almost equally large percentage gave responses which were classified as reflecting a quest for community—"good neighbors," a sense of belonging, and the like (respondents could give more than one reason). Almost none of the respondents gave reasons which could be classified as expressing an orientation toward upward mobility, although approximately half gave "consumption" or neutral responses. No controls, other than the rough control over social class, were introduced into the analysis (compare Dewey, 1948; Martin, 1953).

Finally, the theses of Riesman (1958) and Wood (1958) are generally relevant. Riesman feels that as work becomes increasingly routine and monotonous there is a strong tendency to reject it, placing emphasis on consumption. The suburb, according to Riesman, is a place where consumption has become the principal concern. Thus, "our suburbs are an effort to build a life not based on work, but instead on the family and on voluntary associations" (Riesman, 1958: 401). In short, the suburban way of life represents a rejection of the industrial order—of rationality and production —which is symbolically represented by the city.

Similarly, Wood (1958) argues that people who move to the suburbs are seeking old values of personal efficacy, especially in politics. In the suburbs, it is possible to become directly involved in local politics, or at least to be personally acquainted with many elected officials. Associated with this is the traditional value that cities are corrupt and evil, that bigness is bad, that it is good to be little. In essence, Wood's hypothesis is that the move to suburbia represents a search for the "solid" values of an earlier period in

American history—for friendly people, manageable size, cleanliness, closeness to nature, and the like (Wood, 1958: esp. chs. 2-5).

However, a study by Hawley and Zimmer (1970) based upon a sample of 3,000 residents of selected cities of various size classes, including suburbs, casts some doubt upon Wood's hypothesis. They found that the political involvement of the residents of large cities was essentially identical to that of suburban residents. This was true if the comparison was on the basis of voter registration or proportions voting. Furthermore, residents of large cities appear to be considerably *more* knowledgeable about local elected officials—if knowing their names is taken as a criteria. On the other hand, it did appear that suburbanites were less likely than their large-city counterparts to feel that their elected officials were controlled by special interests or involved in graft (Hawley and Zimmer, 1970: 65-69).

In any case, there is some uncertainty in placing Riesman *or* Wood under the selective migration rubric. It is by no means certain that they feel that, if class and family composition are held constant, life in the suburbs is necessarily different from that of the metropolis. Rather, there is some suggestion that they feel that virtually all middle-class persons are in the process of moving to the suburbs. Hence, they may be attempting a description of the changing life style of the middle class—changes which are bound up with the city-suburban differential.

Finally, Dobriner (1963: 64-67) has observed that primary emphasis has been given to factors which "pull" suburbanites from the city, with less concern with those which may "push" them, such as the city's long association with violence, racial problems, dirt, and so on. In fact, he suggests that "flight to the suburbs may be a polite assertion of the principle of white supremacy" (Dobriner, 1963: 64).

In this connection, he maintains that reliance on verbal responses may be misleading, because people tend to give socially approved answers. In particular, he feels that respondents may say that the suburban move is "for the children" when the real reason is less socially acceptable. To support this contention, he described a series of his own interviews. After receiving the typical response that the move was for "better schools," he pointed out that there were other communities with equally good schools. To this probe, his respondents often listed ethnic or religious reasons for why they did not move to these particular neighborhoods (Dobriner, 1963: 65-67).

This brief review of the hypothesis that suburbanization is a response to psychological "needs" reveals broad outlines. Thus, there is some evidence to support the view that families move to suburbs to acquire opportunities for more extensive and intensive interaction, participate in local government, maximize values of family living, and perhaps to participate in outdoor activities and recreation.

However, it is necessary to qualify the results presented in this section, since all the studies suffer from two common problems. First, in all of them, a number of important variables are not controlled. This is especially important with regard to family composition. Since families with young children are concentrated in the outer zones, this could be responsible for urban-suburban variation in such factors as extent of neighboring; parents may be drawn together in the supervision of children as well as on the basis of other problems and interests resulting from parenthood. Second, even if observed differences do not result from variations in family size, or some other uncontrolled variable, none of the studies separates the presumed effects of selective migration from the possible impact of "structural" differences between city and suburban residential patterns; it may be that "effects" being attributed to selective migration actually reflect structural variation. It may also be that perceptions of the nature of suburban structure, whether initially correct or not, encourage the migration of certain population types, resulting in modifications of structure.

CLASS AND LIFE-CYCLE STAGE

The basic rebuttal to the structural and selective migration hypothesis is that suburban life styles are expressions of class position and life-cycle stage, and the underlying causes of suburbanization are impersonal, related to changes in transportation and communication technology, as well as to the exhaustion of land suitable for housing in the central city (compare Bogue, 1949; McKenzie, 1933; Schnore, 1957a, 1957b, 1959). Among the first to systematically concern themselves with this hypothesis in the context of the debate over suburban life styles were Ktsanes and Reissman (1959-1960). They argue that "suburbia is no different in its social forms than the city to which it sociologically belongs"

(Ktsanes and Reissman, 1959-1960: 187). Because suburbanites come from heterogeneous city environments, their interests and characteristics are highly differentiated. "Behind the unvarying façades of split-levels, ramblers, and ranch houses," they contend that there is considerable variation in occupation, religion, mobility aspirations, and ethnicity (Ktsanes and Reissman, 1959-1960: 190). To support this argument that the class variable underlies city-suburb differences, the authors present data from two New Orleans suburbs which indicate that middle- and working-class people in suburbs behave pretty much the same way as do people of similar status in the city.

An almost classic challenge to the view that differential migration or structural factors are producing a distinct life style was presented by Berger (1960). Berger's sample consisted of one hundred working-class families in a San Jose, California, suburb. The heads of these families were almost all employed as blue-collar workers at a nearby Ford factory and had moved to the suburb because their factory had been relocated there. This is an important point, since the possibility of selective migration was controlled.

Berger interprets his data as indicating that the suburban experience had not changed the life styles of his sample significantly. Thus, there was no evidence of changes in attitudes toward social mobility, class consciousness, religion, or political affiliation. These workers also remained family-centered and frequently visited nearby relatives. While there may have been some increase in "neighboring," there were almost no semi-formal exchanges such as parties or dinners. Finally, they spent much of their leisure watching television; specifically, they indicated preference for such shows as westerns and prizefights rather than variety and domestic comedy, which he attributes to rejection of the middle-class themes prevalent in the latter.

Yet, Berger's data are somewhat ambiguous, especially for the women in his sample. As indicated above, he reports that it is the women who are most likely to engage in informal visiting, identify themselves as middle class, discuss child-rearing, and belong to informal organizations. Equally important, however, are the data on informal participation in general: about half the families report making "many" new friends, and about three-quarters indicate that they entertain friends in their home at least once a month. Although Berger's general interpretation of his data is that interaction and friendship levels are low, he does suggest that women seem more affected by suburban living than do men. Moreover, among those subgroups which had previously been most deprived, there was evidence of substantial shifts in the direction of more middle-class behavior (compare Wilensky, 1961).

Dobriner (1963: 29-55) also suggests that the most adequate approach to understanding suburban life styles is in terms of class variables. He maintains that "suburb" is not a homogeneous concept and, in particular, that generalizations to it must refer to a particular kind of suburb—residential, industrial, mixed, old or new, working-class, middle-class, and so on. Since the working class and the middle class are both moving to the suburbs, it is necessary to understand both if we are to understand the kinds of lives being lived there. "For in the class structure of the suburbs in general, and any particular suburb, we have a major key with which to unlock the suburban riddle" (Dobriner, 1963: 38).

The fact that both classes are populating the suburbs means that there are *two* styles of life even though the middle class is numerically the dominant group (compare Lazerwitz, 1960; Taeuber and Taeuber, 1964). In his study of Levittown, New York, Dobriner (1963: 106-107) found the following themes expressed by his middle-class respondents: stress on privacy, initiative, moderation, and upward mobility. Similarly, they are likely to join voluntary organizations, be independent of the extended family—relying instead on friends—and tend to feel that Levittown is a tempory residence. The working-class residents, on the other hand, lead distinctly different lives; they focus less on privacy, appearances, and the pursuit of distant goals and more on expressiveness. There is also a clearer distinction between male and female roles than in the middle class. Home entertainment was likewise differentiated by class: the working-class groups tending to entertain relatives from the city, while their middle-class counterparts entertain friends, business associates, and neighbors in their homes. These various themes are apparently characteristic of middle- or working-class persons, regardless of location.

Finally, Gans (1962a) has argued that, if differences exist between city and suburban life styles, they are largely a function of class and life-cycle stage. He suggests that in the metropolis there are several life styles—not one. He isolates "cosmopolites," the unmarried or childless, "ethnic villagers," and "deprived," the "trapped," and the downwardly mobile. The first three types develop distinctive ways of living which effectively *isolate* them from their surroundings: they "set up social barriers regardless of the physical closeness or the heterogeneity of their neighbors" (Gans, 1962a: 631). Only the trapped and deprived seem significantly affected by these factors. For example, cosmopolites are relatively well educated, do not have children, and consequently lead certain kinds of lives. Similarly, the unmarried and childless live as they do in part because of their life-cycle stage.

DISCUSSION

The studies reviewed in this chapter suggest that urban-suburban life style differences are neither as clear-cut nor as sharp as commonly supposed. Those that persist when class is controlled appear to be a matter of degree rather than kind. In particular, there is no basis for the view that suburbanites are more involved than their counterparts in the city in intense status competition with their neighbors, excessive conformity, or compulsive concern with social mobility. There also does not appear to be any strong evidence that the elements of suburban life style form a distinctive interrelated constellation organized around a dominant theme—at least, no more so than is the case with residents of large cities who possess comparable socioeconomic status and family size characteristics. Rather, the differences seem quite modest; it appears that the suburban residents are somewhat more involved with their family, neighbors, and community. Whether such differences suffice to define a distinct "suburban" life style—even in the "specific" sense defined in the introduction—is moot. It seems likely that these differences are, as Dobriner (1963: 59) persuasively argues, only minor variations upon more fundamental class themes.

Even the sources of apparent differences are not clear. No study, with the possible exception of Berger's analysis of transferred factory workers, adequately partials out the presumed effects of selective migration from those of suburban structure. While there is considerable speculation about the likely effects of homogeneity, visibility, density, and so on—it is by no means clear that these would operate either with the same force or in the same way (if they operate at all) upon a truly randomly selected group of suburbanites. Indeed, in view of the methodological shortcomings of most of these studies, it is not certain that even the small observed differences are "real" in the sense that they can be attributed either to selective migration or to structural characteristics.

As an example of problems in interpreting data on differences, the fact that suburban residents have more children than their city counterparts is sometimes taken as supporting the selective migration hypotheses—that is, as consistent with the hypothesis that families for whom children are a central value move to the suburbs. But, as is often true of urban-suburban comparisons, these data can be interpreted in more than one way. While these gross comparisons indeed show higher fertility for suburban residents, if age composition is controlled, the differences virtually disappear (Hawley and Zimmer, 1970: 17-19). In other words, there are more children in suburbs because young married couples in the suburbs are more fertile.

This indicates that the causes of family size variation between cities and suburbs lies in the forces underlying concentration of different population types into different residential areas. While it may be that selective migration in psychological terms is one of these causes, this is by no means established. In fact, it is not clear that the reasons given by people for moving to suburbs—at present, the principal technique for evaluating the selective migration hypothesis —correspond to their "real" reasons. While respondents tend to give "familistic" answers such as "better for the children," "more space," and so on (compare Dobriner, 1963: 65-67), they may in fact be providing *themselves* with an ad hoc interpretation of behavior, the causes of which involve such factors as availability of suitable housing.

Moreover, such psychological motivations have probably *always* been involved in movement to neighborhoods characterized by a predominance of owner-occupied single family dwelling units (compare Rossi, 1955); the basic difference today may be that these areas are more likely to be located in the "suburbs" rather than within the political boundaries of large cities. In this vein, Schnore (1957b; 1959) suggests that the exhaustion of central-city land, coupled with the shift to mass production technology in the construction of housing, has led to construction of new homes on the periphery of large cities in "suburban" areas; individual purchasers acquire these homes, but have little to say in the selection of the sites upon which they are erected (compare Duncan, 1959: 697). Schnore also argues that land-use patterns in large cities are changing, and a variety of activities, including the residential function, are being forced out of the center into peripheral ("suburban") areas. If this is the case, then selective migration is largely irrelevant as an explanation for the rapid growth of suburban communities.

Another perplexing methodological problem centers around the way "suburbs" are typically selected for analysis. One of two strategies is commonly employed: the first is to select an area which is politically separate from, but physically continuous with a central city (e.g., Bell, 1958); the second is to define a series of zones according to distance from some central point (e.g., Tomeh, 1964). The resulting comparisons often show that people in less-central areas differ from those in central cities or zones in terms of some aspect of behavior. Unfortunately, even if we assume that an uncontrolled variable such as family composition is not responsible, such claims are exceedingly difficult to evaluate in the context of the theoretical "nature" of suburbs explicated above. These areas are physically a *part* of the city except for the occasional fact of political independence—they may or may not be suburbs as the term is used by ecologists. They serve essentially the same function as other low-density residential areas on the outskirts of large cities and probably reflect little more than the well-known tendency for expansion to occur on the periphery, as suggested above. The fact that they have not been annexed often reflects historical factors; these factors include the extensive development of land in previously existent towns which were formerly located at some distance from the larger center or the creation of "large-scale housing developments" which are politically defined from the start.

How do we justify, either logically or theoretically, distinctions between peripheral areas which have been annexed and those which have not? Indeed, even if we compare central cities with physically and politically separate residential areas, an identical problem is encountered—these areas are functionally the same as peripheral residential ones. This issue is similar to the question of urban-suburban fertility differences raised earlier: while gross comparisons may reveal behavioral differences, more specific comparisons may not. As Gans (1962a) observes, it is only generally true that people live either in "cities" or in "suburbs"—they live in particular neighborhoods and types of dwelling units. Since suburbs are typically less densely settled, is it possible that the differences observed in gross city-suburb comparisons reflect a tendency for people in low-density neighborhoods to lead somewhat different lives from comparable persons in the more densely settled core?

Of course, if small size and political independence are taken as factors either in the selection of migrants or as an independent variable in its own right (compare Wood, 1958), then such an approach makes theoretical sense. But if this is the theoretical strategy, then "suburbanization" is not the variable, but, instead, variations in the size of political units.

Perhaps even more serious is the failure of any of the studies reviewed here to provide *simultaneous* controls over socioeconomic status, life-cycle stage, employment of wife, and other variables which may be involved in the selection of families for suburban living *and* which influence their involvement with neighbors and neighborhood. Until such analyses are completed and replicated, we should regard the present data as largely suggestive rather than definitive.

In particular, what appears to be needed is research which sorts out the independent, joint, and interactive effects of structure, social-psychological selection, class, and family composition. For example, it appears likely that the social-psychological dimensions along which blue-collar workers are selected are different from those for the middle class, and variations in residential areas may affect them differently—that is, there may be an interactive effect between structure and class upon life style (compare Tallman, 1969).

Indeed, it may be that we are asking the wrong questions—that the question of urban-suburban variations in life styles is irrelevant. Given our interest in determinants of particular life styles, perhaps we ought to be more concerned with specifying and evaluating the types of variables which affect them, rather than with the simple comparison of "suburban" and "urban" behavior—a comparison which inevitably confounds a variety of possible causes.

For example, commuting may affect neighborhood involvement, regardless of whether or not the commuter lives in a "suburb." In fact, it is puzzling that commuting has become so completely identified with suburbs (compare Berger, 1960: 8-9). The specialization of land use characteristic of urban areas implies separation of work and residence, and the degree of this separation appears to have been increasing since 1900 (Duncan, 1956; Hawley, 1971: 191-192). Of course, this implies increasingly longer commutes for urban populations in general. Considering the relative importance of commuting for city and suburban populations, the results of a journey-to-work survey for the New York metropolitan region reported by Hoover and Vernon (1962: 138-195) are suggestive. For intracounty commutes (those that begin and end in the county of residence), the average length of the work trip in the "inner suburban" counties (those surrounding the highly urbanized core of counties centering on Manhattan) was considerably shorter than those in the core counties. Moreover, people who commuted to Manhattan from another borough of New York typically took almost twice as long to get to their jobs than was the case with intraborough commutes within the city. Of course, workers who commuted from either the inner or outer ring of counties to the core spent considerably more time on the trip. But since 56% of the region's population live and work in the same county while only 14% commute from the rings, the former is clearly the dominant pattern. In other words, if commuting is an important determinant of elements of life style, then it appears to act with greater force upon residents of large cities than upon workers who live in the suburbs. Other data are consistent with these; for example, Gans (1967: 221-222) shows that residents of the suburb studied by him who had previously lived in Philadelphia reported *less* time spent in commuting than formerly.

Another variable which is probably associated with level of neighborhood interaction is homogeneity. In addition to the study by Caplow and Foreman (1950) reported above, there is a considerable amount of data indicating that neighboring is greatest during the earliest, and presumably most homogeneous, stage of community settlement. This appears to be the case regardless of settlement type-apartment complexes in the center of large cities (Wolf and Ravitz, 1964), English "new towns" (Willmott, 1963), and public housing projects (Wilner et al., 1962: ch. 12), as well as suburbs (Hoover and Vernon, 1962: 185-186). But as Gans (1967: 262) points out, as long as people remain compatible, aging per se will not reduce interaction—only if it is accompanied by changes in population composition and presumably in the bases for interaction.

As with commuting, there is some question about the degree to which the variable of homogeneity applies only to suburbs. As Gans (1962a) notes, the homogeneity of a neighborhood is determined largely by its age and value. Regardless of whether located in cities or suburbs, newer neighborhoods are likely to be relatively homogeneous simply because they have less exposure to population turnover. Commenting on a particular suburb, Dobriner (1963: 12-15) says that, while Levittown may have once been homogeneous, this is no longer the case: incomes, religious affiliations, and the like differed considerably at the time of his study. Neighborhoods and suburbs with homes in the low and medium price ranges are also likely to be relatively heterogeneous regardless of age, because they tend to contain families at different career stages—some are young executives whose incomes are likely to rise steadily, producing at some point a move to a different and better home; others are working class at the peak of their earning power, and hence likely to remain in their present home (see Gans, 1967; Mowrer, 1958; Wattel, 1958). Similarly, Hoover and Vernon (1962: 183-198) point out while initially urban neighborhoods are characterized by low densities and single family dwelling units, over time both population and housing characteristics change markedly. And there are urban neighborhoods which are quite homogeneous with respect to such variables as ethnicity and social class (e.g., Gans, 1962b)—in short, suburbia has no monopoly on homogeneity or lack of it (compare Ktsanes and Reissman, 1959-1960).

In this context of isolating variables which may affect life styles, Hawley (1971: 195-196) suggests that the importance of neighborhood as an associational unit is closely linked to social class. Relatively high-status people are much less attached to a local area—either in terms of location of friends or satisfaction of shopping and recreational needs. But there is some reason to believe that locality *is* an important basis for association for working-class persons; their low income and education may act to limit involvement in the wider metropolitan area. This may be especially true for working-class women.

Focus upon specific determinants of life style may permit sociologists to avoid the troublesome question of why low-density city residential areas in which homes are primarily owner-occupied single family dwelling units, indistinguishable from "suburban" residential areas in appearance and ecological function, are expected to differ from the latter in terms of "life style." Many urban areas are physically identical to many suburban areas, and it is by no means clear that their residents *do* differ in life styles. In other words, if structural variations in settlement patterns affect life style, we ought to ask why and how much—regardless of location. Of course, it may *also* be true that location relative to the city center is a factor whose effects we could examine.

When all is said and done, it appears that there has been an almost hopeless confusion of definitions. Sociologists have allowed the entire area to be defined by nonsociologists in nonsociological terms. The popular media usage of terms like "suburbia," and the associated mythology of distinct suburban life styles, has caused them to become so firmly entrenched in our vocabularies that sociologists have been led to ask questions centered around them—ignoring questions about the kinds of variables which can affect life style.

What often appears to happen in discussion of urban and suburban life styles is confusion of two kinds of analysis. On the one hand, there is the question of spatial differentiation and interdependence within metropolitan communities. On the other is the question of the kinds of life styles being lived in metropolitan areas. The questions are empirical, and theoretical variables relevant to the one are by no means relevant to the other. While it is appropriate to ask whether or not variations in homogeneity or density in neighbor-hoods affect interaction, and it is also appropriate to ask whether or not specialized areal units vary in density and homogeneity within the metropolitan community, it is *not* appropriate to assume that all suburbs are low density or homogeneous—some are and some are not, just as urban neighborhoods differ in these terms. To compare "suburban" and "urban" life styles is thus an almost hopeless task—one in which a variety of possible determinants of life style are so confounded that it may be impossible to disentangle them.

In conclusion, if we approach the question of life styles in terms of isolating the variables which determine them, the rich existing literature on such topics as participation in formal and informal organizations could well be incorporated within this broadened conception of the problem. In this way, it may be finally possible to resolve the complex question of the nature of urban-suburban differences in a way that would contribute to a broader theory of metropolitan life styles.

REFERENCES

BELL, W. (1958) "Social choice, life styles, and suburban residence." pp. 225-247 in W. Dobriner (ed.) The Suburban Community. New York: G. P. Putnam.

BERGER, B. (1960) Working Class Suburb: A Study of Auto Workers in Suburbia, Berkeley and Los Angeles: Univ. of California Press.

BOGUE, D. (1949) The Structure of the Metropolitan Community. Ann Arbor: Horace H. Rackham School of Graduate Studies.

CAPLOW, T. and R. FOREMAN (1950) "Neighborhood interaction in a homogeneous community." Amer. Soc. Rev. 15 (June): 357-365.

CARLOS, S. (1970) "Religious participation and the urban-suburban continuum." Amer. J. of Sociology (March): 742-759.

DEWEY, R. (1948) "Peripheral expansion in a Milwaukee county." Amer. Soc. Rev. 54 (September): 118-125.
DOBRINER, W. [ed.] (1963) Class in Suburbia. Englewood Cliffs, N.J.: Prentice-Hall.
——— [ed.] (1958) The Suburban Community. New York: G. P. Putnam.
DOUGLAS, H. (1925) The Suburban Trend. New York: Appleton-Century-Crofts.
DUNCAN, B. (1956) "Factors in work-residence separation: wage and salary workers, Chicago, 1951." Amer. Soc. Rev. 21 (February): 48-56.
DUNCAN, O. D. (1959) "Human ecology and population studies," pp. 678-716 in P. Hauser and O. D. Duncan (eds.) The Study of Population. Chicago: Univ. of Chicago Press.
ENNIS, P. (1958) "Leisure in the suburbs: research prolegomenon," pp. 248-270 in W. Dobriner (ed.) The Suburban Community. New York: G. P. Putnam.
FAVA, S. (1958) "Contrasts in neighboring: New York City and a suburban county," pp. 122-131 in W. Dobriner (ed.) The Suburban Community. New York: G. P. Putnam.
——— (1956) "Suburbanism as a way of life." Amer. Soc. Rev. 21 (February): 34-37.
FELDMAN, S. and G. THIELBAR [eds.] (1972) Life Styles: Diversity in American Society. Boston: Little, Brown.
GANS, H. (1967) The Levittowners: Ways of Life and Politics in a New Suburban Community. New York: Vintage.
——— (1962a) "Urbanism and suburbanism as ways of life: a re-evaluation of definitions," pp. 625-648 in A. Rose (ed.) Human Behavior and Social Processes: An Interactionist Approach. Boston: Houghton Mifflin.
——— (1962b) The Urban Villagers. New York: Free Press
GRAS, N.S.B. (1926) "The rise of the metropolitan community," in E. W. Burgess (ed.) The Urban Community. Chicago: Univ. of Chicago Press.
——— (1922) "The development of metropolitan economy in Europe and America." Amer. Historical Rev. 27: 695-708.
GREER, S. and E. KUBE (1959) "Urbanism and social structure: A Los Angeles study," in Marvin Sussman (ed.) Community Structure and Analysis. New York: Thomas Y. Crowell.
HAWLEY, A. (1971) Urban Society. New York: Ronald Press.
——— and B. ZIMMER (1970) The Metropolitan Community: Its People and Government. Beverly Hills: Sage Pubns.
HOOVER, E. and R. VERNON (1962) Anatomy of a Metropolis. Garden City, N.Y.: Doubleday.
KISH, L. (1954) "Differentiation in metropolitan areas." Amer. Soc. Rev. 19 (August): 388-398.
KTSANES, T. and L. REISSMAN (1959-1960) "Suburbia—new homes for old values." Social Problems 7 (Winter): 187-195.
LAZERWITZ, B. (1960) "Metropolitan residential belts." Amer. Soc. Rev. 25 (April): 245-252.
LUNDBERG, F., M. KOMAROVSKY, and M. McINERY (1934) Leisure: A Suburban Study. New York: Columbia Univ. Press.
McKENZIE, R. D. (1933) The Metropolitan Community. New York: McGraw-Hill.
MARTIN, W. (1956) "The structuring of social relationships engendered by suburban residence." Amer. Soc. Rev. 21 (August): 446-453.
——— (1953) The Rural-Urban Fringe. Eugene: Univ. of Oregon Press.
——— (1952) "A consideration of differences in the extent and location of the formal associational activities of rural-urban fringe residents." Amer. Soc. Rev. 17 (December): 687-694.
MEYERSOHN, R. and R. JACKSON (1958) "Gardening in suburbia," pp. 271-286 in W. Dobriner (ed.) The Suburban Community. New York: G. P. Putnam.
MOWRER, E. (1958) "The family in suburbia," pp. 147-164 in Dobriner (ed.) The Surburban Community. New York: G. P. Putnam.

MUELLER, S. A. (1966) "Changes in the social status of Lutheranism in ninety Chicago suburbs, 1950-1960." Soc. Analysis 27, 3: 138-145.

RIESMAN, D. (1958) "The suburban sadness," pp. 375-408 in W. Dobriner (ed.) The Suburban Community. New York: G. P. Putnam.

ROSSI, P. H. (1955) Why Families Move. New York: Free Press.

SCAFF, A. (1952) "The effect of commuting on participation in community organizations." Amer. Soc. Rev. 17 (April): 215-220.

SCHNORE, L. (1965) The Urban Scene. New York: Free Press.

——— (1959) "The timing of metropolitan decentralization: a contribution to the debate." J. of Amer. Institute of Planners 25 (November): 200-206.

——— (1957a) "Metropolitan growth and decentralization." Amer. J. of Sociology 63 (September): 171-180.

——— (1957b) "The growth of metropolitan suburbs." Amer. Soc. Rev. 22 (April): 165-173.

SHUVAL, J. (1956) "Class and ethnic correlates of casual neighboring." Amer. Soc. Rev. 21 (August): 453-458.

TAEUBER, K. and A. TAEUBER (1964) "White migration and socio-economic differences between cities and suburbs." Amer. Soc. Rev. 29 (October): 718-729.

TALLMAN, I. (1969) "Working class wives in suburbia: fullfillment or crisis?" J. of Marriage and Family 31 (February): 65-72.

——— and R. MORGNER (1970) "Life style differences among urban and suburban blue-collar families." Social Forces 48 (March): 334-348.

TOMEH, A. (1964) "Informal group participation and residential patterns." Amer. J. of Sociology 70 (July): 28-35.

WATTEL, H. (1958) "Levittown: a suburban community," pp. 287-313 in W. Dobriner (ed.) The Suburban Community. New York: G. P. Putnam.

WHYTE, W. (1957) The Organization Man. New York: Doubleday.

WILENSKY, H. (1961) "Review of Working Class Suburb by Bennett Berger." Amer. Soc. Rev. 26 (April): 310-312.

WILLMOTT, P. (1963) The Evolution of a Community. London: Routledge & Kegan Paul.

WILNER, D., R. WALKLEY, T. PINKERTON, and M. TAYBACK (1962) Housing Environment and Family Life. Baltimore: Johns Hopkins Press.

WINTER, G. (1961) The Suburban Captivity of the Church. New York: Doubleday.

WOLF, E. and M. RAVITZ (1964) "Lafayette Park: new residents in the core city." J. of Amer. Institute of Planners 30 (August): 234-239.

WOOD, R. (1958) Suburbia: Its People and Politics. Boston: Houghton Mifflin.

ZIMMER, B. and A. HAWLEY (1959) "Suburbanization and church participation." Social Forces 37 (May): 348-354.

5

The Family in Suburbia

SCOTT GREER

☐ THERE IS A NECESSARY PROCESS OF TRANSLATION demanded when a social scientist makes a folk term, such as "family," a subject for description and analysis. He must face such an unanalyzed abstraction (and we live most of our hours using what symbolization the tribe provides) and get at its constituent concepts. In this discussion, I will adopt an organizational point of view, seeing the family as a certain kind of group, produced by a combination of kinship and contractual relations. This group, like all groups, can maintain itself over time only if it functions so as to provide rewards which (1) induce the members to continue to relate in "family patterns," and (2) therefore allow this particular group to survive in the organizational environment of this society.

The functions of the family group, then, are "Janus-faced," in Robert Winch's (1971) term. First, for the individual, the family provides socially legitimated sexual relations, legitimate children, and usually the advantage of cohabitation in a single household with whatever social and emotional values are thereby generated. For many individuals in our society, as we shall see later, these functions of the family are central to their life style. Second, however, for the *society*, the family group performs the functions of reproduction, socialization, and the placement of the progeny in the wider social

[149]

system. (I am thinking of contemporary society in the United States and of most families; family groups have served additional functions of a religious, political, and productive nature in the past and still do so to some degree in various contexts.)

Legitimation is thus conferred upon the family by the society; in return, the family, conforming roughly to societal requisites, functions to perpetuate that larger whole. Legitimation takes two forms: marriage is contracted between individuals from differing lineages, and the offspring are ascribed a kinship relation with the parents and with each other. It is important to note that both kinds of relationships are involved in legitimating the social order of the family group. It violates the folk usage, but is correct to say that husband and wife are "kin" only through their common progeny. They are not kin to each other but have, instead, a contractual union. (Father-daughter incest does occur, of course, but is not liable to legal contract in most contemporary societies.)

The principle of kinship in human societies is universal, transitive, creative, and conservative. Why it is universal we do not know, but most analyses emphasize the importance of the lineage, the line of descent. Thus the mother-child relationship may be seen as the primordial bond from which all kin systems originated. Certainly it is a brute fact of human existence that the newborn *must* have adult sponsorship to survive, and that, from such sponsorship, some of the closest human relationships develop, as well as many aspects of what we call "identity." In any event, kinship places the individual, long before he is conscious of it, in a social network where he has claims on others and they upon him. The importance of these claims varies with the organization of the society; in some they are critical for his survival, in others, marginal with respect to most of his life.

Kinship networks are also, to some degree, transitive. That is, the organizational tools provided by kinship—loyalty, common identity, legitimacy—may be used in other institutional areas. Kin may be the basic organizing unit for economic enterprise (as in the E. I. DuPont de Nemours Company through its first century of existence); it may also be the organizational basis for political power, whether in a monarchy or an Irish "political family" in Chicago (Chandler and Salsbury, 1971; Levine, 1966). From Dahl's point of view, kinship networks may constitute political "slack," opportunities to organize influence which are not used fully, or, on the other hand, they may be the controlling political systems in the society (Dahl, 1960; Mair, 1964). This is the sense in which I spoke of the creative aspects of kinship; it is a component, an organizational bond which may be used in many ways, to produce many outcomes.

But kinship, as a principle, is also conservative. It is, ideally, completely limited to those related by genetic descent, the true ground of tribalism. Even modified by the inclusion of "in-laws" —those related only by contract or as side-effects of someone else's contract—its exclusiveness, which is its strength, limits it as an organizational binder. Again, to cite the case of DuPont, the firm could not have become "the first modern corporation" without yielding the claims of kin to the demands of competence (Chandler and Salsbury, 1971). Thus, kinship as a general principle of social organization is possible in almost any society imaginable (at least any which reproduces itself), but the probability that it will be important varies enormously.

KINSHIP AND THE NUCLEAR FAMILY

The nuclear family, a unit consisting of the married couple and their offspring, may thus be merely one knot in an extended web of kinship. I should say, a knot between *two or more* extended webs, for the marital contract binds two different systems. The more functions the kin groups perform for the total society, the less autonomous will be the nuclear family; at an extreme, it may be totally dependent upon kin for access to the economy, the polity, and the gods. At the other extreme, it may be virtually free of all kin dominance, owing little even to parents, gaining little from anyone in the network. The first extreme is exemplified in such societies as the Trobrian Islanders, as Malinowski (1929) describes them. The second fits the stereotype of the "isolated nuclear family" common in contemporary urban sociology, if not in contemporary urban society.

Anthropologists have pointed out the vulnerability of the marriage contract in societies where extended kin networks are strong. Each marriage is, in a sense, a treaty between different and competing powers. Yet the marriage contract is far from invulnerable in societies such as ours, where extended kinship is much less dominant. Indeed, as we approach a time when there is a divorce for every two marriages, we must consider the possibility that other forces in the society which detract from political, economic, and emotional autonomy may also be antithetic to the persistence of the marriage contract (on the divorce rate in the United States, see Winch, 1971). Perhaps the simplest way to put it is this: in kin-dominated societies the kin network provides alternative supports for the marriage partners and thus allows them freedom to leave the nuclear family, while in corporate societies such as our own, individual citizenship and access to the labor market have the same consequence. Only as extended kin or corporate societies insist on the permanence of marriage is the strain toward dissolution resolved.

For clarity's sake, it is important to distinguish between family as one type of social group and household as another. For, though most families live in a common household and most households are organized by the principles of marriage and kinship, the two groups are analytically distinct and, on occasion, empirically so. Thus it is useful to think of "family-organized households" in considering the statistical distribution in the United States today.

There are a number of types of family-organized households. The norm is, of course, the "complete nuclear family" of parents and their children. Variations are important, however; the "stem family," which includes several generations of the same lineage, is rare in the United States, but more common elsewhere. More important is the "incomplete nuclear family," in which one of the parents is missing. This is a large and growing proportion of all American family-organized households. And, finally, we should probably consider separately the married couple without children. Here only one of the principles of family formation is found—the marriage contract. There is some evidence that such family groups are less stable than those where children are produced. (Indeed, it is more accurate to call such households "marriages" rather than families—even though the participants may be looking toward eventual reproduction.)

Finally, we might note the fairly common incidence of "fictive kinship" and "fictive families." Fictive kinship is a way of extending the rights and duties of kinship (or some of them) to persons connected by neither descent nor contract. A common form among Spanish-speaking Americans is the *compadrazgo,* or Godfathership, while among some American Blacks we are told that "going for brother," and "going for sister" are common and apparently useful arrangements. Fictive families range from the extramarital liaison between two persons, to the informal menage including several persons, to the formally constituted "commune," or communal family. None of these forms satisfies the definition of family I have used, for they do not have the organizational bonds that promote legitimacy, order, and persistence. However, they *are* forms of household and, as such, have an importance as possible alternatives to the family, if that harassed institution finally gives up much of the ghost.

SUBURBANIZATION AND SUBURBS

The root cause of suburbanization, the horizontal extension of urban settlement far beyond the older center of the city, lies in a chainging relationship between social organization and space. Transportation and communication technology set the limits to the coordination of human behavior over space; as that technology has improved, it has become possible for households and enterprises to be separated by greater distances and yet coordinated well enough to keep the world going (for an expanded discussion of suburbanization, see Greer, 1962).

There are a number of definitions of suburbs current, their utilities depending upon one's interests in that horizontal extension of the urban place. The simplest is the political: suburbs are outside the political boundaries of the older city. Related to this, but analytically separate, is the definition of suburbs by their distance from the core of the city; this is the ecological definition; A third approach emphasizes the social characteristics of the population. Since these attributes are not highly correlated, it is clear that the folk term suburb is one word standing for many, sometimes contradictory, aspects of the outer metropolis. Each dimension generates a different typology.

Politically, suburbs have in common only the negative—they are not in the central city. However, some are incorporated—in villages, towns, even cities, while some areas are unincorporated, and, for them, the county government and various special districts must provide what local government they have. The municipal corporations were created for a host of reasons—fear of annexation by the city, protection of "property values"—e.g., protection from the "invasion" of the poor and the other ethnic populations, the creation of licensing bodies to issue liquor permits, tax shelters for the rich, freedom from control for industries, and undoubtedly a host of others. Each was, significantly, bent upon escaping the dominance of a larger-scale polity. The areas that remain unincorporated are often kept in this state for the same reason—they are escapes from zoning, building codes, and taxes. Often they are areas which are simply unwanted by any municipal corporation.

The ecological dimension, distance from the center of the city, would appear to be simpler. However, it is complicated by the space-time ratio—the cost in time and money of getting from here to there. Expressways and toll roads of recent vintage bring neighborhoods near them the advantage of very low time costs for travelling great distances; such expressways are usually around the circumference of the city, with spokes to the central interchange. They serve the same purpose in the suburbs as do mass transit systems in the older city. Thus distance from the center *and* distance from the expressway grid must be combined in measuring the effective ecological distinctions among suburbs. That kind of distance of the old, densely settled inner suburbs from downtown may be greater than that from the far-flung exurbias.

Suburbs differ socially along many dimensions; I will emphasize those related to the characteristics of the residential population. Four seem especially important: social rank, life style, ethnicity, and migration status. Social rank indicates the place of a population in the rank order of occupations in the society, its educational level, and its share of the material rewards, including residences. Life style indicates the uses of these resources in choosing a way of life, which may range from the employed, childless adults, urbane dwellers in the towers of the Gold Coast and surrounded by the great market, to the child-oriented, home-obsessed, familistic population where few women are employed outside the home and the home is a free-standing single-family unit. Ethnicity indicates the degree to which a population includes those who are given special status due to their original nationality; through lineage and appearance, they are seen as belonging to another tribe. When things get nasty, they are seen, in Erikson's (1968) phrase, as "pseudo species." Migration status refers to the proportion of the population who are recent migrants to the area. The meaning of recency varies, of course, from the proverbial New England village where your family is recent if it has only been around two generations, to the new developments described by Gans (1970) where the "old settlers" were those who had been around two years.

The ideal-typical suburban population of folk thought would be middle to high social rank, with a familistic life style, few or no visible "ethnics," and a great many recent migrants. Of course, this is not true; as suburbia has come to include a majority of the population in the metropolitan areas, it has dipped down into the working class; as the older, inner suburbs decline in social rank, they have attracted the more "urbane" populations—particularly if they are university towns—and visible ethnic populations; and as those housing developments Gans described grow older, so does the stability and length of residence of their populations.

Suburbs are not new; in the revival of cities following the Dark Ages, as Pireane notes, they were the settlements of merchants, artisans, and laborers outside the fortifications of the city. They occur wherever you find fixed political geographical boundaries combined with economic and population growth. They do, however, present something of an esthetic, if not ethical problem to a people whose metaphors for the hallowed ground are such as "the heavenly city" and "the city on a hill."

Yet, with the transportation revolution of the nineteenth century which opened great new frontiers for urban expansion, the scorn of the city dweller for the suburbs was mitigated by two forces. One was that native pastoralism, often miscalled anti-urbanism, which had marked American society from its beginnings. The other was the increasing custom of the very rich to create sumptuous estates along the commuter railroad lines in Westchester County, along the Main Line in Philadelphia, and up the Chicago North Shore. Thus, from a promotional pamphlet of the 1870s:

> The controversy which is sometimes brought, as to which offers the greater advantage, the country or the city, finds a happy answer in the suburban idea which says, both—the combination of the two—the city brought to the country. The city has its advantages and conveniences, the country has its charm and health; the union of the two (a modern result of the railway), gives to man all he could ask in this respect.

Such was the general tenor of commentary on the surge of suburban development by boosters, real estate developers, planners, and social philosophers. To be sure, some esthetes of an urbane cast of mind ridiculed suburban neighborhoods, particularly those built in mass developments. A good example of this sort of thing are these lines of Wallace Stevens:

> Oxidia, banal suburb, one half of all its installments paid,
> Oxidia is Olympia.

But the general definition was a friendly one.

It was only after World War II that suburbia came to be defined, in the popular press and "pop sociology" as, in some sense, a problem. A spate of books excoriated the suburbs (and presumably the people who lived in them) for their dullness, cultural poverty, and conformity. As Gans has noted, this was really a disguised class warfare. The new suburbanites were there as a result of the housing industry's solution to the problem of a huge housing deficit, the result of a decade of economic depression and five years of wartime economy. They were predominantly young couples, working class and lower-middle class; they behaved in suburbia pretty much as they would have in the central city—as can be seen by examining the newest outer wards of most central cities. To the criticism of the quality of life in the suburbs was added a different and far more basic criticism: the city was suffering from the loss of this population, at the marketplace and in tax income—as well as leadership cadres for public action. Thus, the suburb "leached" the class structure of the city. This, in turn, created decay, and a vicious circle was set up, one which guaranteed further scatteration.

At the same time the suburbs were being so violently criticized, a number of sociologists carried out studies that included them in the focus of attention. They discovered that, when controlled for social characteristics, the chief difference was in housing. Similar populations behaved about the same in city and in suburb, but the overall mix was different, with more suburbanites familistic in life style. In a sense, the lively, growing edge of the metropolis *was* suburbia. And, more profoundly, it was young America, the cradle of future generations. Those who criticized suburbia were simply depressed with conditions in a mass democracy whose affluence was expressed, not by policy, but through the marketplace.

KIN AND MARRIAGE AS THEY ARE
AFFECTED BY SUBURBIA

Since suburbia is not a unitary concept, but embodies distinctions in political structure, ecological position within the metropolis, and the social characteristics of the populations, I shall examine interactions of each with the functions of kin networks and marriage.

The data are sparse and much of what follows is necessarily speculative.

THE POLITICAL DEFINITION

Since the movement to suburbia is a response to the needs of conjugal families with children, we frequently find that the small suburban municipalities are committed to a child- and home-centered polity. Schools are a critical subject of aspiration, contention, and rancor. The school system of the suburbs is very much "our thing" and is taken very seriously. Masotti's (1967) study of school politics in Chicago's suburbs underscores the importance of the quality systems as nodes of migration, while older studies by Bell (1958) and Dewey (n.d.) underline the decision of many to locate in suburbia "for the sake of the children."

Yet we must also remember that, with the enormous population boom of the last two decades, many chose suburbia as the only housing available at a price they could afford. New family formation contributed a great deal to the settlement on the fringes, particularly the movement of young families of lower-middle social rank. "Rejection of the central city" was more often a move to what was available by young families, and a move upward in status for older families who had "made it" in the economic system. As I have noted elsewhere (Greer, 1972: 103),

> many persons might settle for equivalent lodgings in the middle of the city. But the point is moot—until new, single-family dwellings, rather than high-rise public housing developments replace the tenements and row houses near the centers we will not know how many suburbanites are fleeing the city and how many are forced to move outwards because no other acceptable housing is available.

What are the effects of the suburban polity upon the children? One thing to bear in mind is the enormous inequality of resources among suburbs, reflected in the amount of local taxes which can be raised which, in turn, determines most of the budget of the suburban school. The political fragmentation of suburbia results in "the divorce of resources from needs," in Wood's (1958) phrase; some suburbs include enormous shopping centers serving a wide area but paying taxes to one small municipality—others include mostly housing, and cheap, aging housing at that. Kinloch, an all-black suburb of St. Louis, had the highest tax rate in the suburbs and a school system in danger of disaccreditation at the time we studied it; Clayton, containing the county seat and a huge shopping area, had low taxes and a fine school system. Clayton had taxables per student that were 28 times those of Kinloch (Metropolitan St. Louis Survey, 1956-1957).

Of course there is some differential in spending per pupil among the inner-city schools; it is, however, miniscule compared to the variance in suburbia. A recent study by the Center for Urban Affairs of Northwestern University found that the differential in per pupil staffing expenditures for the Chicago public school system was approximately ten percent and was slightly greater with respect to class than to race (Berk et al., 1971).

There is a heated debate over the importance of school funding, compared to such background variables as family culture and genetics in the development of children (Coleman et al., 1966). There is even some question of the importance of education at all (Jencks, 1972). Nevertheless, the notion of equal opportunity is a deeply embedded norm in the political culture of the United States, and it is flagrantly violated in the school system of suburbia. Thus, a family's knowledge that its children are being offered substandard education, by whatever criteria, is a blow to family pride and confirms a sense of inferiority. (This may explain some part of the aggressiveness of inner-city Black parents against the citywide system.)

The political system of suburbia has other consequences. It is, for one thing, small-scale and a matter of common experience—many more people know who is responsible than in center city (Bollens, 1960). Wright and his associates, in studying children through real-life observation, discovered sharp differences in the child's awareness of the adult world, varying with the size of the community. Village children had a better cognitive map of that world than did small-city progeny. This may be generalizable to governmental suburbia.

The divorce of needs from resources has other consequences. Many kinds of service for solving family problems are less developed in most suburbs than in the central city; these include, especially, private service agencies. The discrepancy ranges from suburban areas with no facilities to handle unmarried mothers by aiding in adoption to those which have no babies and must import them from other municipalities, including the central city. (The Cradle, an agency in Evanston, Illinois, has such a problem.)

Finally, to return to the basics, most suburban municipalities express the politics of a child-oriented society, within the ghost of the "republic in miniature" (Wood, 1958). At the same time, the trinity of family, home, and neighborhood is reinforced by the almost exclusive concern of the local polity with providing and defending amenities for these things. The broader spectrum of problems facing metropolitan America—race schism, organizational ineptness, economic inequity, and the integration of the continental nation—tend to go by the board. Familism and homeownership preempt the center stage of the polity, which, in turn, reinforces the deification of nuclear familism.

The small scale of the polity is a value to many; the cliche "keeping government close to the people" is still widely honored. And most suburban polity is small-scale, compared to the enormous polity of the central city. Of course, it may be small-scale bad polity or no polity, yet it provides an illusion of a familiar and trustworthy place—"our town." Suburbanites do tend to know the local officials responsible for their interests and to use the channels in complaints. Suburban wives, usually confined to the area during the day, are likely to become local community actors, while the children in their school, play, and juvenile delinquency are apt to be supervised by adults who are accountable to the parents and the community. The network of adult relationships diminishes the isolation of the single household and acts to protect its members from unheeding bureaucratic pressures.

THE ECOLOGICAL DEFINITION OF SUBURBIA

We have opted for low-density neighborhoods of single-family housing; this requires geometrical expansion of the metropolis as population is added. This further leads to increasing average distance from the center. I will first discuss the consequence of this latter condition—the macro-ecological—then the nature of suburban neighborhoods—the micro-ecological.

One consequence of the expansion of the metropolis is the scatteration of extended kin groups. Thus, in metropolitan St. Louis, the central-city sample reported in 45% of the cases that they visited extended kin living in the same "named place" once a month or more; the comparable statistic for suburbia was less than 39% (Metropolitan St. Louis Survey, 1956-1957). In an upper-middle-class suburb of Chicago, however, over half the sample had regular interaction with extended kin, ranging from 71% of Jews through 33% of Catholics to only 16% of the Protestants. This may be, in part, because the suburb is overwhelmingly Jewish, yet it underscores the fallacy of regarding all suburbanites as equally affected by the scatteration of kin ties (Winch et al., 1967).

Thus, the suburban housewife, almost certain to be separated from her husband for a long commuting-working day, is also more apt to be isolated from relatives. This means it is virtually impossible for her to carry on a career while the children are young—she has no "built-in" or easily accessible babysitters. She is a long way from most job opportunities and, in the probable absence of adequate public transportation, has the added duty of chauffering the children. (One matron observed, as her last child graduated, "Thank God! I've lived over twenty years behind the wheel of an automobile.")

The suburban family is also separated from such functions of the center as parks, museums, great libraries, and the like, many of which could be used even by (or perhaps especially by) women in the child-rearing stage. The local playground and a probably inadequate village library must be her compensation. (More important, as we shall see later, are the neighbors.) Yet the metropolitan press and television are available; what she faces then is a very parochial span of action, but a metropolitan span of attention.

Lest the picture seem too grim, let us remember that it differs only in degree from the situation of most women, at least white, child-bearing, middle-class women in this society. It is evident from the Skokie data that if you transplant the ghetto to the Chicago North Shore you carry a lot of kinfolks with it (Winch et al., 1967). And if you live near one of the great expressways, you and your more distant kin can still get together. As for working, the traditional norms sanction the woman's domestic role as prime, though it is possible that the statistical norms are changing for some classes of women. Still, recent studies indicate that only one-third of a national sample of women would want to work if they did not have to. The proportion who do work is just about the same—one-third (Louis Harris et al., 1970). As for the cultural amenities of the center, the vast bulk of the present suburban population did not use them when it lived in the center city; it is likely that most are satisfied with the shopping mall as a cultural center.

From a slightly different angle, the male head of the household pays his dues to macro-ecological fact. In a small sample survey conducted by Vance Packard (1972), in which he compared a suburb of Los Angeles with a small city in Pennsylvania, he found that, in the suburb, 42% of the sample said nobody locally would know enough to judge his work, life style, or community participation. The proportion in the small city was 12%. And, of course, the commuting male is much less likely to participate in the community; his job is his main arena outside the home. Separation of work and residence showed up in another way: 32% of the suburbanites said their children had never seen their place of work, compared to 14% in the city. (While these findings are striking, one must bear in mind the small size of the Pennsylvania city—which is bound to make everything more accessible. There are many husbands and children within centers of large metropolitan areas who are probably just as affected by the separation of work and residence.)

At the micro-ecological level, the most impressive difference between an older-city form and the suburbs is the sheer increase in the household's space and privacy. This allows for surveillance of children while they have that freedom to navigate which we value, nostalgically perhaps, from our memories or fancies of an earlier America. It is probably a much better scene for child-rearing than the dense neighborhoods of the older city; at least the data indicate that parents think so (Bell, 1958; Dewey, n.d.).

In these low-density neighborhoods, the children tend to bring parents together in, at least, a mutual aid and policing function. Neighboring is very common in suburbia and is frequently instigated by the youngsters. Thus, in one study, neighboring did not vary by any social attribute except age. It was most common among the age category thirty to forty-four years, the place in the family cycle where the children are old enough to be up and around, yet young enough to be still chiefly confined to the small-scale world of neighborhood and neighborhood school. This child-induced neighboring is, to some degree, substitutable for the weakened extended kinship system.

While such a milieu may be by far the preferable scene for raising children, some commentators have remarked upon the relative poverty of opportunities for teenagers to observe the larger world, to participate, to organize their activities. Put simply, in a low-density residential neighborhood, there are apt to be few "hangouts." About all you may have is the high school and the shopping center, and authority may forbid you the first after-school hours, while those who operate shopping centers will probably regard you as a nuisance, bad for business; and you probably will be. Thus, lack of local institutional structure leads to early and frequent use of the automobile and drive-in restaurants as teenage clubs.

As Gans (1970: 305) puts it, "Most of the differences boil down to those of scale: one's private housing space is bigger, while the public social organizations and areas are smaller; it is easier to move around in social and physical space." This is perfectly congruent with an American civic culture which is highly individualistic, tolerates (if it does not encourage) the privatization of social life, and suspects the public interest.

I have been discussing the metropolitan dichotomy of suburb and central city as though we had distinct, bounded, and static types of milieu. And indeed, most serious discussions of suburbia rule out industry and concentrate on the residential function. However, in looking at the changing map of suburbia overall, it is clear that industry is decentralizing, too. The chief institutions of most governmentally bounded suburban areas may be, as Winch notes, the school and the shopping center, but not far from them we find the suburban adaptation to work—the industrial park. A planned development, it capitalizes on open space and the increasingly efficient circumferential expressway system. At an extreme, we have the completely (though privately) planned replica of old-style downtown, with a mixture of production and consumption facilities as well as malls, meeting rooms, gardens, and theatre.

The implications of those ecological shifts for the family are important in two ways. First, they tend to result in a shorter commute for the husband-father; the typical automobile commute is not from the far peripheries to the center of the city; instead, one goes onto the access ramp, drives for perhaps fifteen miles (about twenty minutes), leaves the expressway by the exit ramp, and enters a company parking lot a few minutes later. Thus the space-time ratio alters the meaning of geographical scatteration and brings the home and work place much nearer each other than the map suggests. The husband presumably has more time and energy for interacting with his family, and it becomes a simple matter for them to familiarize themselves with his place of work, if not with the work itself.

A second implication of the decentralization of *all* urban functions is the emergence of apartment clusters near the shopping centers and industrial parks. Such urbane accommodations attract many families in process of settling in after immigrating from other suburbs, giving them time to shop around for their single-family house. But they also attract older couples whose children have left home. Such people want to stay in the area because of familiarity with the scene, local institutions, friends, and, frequently, the nearness to their children and relatives. Thus, one study in Philadelphia, dealing with the possibility of center-city renewal through apartment house clusters, concluded that the demand was slight and was dwarfed by the apartments going up in suburbia (Rapkin and Grigsby, 1960).

A word is in order on experiments with the micro-ecology of suburbia. Whyte's (1956) *Organization Man* was based in part on his study of Park Forest, a planned suburb in Chicago. He describes a world in which the neighbors were omnipresent and the conjugal family was tightly intertwined with many other families, leading to a kind of bland affability which many readers found repugnant. The suburb had all its back doors opening on jointly used courtyards, and privacy was practically impossible. Community, of a sort, was inescapable. but, as this occurred, the singularity and integrity of the conjugal family was threatened. It was a free enterprise version of the kibbutz.

The style did not catch on, except perhaps among some planners of public housing complexes. Instead, planned suburbs were developed on the order of "complete" communities, with great attention to the usual suburban amenities—vegetation, open spaces, shopping and recreation centers, and, as in Reston, Virginia, central features of interest and beauty such as man-made lakes. Such planned suburbia is merely the rationalization of what is frequently achieved by the forces of the market; perhaps their main innovation is the provision of a variety of house types, from single-family units through semi-detached town houses to regular apartment houses. Again, the preferences of some for the convenience of multiple-family dwellings *and* life in the suburbs is recognized, and a place provided for the childless couple and the single person. Such developments make it convenient for the parents and children to remain in proximity after the children have formed new households, and in the way older people prefer—near enough for visiting, far enough for privacy.

In our society, rewards are gained on an individual basis and distributed on a marital and kinship basis. The strain involved can be minimized through bringing work and residence together, but it can never be totally eliminated, for the husband (the income provider) has a disproportionate amount of freedom. He owns his means of production. Perhaps this is why husband domination is more common in suburbia than in the central city (even controlling for race and class); as Winch (1971: ch. 6) suggests, there may be sympathy and gratitude on the part of the dependent wife. I would suggest that she is also trying to protect *her* means of production —the marriage contract.

THE SOCIAL DEFINITION OF SUBURBIA

As I noted earlier, the stereotypes of suburbia are less than complimentary. Suburbanites are seen as cowards, or at least as irresponsible, fleeing the various crises of the central city. They form homogeneous neighborhoods of single-family dwelling units, monotonously alike, and within these neighborhoods generate a culture of bland conformity, untinctured with the cultural and political excitement of life at the center.

More to the point, of course, is the matter of housing. People usually move to gain more space within their income limits (Rossi, 1957). A great deal of suburbia was settled by young families in the child-bearing and -rearing stage who found the best buy for their income. They tended to be homogeneous with respect, then, to age, family stage, ethnicity, and income. Although the given suburb might exhibit a high degree of similarity within, there remain enormous differences between suburbs. The suburb of the stereotype, however, is not those enclaves of stately houses on the Main Line or the North Shore: it is the hundreds of acres of relatively new, relatively modest, three-bedroom houses, each with its shred of lawn and newly planted trees.

The existing housing market, then, "selects" populations for specific areas. Through pricing, it eliminates some households and, through a complex of social and political mechanisms, tends to reject some ethnic groups (particularly darker types) and, in some cases, attract others—for example, Jews. Within these limits, there are still very definite preferences among the existing populations as to just whom they would like for neighbors. In the St. Louis suburbs, we found the most common desideratum to be neighborliness (30%), followed by moral respectability (18%), and good housekeeping and care of property (13%). (Interestingly enough, almost twice as many city dwellers mentioned morality—31%.) In short, they want people of their own social class who are easy to get along with. When social characteristics were mentioned, the most important was family type (18%). And, predictably, 93% of our suburban respondents thought theirs a good neighborhood in which to raise children, compared to two-thirds of the central-city sample (Metropolitan St. Louis Survey, 1956-1957).

Within these limits, ranging from absolute to tenuous and not very constraining, the individual household selects its site. A prime consideration is house type and size for the money; another is convenience. Even in the St. Louis suburbs, a third of the respondents gave convenience of location as the *first* reason for liking the neighborhood. There is also selection by the suitability of the neighborhood for the general life style preferred by the household. This can range from apartment-dwelling near the commercial centers to the single-family household in a spacious neighborhood far from business and heavy traffic. And, even in the suburbs, a substantial number of the urbane, the single swingers and bohemians (not to mention their audiences) will be found in the older, inner areas.

The importance of the suburbs for the family lies in the circumstance that most of them were created *for* the nuclear family. They are convenient for a small group, attached by contract and lineage, to household, garden, do-it-yourself, and hobbies. And they provide private space, indoors and out, in which these activities can be enjoyed in quiet and (depending on the family) peace and harmony.

The suburban move is frequently an upward move in social class and an outward move from extended kin. Schorr (1970: 721) remarks, "As locality and the extended family are no longer coterminous, it is no longer possible to be neighborhood centered and extended family centered at once." This may be a serious loss to the individual or the group as a whole.

Yet, paraphrasing Mogey, Schorr argues that there may also be significant gains in separating conjugal family and extended family (or, to be more precise, parents and siblings of the married pair, since they are clearly the most important extrahousehold kin).

> Younger family groups who move away develop a family-centered society in the place of the neighborhood-centered society they have known. The change is not only expressed in the degree of the contact with one person or another but, as one might expect, in the kind of contact. There is a shift from spending time with collateral

relatives to husband and wife or children. Marital disagreement diminishes and sharing of household chores increases. Less contact with distant relatives is accompanied by more contact with neighbors.

He then goes on to quote Mogey, "There is a tendency for the conjugal type of family to discover itself and for the obligations of kinship to be relaxed." He even suggests that less is more, that escaping the omnipresence of kinfolk increases the value of inter-action. Certainly we have found in our Skokie studies that multiplying the functions of kin, as evidenced in family-owned enterprise, is not conducive to harmonious relations; our respondents generally agreed it was everything from a nuisance to a disaster.

Whatever their life was like in the central city, the conjugal families of the suburbs do tend to be engrossed with their neighbors, community organization, and local institutions. They are more apt to be community actors as well. Further, the wife is almost as likely to be a community actor as her husband and, if she is involved in community organization, is just as likely as her husband to be involved in local politics. The sexual egalitarianism in the suburbs is probably greater than it has ever been in any other part of the society (Greer and Orleans, 1962).

Out of the network of households sharing common interests and identified with a common turf develops the communication and control system of the suburb through the multifarious voluntary associations. Of course, this varies by social class and is especially intense in the upper-middle-class suburbs, but it is almost always more common than in comparable city neighborhoods. Then, if there is an incorporated government, the husband-wife teams are "com-munity binders," a net resource for the suburb. For voluntary associations are usually sex-specific, related to the round of everyday life—PTA, service clubs, church- and child-related organizations for women; work, business, service clubs, veterans, and sports organi-zations for the men. (I should except those sports which have gone unisex, such as bowling and, increasingly, billiards: soon there will be no hiding place down here.) The two kinds of associations meet in the matrix of the conjugal family.

CONCLUSION

I have discussed the effects of "social suburbia" upon the nuclear family, the extended family, and the interactions between them. I should emphasize its effects upon that most important aspect the family—indeed, its raison d'etre—the continuing family, the children. With over half the metropolitan population living in suburbia now, and that the most fertile fraction, the "old home town" will be a suburb for a very large and influential part of the metropolitan population in a very few years.

Children will grow up to be adults with no memories, nostalgic or otherwise, of life in the old central city. They will have learned the nature of the world and their place in it through the conjugal family, neighborhood, and local institutions of the suburb. Their peers in the inner city may cling to the scenes of their childhood, cultural conservatives of the declining neighborhoods; my guess is that, given the chance, they also will make it up and out. After all, suburbia and exurbia are where the jobs are going, just as they are where the new housing is being built. Thus, banal or not, suburbia is the future.

Unless there is a massive shift in preferences, most of these second-generation suburbanites will not evidence either the nostalgia or the repugnance of their parents for the central city. Instead, they will probably be usually indifferent to it, in competition with it when there is conflict of interest. Thus, the job of "saving the central city" will have fewer sympathizers outside the center. The job will have to be done by the city itself, and state and federal government aid will be tempered by the lack of political support in the suburbs for any massive redistribution of income at the expense of suburbanites.

Meanwhile, suburbs will grow increasingly differentiated as they and their populations age. This differentiation is already manifest in the intense rivalry among the youth of different areas, organized around their high schools and, especially, their high school sports. Given the tendency of adolescents to form gangs and to then define and defend turf, one would expect the same territorially defined conflict as obtained in the older central city. Homogeneity within the suburb and heterogeneity between suburbs are analogous to the ethnic mosaic of the older city. But whereas, in that city, raid and reprisal were carried out on foot, so to speak, and thus limited to a few blocks, the automobile-expressway system of the suburbs allows gang conflict over many miles. In a society of expanding scale, the gang behavior of adolescents also increases in scale. Because of this, it is apt to be, more often than not, beyond the purview of the parents, an autonomous activity which separates the generations in the family.

But the future must not be regarded as a straight-line extrapolation of the past. I think we will see the emergence of congeries of centers, each as large as the old central city was not so long ago. They may some day find an appropriate governmental form, one that could allow for civic patriotism, a civic culture, and amenities of which the average citizen of the old railroad-centered city could never have dreamed. If we can bring it off, we can convert the apparent chaos of so many of our great metropolitan areas into families of cities. In most of them, most people will probably live very much the sort of life I have been sketching, but with enormous increments of value.

The reader may resent my moderately optimistic conclusion, especially if he is a patriot of the old, centralized metropolis now in such disarray. And I have hardly noted the problems of the central city in all their dangerous complexity. I can only say that the assignment was otherwise, and this is no place to state my own prejudices and policies for the metropolis as a whole.

A NOTE ON ETHNICITY AND KIN IN THE SUBURBS

As I noted at the beginning of the article, the family is a social group organized by the principles of contract and lineage or descent. A group of lineages claiming common descent is a tribe or nationality. In the United States, such tribalism, or ethnicity, has been maintained through spatial concentration, ethnic endogamy, the functional value of the extended kin structure, and the pressure of the surrounding community.

With the social mobility of second- and third-generation ethnic Americans, up in social rank and outward in space, this entire set of supports is threatened. The suburbs, homogeneous in social rank, life style, and class, are heterogeneous with respect to ethnic lineage. Yet the cultural differences are not great, and their children, exposed to the same socializing agencies, bid fair to resemble each other more than their parents do. The probability of intermarriage and assimilation is very great. Such assimilation is not conducive to the *ethnic* support for the extended kin system and mitigates (if it does not eliminate) pressure from the surrounding community.

With spatial scatteration of kin networks, the neighborhood and local community of the suburbs takes over the function of the old ethnic neighborhood. And this neighborhood is bland, polyglot, assimilated, or assimilating. This shift of attention, in turn, militates against the continuing utility of extended kin. In short, it all points to further assimilation.

The end product of that assimilation is already here, in the "all-American" citizen. His attachment to his nationality of origin is confused by the multiple nature of that origin—the many nationalities that came together and produced him. Thus, in a paradoxical way, as the offspring of "mixed marriages" contract with the offspring of other mixed marriages, some vestige of common origins becomes broadcast.

With a new "myth of common origins," the suburbs begin to look like different tribes of a new nation—white, suburban America. They are separated by space and institutions from the various tribes of another new nation—black, urban America. The danger is that this new, assimilated, suburban population, instead of foregoing ethnocentrism in light of its own recent experience as a *blender* of ethnic lineages, may simply extend the myth of common origin to all white suburbanites. In that case, there is no point in talking of white ethnics; all whites will be ethnic, and white ethnocentrism, or "racism," will complement black ethnocentricism, or "separatism," in a truly dangerous symbiosis.

But we may be able to remember and learn. The days are not far gone when the Irish Catholic and the Yankee hated each other's looks, religion, manners, politics, and souls. Today, this has largely disappeared. Most of the little ethnocentrisms have disappeared, but a greater ethnocentrism can only be prevented through a continual emphasis upon the dangers of creating pseudo-species in the face of the common nation of man. It will be difficult, for that political organization, the state, profits greatly when it can cloak itself in nationalism.

REFERENCES

BELL, W. (1958) "Social choice, life style, and suburban residence," in W. Dobriner (ed.) The Suburban Community. New York: G. P. Putnam.

BERK, R. A., R. W. MACK and J. L. McKNIGHT (1971) Race and Class Differences in Per Pupil Staffing Expenditures in Chicago Elementary Schools, 1969-1970. Evanston, Ill.: Northwestern University Center for Urban Affairs.

BOLLENS, J. C. [ed.] (1960) Exploring the Metropolitan Community. Berkeley and Los Angeles: Univ. of California Press.

CHANDLER, A. D., Jr. and S. SALSBURY (1971) Pierre S. du Pont and the Making of the Modern Corporation. New York: Harper & Row.

COLEMAN, J. S. et al. (1966) Equality of Educational Opportunity. Washington, D.C.: Government Printing Office.

DAHL, R. A. (1960) "The analysis of influence in local communities," in C. R. Adrian, Social Science and Community Action. East Lansing: Michigan State Institute for Community Development and Services.

DEWEY, R. (n.d.) "Peripheral expansion in Milwaukee County." Amer. J. of Sociology 53: 417-422.

ERIKSON, E. H. (1968) Identity: Youth and Crisis. New York: W. W. Norton.

GANS, H. J. (1970) "The suburban community and its way of life," in R. Gutman and D. Popenoe (eds.) Neighborhood, City, and Metropolis. New York: Random House.

GREER, S. (1972) The Urbane View. New York: Oxford Univ. Press.

——— (1962) The Emerging City. New York: Free Press.

——— and P. ORLEANS (1962) "The mass society and the parapolitical structure." Amer. Soc. Rev. 27 (October): 634-646.

Louis Harris et al., Inc. (1970) "The Virginia Slims American women opinion poll."

JENCKS, C. (1972) Inequality: A Reassessment of the Effect of Family and Schooling in America. New York: Basic Books.

LEVINE, E. M. (1966) The Irish and Irish Politicians. Notre Dame, Ind.: Univ. of Notre Dame Press.

MAIR, L. (1964) Primitive Government. Baltimore: Penguin.

MALINOWSKI, B. (1929) The Sexual Life of Savages in North-Western Melanesia. New York: Horace Liveright.

MASOTTI, L. H. (1967) Education and Politics in Suburbia. Cleveland: Western Reserve Univ. Press.

Metropolitan St. Louis Survey (1956-1957) Previously unreported data. (unpublished)

PACKARD, V. (1972) A Nation of Strangers. New York: David McKay.

RAPKIN, C. and W. G. GRIGSBY (1960) Residential Renewal at the Urban Core. Philadelphia: Univ. of Pennsylvania Press.

ROSSI, P. H. (1957) Why Families Move. New York: Free Press.

SCHORR, A. L. (1970) "Housing and its effects," in R. Gutman and D. Popenoe (eds.) Neighborhood, City, and Metropolis. New York: Random House.

WINCH, R. F. (1971) The Modern Family. New York: Holt, Rinehart & Winston.

——— S. GREER, and R. L. BLUMBERG (1967) "Ethnicity and extended familism in an upper-middle-class suburb." Amer. Soc. Rev. 32 (April): 265-272.

WOOD, R. S. (1958) Suburbia: Its People and Their Politics. Boston: Houghton Mifflin.

WHYTE, W. H., Jr. (1956) The Organization Man. New York: Simon & Schuster.

6

Blacks in Suburbs

WILLIAM W. PENDLETON

☐ THE PROCESS OF SUBURBANIZATION has characterized the redistribution of the residential patterns in the United States increasingly during the past two decades. This redistribution has had political and social consequences of enormous dimensions, while influencing the nature of urban life in central cities (Hawley, 1956; Goldston, 1970; Bell, 1968). Though some of the problems associated with suburbanization can be traced to the dispersion of the urban population over a much greater area and at a lower density, the *selectivity* of suburbanization has been most frequently identified as the source of greatest concern (Bell, 1958). Though interpretations vary and studies have given contradictory results, suburbs have been generally viewed as receiving the more affluent, better-educated, younger, conservative, mobile, white parts of the population (Goldstein and Mayer, 1965). Those conditions are well documented, but the resulting interpretation of a homogeneous suburb, at least in comparison to the heterogeneity of the city, has been disputed (Boskoff, 1966; Schnore, 1957, 1958). While individual suburban communities may be quite homogeneous, the suburbs taken as a whole are not. A consequence of that fact is that suburbanization cannot be thought of as a single process with a single set of consequences; it is differentiated, consists of several types and can be expected to vary for different groups and different places at different times.

Before examining the differences between the process of suburbanization for blacks and whites, the variety in the concepts of suburbanization needs some evaluation. Two definitions of suburb are used in this paper, though several others are available. These two definitions are frequently employed because of the ease with which census data can be related to them. They are both ad hoc definitions and subject to the difficulties attendent thereto (Hadden, 1969). One definition is based on the Standard Metropolitan Statistical Area and uses as suburban all of the SMSA that is not contained in the central city. This area is generally called the "ring" of the SMSA. The other definition is based on the urbanized area and equates suburban with the part of the urbanized area that lies outside its central city. This suburban part is often called the "urban fringe." Though central cities of urbanized areas and of SMSAs have a similar definition, all SMSAs do not contain an urbanized area, nor do all urbanized areas lie in a central city. Generally, where a place is the central city of both an SMSA and an urbanized area, the urban fringe will be smaller than the ring both geographically and in population. Though that is not the case for every such city in the country, it is for those that are examined in this study.

Though both of these definitions have the disadvantage of assigning to the central city some areas or neighborhoods that would be generally recognized as suburban and including in the suburban area places that would popularly not be classed suburban, they have the advantage of relying on the political boundary of the city as the demarcation line between city and suburb. For that reason, movement from central cities generated by their political structure, tax rates, school systems, or racial balance is reflected, albeit imperfectly, by the definition.

The suburbanization of the nation is reflected in the differential rates of growth of central cities and their rings between 1960 and 1970. Central cities added 3,194,453 people to their population, though nearly half actually lost population. That represented an increase over the decade of 5.3%. Their rings, however, added 16,598,213 to their population, an increase of 28.3%. Moreover, most of the increase in the populations of central cities was due to annexation rather than migration or natural growth. At the same time, central cities experienced a decline in density that had begun in

1950. The urban fringes of urbanized areas also experienced a decline in density during the period 1950-1960, but relative stability in density during the period of rapid growth between 1960 and 1970. This decreasing concentration of the population in the face of urban growth may indicate that suburbanization does not include an increase in crowding or an extension of the traditional urban patterns into adjoining political units.

Black participation in the suburbanization process began as a small growth of traditionally black areas in the ring or urban fringe (Blumberg and Lalli, 1966; Farley, 1970). In that sense, the initial pattern could not be called a suburban movement. The decade of 1960 to 1970 saw some changes in the pattern. By 1960, 4.8% of the population residing in the rings of SMSAs was black. During the following ten years, while suburbanization was setting a rapid pace in the nation, blacks maintained their proportion of the suburban population; they were 4.8% of the population of the rings in 1970. In the aggregate, blacks would seem to have participated in the suburban movement as rapidly as whites. In fact, blacks increased their share of the suburban population in every region except the South, though the increase was in every case small. In the South, blacks declined from 12.6 to 10.3% of the population in the suburban rings. Yet the proportion of the black urban population residing in the suburbs remained relatively small. Of all blacks residing in the SMSAs, 15.1% lived in rings in 1960 and 16.2% did so in 1970. Regional variation is also important in this context: In the Northeast, 18.7% of blacks living in SMSAs lived in the rings in both 1960 and 1970. For the other regions, changes between 1960 and 1970 are as follows: North Central, 11.6 to 12.8; South, 14.2 to 14.8; West, 25.6 to 29.5.

These figures taken together suggest that blacks have increasingly moved to suburban areas and that their movement is taking place at a slightly increasing rate in comparison to whites everywhere except the South. One question remaining is whether the consequences of this suburbanization of blacks is comparable to those of the suburbanization of whites. A second question is suggested: Is there any decrease in black-white differentials as a consequence of suburban residence? A partial answer to these questions is offered in the rest of this chapter.

Because suburbanization is not purely a national phenomenon, but a product of processes taking place in several communities, the analysis that follows focuses on several different cities. The cities were selected because they generally represent the variety of American cities and should reflect most, if not all, of the regional, size, type, and structural factors that are expected to influence the nature of suburbanization. The selection was subjective, but the cities are coastal and inland, large and small, relatively distant from other cities and close to them, old and new, from every part of the country and differ significantly in their regional composition. They are: Albuquerque, New Mexico; Atlanta, Georgia; Cincinnati, Ohio; Cleveland, Ohio; Detroit, Michigan; New Orleans, Louisiana; Philadelphia, Pennsylvania; St. Louis, Missouri; San Francisco, California; San Jose, California; and Washington, D.C. General descriptive statistics of the size and racial composition of these cities are shown in Table 1.

These data show that the general trend is for blacks to be a much smaller part of the urban fringe than of the central city. The sole exception is Albuquerque, which is exceptional in many ways, as will be seen. The data also show that considerable variation exists among the cities. In addition, the association between the percentage of the population of the central city that is black and the percentage of the urban fringe that is black is moderate (tau = .52). These data indicate the enormous complexity of the process of black suburbanization. Neither the size of the central city nor its suburban fringe is closely associated with the degree of black suburbanization.

TABLE 1

SIZE AND PERCENTAGE BLACK OF THE CENTRAL CITIES AND URBAN FRINGE OF ELEVEN SELECTED AMERICAN CITIES

City	Central City		Urban Fringe	
	Population	% Black	Population	% Black
Albuquerque	243,751	2.2	53,700	2.2
Atlanta	496,973	51.3	671,829	5.8
Cincinnati	452,534	27.6	673,744	3.7
Cleveland	750,903	38.3	1,168,161	3.7
Detroit	1,511,482	43.7	2,466,752	3.7
New Orleans	593,471	45.0	370,226	10.9
Philadelphia	1,948,609	33.6	2,115,970	6.8
St. Louis	622,236	40.8	1,249,559	9.0
San Francisco	1,077,235	20.5	1,824,748	5.6
San Jose	445,779	2.5	581,323	1.2
Washington	756,510	71.1	1,724,979	7.7

SOURCE: These data and those in the other tables in this chapter were tabulated by the Urban Research Section of the Oak Ridge National Laboratory.

There are some limitations imposed by the limitations of the census data available at this time. Analysis is limited to the First Count materials. These are the data based on the questions that were asked in every household in the nation. These materials do offer several variables that bear on the problem: the age and sex composition of the population, which reflects both dependence and fertility; household structure, which, especially in the case of female-headed households, shows the differences in patterns of family living; the value of rented and owned units, which reflects the standard of living of the occupants; and the crowding of housing units, which shows the conditions under which the occupants live at least in some degree. A full exploration of the Census of Population and Housing of 1970 will prove more complete than this analysis, though there is no reason to suppose that it will change the results of the present study.

INDICES EMPLOYED IN THE ANALYSIS

The data available allow several different measures or indices to be employed in the analysis. The indices employed are mostly standard in demographic research, but a brief examination of them is offered in the context of suburbanization.

YOUTH AND AGED DEPENDENCY

Dependency indices or ratios are defined as the relation between the dependent and working populations. When only the age structure is used to define the dependent and working populations, a degree of arbitrariness is introduced, since such factors as morbidity, training, willingness to work, different ages for entry into the working population and differences in sex roles remain uncontrolled. Even so, the population that is young and old relative to that in the intervening years does reflect, even though imperfectly, the burden the population carries in terms of nonproductive members. Of course, such a burden is necessary because of the nature of human biology, but when the dependent population is relatively large, special problems are posed in terms of community services and life styles. The dependency ratio is defined as the number of persons 0-14 years of age plus those 65 years and older per 100 persons aged 15-64. The two components of this index can be used to define youth and aged dependency. Youth and aged dependency are appropriate to the study of suburbs because they should vary differently between city and suburb. In general, the central city, being older and an area of fewer new households, should have relatively high aged dependency and relatively low youth dependency. Two questions can be raised about the dependency indices with respect to black suburbanization:

(1) Do blacks and whites both show the pattern of lower central-city youth dependency and higher central-city aged dependency?

(2) Are the suburban differences between the dependency measures for blacks and whites less than those in the central city?

Two questions analogous to these will be raised with respect to each measure employed because they are basic to understanding the process of black suburbanization. The first determines if black and white suburbanizations have the same type of consequences, the second determines if suburbanization reduces racial differences. If the consequences of suburban as opposed to central-city residence for blacks follow the same pattern as for whites, then the hypothesis that black suburbanization is suburbanization in the traditional sense is supported. Supporting that hypothesis does not mean that a black suburban trend will lead to diminished interracial differences in living patterns or in social characteristics. For that reason, the racial differences in the central cities must be compared with the differences in the suburbs.

With respect to dependency, the pattern of lower youth and higher aged dependency in the central city is borne out by ten of the eleven cities. San Jose, which has some characteristics of a suburban central city, is the only exception. Both blacks and whites show the same kind of differences, but there is some tendency for the city-suburban differences to be greater for whites than for blacks (see Table 2). Though blacks have generally higher youth dependency and lower aged dependency than whites, their total dependency is greater than that of whites. The evidence for a smaller black-white differential in the suburbs is clear in eight cities, but Albuquerque, Cleveland, and San Jose follow a different pattern. In those cities, the black-white differences are smaller in the central city than in the suburb.

TABLE 2

YOUTH AND AGED DEPENDENCY IN CENTRAL CITIES AND RINGS OF ELEVEN CITIES, 1970

City	Youth Dependency				Aged Dependency			
	Central City		Urban Fringe		Central City		Urban Fringe	
	Black	Non-black	Black	Non-black	Black	Non-black	Black	Non-black
Albuquerque	68	48	70	64	7.0	10.3	4.8	8.8
Atlanta	55	30	65	47	10.1	18.0	9.0	7.3
Cincinnati	58	37	61	52	13.8	24.0	14.5	13.9
Cleveland	56	40	54	44	10.9	21.1	5.3	13.8
Detroit	56	35	59	51	9.8	25.3	8.9	10.1
New Orleans	64	34	73	52	12.7	20.8	9.3	7.8
Philadelphia	55	36	61	47	11.6	22.3	11.4	13.8
St. Louis	61	34	71	47	14.4	31.7	11.5	13.1
San Francisco	52	25	55	41	8.1	23.6	6.9	11.6
San Jose	69	54	56	44	4.1	9.2	4.1	9.7
Washington	49	14	62	43	8.7	26.4	5.0	7.8

THE CHILD-WOMAN OR FERTILITY RATIO

The child-woman ratio is a crude measure of fertility. It is the number of children under 5 years of age per 1,000 women aged 14-44 years of age. Since it represents the relation between children and women of childbearing age, it reflects the level of fertility in the population, though it is not immune to the influence of child-spacing practices and migration. The child-woman ratio is related to the life cycle of the family, in that newly established families, if they are concentrated in an area, should greatly increase the child-woman ratio of that area. Specifically, if child-bearing whites are concentrated in the suburbs and child-bearing blacks are concentrated in the central city, the suburban-city differentials should be in opposite directions for the races. Such would be the result, for example, if educated, occupationally mobile blacks with their relatively low fertility were moving to the suburbs leaving poorer but higher-fertility blacks in the city (and in the countryside); and, at the same time, whites were leaving their elderly in the city while bearing their children in the suburbs.

In general, the child-woman ratio shows higher fertility for blacks than for whites, though the reversal in the central city of Cleveland is an important exception. Nationally, the fertility ratio for blacks is higher than that of whites by about 25%. This difference is true in both city and suburb. However, there is a tendency for both races to have higher fertility in the suburbs, with two exceptions for blacks and three for whites. The trend for these cities does not suggest that suburban residence reduces black-white fertility differentials. In six of the cities, the racial difference in the suburbs is less than the difference in the central cities, but in five the trend is reversed (see Table 3).

FEMALE-HEADED HOUSEHOLDS

The Census Bureau defines the head of household as being the male member of the husband-wife pair provided he lives in the household. Thus, women qualify for the status of head of household only when widowed, divorced, abandoned, or otherwise without a resident husband. Nationally, the percentage of female-headed black households is three times that for whites, 28 versus 9, though the percentage of widows is about the same for both races, 13 versus 12%. In every city, the percentage of female-headed households is greater for both races in the central city (see Table 4). The value for whites is always lower than that for blacks. Moreover, the value for whites is less variable from city to city than that for blacks, indicating either a more homogeneous population among the cities for whites or a greater variation in the forces leading to family disruption for blacks in different cities.

TABLE 3
CHILD-WOMAN RATIOS FOR ELEVEN CITIES, CENTRAL CITY AND
URBAN FRINGE, 1970

City	Central City		Urban Fringe	
	Black	Nonblack	Black	Nonblack
Albuquerque	458	373	631	498
Atlanta	429	311	500	378
Cincinnati	463	382	492	433
Cleveland	417	439	391	379
Detroit	481	392	472	419
New Orleans	495	319	573	429
Philadelphia	442	373	470	396
St. Louis	476	353	545	378
San Francisco	428	266	436	347
San Jose	508	442	458	356
Washington	393	160	488	364

TABLE 4
PERCENTAGE OF FEMALE HEADED HOUSEHOLDS IN
ELEVEN CITIES BY RACE, 1970

City	Central City		Urban Fringe	
	Black	Nonblack	Black	Nonblack
Albuquerque	31	12	8	11
Atlanta	30	14	26	7
Cincinnati	31	14	20	8
Cleveland	30	13	18	7
Detroit	26	13	25	7
New Orleans	32	14	23	7
Philadelphia	33	13	28	8
St. Louis	32	15	28	7
San Francisco	30	14	26	9
San Jose	19	9	16	9
Washington	29	16	17	9

The evidence in this case is strongly against a smaller black-white difference in the suburbs than in the city. Only Albuquerque and San Jose have smaller differences in the suburbs and both show unusually low percentages of black female-headed households rather than very high values for nonblacks. Probably the slightly higher values for nonblacks in the cities are due to the age structure of the white population in the central city. Blacks living in the ring are subject to the same disruptive influences on the household as blacks in the central city, though to a smaller extent.

VALUE OF OWNER-OCCUPIED UNITS AND CONTRACT
RENT FOR RENTED UNITS

The best data referring to social status or class available from the early census returns are those referring to the cost and value of housing. Although self-reported evaluations of houses probably suffer from many errors in reporting, they should reflect the general level of values in an area, if not the absolute dollar values. In the country as a whole, blacks owned housing units with an average value of $12,294, while the figure for nonblacks was $20,884. For rented units, the blacks paid an average of $77.00 per month while nonblacks paid $106.00. Though the percentage difference in the rents is somewhat less than that in the value of owner-occupied units, both figures show the well-recognized socioeconomic differences between the races. Neither value is directly reflective of the quality of housing involved. Within and among cities, the same rent or purchase price can apply to housing of different quality. Blacks may not only have less valuable homes, they may obtain less for their money. Nonetheless, in each case and for both central city and suburb, blacks had a lower value of housing unit and, with the exception of Albuquerque, a lower monthly contract rent (see Table 5).

These data confirm, though not without exceptions, the generally held notion of a more affluent suburban area. The exceptions occur more frequently for blacks than for whites, and the suburban-city differences are greater for whites than for blacks. This result argues strongly that suburban residence is less distinctive for blacks than for whites, even though it acts on blacks in the aggregate the same way it does on whites; that is, it increases the value of their housing.

The greatest disparities in both rent and value of housing are found in the suburbs, again indicating that black movement to the suburbs does not indicate blacks live in areas where they are more like their neighbors. An interesting comparison is found in the central city of Washington, D.C., where the disparity is greatest between the races in terms of the value of owner-occupied units and is as great as is found in the suburbs showing the largest suburban difference. Washington has the highest percentage of blacks in its

TABLE 5
AVERAGE VALUE OF OWNED AND CONTRACT RENT FOR RENTED HOUSING UNITS IN ELEVEN CITIES, 1970

| | Owned by Race | | | | Rented by Race | | | |
| | Central City | | Urban Fringe | | Central City | | Urban Fringe | |
City	Black	Non-black	Black	Non-black	Black	Non-black	Black	Non-black
Albuquerque	12,177	18,632	8,601	14,531	66	102	92	70
Atlanta	16,273	24,902	13,732	24,339	68	116	66	137
Cincinnati	15,210	19,369	15,404	21,734	67	98	64	94
Cleveland	16,403	17,685	20.781	28,275	77	88	99	143
Detroit	14,828	16,641	15,633	25,316	78	99	80	143
New Orleans	17,273	27,420	13,725	23,361	59	93	56	117
Philadelphia	9,101	12,838	12,228	21,644	70	105	80	129
St. Louis	11,665	14,870	11,710	20,775	64	83	65	122
San Francisco	21,774	29,352	20,950	31,223	103	134	108	151
San Jose	23,075	27.191	25,848	32,171	129	139	148	158
Washington	20,544	36,710	22,115	32,593	104	155	129	159

population but relatively small percentages of owner-occupied housing units (see Tables 1 and 6). The racial differences in rent, however, are not so great in Washington, and nearly 70% of the units in that city are rented.

QUALITY OF HOUSING

Two measures of the quality of housing are available: persons per room and units lacking plumbing facilities. Here the percentage of units with more than 1.01 persons per room is used as an index of crowding, and the percentage of units lacking a major plumbing facility is used as an index of the lack of services in the household. These measures, like others employed, have some weaknesses. Crowding is related to the age of the persons concerned and the size of the rooms available. A man and wife with two children in a four-room house, for example, might be less crowded than a man and wife with a grown son and a mother-in-law in the same house, though both households would have the same number of persons per room. In general, however, persons living with more than one person per room can be regarded as at least moderately crowded. In addition to varying by race and suburban versus city residence, crowding varies by tenure. Regardless of race or place of residence, renters are more likely to be crowded than owners.

TABLE 6
PERCENTAGE OF HOUSING UNITS OWNED BY RACE IN
ELEVEN CITIES, 1970

City	Central City		Urban Fringe	
	Black	Nonblack	Black	Nonblack
Albuquerque	48	65	25	64
Atlanta	37	44	53	65
Cincinnati	27	42	62	73
Cleveland	38	51	55	72
Detroit	51	66	62	81
New Orleans	27	46	51	72
Philadelphia	47	65	55	74
St. Louis	31	45	55	75
San Francisco	33	36	46	62
San Jose	45	64	45	61
Washington	27	30	43	47

There are, however, some exceptions such as Atlanta and Detroit where suburban nonblack renters are less crowded than suburban nonblack owners. In all such cases, the differences are small. There is no clear pattern of benefit from suburban residence for either race in terms of crowding. There is a tendency for black owners to be less crowded in the central city than in the suburb, though there are some exceptions as well. The major difference seen in Table 7 is that the racial factor seems to dominate everything else. Only in the case of renters residing in the suburbs of Albuquerque are blacks less crowded than nonblacks. This difference is probably due to the city's very small black population (2.2%) and the presence of a significant number of disadvantaged Mexican-Americans. This evidence suggests that, while demographic factors such as dependence and fertility have a similar pattern for blacks and whites in terms of suburban or central-city residence, more socially related factors such as housing respond more readily to racial classifications. The more demographically related variables are in some degree a function of the life-cycle which is influenced by the age of the neighborhood. Hence, blacks living in relatively new neighborhoods might be expected to be demographically like whites in newer parts of the city. Socially, however, they show the pattern of racial differences in education and income that characterize the nation.

TABLE 7

PERCENTAGE OF DWELLING UNITS WITH MORE THAN 1.01 PERSONS PER ROOM IN ELEVEN CITIES BY RACE AND TENURE, 1970

| | Central City | | | | Urban Fringe | | | |
| | Owners | | Renters | | Owners | | Renters | |
City	Black	Non-black	Black	Non-black	Black	Non-black	Black	Non-black
Albuquerque	18	8	22	9	35	20	14	24
Atlanta	14	3	23	6	23	3	15	1
Cincinnati	13	6	17	9	13	7	19	10
Cleveland	9	5	11	6	7	4	5	3
Detroit	12	6	11	5	17	8	17	5
New Orleans	19	4	30	8	28	8	42	11
Philadelphia	7	4	14	5	10	4	17	4
St. Louis	18	7	23	9	19	8	25	6
San Francisco	14	5	15	7	15	4	16	6
San Jose	20	6	24	10	17	5	9	6
Washington	10	2	20	5	16	3	16	4

The pattern of the percentage of dwelling units that lack a major plumbing facility is like that for housing. However, an important suburban factor is shown in Table 8. Though blacks are more apt to have plumbing facilities in the central city, there is no consistent suburban-central city pattern for nonblacks. However, in the central city, the black-nonblack differences are mostly trivial. Exceptions which deserve attention are renters in San Francisco where nonblacks are more likely to be without facilities than blacks and New Orleans and St. Louis where the reverse is the case. San Francisco may have disadvantaged Oriental populations in sufficient number to account for its pattern, while New Orleans and St. Louis are both cities with very large black populations. Probably enforcement of housing codes in some parts of those cities is less strict than in some other cities. One important finding does emerge from these data. There is every reason to believe that legal protection such as that offered in housing codes can reduce black-nonblack differences. At least in some cases, the central city might be more desirable to blacks than the suburbs for such reasons. The pattern of movement of blacks to central cities and nonblacks to suburbs should be examined in terms of the pulls to those areas as well as to the pushes from them.

TABLE 8
PERCENTAGE OF DWELLING UNITS WITHOUT ONE OR MORE MAJOR FACILITIES IN ELEVEN CITIES BY RACE AND TENURE, 1970

| City | Central City | | | | Urban Fringe | | | |
| | Owners | | Renters | | Owners | | Renters | |
	Black	Non-black	Black	Non-black	Black	Non-black	Black	Non-black
Albuquerque	.95	.92	2.2	2.5	8.7	5.6	35.7	8.8
Atlanta	1.3	.94	3.7	1.9	9.1	.64	15.2	1.4
Cincinnati	1.9	1.5	5.6	5.5	5.8	1.1	13.3	4.9
Cleveland	.89	1.1	3.9	3.7	1.4	.67	1.6	1.4
Detroit	.91	.96	3.5	5.5	1.4	.79	4.5	2.8
New Orleans	3.0	.95	7.2	2.2	9.0	.48	19.7	1.1
Philadelphia	1.3	.92	3.9	3.4	2.3	.82	4.6	1.8
St. Louis	2.7	2.1	9.6	6.9	9.3	1.2	13.7	3.0
San Francisco	.86	.87	3.7	7.9	.50	.48	2.3	1.6
San Jose	.16	.40	.78	2.4	.31	.36	.96	1.0
Washington	1.0	.90	2.8	2.4	2.4	2.3	.64	.83

SUMMARY AND CONCLUSIONS

Suburbanization has both demographic and social dimensions. Blacks appear to participate in the demographic dimensions, but not in the social dimensions. Black suburbanization is not unequivocally an indicator of social integration. Though some blacks in some cities are engaged in upward mobility, as indicated by moves to the suburb, large parts of the suburbanization of blacks are probably due to growth in traditionally black areas such as East St. Louis and Decatur, Georgia.

REFERENCES

BELL, W. (1968) "The city, the suburb, and a theory of social choice," pp. 132-168 in S. Greer et al., The New Urbanism. New York: St. Martin's.

——— (1958) "Social choice, life styles, and suburban residence," pp. 225-247 in W. M. Dobriner (ed.) The Suburban Community. New York: G. P. Putnam.

BLUMBERG, L. and M. LALLI (1966) "Little ghettoes: a study of Negroes in the suburbs." Phylon 27 (Summer): 117-131.

BOSKOFF, A. (1966) "Social and cultural patterns in a suburban area." J. of Social Issues 22 (January): 85-100.

FARLEY, R. (1970) "The changing distribution of Negroes within metropolitan areas—the emergence of black suburbs." Amer. J. of Sociology 75 (January): 512-529.

GOLDSTEIN, S. and K. B. MAYER (1965) "Impact of migration on the socio-economic structure of cities and suburbs." Sociology and Social Research 50 (October): 5-23.

GOLDSTON, R. C. (1970) Suburbia: Civic Denial. New York: Macmillan.

HADDEN, J. K. (1969) "Use of ad hoc definitions," in E. F. Borgatta (ed.) Sociological Methodology. San Francisco: Jossey-Bass.

HAWLEY, A. H. (1956) The Changing Shape of Metropolitan America. New York: Free Press.

SCHNORE, L. F. (1958) "Components of population in large metropolitan suburbs." Amer. Soc. Rev. 23 (October): 570-573.

——— (1957) "The growth of metropolitan suburbs." Amer. Soc. Rev. 22 (April): 165-173.

Ethnic Minorities: Politics
and the Family in Suburbia

HARLAN HAHN

☐ AMONG THE MOTIVES that provided an impetus for the origin and growth of American suburbs, perhaps none ranked higher in the minds of potential suburbanites than the desire to locate a suitable environment in which to raise their children. The factors that impelled most suburban migrants to turn their backs on the cities, such as population density, social disorganization, and the fear of alien racial or ethnic minorities, seemed to focus on problems associated with rearing children. In fact, the move to the suburbs was justified by many as an effort to find space in which their children could play, acceptable friends and playmates for their sons and daughters, and immunity from exposure to social temptation or turmoil. The consideration of family issues, therefore, probably was more salient to the people who created and settled suburbs than other social, economic, and political themes that have been subjected to greater attention by social scientists.

The examination of issues related to child-rearing appears to be a crucial element in the appraisal of urban affairs. The family is the basic foundation of society, and the domestic milieu in which individuals are raised may have a critical impact upon their subsequent development. How children reach maturity can play a determinative role both in affecting the nature of the social problems that emerge in metropolitan areas and in shaping the communities where people live.

In addition, the assessment of family characteristics seems to be an especially appropriate approach to the study of suburbanization. In many respects, the suburbs appear to reflect a general consensus that the nuclear family model represents an ideal form of family organization. Single-family homes, small lots, and architectural uniformity provide visible testimony that many American suburbs are the exclusive preserve of nuclear families and that any other residential unit—be it single, extended, or communal—is socially undesirable. Those communities were designed to offer a favorable setting in which children could be raised and in which the small family unit allegedly would flourish. Thus, family attributes cannot be ignored in the evaluation of social problems confronting suburban areas.

Perhaps few functions are more basic to the creation and preservation of a viable community than socialization. Although the performance of this task is supported by government efforts in education as well as by corollary programs in housing, recreation, law enforcement, and other policies, the socializing of young people is usually considered a private responsibility. Especially in suburbs, the family structure within which this socialization occurs may be sharply circumscribed. Any departure from the conventional or customary context for raising children can be the subject of public disapproval as well as private censure. A major purpose of this essay, therefore, is to explore the interplay between public and private values designed to perpetuate the dominance of the nuclear family model in American suburbs.

THE DEFINITION OF WHITE
ETHNIC GROUPS

Although the investigation of family characteristics may be especially relevant to the entrance of white ethnic groups into American suburbs, relatively little research is available on this subject. Perhaps a principal reason for this omission is the problem of definition. Ideally, the criterion for determining ethnicity might be based upon identification rather than upon group membership as delineated by birth, language, or other characteristics. A person who identifies himself as an Irishman, thus, could be classified as Irish, regardless of the number of Irish progenitors in his genealogy.

Even more fundamentally, however, affiliation with an ethnic group usually implies identification with cultural institutions that do not comprise an integral part of the predominant society. Ethnic groups might be recognized simply by observing that they do not share in major aspects of the attributes and behavior that are prevalent in the general population. From the perspective of the total society, people who identify with ethnic groups may be labeled as persons who are marginal to that society, individuals who display deviant characteristics, or those who are classified as "outsiders." As a result, a member of an ethnic group could be defined as a person who—by reasons of his ancestry or cultural background—exhibits characteristics or behavior that appear inconsistent with the predominant values and conventions of the society.

This definition has several important implications. Initially, it makes the precise measurement of white ethnic groups in suburbs—or elsewhere—extremely difficult. Although the U.S. Census contains extensive information on the size and location of both the foreign-born and nonwhite or Spanish-speaking groups, there are no available enumerations of persons who fail to identify with the prevailing values and institutions of the indigenous society. In addition, such a definition seems to imply that the "melting pot" is a myth. Many persons—especially those of foreign extraction—have not been successfully absorbed into the mainstream of American life. In fact, their resistance to assimilation may be the principal feature that has made them a salient and significant object of study.

Furthermore, this definition of ethnic groups is closely related to family characteristics. Perhaps in no other sphere of life is the tension between ethnic identification and the indigenous society more apparent than in the nature of family life. Most of the customs and values of ethnic groups are perpetuated within the confines of the family unit. While adherence to prevailing social values might be a necessary prerequisite to social, economic, or political success in the community, the observance of ethnic traditions is often permitted in the private world of the family. Hence, the family may be a primary agent not only for socialization into predominant cultural norms, but also for the preservation of ethnic styles of life. Although many ethnic homes bear a superficial resemblance to the dominant American model, families also comprise a major vehicle for the maintenance of ethnic distinctions. The processes of socialization that occur within families reflect major differences between ethnic and indigenous households.

ETHNIC GROUPS AND
FAMILY CHARACTERISTICS

In addition, the organization of ethnic families may depart from the classic American—especially suburban—model in several critical respects. In comparison with prevailing standards, many ethnic groups are more likely to include large families, they are more apt to maintain extended kinship systems, and they may consist of a disproportionate number of single or unattached males who have come to industrial areas to "seek their fortunes." Each of these features of family structure can be traced to the traditional rural origins of many immigrants. In an agrarian society, large or extended families not only constituted an economic advantage and an important method of caring for the weak and the elderly, but they also comprised a means by which both spouses asserted their observance of traditional sexual roles. Similarly, migration from the farm to the city provided a route by which young single males could obtain the economic resources and the social skills needed to begin a family of their own, either in native communities or in the place of their destination. Although these traditions seemed particularly appropriate to an isolated rural society in less-developed nations, they became significant aspects of a distinctive life style for many ethnic groups transplanted to urban America.

Perhaps the two most critical features of ethnic family organization are the relatively high frequency of large or extended kinship systems and single or unattached males. In either case, many ethnic families do not seem to fit the model of the nuclear suburban family. There is a critical divergence between the family structures maintained by a large number of persons who identify with ethnic groups and the indigenous suburban model. The conflict between those segments of the population is exacerbated by the relatively narrow range of family organizations accepted by such communities.

The tolerable limits of family structure in America—particularly in suburbs—are sharply constricted. American society appears to prescribe both a minimum and a maximum family size for its population. While the minimum consists of two adults, the maximum may be variously defined as four or five—but probably no more than six—children. An unattached individual or a single person with children is at variance with pervasive social norms concerning the minimum size of a family. Parents with a single child even might be regarded as departing from prevailing attitudes regarding desirable family size. Similarly, households containing two adults and a large number of children and dwelling units that include several related or unattached adults defy the maximum standard set for the family. Undoubtedly, the mounting divorce rate and growing efforts to limit population growth may expand the permissible range of family size, but divorced persons who fail to remarry and individuals who persist in conceiving an increased number of children also may be subject to social sanctions.

There are a variety of means by which social censure is conferred upon individuals or families that do not conform to the dominant model of family organization. Humans usually seek to associate with others whom they see as possessing similar attributes, and they tend to shun contact with those whom they regard as deviant or different. Among the attributes that are considered critical in screening potential friends and associates, the family may be of prime importance. In personal interactions, for example, one of the first questions addressed to a stranger frequently is a polite but probing inquiry about his family. Moreover, much social conversation tends to revolve around family problems and subjects associated with the raising of children. Since the family is regarded as fundamental to society, refusal to respect either the maximum or the minimum limits on family size can evoke stronger social condemnation than many other forms of behavior. The stigma attached to such violations may often be imposed privately or implicitly through means such as avoidance, resistance, or the withdrawal of social acceptance. By denying basic human instincts for approval and sociability, such private and frequently unspoken forms of censure might be even more severe than other penalties. The large family with unruly children or with several other adults living in the household may not only encounter criticism for creating noise and disturbance in the neighborhood, but, perhaps more seriously, they could be barred from many forms of social interaction with their neighbors. Similarly, single persons may find themselves excluded from some social activities, and the unmarried parent might become the target of as much gossip and disapproval in the modern middle-class suburb as in the eighteenth-century puritanical village.

Although social interactions are usually considered private and unrelated to local issues, they may provide the basis for informal groupings and networks of broader import to the community. Personal contacts based upon perceived similarities or differences can form the foundation for intense and persistent community cleavages that are reflected in public or political controversies, as well as in private conversations. Both the relatively narrow range of acceptable family structures imposed by the suburban milieu and the salience of family characteristics to most suburbanites, therefore, tend to promote a sharp division between many ethnic groups and indigenous settlers who subscribe to the nuclear family model.

Perhaps one of the primary effects of this conflict is to enhance the tension between ethnic groups and other suburban residents. Although this clash is often ascribed almost exclusively to socioeconomic origins, it may also have its source in family and cultural characteristics. For the entire community, departure from the norms governing family size and organization can be even more salient than alien cultural traditions perpetuated within family units. While many of the latter traditions may not interfere with existing social routines, variations in family structure seem to have an impact upon the organized life of the community. As a result, discrepancies between the type of family organization adopted by many ethnic groups and the suburban model have important political, as well as social and economic, connotations.

PUBLIC SUPPORT OF THE

NUCLEAR FAMILY

Although there has been relatively little discussion of this phenomenon, most government policies in suburban as well as other communities are designed to primarily serve the nuclear family. Perhaps the principal means by which public institutions seek to inculcate the cultural values that dominate American society is through the educational system. A convincing case, in fact, could be made that one of the major functions of public schools is to socialize children concerning the conventions that regulate behavior in society and perhaps thereby to counteract or mitigate the alien traditions nurtured within ethnic family units.

Perhaps even more important, however, are the policies that contribute—directly or indirectly—to the perpetuations of socially prescribed models of family size and structure. In many respects, this pattern may not be unexpected. Most consumers of government programs are members of nuclear families; political processes are designed to be responsive to the will of the public. Since the family forms the basic structure of society, it may be natural that other social and political institutions are organized in conformity with this structure rather than at variance with it. Although this tendency may fail to meet the needs of critical segments of the population—including many ethnic groups—that do not subscribe to the strict nuclear family model, an examination of public policies in housing, transportation, recreation, law enforcement, and other areas illustrates the extent to which government programs generally conform to the contours of this model of family organization.

SUBURBANIZATION AND SUBSIDIZATION

Perhaps the clearest example of the way in which public agencies have supported the nuclear family unit can be found in housing policies which were responsible for the birth and development of suburbs. Not only was the quest for a better environment in which to raise children a major impetus for the suburban trek, but it was also officially sanctioned by the federal government. In fact, a major legislative rationale for many programs enacted prior to the growth of suburbanization was the desire to provide an improved environment for the maintenance and protection of the family as a basic social institution. Largely through FHA and VA programs, which facilitated the purchase of homes by providing low-interest loans, millions of dollars were spent to support the process of suburbanization. Moreover, suburban developers, cognizant of the opportunities created by those federal funds and anxious to cater to the whims of potential homebuyers, responded by erecting vast numbers of houses planned exclusively for the nuclear family.

In the process of suburban growth, many persons found that they were excluded from the suburbs. While the federal government had been instrumental in fostering the suburban enterprise from its inception, it apparently did not intend that the gates of suburbia would be thrown open to the general population. People who were single, those with large families, and ethnic or racial minorities frequently found it difficult to acquire loans on favorable terms for the purchase of homes. Moreover, even a successful loan applicant from these groups might confront almost insurmountable obstacles in locating a suitable structure in the suburbs. Unlike public housing projects, which were restricted almost exclusively to central cities and which consisted of multifamily structures, the overwhelming number of houses built in the suburbs were single-family dwelling units. Families with a large number of children or with relatives or unattached adults living in the home found that the three- and four-bedroom dwelling units constructed by tract developers did not meet their requirements. Suburban houses were built for the needs of the "average" nuclear family. For many large or extended families, the houses in the suburbs were simply too small, and for the single individual they were usually too large. Hence, the suburbs became an almost exclusive preserve for groups that could satisfy three essential requirements: color, affluence, and family size. The emergence of the suburb as community populated by prosperous, Anglo, middle-class nuclear families was a phenomenon promoted and subsidized by the federal government.

Those programs were augmented by other government policies that enhanced the growth of suburbs as middle-class enclaves. The allocation of massive federal resources by the government to highway construction, for example, facilitated the migration of the middle-class white-collar worker, who could afford the costs of commuting; the escape route from the problems of the core city was paved by federal funds. In addition, the delegation of many critical public functions to local governments enabled homogeneous communities to promote their own values and life styles in other fields of public policy. Although housing was the principal means by which government subsidized the white middle-class model of the suburb, official support of this type of suburbanization extended through a wide range of government activities.

A HIATUS IN
GOVERNMENT SERVICES

Another indication of the means by which public institutions sustain the nuclear family can be observed by noting that government policies seem to have a major impact upon average citizens only during certain initial periods of their life cycle. From the time of birth until a person leaves his parental home, many children and young people are supported by public programs in housing and in the provision of services such as education. During those critical years, many aspects of a person's existence are nurtured by government actions designed not only to enhance his socialization, but also to bolster the family unit of which he is a member. Subsequently, many individuals receive similar government benefits as they purchase homes and form new family units of their own.

There is, however, usually an important gap between the time that an individual leaves his family and reconstitutes a family of his own. After a person has left his parental home and completed his education, there are relatively few government programs of which he is a direct or indirect recipient. This period of time, which extends chronologically from the late teens to the twenties, seems to represent a crucial hiatus in the provision of government services.

Since most citizens are affiliated with a family unit, releatively little attention is devoted to the needs and aspirations of persons who lack that attachment. Yet, this is precisely the age category particularly susceptible to crime and other manifestations of social deviance. Most violations of the law are committed by young people, especially males, in their late teens or twenties. This also seems to correspond with a period in the life cycle of most individuals characterized by personal disorganization and psychological malaise. While the definite causes of this phenomenon undoubtedly are numerous and difficult to unravel, they may be related to psychological and environmental factors as well as to social and economic circumstances. (Parenthetically, it also seems ironic that government policies designed to support the nuclear family are not matched by equivalent efforts to create families. If the nuclear family is critical to the maintenance of social stability, a logical extension of this argument might be to put government in the dating, mating business.) Nonetheless, family considerations and related issues seem to form an integral feature of many social problems. Perhaps the withdrawal of a person from the comfortable environment of his parental family and the absence of the responsibilities imposed by a new family unit might contribute to a sense of personal estrangement that is undesirable not only for the individual but also for the society as a whole. Among the factors that promote the growth of social problems, therefore, the removal of family support and the lack of public policies to meet the needs of persons in those circumstances might deserve serious attention.

The absence of government programs to serve young people who do not have stable family attachments may have a particular impact upon single migrants seeking work in metropolitan areas. The provision of services for this segment of the community usually has been consigned to private rather than public institutions. The emergence of many of those services not only seems to be a blight upon the community, but the services may fail to meet the needs of those persons adequately. Inexpensive hotel rooms, for example, might offer a temporary solution to the housing needs of transient workers, but they can hardly provide the supportive environment of a family unit or an ethnic community. Similarly, bars and taverns might serve a social purpose as a meeting place for lonely migrants; but they may also strengthen the familiar effects of economic deprivation, differential association, and crime. The social and recreational facilities available in suburbs do not seem adapted to the needs of single migrants. In a community dominated by nuclear families, an individual hoping to achieve the financial resources to start his own family or to bring his family from a foreign country may appear to be a deviant and an "outsider." Although the tension between nuclear and extended or large families may provoke a major source of political conflict, perhaps the primary source of concern in many communities will be focused on the growing population of single migrants. As increasing numbers of unattached males migrate to outlying areas in search of employment, they may promote controversies which local governments will find it necessary to heed. Since those groups might appear to be primarily responsible for increases in crime and other forms of social deviance, they could become a major target of public hostility and antagonism.

THE CITIFICATION OF SUBURBS

There are a number of trends which indicate that a new and diverse breed of urban migrants is seeking shelter in suburban communities. The growth of large population concentrations on the fringes of the central city has promoted both an expansion of service industries catering to the suburban consumer and a demand for people to provide those services. In addition, the relocation of industrial plants in suburbs has encouraged many manual as well as white-collar workers to seek residences near their places of employment. As a result, entry has been provided for groups which previously lacked access to the suburbs. Suburbs are experiencing a "second wave" of migration radically different from the movement that gave them birth.

In general, major metropolitan areas in the United States seem to be undergoing a citification of suburbs as well as a suburbanization of cities. Perhaps the former phenomenon is an unavoidable corollary of the latter. As an expanding share of the population has achieved middle-class status, they have sought many of the same amenities that inspired the first wave of migration to the suburbs. As a result, population growth in suburban areas has promoted the development of satellite core cities. The original characterization of metropolitan regions as a relatively uncomplicated concentric circle consisting of a central city surrounded by suburban rings has been modified by the emergence of suburban core cities that have become secondary centers of commercial and industrial activity. The emigration of new and culturally heterogeneous groups to suburban areas, therefore, has introduced massive social problems not unlike those that plagued the original central city.

However, there may be two alternative consequences of this process. In some metropolitan regions, Anglo, middle-class enclaves may succeed in preserving their homogeneity through careful planning, zoning practices, and other political techniques, thereby forcing other suburbs to accept the new wave of migrants flocking to the residential fringe surrounding central cities. In those areas, there may be a "specialization" of suburbs. While some communities maintain their predominantly Anglo, middle-class character, others might become primarily ethnic and industrial.

In other localities, however, the trend toward urbanization—especially in large satellite core cities—may be irreversible. The downtown areas of many suburban cities might begin to resemble the decaying core of the urban area. Whereas the heart of the central city previously was the magnet drawing migrants who sought marginal or unskilled work, this attraction could shift to large suburban communities. As a result, many suburbs not only may encounter an influx of cut-rate hotels, poolrooms, and bars to serve the new migrants; they might also confront the emergence of severe over-crowding as large or extended families squeeze themselves into existing suburban houses that they can afford.

The influx of new migrants to suburban communities may introduce a host of trying and unfamiliar social problems to those localities. But it also seems to present suburban communities with some interesting and important challenges in formulating public policies to solve those problems. Although government programs traditionally have devoted almost exclusive attention to the role of socioeconomic factors in the amelioration of social ills, the migration of many heterogeneous groups to suburban areas might tend to emphasize and underscore the corollary importance of family and cultural characteristics in the resolution of community issues. Since the suburbs were founded primarily as a haven for the raising of children and for the preservation of the nuclear family, the movement of groups with different life styles and family charac-teristics to those areas seems to arouse new issues and new forms of conflict which the political process must handle. The ability to cope with those questions, however, also appears to require a major reevaluation of the role of government and private institutions in the community.

THE DELINEATION OF PUBLIC AND
PRIVATE RESPONSIBILITIES

Traditionally, the distinction between functions that are regarded as a legitimate responsibility of government and duties that are viewed as appropriate to the private sector of society has not been clearly defined. In fact, a complete enumeration either of public activities that affect private institutions or of private acts that impinge upon governments probably would be impossible. None-theless, there seem to be generally recognized boundaries between public and private obligations that are occasionally obscured. Although government programs in housing and other areas may contribute to the support of the nuclear family, for example, family problems usually have been considered solely a private concern. In general, social and economic factors usually have been included within the purview of public agencies, but cultural and psychological concerns have been considered private matters.

One method of assessing the role that government plays in society may be provided by examining those government agencies or programs that have a direct and immediate impact upon the everyday lives of most individuals. Although such a list is destined to be incomplete, a partial count of public representatives that the average citizen encounters or observes in his daily life might include postmen, law enforcement or traffic officers, and, for persons in younger age categories, school teachers. In addition, some other government programs such as public streets or transportation and the monetary system may impinge upon the everyday life of most people. While there are numerous other government activities that have a direct effect upon certain segments of the population as well as actions that have an indirect impact on the daily lives of most persons, such functions appear to be a principal means by which people become conscious of government in the course of their everyday lives. By promoting transportation, commerce, and a stable social order, and by providing a means for the socialization of young people, these government programs seem designed to facilitate mutual accommodation and interaction among different citizens rather than to promote life in isolation from society.

Perhaps the principal feature of government actions that impinge on the daily lives of most people, however, is that they are limited both in their number and in their significance. Ironically, none of those activities appears to be entirely relevant to the basic needs of people for the maintenance of life. Regardless of how the requirements for personal survival are defined, government programs that influence everyday life seem unrelated to such fundamental human needs as food, clothing, shelter, sleep, or sex. Although the distribution of costs and benefits through some government programs may influence the quality of life for certain segments of the population—and other programs may have a similar indirect impact upon larger groups of people—the public activities that have the most immediate and direct effect upon daily life seem essentially unrelated to the basic needs of man.

The absence of public programs to meet the essential requirements of human survival may have an especially critical impact upon groups that display unusual or atypical characteristics. In a modern urban society, for example, a large family is often an economic burden rather than an advantage; and the effort to feed, clothe, and house many children or other persons may emerge as a consuming purpose in life. Similarly, for many individual migrants, the problem of locating a nourishing meal and a place to sleep might become an aim that overshadows all other ambitions. In either event, the attainment of those objectives can be impeded by a lack of familiarity with the customs and life styles of the indigenous society. Many basic human needs have been consigned to the private marketplace rather than to public agencies, and they have remained unfulfilled. As a result, many members of ethnic groups—as well as other persons—have been forced to concentrate upon the satisfaction of those needs to the neglect of the goals that might contribute to a more meaningful existence.

Yet the assignment of the obligation of meeting basic human needs to the public rather than to the private sphere of society also may comprise an imperfect mechanism for the satisfaction of those needs. Perhaps the fundamental difficulty is that both government programs and private institutions are based upon majoritarian principles. In the public sector, the adoption of government policies requires the approval of more than fifty percent of the people responsible for making decisions. As a result, a plurality of the population that has succeeded in fulfilling the criteria for personal survival in the private sector of society may be unwilling to recognize the needs of others who have not gained this success. Just as public policies have been designed to sustain rather than to undermine the nuclear family model, they have usually favored the interests of persons whose aspirations surpass the basic requirements of survival rather than those who have not yet achieved that threshold. To the extent that government programs are an accurate reflection of popular desires, therefore, the lack of public policies affecting some of the most basic goals of man may be indicative of a conviction among many people that private sources are a more appropriate locus for the satisfaction of those needs than is government.

Similarly, the private sector of society is founded upon an analogy to the principle of majority rule. In economics, however, the basic unit of influence is dollars rather than people. Unlike the public sector, where voting strength is allocated equally to all people, power in the private marketplace is distributed in proportion to the amount of resources that one has available to invest in his choice among competing products. Hence, persons possessing both extensive financial resources and relatively exotic tastes may dominate the economy at the expense of groups with little money and few basic needs. Whereas a small portion of the private market may exist to serve inherent needs for such items as simple food and clothing, larger organizations have emerged for the purpose of catering to more sophisticated preferences. The implications of this trend may become increasingly critical as America achieves the status of a middle-class society. The concentration of economic power among persons whose desires extend beyond basic survival needs could inhibit the capacity of the private market to satisfy those segments of the population with more fundamental demands. Undoubtedly, enterprises designed to serve the minimal needs of people will continue to exist, but they may be overshadowed by industries catering to the more esoteric tastes of the wealthy.

PUBLIC POLICY AND

SOCIAL PROBLEMS

In both the public and the private sector of society, institutions are organized primarily to satisfy the wishes of the largest share of the market, whether it consists of persons with a large amount of political influence or people with a bulk of the financial resources. Relatively little attention is devoted to individuals who are unable to fulfill the basic needs of human survival. Yet the satisfaction of those needs may constitute one of the most important values of both the public and the private sector. Clearly, private enterprise cannot be considered either humanitarian or efficient if it does not satisfy the needs of the hungry or of those who are inadequately protected from the elements. Similarly, government policies may not be considered successful if they neglect the needs of deprived portions of the population.

The adoption of the goal of meeting basic survival needs may require a major redefinition of the private and the public sectors of society. In the private sphere, it seems unlikely that changes could be made to ensure that the fundamental requirements of people with few monetary resources are satisfied. In the public sector, however, the goal of governmental policy is less the fulfillment of individual desires than the satisfaction of the collective interests of the whole population. Even though governments may be designed primarily to provide "the greatest good for the greatest number," the amelioration of the problems encountered by small segments of the population who produce large amounts of social disorganization might contribute more to the general health of the society than the pacification of the will of the majority.

Perhaps one method of achieving important social objectives might entail the adoption of a new type of cost-benefit analysis for government policy. Instead of seeking to satisfy the demands of the majority—or of those with a large amount of political power—perhaps the primary objective of government should be the alleviation of the needs of people who account for the largest proportion of "social problems" in the society. By adopting this purpose, the public sector might be able not only to ensure that all persons are provided with the minimum requirements for survival but also to reduce the costs imposed upon other segments of the population through crime and other forms of social disorder. Segments of the population encountering extensive "social problems" would seem to constitute a more logical source of concern for policy makers than other groups that comprise larger numbers of people but fewer problems for the society as a whole. Although this approach may defy the majoritarian premises implicit in the formulation of policy, the long-range effects of such an orientation could produce substantial benefits for the entire society.

SOCIOECONOMIC AND CULTURAL CHARACTERISTICS

The implementation of this change in the development of public policy, however, might extend beyond the traditional social and economic concerns of behavioral scientists and policy makers. The fulfillment of personal aspirations for people entrapped by endemic social problems may include more than the acquisition of an adequate education, a decent job, and a reasonable income. Individual aspirations also may encompass a sense of psychological well-being and a supportive environment in which to confront the manifold problems of life. Although the latter objectives have traditionally been defined as lying outside the legitimate purview of government, they may even be more critical than the concerns that have traditionally preoccupied policy makers.

Studies of population characteristics and social problems in urban and suburban areas usually focus on socioeconomic variables. While other variables clearly have not been excluded from consideration, the principal thrust in numerous investigations is devoted to the explanatory power of these attributes. In the formulation of public policy, presumably there is at least some relation to the analyses performed by social scientists, and the same assumptions and orientations tend to permeate many governmental programs. As a result, political decisions frequently may be based upon economic and sociological assumptions to the neglect of psychological and anthropological considerations. Although few would deny the importance—perhaps even the primacy—of socioeconomic factors in the creation or the resolution of social problems, the psychological impact of instability, disorganization, and disorientation in family structures seems worthy of equal—or at least of increased—attention.

Perhaps more importantly, the approach adopted by most social scientists represents an intellectual perspective containing the seeds of an enduring conflict that could emerge with renewed vigor. The emphasis on socioeconomic attributes seems to detract from a stress on cultural characteristics in the investigation of social problems and trends. Without denying the utility or the efficacy of socioeconomic variables, they may not provide a total explanation for many areas of human behavior. A large number of individual actions perhaps can only be explained by examination of cultural or life-style features rather than by the education he attains, the income he acquires, or the occupation in which he is employed.

Socioeconomic attributes may emerge as a major focus not only because of their utility in the construction of expansive categories of people but also because of the implicit concern of many social scientists with the formulation of social policy. Efforts to improve the social status or the economic conditions of deprived segments of the population clearly seem more attainable through political action than attempts to change the state of mind or the life styles of those groups. Perhaps the emphasis on socioeconomic considerations both in research and in public policy, therefore, reflects an implicit acknowledgment of the constraints imposed by existing social and political institutions and arrangements. Programs to raise individual incomes, to provide expanded occupational opportunities, and to offer educational advantages, enjoy a higher probability of success than plans to improve the mental health of a society or to offer a more meaningful life for individuals. This does not necessarily imply, however, that socioeconomic factors are of greater importance to the persons affected by those programs. In fact, personal happiness and a satisfying life may be of greater significance than income, education, or occupation to many people. The emphasis upon the latter considerations may simply reflect the fact that governments traditionally have been better equipped to handle socioeconomic subjects than psychological matters.

TOWARD A NEW SUBURBAN SOCIETY

Although the introduction of new and unfamiliar elements into the suburbs may confront those communities with a bewildering array of problems, this trend also seems to provide metropolitan areas with a critical second chance to rescue urban society from the danger of total collapse or decay. In many respects, public efforts to improve the conditions of life in central cities appear to have failed. Government programs to raise the socioeconomic level of urban residents not only have been inadequate, but they also seem to ignore other crucial aspects of everyday existence. As a result, many of the largest cities in the country appear to have reached a stage of development in which they are inhabited primarily by a confused and disoriented population that is incapable of sustaining a viable sense of community.

Perhaps the principal task—and the major challenge—confronting suburbs is the accommodation of increasing heterogeneity. For many years, the image of the tranquil suburbs could be maintained because there were few alien or distinctive elements of the population that threatened to disturb the orderly life of the community. Moreover, many suburbs sought to promote this image through techniques such as housing standards, zoning regulations, and other methods designed to prevent the entrance of disruptive groups into the community. As suburban areas undergo the same processes of population growth and industrialization that previously affected central cities, however, they may find it increasingly difficult to bar their doors to outsiders. Perhaps more importantly, the movement of new and divergent groups to the community could add a variety and a richness that have often been absent in suburban localities. Rather than a liability, the growing heterogeneity of suburban populations might be an asset that could contribute significantly to the diversity of community life.

The changing character of American suburbs may necessitate an increased acceptance of divergent life styles and customs. As mounting numbers of ethnic groups move to the suburbs, it seems unlikely that they will be easily assimilated into the prevalent values of the community. In most parts of America, the "melting pot" seems to have produced a parfait rather than a blend; and dominant segments of the population frequently find that they must adjust to distinctive ethnic traditions rather than forcing those groups to adopt the standards of the indigenous society. This tendency could have a particularly strong impact upon family characteristics. Perhaps the major priority in suburbs is the need for an increased tolerance of individuals such as single migrants and persons in large or extended families who defy the usual conventions of the nuclear family. Although the informal pressures exerted upon ethnic suburbanites ultimately might induce increased compliance with the social norms that regulate family size, the first major wave of white ethnic groups that migrate to the suburbs may include a disproportionate number of persons with unusual or atypical family structures. The accommodation of those groups would seem to require some critical changes in community attitudes.

Even more importantly, however, the movement of heterogeneous groups to the suburbs might involve important alterations in prevailing institutional arrangements. Perhaps an appropriate basis for this effort would be a major reappraisal and revision of the traditional distinction between public and private spheres of responsibility. Although family matters are often defined as an intrinsically private concern, many government programs serve the purpose of sustaining and promoting the nuclear family model in American society. Similarly, while major attempts are made to meet socioeconomic needs through public policies, the satisfaction of the basic requirements of personal survival is usually relegated to the private sector of the economy. One method of resolving this apparent inconsistency could entail a candid admission that government as well as private institutions have a crucial responsibility to meet fundamental human needs. The obligation to fulfill such critical requirements as food, clothing, and shelter may not be an end or an ultimate objective, but it could be a means of achieving even more significant goals, such as the enrichment of life and the maintenance of the family unit as the foundation of society.

The ability of political leaders to meet the challenges that confront them, therefore, may require a major reorientation of government plans and priorities. The implications of this redefinition of government responsibilities might be illustrated by consideration of another major area of public policy. In addition to the programs enumerated earlier, one of the most direct means by which government attempts to sustain families is through welfare activities such as aid to dependent children, assistance to persons classified as disabled, and support for the indigent. The acceptance by public institutions of the responsibility to meet the fundamental needs of personal survival might involve the subsidy not only of groups that conform to those definitions but also of any person who is unable to fulfill the minimal requirements of existence due to social, economic, or cultural circumstances. Perhaps more importantly, however, efforts to improve the quality of life might encompass government actions that stress techniques such as psychological support and emergency aid as well as the provision of financial assistance. Many of the most significant obstacles that people encounter in their lives are not economic problems, and they may require more than monetary resources for their solution. Although it may be difficult to demonstrate the effectiveness of noneconomic assistance, the importance of this aid might justify an increased emphasis on this approach.

While many programs may require the participation of state and national governments, there are also numerous actions that communities might initiate to solve local problems. In particular, the increased admission to the suburbs of ethnic groups—including both large or extended families and single migrants—could impose more strains upon some community resources than upon others. Since the suburbs were designed primarily for raising children, the impact of children from large families upon the educational system and other government services might be minimal. But the increased entrance both of elderly persons in extended families and of single young people in their late teens and twenties may pose a severe challenge to those localities. In the field of housing, for example, there may be a growing need both for dwelling units with numerous rooms to accommodate large or extended families and for small, inexpensive apartments to house single migrants. The provision of this housing could entail not only the encouragement of private developers but also the revision of municipal planning and zoning regulations. Similarly, suburban areas might find it necessary to develop new recreational programs both for the aged and for young migrants. The range of those suggestions might be indicated by the recommendation that pool tables as well as playgrounds be installed in city parks. In addition, however, some method might be found to fill the critical gap in government services that exists between the time that a person leaves his parental home and forms a family of his own. Since persons in this category account for a disproportionate amount of crime, delinquency, and other social problems, this effort could produce substantial benefits for the community. Perhaps one approach to this objective might entail a reduced stress on the punitive aspects of law enforcement and a correspondingly increased emphasis on the provision of community services.

Suburban areas not only have an opportunity to profit from the earlier mistakes of central cities, but they also may be able to renew and expand some of the values that were responsible for their growth. Among those values might be included the desire for grass-roots participation, community control, and the creation of a favorable setting for families. The extension of those features to ethnic groups reflecting different cultural traditions and family structures could be advantageous both to those groups and to the remainder of the community. To the extent that suburbs have succeeded, they have achieved that goal by fostering a genuine sense of community among their residents. By drawing upon the values that gave them birth and by extending those values to new and diverse groups, the suburbs could acquire the resources and the will necessary to cope with the problems that may increasingly confront them.

Part III

GOVERNMENT AND POLITICS

GOVERNMENT AND

POLITICS

Governments and the political process determine much of what happens in suburbia. Unlike the central city, suburbia consists of large numbers of governmental units which are poorly coordinated. The structures of these individual governments, their relationships with each other and the central city, and the impact of suburbia on state capitals and Washington are all undergoing rapid change. These changes have significant implications for the coordinated growth and development of the metropolis, the quality of life in individual suburbs, and the nature of state and national suburban policy. The three contributions to this section address these issues.

In the first essay, Scott traces the trends in the development of suburban governmental structures, including increased bureaucratic professionalism and the growth of special district authorities which infringe on municipal prerogatives. He also examines the process by which change and development occur in suburban governmental structure and identifies the four most important factors: state constitutional/statutory provisions, regional and local custom, community idiosyncracies, and local socioeconomic characteristics.

Friesema reexamines the contemporary utility of three well-known models of metropolitan politics (Holden's "diplomatic system," the "market model" of Ostrom et al., and Williams' "life-style/social access" hypothesis) in light of recent developments such as the emergence of metropolitan councils and regional planning authorities, increased state and federal intervention, the organization of metropolitanwide interest groups, and a series of legal challenges to established jurisdictional prerogatives.

Using 1968 presidential election data, Zikmund offers a compelling case for the absence of a national suburban political pattern. His analysis suggests two distinct levels of suburban differences: (1) within any given metropolitan area, there are significant differences among suburbs in the support given to particular presidential candidates and (2) the suburbs of any particular SMSA, despite individual differences, tend to respond electorally in a patterned way which is significantly different from the suburbs in other SMSAs.

8

Suburban Governmental Structures

THOMAS M. SCOTT

□ ONE MUST CONFRONT AT LEAST TWO PROBLEMS when one sets out to comment meaningfully on suburban governmental structure. First, there is remarkably little research on which to draw. This dearth of a strong data and research base is due to several factors, including the fact that there are more than 20,000 units of local government in U.S. Standard Metropolitan Statistical Areas, most of which are in some sense suburban. In addition, the concept "suburban" has not been defined with the kind of precision and clarity that facilitates the collection of data, so that aggregate analyses of distinctly suburban phenomena have not been encouraged.

The second problem comes from the fact that local governmental structures, whether suburban or nonsuburban, are defined in the constitutions and statutes of each of the fifty states, and the variation in such laws and provisions from state to state is considerable. To date, no one has developed a general description and taxonomy of local governmental structures in all the states. Without such a tool, general, systematic, and comparative analysis of local governmental structures is impossible. As a consequence, most scholars and commentators on local governmental structures deal in very general and broad categories, such as those found in standard government texts, or focus specifically on the structures in the limited number of cases they happen to know well. Our approach in this report will necessarily include both these strategies and will, consequently, rely heavily on the experiences in a few states—particularly Minnesota—as well as on the more general materials on local government.

At the outset, there are several definitional matters that should be clarified. First, the term "suburb" has many meanings, but for our purposes we shall use it simply to refer to the collectivity of local governments within Standard Metropolitan Statistical Areas in the United States, with the exception of governments in the central cities of those SMSAs. While much of the writing and comment on suburbs is limited to municipal government, a full analysis should include counties, towns and townships, school districts, and other special-purpose government, as well.

As Table 1 indicates, American metropolitan areas contain a large number of suburban governmental units. According to the 1967 Census of Governments, U.S. SMSAs included more than 20,000 units of local government, all of which (except for the 225 central cities) can be classified as suburban by our definition. Table 1 also compares the numbers of units and the numbers of elected officials inside and outside metropolitan areas.

It is interesting to note that there are considerably fewer units of local government in SMSAs than there are outside, despite the fact that the majority of the U.S. population resides within SMSAs.

TABLE 1
COMPARISON OF METROPOLITAN AND NONMETROPOLITAN UNITS OF LOCAL GOVERNMENT AND NUMBERS OF LOCAL ELECTED OFFICIALS; 1967

	Units in SMSAs		Units Outside SMSAs	
	Units	Elected Officials	Units	Elected Officials
Local Units	n	n	n	n
Counties	404	12,476	2,645	61,723
Municipalities	4,977	45,142	13,071	98,785
Townships	3,255	30,480	13,850	99,123
School districts	5,018	28,895	16,764	78,678
Other special districts	7,049	17,668	14,125	32,275
Total	20,703	134,661	60,455	370,584

SOURCE: U.S. Bureau of the Census, 1967.

Similarly, the data in Table 1 indicate that there are three times as many elected local officials outside SMSAs than there are in SMSAs.

If one examines the number of local units in SMSAs of various sizes, one discovers that the number of local governmental units per capita increases as the total population size of the SMSA decreases —i.e., the largest SMSAs have fewer numbers of local governmental units per capita than do the smaller SMSAs. While we do not know enough to comment extensively on the dynamics of the process which has created this result, it suggests that the oft-cited problem of local governmental proliferation and fragmentation may be greater in non-SMSAs and in smaller SMSAs than it is in the larger urban complexes, although, of course, in actual numbers there are more units in the larger SMSAs.

The second definitional problem concerns the term "structures of government," which is also loosely used to include a wide range of phenomena. For our purposes, however, we shall refer to two aspects of the "structures of suburban local government" concept. First are intra- or internal suburban governmental structures, and second are intersuburban forms and structures that exist within particular metropolitan regions, which can and do affect the nature of governance both in the metropolitan area at large and *within* particular suburban units.

"Internal local governmental structures" may refer to the basic forms of government, usually some variation of either the council-manager or mayor-council system. It may refer to the electoral system, including partisan or nonpartisan elections, district or at-large representation systems, staggered terms of office, the size of legislative bodies, long or short ballots, and so on. It may refer to the extent of professionalization and specialization in local administration, including the degree to which merit considerations are a requirement for government employment.

Although there is no general descriptive analysis of the total range of internal local governmental structures in the United States, there are two facts about which we can be certain. First, there is considerable variation in the structures of local government from place to place. Indeed, one of the remarkable features of American local government is the extent to which virtually no community looks exactly like any other in terms of its forms and structures. Even in situations where state law and practice is very specific and highly prescriptive, considerable local variation occurs. Thus, systematic analysis of local governmental structures is very difficult, and generalizations can at best only be tentative.

The second fact is that, for the most part, state laws do not distinguish between suburban and nonsuburban local governments, with the result that suburban local structures do not differ markedly from nonsuburban local structures, although the distribution of various forms and practices may vary. There is some evidence, however, that this situation is changing and that special forms and structures are now being made available to urban and suburban local units. We will consider this development more fully in a subsequent section of this essay.

"Intersuburban governmental structures" refers to the variety of formal and informal governmental arrangements created to achieve some measure of coordination, integration, regionalization, and so on of governmental activities and services in metropolitan areas. Normally, these arrangments are of greater scale than traditional local units of government, and their structures and powers are often established on an ad hoc basis. Their existence is by no means limited to metropolitan areas, but they have been more predominant in the urban context, particularly in some of their more recent manifestations—e.g., Councils of Government (COGs), areawide planning agencies, and the like. Intersuburban structures also includes the more traditional, regionally based, special district governments, most of which are limited to a single function and most of which cut across and overlap jurisdictions with existing traditional units of local government. Special districts are perhaps as common in nonurban regions as they are in metropolitan areas, but their use in the urban and suburban context continues to grow, and they represent a very significant dimension for the structuring of suburban local government.

In this essay, we shall focus on three major topics. First we will describe the significant recent *trends* in the development of suburban governmental structures both internal and intersuburban and examine some of the more obvious implications of these trends. Second, we will discuss the *process* by which change and development in suburban local governmental structures occurs. Third, we shall review some of the reasons *why* local structural change is an important part of suburban political processes and suggest some implications for the future.

DEVELOPMENTS IN
SUBURBAN GOVERNMENTAL STRUCTURES:
INTRALOCAL

Because aggregate data on suburbs per se are not readily available, one can only estimate the kinds of developments that are occurring in governmental structures on the basis of limited observation. On this basis, three intralocal structural developments seem significant. The first is the increased use of some variant of the council-manager form of local government whether at the municipal or county level. The studies of Kessel (1962) and Scott (1968) reported this phenomenon in a few states earlier in the 1960s, and the update of some of these data indicates that the trend toward the manager form continues in the suburbs of Illinois, Ohio, and California (although at different rates) and in Minnesota, where direct observation has been possible.

Normally, the manager plan is adopted at the point in its development when the community feels that its problems and resources are sufficiently large to warrant the benefits and costs of this governmental structure. For wealthy communities, this may occur when they are relatively small (less than 5,000 population), whereas in less-affluent situations the population required to support the managerial form will necessarily be larger.

The second major development is related to the first and involves a general increase in the degree of professionalization of suburban local government. This professionalization is due to several factors, most important of which, probably, is the greater levels of affluence in suburban areas, which permits a relatively greater array of functions and a relatively greater capacity to support those functions through higher personnel costs.

The trend toward professionalization in suburban governmental employment is complicated, however, by two other factors: the role of party patronage and the role of public employees' organizations and unions. Patronage employment has generally been less common in suburban areas than in the older central cities, partly because parties often play a less-direct role in suburban local politics and partly because many suburbs have adopted some form of merit employment. Governmental employees' unions are also less predominant in suburbs than in central cities according to recent surveys published by the International City Management Association (although more common than in rural communities) and are least predominant in communities with the city-manager form of government (Stieber, 1969). These data indicate that suburban city-manager communities have probably come closer than other types of communities to the personnel practices utilized in large parts of the private sector, although it is probably fair to say that unionization is on the increase. In the meantime, however, city managers maintain considerable discretion in personnel practices, and, for the most part, this has resulted in the utilization of professional criteria in hiring and firing. Likewise, the same observations hold true for the educational systems in suburbia, which in most instances are governmentally distinct from municipal, township, and county government.

Increased professionalization in local government has several implications. For one thing, it is often associated with increased specialization, and, thus, it tends to encourage the development of new and often specialized functions in the community. For more-affluent communities, this may not create problems, but where financial circumstances are more difficult it can generate frustration on the part of well-trained but highly specialized personnel who are forced to undertake a broader range of responsibilities. This frustration can, in turn, produce high rates of turnover as local personnel eventually find positions in larger or more-affluent communities, where their assignments are more appropriate to their training and inclinations.

Also, professionals tend to associate with other professionals of similar backgrounds and expertise. Indeed, they may find such interprofessional associations more rewarding and interesting than the relationships connected with their responsibilities within their own communities. To the extent that such external relationships reduce the professional's commitment to and vital interest in the local community, its citizens, and its elected officials, such tendencies may result in real tensions between the professional and his constituents.

In short, increased professionalization in suburban governmental personnel can provide real benefits to the local community in the form of more efficient and more competent administration and management. But it may come at a cost, in addition to the direct financial cost of employing such people. It may encourage over-specialization, it may result in relatively high rates of turnover, and it may foster a gulf between the citizen and the community's management staff.

The third major intrasuburban development involves a reduction of local responsibilities at the municipal and county level resulting from the creation of special distinct governmental units, most of which exist to perform a single function. This process has been under way at various rates in various locales for many years, and it occurs in rural as well as urban settings. But it has clearly been an important ingredient in the shaping of suburban structures and functions.

There are, of course, many reasons for the popularity of the special, single-purpose governmental district, especially in metropolitan areas. For one thing, such governmental units are often exempt from state legal restrictions on taxing and borrowing and thus provide the only way to manage functions requiring large capital outlays. Second, they are often the most politically attractive solution to particular governmental problems in that they do minimum violence to existing government structures and political arrangements. Certainly, this has been the case in the Twin Cities area, where special districts have been utilized for airport planning and construction, for sewage collection and treatment, for public transit, and for health planning.

Special districts have been, by and large, the most successful means by which some measure of metropolitanwide-governmental coordination occurs. For this reason, their development of traditional local governments in metropolitan areas. Thus, their utilization has probably contributed as much as any other factor to the assurance of the continued maintenance of existing suburban local units.

For some, this seems most unfortunate because it postpones the day when, from their standpoint, governmental fragmentation and atomization in metropolitan areas inevitably must be eliminated. For others, special districts provide a means by which governmental scale is increased *only* in those instances where it is absolutely required, thus assuring the continued local control of those functions most appropriately served at that level. The argument between these two positions will doubtlessly continue in one form or another, as will the increased use of special districts.

The reason for introducing the special district issue at this point, however, is that it raises two more basic, intralocal structural questions. First, special districts create particular problems for the citizen and his relationship to local government. Special districts are often governed by obscure commissions or boards of trustees, usually selected in equally obscure fashion—perhaps by areawide election, but more commonly by some mysterious and ad hoc appointive process. Usually they are "citizen legislatures," which means that they meet infrequently and leave much of the policy development and day-to-day management to the professional staff.

At best, special districts have low political profiles. Media coverage is usually limited. Taxes and assessments are often determined independently and collected through municipal or county units. It is the rare citizen, indeed, who knows his representatives on the various and sundry special districts operating in his jurisdiction. Governmental accountability is extremely difficult under such circumstances, and the whole system often reinforces the citizen's sense that government, even local government, is little more than a huge, impersonal, unresponsive bureaucracy.

A second intrasuburban structural issue is that the increased use of special districts has gradually whittled away at the set of functions performed or potentially performed by traditional suburban governments. As a matter of fact, the special district issue is only one part of the broader question, raised especially for suburban governments. What are the appropriate levels for decisions about, payment for, and management of a wide range of domestic governmental functions? It has been our custom to think of municipal governments as general purpose governments while, as a matter of fact, municipalities, particularly suburban municipalities, are now only one of several units of government in metropolitan areas performing a relatively narrow range of tasks.

It has also been our custom to treat counties as fairly specialized units of government, limited to their rural traditions and concerned primarily with roads and judicial functions. We now find considerable vigor on the part of many urban counties as they endeavor to carve out for themselves an increasingly significant chunk of the metropolitan governmental pie. In parts of the country, the same can be said for township governments, which until recently have been treated by many urbanists as little more than the vestigial and anachronistic remains of a preurban condition.

During the 1950s and early 1960s, there was considerable academic discussion of the role of the various types of local units in metropolitan governance accompanied by a set of standardized structural prescriptions that would have, by and large, consolidated all such units and submerged them in metropolitanwide governmental structures. Since that time, and perhaps in partial response to those proposals, the various types of urban local governments have been actively competing with one another to ensure themselves a viable role as the future metropolitan governance system evolves. The state and federal governments are increasingly active participants in this competition, and the eventual shape of the structure is, at this point, by no means clear.

DEVELOPMENTS IN

SUBURBAN GOVERNMENTAL STRUCTURES:

INTERLOCAL

There are three broad categories of change in intersuburban governmental arrangements occurring in metropolitan areas affecting suburban structures, two involving new developments, the third involving changes that were anticipated but did not occur. The first has already been alluded to: namely, the increase in the use of larger-scale units of government to perform functions that otherwise would be municipal or, perhaps, county in nature. The simple fact that a particular function is no longer performed locally does not lead, necessarily, to the lack of local interest in or direct involvement with the function. However, it does mean that the local unit shifts from an operating role to an "interested party" role. This shift involves a whole new set of political and administrative costs.

The second major intersuburban governmental change has much the same net effect. This is the increase in the development of formal and informal relationships between and among officials, usually administrative officials, in several suburban communities in a particular metropolitan area. These relationships may be formally institutionalized in regular meetings of leagues of counties, municipalities, or school officials. They may take the form of regional or statewide associations of city managers, tax assessors, police chiefs, school superintendents, engineers, planners, and the like. They may take the form of a more informal but regularized sharing of information, resources, skills, and so on, such as the group that has developed among several of the highly trained professionalized city managers in the Twin Cities suburban area. These organizations play an increasingly important role in lobbying efforts in state legislatures, much in the way that Browne and Salisbury (1971) describe the situation at the national level for such groups as the National League of Cities, U.S. Conference of Mayors, National Association of Counties, and so on. In addition, these organizations, in the suburban setting, have important policy implications through the sharing and exchange of information of administrative practices, personnel policies (including in the Twin Cities case, joint negotiations with public employees from several communities), ways to handle particular local problems, and the like.

The structural implications for suburban governments of these two developments are very significant. The creation of new large-scale units for certain functions and greater utilization of formal and informal interchange with other local units and officials requires that local decision makers (elected and administrative) spend increasing proportions of their time and energy outside their own community. In the Twin Cities area, some of the more active suburban managers estimate that as much as fifty percent of their working time is spent in dealing with units external to their own borders, so that more and more of the local day-to-day routine is turned over to other personnel. Indeed, in one or two instances, managers and their councils have experienced a parting of the ways over the issue of how much time should be spent by the manager attending to strictly internal affairs.

The main point here is that we have now created a situation where local governments, especially in suburbia, to be effective must allocate increasing portions of their administrative and political resources to external relationships involving other local units, regional units, the state, and the federal government. In a sense, local governments increasingly find themselves in the same predicament that the federal government has found itself in in periods following aggressive isolationism, when it has been necessary to create elaborate legislative and administrative mechanisms to handle external relationships and problems. The difference for suburbia is that the necessity to engage in and maintain external relationships is now a reality, but the appropriate structures and institutions have not yet been built into local legislative and administrative processes. One does not want to suggest that suburban governments necessarily need structures equivalent to the Pentagon, the State Department, or the CIA, but certainly more regularized institutions and practices will be required if the elaboration of external relationships continues.

Finally, the period of the 1960s and early 1970s is as important for what changes have not occurred in suburban governance as it is for those that have. Clearly, the shift toward professionalism and management has occurred more readily in suburbia during this period than in other settings and at other times. Clearly, the general-purpose role of local suburban units has diminished as functions are divided among different units and different levels of government. Clearly, local political administrative resources are increasingly devoted to activities and concerns beyond local boundaries.

On the other hand, some of the things predicted for the 1960s have not happened. Suburban local government does not seem to be going out of business. Indeed, in most metropolitan areas, more units (especially municipalities and special districts) are being created. The major suburban structural "nonevent" of the 1960s, however, has been the lack of major consolidations of local units in the metropolis, either in the forms envisioned for St. Louis or in the more "federated" variations, a few of which have actually been created (e.g., Nashville-Davidson County). Most recently, the major thrust of coordination-integration-consolidation movement has been the Council of Governments (COGs) approach, which strengthens rather than weakens existing local units by permitting their direct representation in the areawide agency.

In this context, the Minneapolis-St. Paul experience is interesting, although certainly not representative of general trends nor necessarily indicative of future directions in other metropolitan regions. In the Twin Cities metropolitan area, local units (especially municipalities) continue to be created although the recent tendency has been toward the incorporation of entire townships thus creating new units with large territory and potentially larger populations than those smaller areas incorporated during the 1950s and early 1960s. Consequently, the suburban municipalities of the future will for the most part potentially, at least, be in a better position to afford their own governance.

Two more dramatic and potentially far-reaching developments have occurred that may well be emulated in other metropolitan areas. First was the creation, in 1967, by the Minnesota State Legislature, of the Metropolitan Council. The council consists of a one gubernatorially appointed representative from each two state senatorial districts in the seven-county metropolitan area and a chairman, also appointed by the governor. The council has the power to review planning decisions of local units and of regional units, as well as coordinating planning for the major development decisions in the region. It also passes on all applications for federal grants. It coordinates the activities of the several semi-autonomous functionally defined boards: the Metropolitan Transit Commission (which also operates the metropolitan bus transit system), the Metropolitan Health Board (which is the official areawide health planning agency), the Metropolitan Sewer Board (which plans, constructs, and operates all sewage treatment facilities in the metropolitan area), and the Metropolitan Airports Commission.

The extent to which the Metropolitan Council concept will be emulated in other parts of the country will depend in large measure on its ability to deal with several critical issues that are not yet resolved. One is the relationship with existing local units which, except for the specific functions mentioned, remain much as they were. The council concept, however, impinges on local government in two ways. First, its review and veto powers over many varieties of local plans reduces considerably local planning autonomy. Second, as the council has developed, it has added functions which were formerly local in nature. The first step was the metropolitanization of sewage treatment. This step was of relatively little political consequence when it occurred, because the sewage situation had become so serious that all parties agreed a radical and regional response was required. Such consensus is much less clear in some of the issues where the council has more recently moved—e.g., transit, health planning, and areawide housing regulations.

The second issue involves council responsiveness and accountability to local governmental interests. Under the present structure with council members, appointed by the governor, representing state legislative districts rather than local governmental units, local officials feel themselves left out of the picture and find limited opportunities for leverage in council decision-making. The problem is a thorny one since, under the present structure, the Metropolitan Council is a quasi-state-level agency responsible ultimately to the governor and the legislature. Under a different selection system—e.g., direct popular election or some kind of local unit representation—the council, representing an area containing more than half the state's total population, would become much more politically independent of the legislature and could become a strong competitor at the state level. Since any change in structure will require legislative action, the future of the council's representation system at this point is not clear.

The second major innovation in the Twin Cities area that will have important long-term effects on suburban development and suburban decision-making was the passage by the 1971 legislature of the so-called "fiscal disparities" bill. This legislation is designed to resolve one of the most difficult problems facing contemporary suburbia —namely, the problem that some suburbs, by virtue of their tax base, are much wealthier than others and can afford higher levels of public service at relatively low cost to residential property owners. This problem becomes more acute as the "wealthy" communities tend to become wealthier, and as many of those communities with a relatively large residential base begin to realize the actual limits on their potential for adding strong commercial resources. The Minnesota plan specifies that the benefits from all new commercial development in the metropolitan area will be divided, with sixty percent going to the community in which it is located, the remainder going into a general fund to be allocated among all communities in the metropolitan area on the basis of a formula that assesses need and tax effort. Initially, the amount of money to be redistricted is relatively small, but as commercial property continues to develop the amounts will grow and the redistributive effect will be of considerable importance.

In the preceding paragraphs, we have considered some of the major recent developments in suburban governmental structures, both those internal to local government and those occurring outside local control when the impact on local decision-making structures is pronounced. Observation of the past decade suggests that suburban local governments are not being eliminated directly through consolidations, annexations, "metro" governments, and so on. Some of their traditional functions are being whittled away by the continuing creation of special-purpose governments, while local discretion and autonomy in some types of decision-making have been reduced by the planning and review requirements of other local, regional, state, and federal jurisdictions. This, in turn, has forced suburban units to devote increasingly larger and larger proportions of their political and administrative energies and resources to concerns outside their local borders. At the same time, greater complexity in the management of local government, coupled with the usual system whereby suburban legislative bodies consist of nonprofessional, part-time "citizen-legislators" has increased the importance of the role of the professional public official and encouraged greater professionalization of local governmental personnel.

These processes have contributed to the creation of a wide range of more or less formally organized groups of professional suburban administrators (and in some cases, politicians, as well), crossing local suburban lines, where ideas, skills, and resources can be shared. These groups, despite the fact that they are not formally structured into local government play an increasingly significant role in policy development and administration in suburbia.

THE PROCESS OF SUBURBAN
GOVERNMENTAL STRUCTURAL CHANGE

Although we are not able to report systematically on the trends in suburban governmental structures described above, two things are clear. First, there is continuous change in local governmental structures in metropolitan areas throughout the country. Individual communities frequently evaluate and reassess their forms and structures and engage in the processes required for their change. At the state and metropolitan levels, too, public and private discussion of structural arrangements occurs continuously, and change is a common event.

Second, the adoption of some of the more recent forms and structures has not occurred uniformly throughout all metropolitan areas, and, within particular metropolitan areas, the rates of change often vary between local units.

In line with the discussion in the preceding section, suburban structural change comes in two forms: change within a particular local unit and change in the structural arrangements between and among local units. This latter form of structural change has not been thoroughly analyzed, although the case studies of change and proposed change in several metropolitan areas provide some insights into the process by which metropolitanwide governmental restructuring occurs.

Two common elements in these studies emerge. First, the nature and character of the local political system is critical in determining the outcome of attempts to restructure local government at the metropolitan level. Since local political systems are idiosyncratic to their locales, this suggests that future metropolitanwide governmental organization will be very much a product of local forces and circumstances. Second, when metropolitanwide change has taken place, it has often been at the insistence of the state or federal government requiring some form of areawide planning and policy-making, and it has been distinctly conservative from the standpoint of protecting the interests of existing local units of government.

There is nothing in these fragments of evidence to suggest that the process of metropolitanwide governmental structural change will not continue to be characterized by the application of some form of the carrot or stick, and that existing local units will not continue to build themselves directly into any new arrangements.

In the following discussion, we will focus on the process of intrasuburban structural change—i.e., the change in local forms and structures occurring within particular suburban communities. The basic premise for this discussion is that internal suburban governmental structures are not created and shaped by accident. Rather, they are the result of several constraining factors, not the least of which are the basic nature and character of the individual communities the forms and structures are designed to serve. Local governmental structures are nothing more or less than "rules of the game," and, in governance, as in other games, rules are designed to facilitate the game for the players. If the rules fail to serve the players, they may attempt to change the rules, or, as is often the case in suburbia, they seek other games where the rules more nearly approximate their skills and interests.

As several studies have demonstrated, however, local governmental structures are more than a simple reflection of a community's socioeconomic characteristics (Wolfinger and Field, 1966). Existing research has not specified all the variables, but in general terms the form of local governmental structures in a particular community is the result of the interplay among four factors: state constitutional and statutory requirements and provisions; regional and local custom and tradition; certain relevant and unique local idiosyncracies; and the social and economic structures of the particular community. Variation in any or all of these factors is largely responsible for the variation in local governmental structures throughout the country. And variation among these factors affects the process by which local units change and adjust their forms and structures.

Our basic argument is that these four factors form a set of increasingly constraining parameters within which the local community may define its own "rules of the game"—i.e., its governmental structures. Figure 1 illustrates this concept and suggests, of course, that the limit for each parameter is determined in part, by the limits imposed by the higher-level parameters.

The figure illustrates one hypothetical situation where the freedom permitted local units by state law is fairly broad (ranging 3.5 degrees on either side of the zero point in the hypothetical illustration), but where the decision latitude of the local unit is increasingly constrained by regional custom, local idiosyncratic factors, and its own social and economic mix. The net result of these increasingly constraining factors is that, despite generous state constraints, the community really is quite limited in its decision latitude (ranging only 1 degree on either side of the zero point in the illustration).

Unfortunately, we are not yet in a position to specify or quantify these four factors. If we were able to do so, we could with considerable precision explain and predict the process of local governmental structural change. Until we can do so, however, the discussion must necessarily be limited to generalizations.

The most obvious and in some cases the most significant factor in determining local governmental structures, whether suburban or not, is the constitutional and statutory provisions of each state. Although there is no systematic survey of the variation in these laws among states, it is clear from casual sampling that there are significant differences. In our research on restrictions in three states (Ohio, Illinois, and California), there were important differences both in the variety of forms local units were permitted to adopt and the ease or

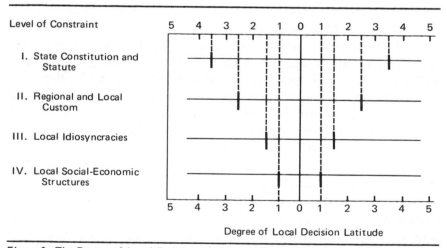

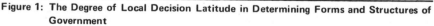

Figure 1: The Degree of Local Decision Latitude in Determining Forms and Structures of Government

difficulty of making structural change. The most obvious example in these data was Illinois, which, until the mid-1950s, did not officially permit the council-manager form as an option for local units (Scott, 1968).

The most interesting current question regarding state legal constraints on local governmental structures, however, is the extent to which some states have begun to differentiate between metro-politan-suburban and nonmetropolitan local government. Again, there are no general data surveying these developments, but scattered evidence suggests that legislatures are increasingly willing to distinguish between metropolitan and nonmetropolitan areas and to permit local forms and structures more appropriate to metropolitan problems and concerns. The Minnesota case provides at least two illustrations. Several years ago, the legislature enacted a Joint Powers Act which permits local units to perform any functions jointly that they are empowered to perform individually. This act applies statewide, but it was designed in part for and has been applied extensively in the metropolitan region, where it has greatly facilitated interlocal cooperative agreements and arrangements. The second illustration is the fiscal disparities legislation discussed previously, which alters the local profitability of new commercial development in the Twin Cities metropolitan area and thus is directed specifically at a peculiarly metropolitan problem.

The second factor in the analysis is regional and local custom and its impact on suburban governmental structures. The argument in support of the importance of this factor comes in part from the research reported by Wolfinger and Field (1966) and by Scott (1968). Wolfinger and Field do not deal specifically with suburbs, but they do conclude that there are important regional differences in the extent to which various forms of city governments are utilized. My study is directed specifically at suburbs in three states and finds variation among the states in the degree to which and the rates at which suburbs have adopted the council-manager and other related structures of local government. The basic argument suggests that communities learn from and emulate each other in many ways including their forms and structures of government and that this learning and emulation process often occurs among units in close proximity, particularly within a given state or metropolitan area. The regionalization of the learning and emulation process is stimulated and encouraged by many of the phenomena mentioned earlier, including state and metropolitan leagues of municipalities, state and local associations of city managers and other local officials, locally based public administration training programs, common news media sources, and so on. There is, to be sure, considerable nationally based learning and emulation, but the extent to which communities, especially suburban communities, observe each other, share ideas, compare their responses with their neighbors, and the like is quite impressive and is clearly a major element in shaping local decisions, including decisions about governmental forms and structures.

The third factor affecting suburban governmental structures is the peculiar and idiosyncratic elements that may exist in any particular community. These are not the kinds of things that social scientists can easily identify and study; rather, they are the grist for the mill of the local historian and journalist. They may involve peculiar bits of local history, or particular personalities and personality clashes. They may stem from a specific configuration of local business interests or the distribution of populations at certain times in the development of the community. One is reminded of a recent merger of two Twin Cities suburbs that occurred more or less simultaneously with the marriage of the city manager in one of the communities with the village clerk of the other.

The problem with this third factor is that it is virtually impossible to define and analyze systematically; yet those involved in the decision processes in local government, including governmental structural decisions, know that such considerations can be very important—indeed, critical—under the right conditions.

The fourth constraining factor is the impact of local socio-economic characteristics and attributes on suburban governmental structures. There is a considerable body of research on this factor, although most of it is not limited to suburbs per se, and the findings are not altogether conclusive. In general, however, these studies show that middle-sized communities characterized by high socioeconomic status, fertility, population mobility, and social homogeneity are most likely to have adopted the council-manager plan and its accompanying structures, while communities with opposite characteristics tend to retain more traditional forms of local government, particularly the mayor-council system. The findings show that the newer or reform structures of local government are more readily accepted where middle-class values and experiences prevail and where they are relatively unhampered by social and economic cleavages that manifest themselves in more traditional partisan practices and structures (Schnore and Alford, 1963).

Since we are interested in the processes of structural change, the findings and conclusions of the Scott (1968) study of suburbs in Illinois, Ohio, and California are instructive, since they include a change or developmental dimension.

Simply stated, the data from the suburbs in those three states showed that new forms of government were adopted *first* by communities with very special characteristics, essentially those with strong, relatively stable, family-based, middle-class populations. Suburbs adopting the newer governmental forms, subsequently, were less likely to exhibit such characteristics. The argument explaining these data is based on the assumption that the adoption of new governmental forms represents a major decision by a community. The decision is especially difficult if the proposed change involves forms and structures about which the community has little first- or second-hand knowledge—i.e., when there are few, if any, reference communities whose experience can serve as a guide. This is the case, of course, if communities in a particular region or locale are considering the adoption of governmental forms that have not been tried by any of their neighbors.

Because of the greater uncertainties the initial adoptions of new structures in a given region are usually accomplished by units with special characteristics that include both the lack of strong cleavages that might prevent concerted community action and the social and organizational resources necessary to sustain the high level of political activity required to accomplish governmental structural change.

Once the first communities (the "pioneers") adopt the new forms, they become the reference group to which other local units can turn for advice and counsel regarding the changes. The existence and availability of the reference group make the decision process easier for subsequent changers (emulators) and tend to accelerate the rate at which the new forms and structures are adopted by other local units, whose socioeconomic characteristics do not need to conform as closely to the pioneer model. This reference group phenomenon is, as was suggested earlier, especially important within metropolitan areas where the sharing of information and experiences is facilitated geographically and organizationally.

Variations in suburban local government structures, then, are the consequence of state constitutional and statutory provisions, regional and local customs and traditions, certain relevant local idiosyncracies, and basic social and economic structures of the particular community. Each factor operates within the constraints imposed by the others in the order illustrated in Figure 1. If state law does not permit a particular option, then the other factors are irrelevant. If regional custom encourages professionalism and "good government" forms, and they are available under state law, then local idiosyncratic partisan structures will be difficult to maintain at the local level, and so on.

These factors affect the structures of government in all kinds of communities, but the fourth factor—the social and economic characteristics of the particular community—is especially critical in suburban units for several reasons. First, suburbs are usually newer and therefore less bound by tradition and long-term political practices and customs. Thus, local idiosyncratic considerations are often less important than in older, more established communities. Second, once they begin to reach full development, suburbs often become more homogeneous than nonsuburban communities, and it becomes easier, therefore, to formulate structures most appropriate to a particular social and economic mixture. Third, suburban citizens have relatively greater freedom to "vote with their feet"—i.e., to move about, changing their place of residence, in order to select the communities that provide the services and qualities they most desire. To some extent at least, these qualities are a result of local governmental structures.

WHY DOES STRUCTURAL CHANGE OCCUR?

The preceding discussion describes some of the more important recent trends in suburban governmental structures and offers a tentative explanation of the process by which change and development takes place. The more basic question, however, is not so easily answered—why are suburban governmental structures changed? Does it really matter that communities have different forms? Or, to put the question as Norton Long would, "What are the differences that make a difference?"

In general, this question cannot be answered precisely. We still do not know, in a systematic way, whether the manager form or the mayor-council form produces better policy, fairer problem-solving, or more efficient use of resources. We cannot compare the responsiveness and sensitivity to public demands of various kinds of local electoral systems. We do not know whether or not a highly professionalized personnel system is preferable to governmental employment by patronage. We cannot really demonstrate that areawide, single-function governments are always more efficient than less-coordinated arrangements, nor can we defend with hard data the proposition that regional planning authorities result in better metropolitan development. We have a considerable folklore and ideology on these and related questions, but our ability to bring hard empirical evidence to bear on them is very limited.

Nonetheless, the processes of "rule-setting" and "rule change" (i.e., adopting and altering structures of local government) are clearly important activities for those involved in the political process, especially in contemporary suburbia. The processes are important for two reasons, in much the same manner that structural reform has always been an important part of the political system. For a few, it is primarily an ideological question. There is a faith that some forms of government are preferable to others because it is believed that they produce better results and provide more desirable decision-making mechanisms. For such persons, the relationship between structures and outcomes is usually not specified except in very general and often vague terms. This kind of reform perspective is often "faddish" and is usually propounded by some of the more articulate elements in the system, traditionally referred to as "goo goos" by their less-reform-oriented counterparts.

To a considerable degree, the recent trends in suburban government described above are the logical extension of a reform tradition that had its birth in the late nineteenth century. It stresses professionalism and management skills in local government. It abhors partisan entanglements at the local level. It discourages local parochialism and emphasizes broad-scale and broad-range planning and regulation. This perspective is most clearly manifest in the council-manager form, in merit systems of governmental employment, in nonpartisan and at-large electoral systems, in areawide planning and regulatory agencies, in regional rather than local decision-making mechanisms, and so on.

Interest in and concern for local structural change is not, however, simply a broad-ranging, nonspecific, ideological issue. Local rules of the game can and do affect particular interests and can facilitate particular goals and objectives. Specific examples abound. Shifting from an at-large to a district election system can change the balance of power in a local legislative body and vice versa. Creating new positions can shift the balance of power in local administrative relationships. For example, several Twin Cities suburbs have been able to circumvent the power of old-time police chiefs by creating a Public Safety Director position and subsuming the police department in the new agency. Countywide tax assessment can reduce local variation in assessment practices and eliminate inequities among suburban communities. Creating local housing and redevelopment authorities can effectively reduce the power of municipal councils to regulate future land-use development.

On balance, governmental structural reform is for most citizens a combination of general ideological persuasions, the roots of which they may neither understand nor be concerned with *and* a means to achieve specific but not necessarily well-thought-out goals and objectives. In this sense, then, much reform activity is in part symbolic and in part goal-oriented, although the symbolism may be vague (and even contradictory) and the long-term implications of change rarely understood.

This combination of the symbolic and the practical is most clearly exemplified in the large and growing number of suburban communities that have adopted the professionalized, nonpartisan planning and regulatory-oriented structures of the twentieth century. For many of the citizens of such communities, their basic interests and concerns are fragmented among a variety of local, regional, and national governmental jurisdictions. The jurisdiction of residence is often not the jurisdiction of work activity, recreation activity, consumer activity, or educational activity. It may not even be the jurisdiction of sewage treatment, water provision, mosquito abatement, and tax assessment. For many of these citizens, physical and social mobility reduces their long-term commitments to and concerns for what is often a temporary residential jurisdiction.

Under these circumstances, it is little wonder that professional, nonpartisan planning and regulatory-oriented governmental structures are so popular. How else could such citizens assure themselves of the reasonably competent provision of local functions and services, along with a guarantee that property values (with which they are critically concerned since the place of residence represents a major investment commitment for many) will be maintained over time; all at very low political cost, because the hired professionals will make or shape most of the important decisions with these goals in mind. It is true, of course, that such a system may be somewhat more expensive financially, but the suburbanite is distinguished by his relatively higher levels of affluence, and the monetary cost is low, indeed, for the benefits he receives.

Such citizens, especially those with a more ideological bent, may be offended by this analysis, but in truth what they have done, perhaps better than any other subset of the population at any other time in our history, is to purchase good government for themselves at a very reasonable cost by insisting on local structures (rules of the game) that reduce political considerations and costs to a minimum and by discouraging the intrusion of persons and interests who do not wish to play the game in this manner.

WHAT ABOUT THE FUTURE?

To what extent can we anticipate that the post-World-War-II trends described above will continue, or are there indications of potential change? To be sure, forecasting the future is a treacherous business at best, but three elements seem destined to affect the further development of suburban governmental structures.

First is the general and pervasive disenchantment with all political and governmental institutions and especially the growing sense of alienation toward governmental professionals, bureaucrats, and technicians. This concern is expressed both by the general public and increasingly by their elected local officials. Suburbs are by no means immune; indeed, the problem may be most serious in the suburban setting, because suburbs have become the most professionalized of our local institutions and are thereby most vulnerable to the criticism and disenchantment.

In many places, this issue has spawned attempts to increase citizen participation. Our own judgment is that the heat of the present distress is temporary. Suburban residents, especially for the reasons cited above, are not fundamentally interested in extensive citizen involvement in local government. On the other hand, profession-alized public servants have, perhaps, overstepped the bounds of their circumstances and need to be reminded that some semblance of responsiveness and accountability is desirable. In many ways, the present level of alienation toward political institutions is the result of the fact that we have moved very swiftly, especially in suburbs, to government by technocrat without raising and resolving the fundamental issues of the relationships between the governmental professional and the part-time, semi-interested citizen.

There is one type of suburban setting where these concerns may have more important long-term implications. Some suburbs have now developed to the point where their social and economic structures resemble their central-city counterparts more nearly than they do their suburban neighbors. Their populations are large, and divers socially, racially, and economically. Their commercial base includes basic industry as well as the more typical shopping centers. Their physical plant is old and deteriorating. Such communities are readily apparent in many metropolitan areas. In others, the first signs of such development are only now beginning to appear. In any case, the maturation and "citification" of what were formerly residential enclaves is usually accompanied by alterations in the political processes. More specifically, one can observe the development of basic political cleavages, the emergence of partisan organizations and activities, and the eventual insistence on governmental structures that more closely reflect and accommodate the politicization of the community.

The second problem is related to the first. It concerns the question of local (decentralized) versus regional (centralized) decision-making and control. There seems to be little impetus left in the metropolitanization movement of the 1950s. Instead, during the 1960s, we have observed a very complicated ad hoc process whereby functions are added, subtracted, and shifted among the various types of metropolitan local governments, while, at the same time, each struggles to ensure its own survival in the emerging urban complex. The issue is no longer local autonomy versus centralized decision-making. Rather, the issue is the shape of the eventual distribution of local powers and functions throughout the metropolis. The tough question is the extent to which and the circumstances under which locally based units may participate in regionwide decision-making. Should they participate directly as individual units? If not, what should be the nature of their input in the decision-making process? We are a long way from resolving these questions.

The third problem comes from the fact that suburbs are, in one crucial way, different from most of the other communities we have concocted throughout our history. Most communities, large or small, have been built with economic development objectives as their paramount raison d'être. This has meant that the vast proportion of fundamental local decision-making has been oriented toward the economic viability and prosperity of the community, and citizen issues have traditionally been of secondary importance.

Most suburbs, on the other hand, have emphasized citizen and residential concerns and in many cases have treated economic development as a lower form of public activity and as a last resort should taxes on residential property become too high. This attitude was understandable at an earlier stage of suburban development, where the costs of public service were low and where prospects for relatively inoffensive commercial development seemed bright. During the late 1960s, however, these circumstances changed. Public service costs, in suburbs as elsewhere, rose very rapidly while the supply of commercial development, inoffensive or not, diminished. Suburbs were, for the first time, in serious financial trouble. Indeed, for many there is no chance for their continued viability without major redistributive policies on the part of the state and federal governments.

The prospect of significant fiscal dependence on other units of government is quite likely to reduce substantially suburban local autonomy and increase the extent to which local units will be required to interact and bargain with other units and levels of government.

It should be clear from the preceding discussion that the three problems outlined are by no means peculiar to contemporary suburbia. Indeed, they are endemic to government, especially local government, across the board. We have simply stressed their suburban dimensions and manifestations.

It should be clear also that suburbs, perhaps more so than other types of local communities, are affected by and involved with the actions, decisions, and policies of other governmental units and levels. Certainly the federal government played a major role in the early development of suburbs through its postwar housing and transportation policies. The federal government is now beginning to exert its influence in a different but equally significant way, through its grants and loans policies and their areawide planning requirements. The state's role, too, has been important, exercised primarily through its power to design various kinds of metropolitan governmental structures, particularly special district governments.

More important, perhaps, than the impact of federal and state levels are the interactions between and among suburban communities themselves. Suburban residents, workers, and consumers are no respecters of suburban boundaries, nor are the problems and issues they create as they move about in the metropolitan region. The very limited ability of suburban units to control their borders forces interaction and interinvolvement with their neighbors, to the point now, where external relationships are often as important to a suburb as are its internal affairs.

Finally, and in conclusion, it should be clear from this entire discussion that it makes very little sense to treat suburban governmental structures apart from all the other elements of the local political system. As we suggested earlier, suburbs, perhaps more than other kinds of communities, have consciously designed and shaped their political structures with the result that their institutions closely reflect local social, economic, and political characteristics. For this reason, and because we anticipate important changes in suburban social structures in the foreseeable future, we can also expect important new directions in local governmental forms and institutions.

REFERENCES

BROWNE, W. P. and R. H. SALISBURY (1971) "Organized spokesmen for cities: urban interest groups" pp. 255-278 in H. Hahn (ed.) People and Politics in Urban Society, Volume 6 of Urban Affairs Annual Reviews. Beverly Hills: Sage Pubns.

KESSEL, J. (1962) "Governmental structures and political environment." Amer. Pol. Sci. Rev. 56 (September): 615-621.

SCHNORE, L. and R. ALFORD (1963) "Forms of government and socio-economic characteristics of suburbs." Administrative Sci. Q. 8 (June): 1-18.

SCOTT, T. M. (1968) "The diffusion of urban governmental forms as a case of social learning." J. of Politics 30 (November): 1091-1108.

STIEBER, J. (1969) "Employee representation in municipal government," in Municipal Yearbook. Washington, D.C.: International City Management Association.

U.S. Bureau of the Census (1968) Census of Governments; Government Organization. Washington, D.C.: Government Printing Office.

WOLFINGER, R. E. and J. O. FIELD (1966) "Political ethos and the structure of city government." Amer. Pol. Sci. Rev. 60 (June): 306-326.

9

Cities, Suburbs, and Short-lived Models of Metropolitan Politics

H. PAUL FRIESEMA

□ IN A RELATIVELY SHORT TIME, SOCIAL SCIENTISTS have developed an extensive and imaginative array of shorthand models of the metropolitan political process. Beginning no earlier than the late 1950s, this outpouring has included attempts to develop orienting frameworks proclaimed to be functionalist in orientation, or as focusing on political culture; a market model; a theory of countervailing forces; a life-style model; a diplomatic system; an international integration analog; a bargaining model; and no doubt some others.

While these constructions of the metropolitan political order have been developed, refined, and subjected to critical review, the sad fact is that their full value is still undemonstrated. The first theoretically based, proposition-testing comparative research on the interjurisdictional politics of metropolitan areas remains to be done. But demographic changes, a few court cases, and the passage of a relatively short amount of time may be combining to make these models obsolescent, even before they are field tested. These models may quickly become particularly flawed guides to understanding crucial issues about large cities and their surrounding suburbs. Some basic reformulations may be in order, even though the recent formulations have seemed rich and promising. On the other hand, these recent (or recently appreciated) changes in the nature of metropolitan politics may give added impetus to at least some parts of the theoretical formulations, by making assumptions of some models seem more realistic. This paper will present a very preliminary assessment of what might be the theoretical consequences of these changes, including among these changes a series of legal cases, currently going through the courts, which challenge the jurisdictional prerogatives of white and wealthy suburbs.

POSTREFORM METROPOLITAN THEORY

The period of conceptual enrichment of the study of metropolitan political processes probably needs to be placed in perspective. Political scientists have, of course, been focusing attention upon the "metropolitan problem" since the 1920s. But a watermark was reached in the late 1950s. Certain general conclusions seemed to emerge from a series of unification attempts and from two exhaustive studies of metropolitan areas (the New York and St. Louis metropolitan surveys).

Probably the most important general conclusion (perhaps "general feeling" would be more like it) was that efforts to reform metropolitan areas through political unification were not going anywhere. From this feeling, a series of other things seemed clear enough, at least to some. It was going to become necessary to come to grips with the extreme complexity of interjurisdictional conflict and cooperation occurring within metropolitan areas. While the metropolitan areas were expanding rapidly, the basic *form* of political organization was fairly stable. The chief political units engaged in metropolitan politics were the (rapidly increasing) municipalities. One more or less permanent political cleavage was between the central cities of metropolitan areas and the array of suburban municipalities around each central city. The suburbs were no longer limited to being the province for prosperous WASPs. Indeed, with the continuing suburban exodus, plus inmigration to the cities, it appeared that the urban population would be socially and economically differentiated (rather finely differentiated, at that) on the basis of the jurisdiction in which people lived. The suburban areas were going to be increasingly important sites for economic activities.

From this broadly shared perspective on some key elements of metropolitan politics, postreform scholarship proceeded along a couple of fronts. A series of very detailed and impressive studies of politics in particular metropolitan areas were conducted (for example, Martin et al., 1967; Mowitz and Wright, 1962). Such studies were far from atheoretical. The Martin et al. (1961) volume developed a whole inventory of propositions; Mowitz and Wright (1962) employed an implicit general systems framework. The other postreform thrust was to develop and explore theoretical simplifying models of metropolitan politics. These models were characteristically presented without reference to any particular field research, except for illustrative purposes (although the richest formulations were stimulated by field research in the Los Angeles, Cleveland, and Philadelphia metropolitan areas). While this thrust to metropolitan studies involved many people and many alternatives, undoubtedly among the most important statements (as judged by critical comment, citations, reprints in collections, and similar indicators in addition to author's subjective judgment) are Holden (1964) and Ostrom et al. (1961). Williams (1968) seems to have provided another important effort, although his work is largely a major refinement of the Ostrom et al. statement. This work is elaborated in Williams (1971).

Holden (1964) argues the utility of treating the metropolitan area as a diplomatic system analogous to the diplomatic system of international relations and subject to equivalent propositions. Ostrom et al. (1961) propose what has come to be called a "market model" of metropolitan politics, with the various municipalities within metropolitan areas serving as service producers, competing for customers who choose municipal location upon the basis of the mix of goods they desire. Williams' (1968) "life-style model" is based upon the premise that jurisdictions will integrate their system maintenance public services, but isolate their life-style services.

These models seem to be based upon certain assumptions and assertions about what are the stable conditions of the metropolitan areas. But these assumed stable conditions may not be stable at all. It may be that some constants of the models have the capacity to become highly variable in the metropolitan areas of the 1970s. While this analysis will focus upon some questions concerning the Holden, Ostrom et al., and Williams formulations, it should be clear that many of the questions which are raised could, as well, be applied to the other postreform models of metropolitan politics, such as Crouch and Dinerman's (1964) theory of countervailing forces, or the effort to apply Karl Deutsch's work on international integration to metropolitan areas by Jacob and Toscano (1964).

Many of the things which suggest the need to reexamine the utility of our models of metropolitan politics for the 1970s (and particularly for use in examining city-suburban issues) are the product of long, observable development. Some of these are: the emergence of metropolitan councils, regional planning authorities, and more direct federal (and state) intervention to stimulate or compel areawide plans and programs; the stabilization and routinization of relations between core-city municipalities and the suburban municipalities which border the core city; the emergence of metropolitan, areawide, interest-articulating groups concerned with such issues as open space acquisition, fair housing, pollution control, and the like. In addition to these developments, one other element may well compel some conceptual reformulations. This has just recently emerged, and its final impact is still quite uncertain. This element is a series of legal suits successfully challenging established jurisdictional prerogatives. Included among this list of cases is the Serrano decision in California and parallel cases in other states, which have held that education is a fundamental right protected by the Constitution. As celebrated and important as the Serrano case undoubtedly is, its impact upon models of metropolitan politics may prove to be less than that of some other legal suits. In the case of Gautreaux v. Chicago Housing Authority, scattered-site public housing may be ordered for the whole Chicago metropolitan area. Judge Roth has ordered a school integration and busing plan for Wayne, Oakland, and Macomb counties in the Detroit metropolitan region. The spate of recent suits challenging suburban beach restrictions which limit recreational use to residents is an indication that the challenge to restricted privileges, based upon membership in some suburban constituency, is likely to continue.

Of course, it is far from clear that any of these landmark cases will ultimately be sustained by the Burger Court. Nor is it certain that the court decisions can be enforced, even if sustained. We may learn what "massive resistance" is really like before the decade is through. But if these decisions are maintained, the whole nature of metropolitan politics will be changed. Already the models seem inadequate. They are only of limited use, in fact, in understanding the metropolitan political process by which these fundamental issues on tax equalization, metropolitan housing, and school integration are being raised, fought, and decided. And these questions are undoubtedly the most important issues of metropolitan or city suburban politics in the next years. These issues make such presumably staple issues of metropolitan politics as competitive annexation, municipal income taxes on nonresidents, or water for Wauwatosa seem old hat and trivial.

THE DIPLOMACY MODEL

If we consider the governance of the metropolis as a problem in diplomacy, we make certain assumptions about the salience and centrality of certain patterns of metropolitan political behavior. Perhaps the most central assumption is that the major actors in the diplomatic system are municipal corporations, with general power, which can be treated as roughly comparable with nation-states. Directly related to this assumption is the notion that metropolitan politics consists, essentially, of bilateral or multilateral relations between and among these municipalities within the metropolis, without involvement by private interests, other than the territorially defined jurisdictions. Holden (1964: 627-628) posits three characteristics of "diplomatic systems" which he asserts are common to the international and metropolitan arena. Two of these are:

> Diplomatic systems are social systems consisting of corporate units bound within an 'ecological community.' These corporate units are the *primary* actors in the system.

> Diplomatic communication is governed by a ritual which proscribes non-governmental initiatives toward the resolution of substantive issues in which governments are partisan.

These presumptions about metropolitan politics seem increasingly narrow. Not only are there a wealth of other nonlocal *governmental* actors involved in metropolitan issues, but the role of nongovernmental, interest-articulating groups and occasionally individuals in the politics of metropolitan areas is now apparent and legitimate. To treat such actors and involvements as remote or secondary seems to place an unnecessary stricture upon the study of metropolitan political phenomena.

The current legal challenges to jurisdictional advantage are clear enough indications of the constrictions of the diplomatic model. Not one of these issues can remotely be conceived as a bilateral or multilateral issue of interjurisdictional relations. In none of these issues are the plaintiffs acting on behalf of municipal jurisdictions. They are private citizens, acting for themselves and people in like circumstances, frequently aided by "public service" interest groups. Holden's model does include a role for the courts as arbiters of metropolitan issues. But legal challenges are conceived as the way to solve interjurisdictional issues which cannot be resolved by municipal diplomacy: the functional equivalent to internation warfare. That characterization seems irrelevant to the current issues of urban-suburban conflict. Of course, it would be possible to save the whole diplomacy analog by simply defining the legal suits over property tax equalization and so on as interesting cases, but outside the scope of metropolitan politics. That would be unfortunate, for these issues have emerged *because* of the distribution of people and values among jurisdictions within metropolitan areas, and their resolution will have major suprajurisdictional impact.

If the diplomatic model is not very useful in understanding the major issues of urban-suburban politics of this new period, its utility in focusing the issues of *interjurisdictional* politics which do exist in metropolitan areas has also been reduced, particularly with regard to the core-city municipalities vis-à-vis the suburban municipalities. One reason for this loss of salience of the model with regard to city-suburban bilateral municipal relations is that the characteristic "issues" of urban-suburban jurisdictional politics have become stabilized, routinized, and almost made technical in recent years. They no longer arouse much emotional interest, nor do they involve major policy issues. As a consequence, there seems to have been a substantial transformation in the nature of interjurisdictional metropolitan politics.

It seems clear that a very large percentage of the intermunicipal bilateral contacts within metropolitan areas have to do with border maintenance and contiguity issues. But for most core cities, at least of the larger metropolitan areas, the borders have been basically stable for a good long while. Competitive annexation or tax-induced, unwarranted municipal incorporation may still agitate relations among jurisdictions on the metropolitan fringe, but it is *not* a very pressing issue of city-suburban relationships. The fights over location of the freeways are mostly over now, and the interjurisdictional politics of mass transit are really only on the horizon. The tax-free use of city municipal services by suburban residents commuting into the city for work or entertainment was, until recently, a passionate underlying source of interjurisdictional conflict and discord. But the two-way, and every which way, flow of traffic throughout the metropolitan area is now apparent even to many municipal pols. Many of the classic nineteenth-century wealthy suburbs, touching the city, have opened up to a more heterogeneous, urbane population, as their housing stock aged. Those suburbs, as well as other working-class suburbs on the core-city periphery, seem to reduce some of the obvious visible social distance between large core cities and the suburbs. They amount to almost a buffer or an irregular transitional zone. Bilateral politics are, of course, mostly among contiguous jurisdictions. The interesting consequence of all this is that one of the recently presumed constants of metropolitan politics, a basic political cleavage between the core-city municipalities and the suburban municipalities, has become far more complex and obscure, while we were looking at other questions.

The network of interjurisdictional bilateral agreements and understandings which lace the metropolitan areas do, indeed, amount to a successful political integrative effort. Their pattern and consequences need further study. But the diplomatic analogy distorts understanding, particularly about the relationships among established jurisdictions. Among other things, it suggests arms-length dealings and a presumption of jurisdictional autonomy (sovereignty). Holden (1964: 628) writes, "In the metropolitan case, the counties, school districts, special authorities, and other corporate actors each seek to maintain their own 'interest.' To this end, each mobilized available resources of many kinds: money, lawyers, retaliatory legislation, 'civic pride,' and public relations." If that was the recent characteristic pattern of bilateral interjurisdictional politics, in the present period of easy adjustment and accommodationist dealings among jurisdictional actors, it no longer fits. The diplomatic analogy also suggests that interjurisdictional relations are very political, involving negotiations among leaders, rather than the interjurisdictional dealings by department heads, professionals sharing interests and commitments, and other lower-level jurisdictional actors. These latter types of relations, in fact, seem to account for must bilateral dealings within metropolitan areas. When Holden (1964: 628) posits, "Under the ritual, issues are raised and formulated by the highest political officials of the corporate units or those specialized functionaries ('ambassadors') who carry and interpret messages between units," he is speaking of a pattern of political behavior which is in no sense characteristic of bilateral interjurisdictional relations of the period we have entered, even if very apropos a decade ago, when he developed the analogy.

Meanwhile, a major arena for metropolitan political decision-making has emerged in the suprajurisdictional planning authorities and their various hybrids. There is still so little detailed comparative study that it is unclear whether the apparent international analogs to these metropolitan agencies (the IGO, NGOs, and so on), and particularly the scholarly literature on international organizations, might lead to fruitful insights. What does seem clear is that the presumed similarities between the behavior of nation-states and that of municipalities, which made the "diplomatic model" (as well as the regional integration literature) seem so useful an analogy, involved presumptions about the nature of metropolitan politics which are no longer tenable, particularly with regard to city-suburban relations. The critical issues of metropolitan politics no longer pit jurisdictions against each other, as the main actors, even though they occur because of the distribution of people and values among jurisdictions. The relations which do occur between political jurisdictions are of quite a different nature than the model posits. As illuminating and exciting as the Holden article has been, and as influential as it has been upon political scientists (including this author), it highlights the point that many of the presumably stable elements of metropolitan politics—the constants—have not stayed stable at all in the post-reform period of study of this subject.

THE MARKET MODEL

If we treat the metropolitan political community as if it can be summarized as a polycentric metropolis, whose essential actors are jurisdictions competing in the same marketplace for desirable residents, we make certain evaluations concerning the things which are important in metropolitan areas. Norton Long and others have pointed out, for example, that a market model presumes that consumer demands should take precedence over citizenship questions, when evaluating governmental outputs. There are other significant assumptions with the market model, and many of them look more problematical in the 1970s than they did in the 1960s.

Some of the concerns about the viability of the market model as a framework for examining and understanding metropolitan political structure are the same ones which arose when examining the diplomacy analogy. As with the diplomacy model, the market model presumes that the critical issues of the metropolis are interjurisdictional, and the chief actors are independent jurisdictions in mixed competition and cooperation with one another. But that presumption does not seem to allow a user of this framework to thereby get a handle on many of the metropolitan issues of the 1970s.

While some critics of the market model have raised questions concerning the freedom of choice of many particular "customers" to pick a jurisdictional entity which maximizes their own desires, perhaps the more critical assumption of the market model is that jurisdictions have the ability to fashion the list of services they offer and the price they will charge for those services, so as to attract or retain desirable "customers." But the metropolitan marketplace is not a free marketplace, as we are increasingly becoming aware.

With the external interventions of state and federal agencies and of courts, plus the emergence of suprajurisdictional programs of a wide variety and of equalization formulas, the ability of the municipal officials of a jurisdiction to even significantly affect the array of services which will be provided to people and firms living within the geographical limits of their jurisdiction is really quite limited. There remains considerable variation in the services and amenities available from jurisdiction to jurisdiction, but only a diminishing fraction of those services and amenities are subject to significant modification and control by the presumed entrepreneurs of the city halls. What is true about jurisdictional control over the mix of services is equally true about control over the prices (taxes and service charges) which can be charged.

While an inability to control and even significantly affect the package of services is characteristic of *all* jurisdictions within metropolitan areas to some degree, it is particularly apparent for the central cities and the first ring of suburbs. Long-established jurisdictions have a very difficult time altering in any significant way the mix of services and charges they can offer. They not only have lost autonomy along with other jurisdictions, but they are largely locked into the land-use patterns, the millage limitations, and the budgetary commitments which have built up over the years. If new jurisdictions on the suburban fringe have some significant ability to fashion the mix of services they offer, in a response to demand, the central cities and older suburbs largely do not. One is hard-pressed to recall any central city or long-established suburb which has significantly altered the package of public services it offers, except under external duress or in response to compelling outside inducements having little or nothing to do with any pressure to compete in a metropolitan marketplace. By contrast, the social and economic composition of a jurisdiction may change rapidly. One might argue that, even if the presumption that jurisdictional actors can freely fashion the package of services they offer will not stand examination, nothing serious is lost, because the jurisdictions do, in fact, continue to provide a great variety of mixes of services and amenities to residents. Moreover, private citizens and firms may still respond *as if* it were a competitive marketplace in which they could choose. That may be true, but the authors of the model have used the presumption of real, conscious competition among jurisdictions as a basis for anticipating other characteristics of metropolitan politics. They did argue, for example, that the conscious competition might lead to economies of service which would exceed the economies of scale that might occur with metropolitan unification. "Patterns of competition among producers of public services in a metropolitan area, just as among firms in the market, may produce substantial benefits by inducing self-regulating tendencies with pressure for the more efficient solution in the operation of the whole system" (Ostrom et al., 1961: 838). Ostrom et al. (1961: 839) also believed that the competition to provide services would lead to a variety of predictable, suprajurisdictional service arrangements such as sales of services from one jurisdiction to another to achieve the appropriate scale of operation for that particular public service:

> Efficient scales of organization for the production of different public goods may be quite independent of the scales required to recognize appropriate publics for their consumption of public goods and services. But competition among vendors may allow the most efficient organization to be utilized in the production, while an entirely different community of interest and scale of organization controls the provision of services in the local community. . . . Under these circumstances, a polycentric political system can be viable in

supplying a variety of public goods with many different scales of organization, and in providing optional arrangements for the production and consumption of public goods.

The necessary presumption of the market model, that jurisdictional actors can and will largely fashion the range of public goods and services that they will exclusively offer their residents, will become even less tenable, to the extent that the Serrano decision, cross-district busing, and other legal efforts erode the list of exclusive services any particular jurisdiction can market.

Whatever the continuing validity of the market model in predicting behavior—and at least some of my concern can be put into empirical, testable form—the model has a larger problem, for use by political scientists in the 1970s. The model presumes a *direction* to the relationship between jurisdictions and the jurisdictional actors, on the one hand, and the groups of contenders (the "consumers") in the metropolitan areas on the other, that simply is not very broadly useful. The model presumes it is desirable to view the jurisdictions as the independent variables and the groups locating themselves throughout the region as the dependent variables. Without denying that there may be some issues in which that presumption is useful, nor denying that in most real world systems, interactions work in both directions, it really is not too broadly beneficial in the 1970s to conceive of the political jurisdictions as independent actors affecting locational choices, group settlements and the like. For anyone who has followed the efforts to break down various suburban walls or seen the relationships between real estate developers and suburban city council, it would seem far more useful to base a model in the direction of viewing the suburban jurisdictions as entities *who are acted upon*, rather than as initiating parties. But the market model presumes that the jurisdictions are the operative agents of change.

The market model, which has seemed to have such a high potential as a framework for analysis of metropolitan politics, is based upon some substantial suppositions about political behavior which seem increasingly unlikely. It may be that the market model will still serve some narrower uses. But it does not seem likely to continue to be looked upon as a basic framework for examining metropolitan politics.

THE LIFE-STYLE HYPOTHESIS

While Oliver Williams (1971) refers to his urban theory of social access as a conceptual model, it is not an elaborated analog, as are the diplomatic and market models. Instead, particularly when Williams is referring to the politics of metropolitan areas, this social access theory can be reduced to a single, powerful hypothesis: "Assuming no outside interventions, policy areas which are perceived as neutral with respect to controlling social access may be centralized; policies which are perceived as controlling social access will remain decentralized" (Williams, 1971: 93).

Williams argues that it is useful to think of policy areas in which people have desired patterns of interactions with others who share that set of desires as *life-style* policies. These patterns of preferred interactions occur in specific territories. One of the salient territories within which preferred interactions will transpire is the political jurisdiction. Life-style policies, for jurisdiction, include preeminently educational policy, and the screening of potential partners in social interactions. Thus, policy areas such as schools, zoning, housing controls, and cultural and recreational programs become clear life-style issues, in that *jurisdictional membership* includes access to a set of policies which are desired, and restricted to others holding such membership. Other public policy issues, which are not manifestations of a preference for desired social access, are labeled *system maintenance* policies. Thus, Williams predicts that, barring "outside intervention," jurisdictions may centralize their system maintenance programs, to achieve economies of scale, and the like, but not centralize their life-style services. While there may be ambiguity at the margins in deciding which particular policy areas should be classified as life-style and which as system maintenance, this distinction may well shed some light on the confusing pattern of centralization and resistance to centralization, of mixed cooperation and conflict, which have become the hallmarks of metropolitan politics. There is considerable face plausibility to the distinction, although it is not without problems (e.g., a large percentage of the suprajurisdictional "centralization" which occurs seems to be between jurisdictions which share common life-style preferences). With slight transformations, the "life-style model" seems to provide a usable framework for examining contemporary issues of metropolitan politics and formulating testable propositions. Some potential values of this framework include:

(1) Variation in jurisdictional activities are, or can be, conceived as dependent upon initiatives, or inputs, from a variety of sources. Metropolitan politics need not be conceptually limited to issues of jurisdictions acting upon each other or upon "consumers."

(2) While Williams seems to treat the potential responses to pressures for change in policy areas as either centralization or continuing decentralization, one does no violence to the theoretical underpinnings to assert instead that jurisdictions will successfully resist efforts and entreaties to alter their preferred social access (life-style) policies, in the absence of outside intervention.

(3) While Williams appears to consider "outside intervention" as outside suprajurisdictional *governmental* intervention, through such means as state legislative mandates or federal financial inducements, it is consistent to consider court interventions, or perhaps even the nonlitigous actions of social movements as outside interventions.

With such an orienting framework, we can develop propositions and expectations which seemingly have some relationship with what is occurring in metropolitan areas. Jurisdictions within metropolitan areas are under considerable pressure to alter many of their public policies. When those pressures deal with questions of removing restrictions upon social access (life-style issues), as they often do, there seems to be significant jurisdictional resistance; there seems to be considerable outside intervention. Thus, the life-style theoretical construct provides a basis for dealing with the metropolitan issues of the 1970s; on one hand, the many-faceted attempts to break down the pattern of exclusive privileges accruing to people on the basis of the jurisdiction in which they reside, and the resistance to such attempts, and on the other hand, the rapidly emerging suprajurisdictional efforts (on institutions) which are also an increasingly important part of the metropolitan mix. While the very promising diplomatic and market models of metropolitan politics now seem to be of less broad utility than they did recently, the life-style model seems applicable and relevant. It seems to be worth elaboration and testing.

The life-style model suggests the profound rupture in established patterns which will occur if the Serrano decision and other legal challenges to jurisdictional prerogatives ultimately are declared to be the law of the land. Such legal interpretations would also scuttle another promising theoretical approach to metropolitan politics.

REFERENCES

CROUCH, W. and B. DINERMAN (1964) Southern California Metropolis: A Study of Development of Government for a Metropolitan Area. Los Angeles: Univ. of California Press.

HOLDEN, M., Jr. (1964) "The governance of the metropolis as a problem in diplomacy." J. of Politics 26 (August): 627-647.

JACOB, P. E. and J. V. TOSCANO [eds.] (1964) The Integration of Political Communities. Philadelphia: J. B. Lippincott.

MARTIN, R. et al. (1961) Decisions in Syracuse. Bloomington: Indiana Univ. Press.

MOWITZ, R. and D. S. WRIGHT (1962) Profile of a Metropolis. Detroit: Wayne State Univ. Press.

OSTROM, V., C. M. TIEBOUT, and R. WARREN (1961) "The organization of government in metropolitan areas: a theoretical inquiry." Amer. Pol. Sci. Rev. 55 (December): 831-842.

WILLIAMS, O. P. (1971) Metropolitan Political Analysis. New York: Free Press.

––– (1968) "Life-style values and political decentralization in metropolitan areas," pp. 427-440 in T. Clark (ed.) Community Structure and Decision-Making: Comparative Analysis. San Francisco: Chandler.

10

Suburbs in State and National Politics

JOSEPH ZIKMUND II

☐ IN A RECENT BIBLIOGRAPHICAL ESSAY entitled "Inner-City/Outer-City Relationships in Metropolitan Areas," Timothy Schiltz and William Moffitt (1971) suggest that, after two decades of research, we can "feel reasonably certain that the typical central city is Democratic-oriented and that the typical suburban area is more likely to be Republican, but it now appears that the disparity between the two has been considerably overstated." Going on, the authors conclude that

> the final analysis of central city-suburban disparities can be only that we do not know exactly what they are. No one seems quite certain as to what the political outlook of 'suburbia' is. Suburban political behavior and how it compares to central city behavior is a vast, largely unexplored area of research which urgently needs attention [Schiltz and Moffitt, 1971: 85].

It is altogether possible that the primary reason why these authors reached the conclusion they did is because the generalized use of the term "suburbia," like the generalized use of the term "central city," is fundamentally meaningless. How disheartening that after those same two decades we still find writers trying to structure their thinking in terms of a single, national "suburban" pattern in state or national politics. Unfortunately, such misconceptions continue to receive some kinds of academic support.

In the late 1950s, Robert Wood began his classic study of *Suburbia* with the heading "Suburbia as Looking Glass." This phrase was used to indicate that suburbia was

> a looking glass in which the character, behavior, and culture of middle class America is displayed . . . the suburbanite is, by statistical definition, the average American . . . the suburban trend should typify our contemporary way of life [Wood, 1958: 4].

Although Wood (1958: 8) went on to reject much of the normative aura which surrounded the looking glass concept of the fifties, he affirmed its applicability to "middle class Americans wherever they live."

Scammon and Wattenberg (1970: 70-71), describing today's dominant electoral force, pick a midwestern suburbanite as their example of "middle voter":

> Middle Voter: A metropolitan . . . middle-aged, middle-income, middle-educated, Protestant, in a family whose working members work more likely with hands than abstractly with head.
>
> Middle Voter is a forty-seven-year-old housewife from the . . . [suburbs] of Dayton, Ohio, whose husband is a machinist.

Seeking to destroy the previous suburban stereotype, Scammon and Wattenberg (1970: 294) argue that these suburban-based, middle-class Americans cannot be typified by wealth and white-collar jobs.

> They live not in the $70,000 homes in Darien, Connecticut, a 40-mile-distant exurb of New York City, but in the $20,000 houses of Parma, Ohio, immediately abutting Cleveland. They are plumbers. They are foremen. They are airplane mechanics. They are small merchants. They are mostly white, but there are some black busdrivers whose wives are schoolteachers who are moving in.

Wood and Scammon and Wattenberg recognize that this typical middle American is merely a statistical abstraction—the authors' way of summarizing their picture of a statistical average. Unfortunately, some have been inclined to believe that middle voter or typical suburbanite is a mass, popular reality.

Our position throughout this chapter will be that it is equally dangerous to believe in the reality of middle voter, as a uniform type of suburban resident, as it is to believe in the stereotyped pro-Republican character of suburbs and suburbanites. People living in suburbs reflect a tremendous diversity. There is no single, suburban vote in presidential elections, just as there is no single, socioeconomic characteristic which pervades all suburbs. We may speak, as the title of this chapter implies, about "suburbs" in state and national politics, but the evidence of suburban diversity is so overwhelming as to make the discussion of a general suburban pattern both empirically inaccurate and theoretically misleading.

> Because individual suburban places possess dissimilar functional, historical, political, and demographic attributes, they also possess dissimilar social environments. . . . From the standpoint of sociology, the term "suburb" has relatively little meaning. It refers only to a locality nearby and somehow interdependent with a city. To employ the term to suggest a single type of place, or to imply that there is an overarching pattern of "suburban" social organization, is to engage in an overgeneralization of serious proportions [Kramer, 1972: xv].

The question at hand is not how suburban voters respond to elections, but rather how the voters in various suburbs in particular metropolitan areas respond. The two statements are not identical.

POLITICAL DIVERSITY AMONG SUBURBS

The single most overriding fact which emerges from the study of presidential elections in both suburban communities and central cities is the diversity of patterns which one finds. Lest we come to feel that the real difference between central-city and suburban patterns is the uniformity of central-city patterns versus the diversity of suburban patterns, it is important to remember that, for example, in 1956, the cities of Baltimore, Buffalo, Chicago, Cincinnati,

Milwaukee, Minneapolis, New Orleans, and San Francisco all cast a majority of their votes for President Eisenhower, while Boston, Cleveland, Detroit, New York, Philadelphia, Pittsburgh, and St. Louis went for Mr. Stevenson. In that same election, again merely as an example, the most Democratic ward in the city of Chicago voted 83.9% for Stevenson, and the most Republican ward in the city voted 70.7% for Eisenhower (Scammon, 1958).

Suburban differences appear at two important, distinct levels. First, suburbs within any given metropolitan area differ greatly on the bases of the socioeconomic characteristics of their residents and their electoral support for the several political parties contesting presidential elections. Second, the suburbs within any particular metropolitan area, despite their individual differences, seem to form a kind of integrated system of suburbs which responds in a regular and patterned way to the year-by-year forces of national politics and which differs significantly from the responses of suburbs in other metropolitan areas.

In 1968, the suburbs around Boston, Cleveland, Philadelphia, and Pittsburgh produced a number of examples of these assertions. See Table 1 for the record of suburban diversity within each of these four areas. What is evident, as it has been evident since the end of World War II, is that there were Democratic suburbs and Republican suburbs and that some suburbs in each area changed little in their voting patterns between 1964 and 1968, while others changed a great deal between these two elections. Thus, we have clear evidence for the first assertion—the diversity of voting patterns among suburbs within any particular metropolitan area.

TABLE 1

SUBURBAN DIVERSITY IN THE PRESIDENTIAL ELECTION OF 1968
(in percentages)

	Boston	Cleveland	Philadelphia[a]	Pittsburgh
Most Democratic Suburb	82.2	54.2	61.0	69.2
Least Democratic Suburb	31.4	6.3	25.5	10.7
Median Democratic Suburb	59.6	37.1	36.7	39.6
Percentage of Suburbs 50% Democratic or More	71.8	9.8	8.3	25.6
Largest Percentage Point Change 1964 to 1968	14.9	32.5	23.2	23.3
Smallest Percentage Point Change 1964 to 1968	6.3	13.5	3.8	5.1

a. Calculated from sixty suburbs in Pennsylvania.

At the same time, the data on Table 1 begin to hint at the fact that the suburbs in each of these four metropolitan areas responded differently to the 1968 election than suburbs in the other three areas. Put another way, despite the diversity among suburban communities within each metropolitan area studied, it seems that the suburbs surrounding a particular central city respond to national political events such as elections as one integrated whole. This does not mean that all the suburbs in the area react precisely the same way—we have already demonstrated that this is not the case. Rather, it argues that similar suburbs in a metropolitan area tend to react in similar ways. This is illustrated when we look at the ways suburbs in each of the four areas moved politically between the 1964 and 1968 elections. By comparing the earlier vote of a group of communities with the later vote, we can see how closely the responses of similar communities correspond. Statistical correlation is the most precise way to make such an analysis (see Table 2). Correlation coefficients above .90, as we have here, indicate an extremely strong relationship between the component parts used in the analysis, thus indicating very similar responses by similar suburbs to the changed political forces between 1964 and 1968.

A CASE STUDY:

THE PHILADELPHIA AREA

Perhaps the best way to test the idea of a system of suburbs within a given metropolitan area would be to look at an area which rests in or overlaps two different states—the New York City area in New Jersey, New York, and Connecticut; the Philadelphia area in New Jersey and Pennsylvania; or the Cincinnati area in Ohio and

TABLE 2

CORRELATION OF 1964 WITH 1968 VOTING PATTERNS OF SUBURBS

	Correlation Coefficient
Boston (39 suburbs)	.96[a]
Cleveland (51 suburbs)	.96
Pittsburgh (47 suburbs)	.95
Philadelphia (60 Pennsylvania suburbs only)	.90
Philadelphia (99 Pennsylvania and New Jersey suburbs)	.87

a. All correlations reliable to .05.

Kentucky. The availability of relevant data makes the Philadelphia area an ideal choice. The information for the Philadelphia area presented thus far in the discussion has come from 60 suburbs in Pennsylvania. There are also 39 communities on the New Jersey side of the Delaware River which qualify as suburbs of Philadelphia.[1] When all 99 suburbs are taken together and tested the result is quite remarkable. The correlation coefficient for the 60 Pennsylvania suburbs, comparing the 1968 with the 1964 presidential vote, was .90; the correlation for the Pennsylvania and New Jersey suburbs taken together was .87. The addition of New Jersey suburbs to the test produced no important change in result.

Now let us return to the question of differences (if any) between the voting patterns of suburbs and the central city raised by Schiltz and Moffitt. Direct city-to-suburb comparisons are not more common, in part, because one seldom finds aggregate vote and socioeconomic data which come from corresponding election districts and census tracts either within central cities or among suburbs and because both election districts and census tracts often have different boundaries from one time to another.[2] One important exception is the Philadelphia metropolitan area for the years 1956 and 1960. During these years, election districts and census tracts were identical—with only minor exceptions—in the Philadelphia city wards, in the Pennsylvania suburbs, and in the New Jersey suburbs. Here we have one chance, at least, to compare the voting patterns of individual suburbs with those of central-city wards and to explore the impact, if any, of state boundaries on the voting patterns. Correlations were made for suburban communities and central-city wards between the percentage of the electorate voting Democratic and the percentage of the area's adults, as recorded by the 1960 census, having any college education.[3]

Several important patterns emerge from these data. First, there seems to be a relatively high statistical correlation between voting and the education level of the communities which make up the Philadelphia suburbs on the Pennsylvania side of the Delaware River. Second, the correlations are not significantly lower when the New Jersey suburbs are added to the sample. Third, there are considerably lower correlations when Philadelphia city wards are added in to the total number of geographic units under investigation (see Table 3). Just how should we interpret this information?

Historically, we have come to recognize the essential social and economic integration of the entire metropolitan region—the central

TABLE 3
CORRELATION OF VOTING WITH EDUCATION IN
THE PHILADELPHIA METROPOLITAN AREA: 1956 AND 1960

	1956	1960
Pennsylvania Suburbs (n = 60)	−.77[a]	−.73
New Jersey and Pennsylvania Suburbs (n = 99)	−.74	−.71
All Suburbs Plus Philadelphia Wards (n = 151 in 1956; 158 in 1960)	−.63	−.59

a. Number in each cell is the correlation coefficient. All coefficients are reliable to .05. The correlations are all negative because education level is correlated to percentage of Democratic vote.

business district, the rest of the central city, the surrounding residential suburbs, and the variety of satellite communities economically interdependent with the center city (see Ward, 1971: chs. 4 and 5; Hoover and Vernon, 1962; on the Philadelphia metropolitan area, see Williams et al., 1965). There is some evidence to indicate that the political attitudes of suburbanites and central-city residents in the same metropolitan area are quite similar, assuming that the socioeconomic status of the respondent is held constant (Zikmund, 1967). The exception, of course, occurs most frequently when local issues are at stake which tend to divide city from suburb or threaten the legal existence and political independence of the suburb (see Sofen, 1966: chs. 1-4). If there does exist a city-suburban attitude continuum based on socioeconomic level rather than merely as a reflection of city versus suburban residence, then we would expect to find evidence of this in the actual voting patterns of individual suburbs and central-city wards in the same metropolitan area.

The statistical evidence reported above clearly supports the assertion that voting patterns follow socioeconomic patterns in the Philadelphia area. The wealthier the suburb, the more likely it is to vote Republican, and so on (for further evidence, see Bell, 1969: 285; for on-the-whole agreement, see Walter and Wirt, 1972).[4] More importantly, the combined Pennsylvania-New Jersey vote data of suburbs strongly indicates that this relationship is not tied to state boundaries. The suburban portion of this metropolitan area seems to respond politically as if it were an integrated residential system even where some of the suburbs lie in one state while others rest in a second.[5] By contrast, the introduction of central-city wards into the picture weakens considerably the relationship between voting and the educational level of the area. The problem now becomes one of

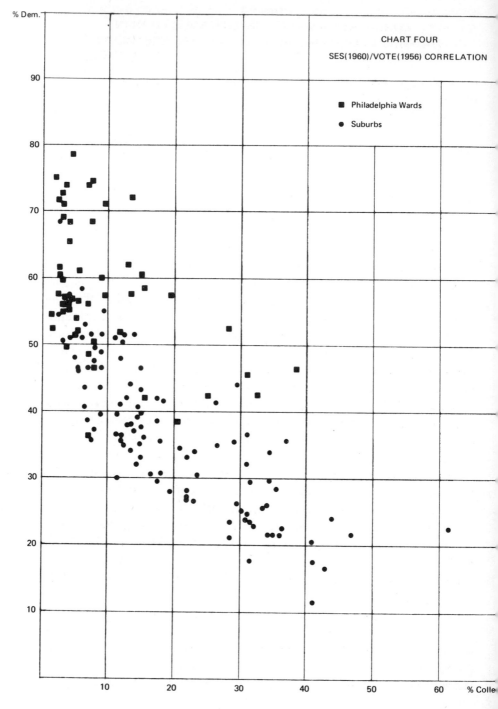

trying to identify and explain the essential difference in the pattern for suburbs versus the pattern for central-city wards.

Two alternative distributions could account for the findings in Table 3. First, the suburbs could demonstrate a high correlation which would focus on one regression line and the central-city wards also could demonstrate a high correlation which would fit a second, parallel, but essentially different, regression line. If this were the case, suburbs at a given socioeconomic level would always be more (or less, as the case may be) Republican than city wards at that same socioeconomic level. Thus, some explanation would have to be provided to account for this general, more (or less) Republican pattern among suburbs. Second, the suburbs could demonstrate a high correlation between political and socioeconomic factors, while the central-city wards would exhibit no such (i.e., a random) relationship. In this latter case, we would have to try to explain why the city wards failed to produce a high correlation between voting and socioeconomic level.

The best way to decide between the two alternatives is to look at the scatter diagrams. Since the 1956 and 1960 distributions turned out to be quite similar, only the 1956 chart is reproduced here for analysis (see Figure 1). Note that the Philadelphia city wards do form a broad line which, generally speaking, is more Democratic than the suburbs—suggesting the first alternative outlined above. However, our attention should focus even more on the spread—the almost random spread—of Philadelphia wards at the lower end of the education scale (0 to 20% college-educated). What kinds of explanations are possible for such results?

Looking at the close adherence of the suburban communities to the vote/education regression line, it is difficult to imagine some special, unique characteristic of suburbs which makes them uniformly more Republican than central-city wards. The pattern for suburbs alone seems both reasonable and satisfying. Suburbs vote roughly on the basis of the socioeconomic levels of their residents (for a qualification of this in a discussion of the impact of ethnic loyalties on suburban vote, see Walter and Wirt, 1972: 120-123). Is there as satisfactory an explanation for the especially pro-Democratic voting pattern of city wards? For years, we have been asking the question: why are suburbs more Republican than the central city? In effect, we have been assuming that the city vote was somehow "normal," while the suburbs have been turning in especially or uniquely pro-Republican "abnormal" votes. Perhaps we have been asking the wrong question and seeking answers in the wrong places—i.e., the suburbs. A more appropriate, and potentially more fruitful, question might be: if the suburbs tend to vote on the basis of the socioeconomic levels of their residents, do the several neighborhoods within the city vote on the same basis and, if not, why not? At this point our evidence from Philadelphia indicates that city wards are, in fact, uniquely pro-Democratic when compared with the surrounding suburbs.

It would be impossible to try to explain completely these findings from the city of Philadelphia without looking at the particularities of every city ward at the time of these elections. Even then, it is inevitable that many of the individual pieces of data would remain beyond our total comprehension. However, it is possible to suggest two generalized, possible explanations for these results. One of these would be the differing levels of political organization among the city's wards. It was at this time that the present Democratic organization in Philadelphia was in the formative stage. The old Republican machine had been thrown out in the late forties by reform Democrats. At the same time, the full workings of the new Democratic organization had not fully emerged throughout the city. Under such circumstances it is likely that the long-range impact of organization—increased turnout and loyal party voters—would produce a situation in which the city would vote more Democratic than its unorganized suburban neighbors but would not, as yet, be able to produce uniform results among all city wards at the same socioeconomic level (historical information from Reichley, 1959).

A second plausible interpretation might focus on the high concentration of unreconstructed—not totally homogenized into the "American" middle-class way of life—ethnic peoples in the city proper (on the persistence of ethnic politics in big cities, see Glazer and Moynihan, 1964). In Philadelphia, with its abundance of Italians, Jews, and Blacks, among others, we would expect somewhat higher pro-Democratic voting trends at this stage of American political history than in suburbs, where ethnic identity is often submerged and where more universal socioeconomic factors come into play (on Philadelphia ethnics, see Banfield, 1965: 108; Binzen, 1970).

On the other hand, it is virtually impossible to develop a reasonable explanation for why suburbs are uniformly more Republican than city wards other than the fact that suburbs tend to vote on the basis of socioeconomic level while city wards do not. To produce an alternative explanation forces us into some kind of artificial "fudge factor" such as "suburbanization," which allows us to take the empirical evidence and mold it into our preconceived theoretical structures. In sum, it seems to make much more sense to treat suburban patterns as if they demonstrate the typical American pattern rather than to treat them as deviants from the historical patterns of central cities.

When we suggest that the vote in suburbs may be more "normal" or representative of the nation as a whole than the central-city vote, the assertion is not that there is some kind of unique suburban man who stands as the archetype of the political American, after the fashion of Scammon and Wattenberg's middle voter, for example. Quite the contrary, it suggests that those geographic entities which we call suburbs contain a variety of peoples who are rapidly becoming more statistically representative of the total American population than either the cities or rural areas ever were—at least in terms of containing people in some numbers from all major social and economic categories—and who are uniquely unencumbered by the special historical factors which made large cities more Democratic than the socioeconomic levels of their citizens would indicate —racial and ethnic identities and tightly organized political machines (for an interesting body of statistical evidence supporting this argument, see Hirsch, 1968). By the time of the 1970 census, a plurality of Americans had come to live in the suburbs (about 36% of the total population, up 11 percentage points from 1950; Scammon and Wattenberg, 1970: 67-68). Barring a radical shift in American population trends in the next two decades, it is quite likely that, by 1990 or certainly by 2000, a majority of Americans will live in the suburbs. Put simply, if the suburbs do not now demonstrate the dominant political patterns of the nation, they soon will. Ultimately, suburban voters are likely to typify the dominant features of American politics because of the size of their numbers and the overall representativeness of their members.

HOW SUBURBS HAVE VOTED IN
PRESIDENTIAL ELECTIONS

Scammon and Wattenberg (1970: 294), at the end of their discussion of contemporary voting trends, seem to reach the same basic conclusion as we expressed above:

> Suburbia will indeed be the major psephological battleground in the years to come but will probably be the major battleground only because so many Americans will be living there. Anyone who automatically deeds that turf to Republicans does so at his peril.

The authors imply that suburbs have been becoming more Democratic in recent years and that the trends represented by this movement are of a permanent or lasting nature rather than the product of short-term fluctuations in the fortunes of the two major parties. Looking back on the suburban vote data which have been accumulated over the past two decades, it is apparent that Scammon and Wattenberg are correct, and the trends they highlight have been going on for some time. Assuming the national Democratic/Republican vote ratio in any given election may be used as a standard to judge the political likeness of any subpopulation to the nation as a whole, we would argue that most suburban communities, or at least those studied here, seem to be moving toward a cluster around this standard and away from an especially pro-Republican or pro-Democratic character which may have existed in the past. A recent analysis (Zikmund, 1968: 253-254) of election trends in presidential contests between 1948 and 1964 reached the following conclusions:

(1) In the period from 1948 to 1964, even disregarding the strong pro-Republican trend in 1948 and the heavy pro-Democratic trend in 1964, a slight pro-Democratic shift seems to have occurred, generally, among suburbs.

(2) This general pro-Democratic pattern is considerably more evident among previously Republican suburbs than among previously Democratic suburbs.

(3) The overall effect of these voting shifts has been to narrow the percentage point range between the most extreme Republican suburbs and the most extreme Democratic suburbs: that is, to decrease the range of political diversity among suburbs within each metropolitan area. . . .

(8) Changes in social-economic patterns from 1950 to 1960 . . . did not account for the changes in suburban voting patterns between 1948 and 1960 or 1964.

In addition, the election-to-election shifts of suburban communities over this period followed national patterns quite closely (Zikmund, 1968: 253). The 1968 election did nothing to alter the fundamental character of these findings. In two of the four areas included in this study—Cleveland and Pittsburgh—1964 to 1968 shifts produced even further diminishing of the political differences between the most Democratic and the most Republican suburbs in the metropolitan region. In the other two areas—Boston and Philadelphia—there was a very slight increase in the range of political extremes, but not enough to reverse the general trend in the opposite direction for the entire twenty-year time span.

Does this imply that the assertion of political diversity among suburbs with which we began this chapter is not correct or is not likely to hold in the near future? The answer to both questions is no. The evidence for political diversity among suburbs in the past and at present has already been spelled out (see again Table 1). The question now is the trend in voting patterns over the long run. Table 4 presents the relevant evidence. Results for 1948 are compared directly with those of 1968 because both elections were very close contests and because suburbs tended to exhibit a pro-Republican bias both years, as contrasted with the 1960 presidential contest when suburbs showed a slight pro-Democratic bias. In all four metropolitan areas, the twenty-year pattern brought the median suburb closer to the national Democratic percentage. (Boston suburbs, of course, produced this same generalized result, but, in fact, seem to be

TABLE 4
TRENDS IN SUBURBAN VOTING PATTERNS 1948 TO 1968

	1948	1968
Boston		
Difference of median suburb[a]	−10.7	9.6
Range of suburbs[b]	56.6	49.8
Cleveland		
Difference of median suburb	−8.5	−5.6
Range of suburbs	60.3	47.9
Philadelphia (60 suburbs)		
Difference of median suburb	−19.6	−6.0
Range of suburbs	46.8	35.5
Pittsburgh		
Difference of median suburb	−14.1	−3.1
Range of suburbs	74.3	58.5

a. "Difference of median suburb" means the difference between the median Democratic vote of suburbs in the particular metropolitan area and the Democratic vote for the nation as a whole. Cell number is expressed in percentage point difference.
b. "Range of suburbs" means the percentage point range between the most Democratic and least Democratic suburb in the particular metropolitan area.

becoming predominantly pro-Democratic in character rather than more typical of the nation as a whole.) Similarly, in all four areas, there was a sizable decrease in the percentage point range between the most Democratic and least Democratic suburbs; in three of the four areas, this occurred as a consequence, at least in part if not entirely, of a pro-Democratic shift by the least Democratic suburb in the metropolitan area. The exception was the Cleveland area, in which suburbs have tended to become slightly more Republican in recent years. Again, the reader must be reminded that the assertion is not that there is emerging some kind of average suburban man who votes like the nation as a whole, but rather that suburbs, when all their individual differences are averaged together, are coming statistically closer to the nation as a whole; thus, the Cleveland and Boston results, while contrary to the patterns of the two other areas studied, do not necessarily contradict the proposition being tested.

The final question regarding the impact of suburbs on national elections concerns the Electoral College. Since the electoral vote is cast in state blocs, the important problem is the impact of the vote in suburban communities on statewide pluralities for one party or another. It is one of the ironies of contemporary American politics that, as suburbs have been moving, generally speaking, in a more Democratic partisan direction—e.g., from 35 to 40% Democratic—it is altogether possible that they are contributing the same, or roughly the same, vote pluralities to the Republican Party. As a hypothetical example, assume that a suburban community of 4,000 total voters in 1948 cast 35% of its vote for the Democrats and 65% for the Republicans. This would have provided Mr. Dewey with a 1,200 vote plurality. Let us say that the same suburb voted 40% Democratic and 60% Republican in 1968. At the same time, let us assume that the voting population increased by only 50% to 6,000. As a consequence, Mr. Nixon also would have a 1,200 vote plurality. This is not a far-fetched example, for a number of suburbs in this study increased by almost 50% during the 1960s alone. However, the discussion cannot stop here, for, during this same time, both the large central cities, generally speaking, and the rural hinterlands have lost population. In effect, by merely staying even, it is quite likely that our suburban victor in 1968 actually turned out to be further along the road toward capturing the state's electoral vote than his counterpart twenty years before.

Thus, the problem of the impact of suburbs on the electoral college vote boils down to a series of more basic questions. How have the population patterns of the state—in the large central cities, the variety of suburban communities, and the rural areas—changed over the past two decades? How have these population shifts affected the partisan character of the various parts of the state? And finally, how have the previous two sets of changes altered the pattern—both qualitatively and quantitatively—of electoral pluralities throughout the state? The answers to these questions will differ from state to state. To answer for the Boston area, where suburbs seem to be becoming significantly more Democratic, will not be the answer for the Cleveland area, where just the reverse trend may be occurring. Nor will it do for the Chicago, Detroit, or Los Angeles metropolitan areas. In the end, we are forced back to where we began; that is, to the diversity of political patterns in suburbs looking at the nation as a whole and to the importance of exploring the question of suburban impact within the context of the particular metropolitan area and the particular state in question. To attempt to generalize at the national level is to render the endeavor virtually meaningless.

SUBURBS AND STATE ELECTIONS:

PENNSYLVANIA CASE STUDY

The argument developed above concerning suburban impact on the electoral college vote applies in the same way to the question of suburban impact on state elections and for precisely the same reasons. Therefore, instead of trying to look at suburbs from a number of different states and metropolitan areas, we will look only at the Philadelphia and Pittsburgh areas and their impact on Pennsylvania state elections. We would not suggest that the particulars of this case study are typical of the nation as a whole. Quite the contrary; however, they do exhibit several patterns which we would expect to find elsewhere in the nation, and they suggest at least one unique idea which deserves considerably more investigation.

Probably the strongest pattern found in these two metropolitan areas from 1954 to 1970[6] is the way suburbs follow the statewide electoral swings in party support (see Table 5). In 1954, the state voted 53.8% Democratic; in 1966, the state went Republican and the Democrats captured only 47.0% of the total vote; in the most recent gubernatorial election in 1970, the Democrats again carried the state

TABLE 5
PARTISAN SHIFTS IN SUBURBAN VOTING IN GUBERNATORIAL
ELECTIONS FROM 1954 TO 1970: PENNSYLVANIA

	Philadelphia	Pittsburgh
Number of Suburbs Showing Consecutive Decreases in Democratic Percentage: 1954, 1966, and 1970	0	0
Number of Suburbs Following the Statewide Pattern: Democratic Percent Down from 1954 to 1966 and up from 1966 to 1970	51	33
Number of Suburbs Showing Consecutive Increases in Democratic Percentage: 1954, 1966, and 1970	9	15
Number of Suburbs More Than 3.2[a] Points Higher in Percentage Democratic: 1954 to 1970	50	38

a. The statewide shift between 1954 and 1970 was plus 3.2 percentage points for the Democrats.

with 57.0%. Thus, the statewide trend for these three elections shows Democrats, Republicans, and then Democrats winning and produces a 3.2 percentage point gain for the Democrats between 1954 and 1970. How do the suburbs compare? Sixty-nine percent of the Pittsburgh suburbs and 82% of the Philadelphia suburbs followed the statewide election-to-election pattern for these three elections. Of the suburbs which did not follow the statewide trend, none demonstrated a pattern which was consecutively more Republican for each of the three elections. Instead, all the exceptions showed a pattern which was consecutively more Democratic for the three elections. The more frequent incidence of exceptions among Pittsburgh suburbs may be due to intrametropolitan influences because the city of Pittsburgh itself went progressively more Democratic in these three elections, while the city of Philadelphia followed the broader, statewide pattern.

In addition, it seems safe to assert that the general pattern for suburbs in both areas was pro-Democratic in state elections (see again Table 5). Not only do the exceptions to the statewide pattern tend to fall in a pro-Democratic direction, most of the suburbs in both metropolitan areas shifted considerably more Democratic than the statewide shift of 3.2 percentage points. Once again, we must be careful not to overestimate the impact of these changes in partisan composition for increased population in most suburbs may still be producing relatively strong Republican pluralities, though these too tend to fluctuate with the statewide shifts from election to election.

Third, the voting patterns of these suburban communities in state elections are just as closely tied to the socioeconomic character of suburbs as were the presidential elections discussed earlier (see Table 6). With the exception of the 1950 correlation for Philadelphia suburbs, the results here are comparable to those found in the earlier section on presidential elections. If there needs to be an attempted explanation for the 1950 Philadelphia data, the answer probably lies in the political history of the city at this point in time. Again, the old Republican machine was collapsing, while the newly emerging Democratic organization had not reached full efficiency.

Looking at these data from the Philadelphia and Pittsburgh metropolitan areas, aware of Walter and Wirt's modest correlations between socioeconomic factors and voting in suburbs across the nation, and strongly drawn by Elazar's (1966: chs. 4 and 5) ideas concerning dominant state political cultures, the temptation to test the Philadelphia and Pittsburgh suburbs together in a Pennsylvania state election was overwhelming. The correlation coefficient for the suburbs in both metropolitan areas taken together was .55, significantly lower than for each of the areas treated separately. As a consequence, we have added evidence pointing to the metropolitan area itself as the structuring force in SES/vote patterns among suburbs. At the same time, the idea of a statewide political culture affecting the voting patterns of suburban communities generally throughout a given state deserves further analysis in states like Ohio, Texas, and California, where more than one metropolitan area exists in the same state.

SUMMARY AND
CONCLUDING REMARKS

The arguments developed in the body of this chapter may be summarized as follows:

(1) There is a great deal of social and political diversity among the suburbs surrounding any particular central city. This diversity produces regular Democratic suburbs and regular Republican suburbs in most metropolitan areas, with the majority of suburbs in most metropolitan areas tending to the Republican side of the ledger.

TABLE 6
CORRELATION OF GUBERNATORIAL ELECTIONS WITH
EDUCATION IN PENNSYLVANIA SUBURBS

	1950	1960
Philadelphia Suburbs	−.55[a]	−.75
Pittsburgh Suburbs	−.79	−.76

a. Figure in each cell is the correlation coefficient relating the percent of the community's population with some college education and the percentage voting Democratic; 1950 figures relate 1950 census results to 1950 gubernatorial election results; 1960 figures relate 1960 census data to 1958 gubernatorial results. Correlations significant to .05.

(2) The social and political patterns of suburbs vary considerably from one metropolitan area to another, even within the same state. As a consequence, it is very misleading to speak of any kind of national suburban pattern.

(3) The partisan movement of suburbs, generally, tends to follow quite closely the nationwide and statewide political trends. Often when suburbs seem to be moving counter to broader nationwide or statewide trends, they are, in fact, responding quite typically to local, metropolitan area trends involving the central city and its suburban ring.

(4) Within particular metropolitan areas, there is a strong relationship between voting and aggregate socioeconomic factors. The strength of this relationship tends to decrease significantly as one tries to lump together suburbs from different metropolitan areas within the same state or suburbs from several metropolitan areas in different states.

(5) If there is a long-term shift in suburban political loyalties, it is probably a multidirectional aggregation of shifts which will tend to move the average partisan division of suburbs across the nation as a whole into even closer alignment with the partisan balance of the entire country. The long-term effect of these many individual trends will probably result in some suburbs becoming more one-party—both Democratic and Republican—in character while the majority of suburbs become more competitive between two major parties.

(6) The cause of this long-term change in average partisan composition of suburbs is not a result of some abstract suburbanization process, but rather is the fact that the socioeconomic character of suburban communities, when averaged across the general American population —both in the sense of the social mix of peoples living in the suburbs and of the proportion of the total American population living in these communities.

(7) Finally, to properly come to grips with the problem of the impact of suburbs on national or state politics, one must analyze the impact of

suburbs on the real constituencies of electoral competition. In effect, there is no national electorate; there are only state and local electorates. Thus, one must seek to answer the question of suburban inpaot primarily through a variety of more basic questions for each state and each metropolitan area separately, recognized and studied as a distinct political arena unto itself. These and these alone give a context for studying suburban impact.

Ultimately, the problem with trying to generalize about suburban impact—whether in state elections, in the Electoral College, or in legislative politics—is that the suburbs are seldom the masters of their own power. In elections, the question of impact revolves around other questions such as the various population shifts in the state and the impact of these numerous shifts on the partisan balance and on the partisan pluralities in the many distinct political areas of the state. The same principle applies in the investigation of suburban impact on legislative politics as well. Not only must we try to explore the total movement of peoples around the state, but also such purely political factors as legislative districting personality of candidate, and a variety of local factors, too.

Several years ago, I concluded an article (Zikmund, 1967: 75) on political activity and political attitude patterns in suburbs with the following sentence: "The key to an accurate understanding of the suburban phenomenon rests with theories and research that concentrate on each particular metropolitan area and on the unique environment it provides for suburbs and suburban residents." Nothing that I have seen or done since that time has caused me to change that judgment. We must look at the impact of San Francisco suburbs on California politics or Philadelphia suburbs on Pennsylvania politics. The further we move from this emphasis on the particulars of each distinct case, the greater our chance of overgeneralization, of inaccuracy, and of just sheer nonsense. There is no national suburbia; there are no uniform suburban patterns or characteristics which distinguish these places from central cities or other kinds of satellite communities—except simple geographic location. It behooves us to stop using the term as if there were a national suburban phenomenon—even as a short cut, when we really know better—because we perpetuate ignorance and misunderstanding when we do. Suburbs must be understood as individual communities within a metropolitan system, not because some ideology of metropolitanism tell us, but because that is the way suburbs actually develop and function (for examples of studies with a specific metropolitan area focus, see Gilbert, 1967; Williams et al., 1965).

NOTES

1. The working definition of a "suburb" used in this study is elaborated in Zikmund (1965: 32-35). The definition focuses on the commuter relationship of a satellite community to the central city. Thus, when we speak of Philadelphia suburbs in New Jersey, we mean communities which hold a considerable working population which commutes daily into the city of Philadelphia to work.

2. For example, the reason we cannot add Philadelphia wards to the 1964-1968 suburban analysis just completed is because the ward boundaries were altered significantly during these four years making meaningful comparisons between the two elections impossible.

3. Other social-economic factors were tested in a similar way with roughly comparable results. These are not included here in order to conserve space.

4. However, Walter and Wirt (1972: 117-119) had lower correlation coefficients than we have found and they express considerable concern over the use of ecological correlations in this way. It is our opinion that the major reason Walter and Wirt found lower SES/vote correlations is that they used 407 suburban communities spread throughout the nation as a whole, while we calculated the correlations separately for each "system of suburbs" within the particular metropolitan areas studied. We assert that the strongest SES/vote correlations occur among suburbs within any single metropolitan area.

5. This applies only to presidential elections. In state elections, of course, it would be expected that suburbs would reflect the political climates of their particular states rather than some general metropolitan political climate.

6. The 1954 gubernatorial election was chosen as the base of this analysis because the Democrats won both that year and in 1970, thus minimizing the likely pro-Democratic shift among suburbs due to the influence of statewide partisan trends.

REFERENCES

BANFIELD, E. C. (1965) Big City Politics. New York: Random House.

BELL, C. G. (1969) "A new suburban politics." Social Forces 48 (March).

BINZEN, P. (1970) Whitetown USA. New York: Random House.

ELAZAR, D. J. (1966) American Federalism. New York: Thomas Y. Crowell.

GILBERT, C. E. (1967) Governing the Suburbs. Bloomington: Indiana Univ. Press.

GLAZER, N. and D. P. MOYNIHAN (1964) Beyond the Melting Pot. Cambridge, Mass.: MIT Press.

HIRSCH, H. (1968) "Suburban voting and national trends." Western Pol. Q. 21 (September).

HOOVER, E. M. and R. VERNON (1962) Anatomy of a Metropolis. Garden City, N.Y.: Doubleday Anchor.

KRAMER, J. (1972) North American Suburbs. Berkeley, Calif: Glendessary.

REICHLEY, J. (1959) The Art of Government. New York: Fund for the Republic.

SCAMMON, R. M. [ed.] (1958) American Votes: Volume Two. New York: Macmillan.

––– and B. J. WATTENBERG (1970) The Real Majority. New York: Coward-McCann.

SCHILTZ, T. and W. MOFFITT (1971) "Inner-city/outer-city relationships in metropolitan areas." Urban Affairs Q. 7 (September).

SOFEN, E. (1966) The Miami Experiment. Garden City, N.Y.: Doubleday Anchor.

WALTER, B. and F. M. WIRT (1972) "Social and political dimensions of American suburbs," pp. 109-111 in B.J.L. Berry (ed.) City Classification Handbook. New York: Wiley Interscience.

WARD, D. (1971) Cities and Immigrants. New York: Oxford Univ. Press.

WILLIAMS, O. P. et al. (1965) Suburban Differences and Metropolitan Policies. Philadelphia: Univ. of Pennsylvania Press.

WOOD, R. (1958) Suburbia. Boston: Houghton Mifflin.

ZIKMUND, J. II (1968) "Suburban voting in presidential elections: 1948-1964." Midwest J. of Pol. Sci. 12 (May).

––– (1967) "A comparison of political attitudes and activity patterns in central cities and suburbs." Public Opinion Q. 31 (Spring): 69-75.

––– (1965) "Suburban vote patterns in the Northeast: 1948-1960." Ph.D. dissertation. Duke University.

Part IV

POINTS OF CONTENTION:

SUBURBAN ISSUES

IN THE SEVENTIES

POINTS OF CONTENTION:

SUBURBAN ISSUES

IN THE SEVENTIES

The points of contention in the suburbs increase in scope and intensity as the area urbanizes and becomes more complex and heterogeneous. Suburban issues such as metropolitan government, which not too long ago appeared to be of crisis proportion, pale by comparison with recent and emerging ones. The older problems tended to be administrative and organizational or related to physical plant; the new breed of issues are more human, more normative, and considerably more emotional. Whether they are resolved, and how, will influence the shape and stability of suburbia and the entire metropolitan area.

The opening essay of this secion, "Problem-solving in Suburbia: The Basis for Political Conflict" by Downes, provides a framework for examining suburban problems. It includes an analysis of the conflicts created by suburban problems and their severity, a way of typologizing these problems, and a discussion of government's role in problem-solving and the political obstacles to meaningful problem-solving in suburbia.

The next four papers focus on specific suburban policy issues which now have, or soon will have, a significant impact on the suburban scene. Babcock and Rubinowitz deal with complementary issues—exclusionary zoning and the inaccessibility of the poor and Black to suburban housing. After tracing recent developments in zoning law, Babcock concludes that the long-established municipal power to regulate and control the use of land within its jurisdiction is shifting to the state as a result of court interpretations.

In the companion essay, Rubinowitz analyzes the obstacles being encountered by the poor on the road to the suburbs and discusses alternative strategies for overcoming them. While Babcock concentrated on the relatively private strategy of challenging exclusionary zoning ordinances, Rubinowitz concentrates on public and quasi-public strategies at the county, regional, state, and national levels.

Wirt plays a variation on Babcock's theme of receding municipal autonomy in his analysis of financial and desegregation reform in suburban education. Local control of educational policy, like land use, has been sacrosanct in American society, but the revolution is on. The strategy for reforming racial and class discrimination in education, as in zoning, has been the legal challenge. But if the challenge is successful, and at this point the outcome is not clear, there is yet no agreement on the alternative to local control.

The last subject treated in this section is a "sleeper." As suburbia urbanizes and crime rates soar, personal security and the effectiveness of suburban law enforcement agencies are rapidly becoming an issue. Ostrom and Parks address themselves to the problem of police function fragmentation. In anticipation of efforts to promote the consolidation of metropolitan law enforcement agencies, they review the evidence on the efficacy of consolidation and report findings from their own research on the relationship between the size of police jurisdictions and police service levels.

11

Problem-solving in Suburbia:
The Basis for Political Conflict

BRYAN T. DOWNES

□ MOST INTRA- AND INTERMUNICIPAL POLITICAL CON-
FLICTS which occur in suburbia are the result (as they are in central
cities and elsewhere) of disagreements over the following kinds of
questions:

> (1) What are our problems? How serious are they? Do any constitute
> crises?
>
> (2) How can we best solve our problems, particularly the ones we consider
> serious or crises?
>
> (3) What role should government (local, state, and federal) play in this
> problem-solving process?

In this essay, I intend to examine why these questions give rise to
controversy and conflict in suburbia.

LOCAL-LEVEL POLITICS

One positive consequence of the political conflicts of the last two
decades has been that political scientists increasingly view *local*

politics as a process which not only involves the making but also the implementation of public policy—as a process of authoritatively deciding who gets what, when, how (Lasswell, 1958; Dye, 1972; Downes, 1971). Researchers have both described and explained this process. They have also begun to be more concerned about the consequences which this process has for meaningful problem-solving by governmental institutions. Not only are they interested in understanding *who gets what and why* but also *what difference it makes* for people and the problems they face (Lineberry and Sharkansky, 1971). For example, with regard to this latter concern, it is becoming apparent that, although increasing expenditures for public education (or any municipal service for that matter) usually has some impact on the *quantity* of educational services local school districts are able to provide, it may have very little impact on the *quality* of education students receive. This is most apparent in ghetto-area schools in our largest and oldest central cities but also increasingly true in many suburban schools.

The major strength of current efforts to understand local politics is, of course, the focus on both the causes and the consequences of governmental activity. A major weakness, however, and it is a critical one, is the *neglect of the causes and consequences of local government inactivity,* particularly the unwillingness or inability of policy makers to solve central-city and suburban problems. In studying politics, we should not only be concerned with who gets what, when, how but who gets left out of this process and why? Who does not benefit from existing governmental arrangements and activities? Whose demands are not responded to in other than a token or symbolic manner? Whose problems do not get solved? And, most importantly, why is this the case?

This concern is analogous to that recently shown by some political scientists in the "second face of power" (Bachrach and Baratz, 1970). Political power is most often thought of as being used to assure that political demands are responded to positively and that certain public policies are enacted and implemented. But those who have political power can and frequently do use their power to assure that demands of certain citizens *are not responded to* by local decision makers and that these same citizens do not participate meaningfully in local politics. In this case, power is being used to confine the scope of local politics to relatively safe problems, demands, issues, and participants. These are ones which do not pose a threat to those who dominate the process of deciding who gets what, when, how. One has only to review the history of demands by black Americans, the poor, and the powerless for meaningful problem-solving by local (state and federal) policy makers to realize the validity of this perspective.

SUBURBAN POLITICS: A
CONTINUOUS CONVERSION PROCESS

The process of authoritatively deciding who gets what, when, how in suburbia (as it is elsewhere) is probably most accurately conceptualized dynamically as a continuous conversion process—a process spanning a given period of time, which, for purposes of study, can be broken down into a series of interrelated stages (Downes and Greene, 1973).

The *first stage* in this conversion process involves the perception of a problem by an individual or group and the formulation and articulation of a demand for a policy change which should lead to the problem's solution. A *demand* is a request for an expansion or contraction in the net scope or way of functioning of local government (Agger et al., 1964: 6-10). Suppose, for example, that a group of people living in one of several inner suburbs which make up a particular school district becomes increasingly dissatisfied with the quality of education its children are receiving in the district's public schools. They decide that, in order to upgrade the quality of education, they, as parents, must be able to participate more directly in educational policy-making, particularly in such matters as cur-riculum, hiring and firing of teachers and administrators, setting standards for evaluating teacher performance, and so on. Hence, they present a demand to the district's school board for community control—for municipal or neighborhood school boards rather than a single, districtwide board (Altshuler, 1970). This demand is for the redistribution of policy-making and implementing authority down-ward from a central to the municipal or neighborhood level.

The *second stage* involves the transformation of the initial demand into an issue. An *issue* is a demand that members of the political system—its authoritative policy makers, for example—are prepared to deal with as a significant item for discussion through regular channels. For example, the demand for community control of district schools becomes an issue when the school district's board agrees to consider such a policy and the demand is either referred to a board committee or placed on the board's formal agenda. This is a very critical stage in the political process, since demands for policy change, particularly by the powerless, are so often ignored rather than acceded to by government policy makers.

The *third stage* in this conversion process involves the transformation of an issue into public policy. An issue becomes a public policy when a city council, school board, or some other authoritative decision-making unit acts on the proposed alteration in the scope of governmental activity. Thus, community control is converted from an issue into school policy when a school board adopts or does not adopt a community control ordinance.

The *fourth stage* in the conversion process involves the implementation of the policy or program. This is a stage that is too often neglected by proponents of policy change. It is a critical stage, since policies and programs which are not implemented do not solve problems. For example, particularly important for the successful implementation of community control would be that meaningful policy-making and implementing authority be transferred to municipal or neighborhood school boards.

The *final stage* encompasses feedback and subsequent learning by members of the authoritative policy-making or implementing unit. What was the policy's impact on the target situation or group, was it intended or unintended, anticipated or unanticipated, symbolic or tangible? Does learning take place and result in the policy being adjusted or fundamentally revised, if feedback from citizens and research is negative? In the case of community control, it is particularly important to evaluate whether direct participation in educational policy-making by municipal or neighborhood school boards does actually result in upgrading the quality of education students receive (Fantini et al., 1970).

By breaking the political process down into a series of interrelated stages, we are better able to see how the outcomes of each stage build on the outcomes of previous ones. For example, the *second stage* begins with an impetus, a demand, for policy change. People, of course, do not engage in demand-articulating activity unless they are involved in their community, feel that something important is at stake, or have achieved a certain degree of political consciousness. Furthermore, demands for policy change frequently give rise to controversy and conflict. For example, demands for the adoption of open housing, zoning changes, capital improvements, expansion of park and recreational facilities, and increases in salaries of city personnel may all generate a great deal of conflict not only among community residents but also among policy makers. As a result, a variety of outcomes are possible, ranging from acceptance of the demand and then its transformation into an issue, at one extreme, to its modification or rejection at the other. If the former, policy makers are predisposed to further change. If the latter, the initial impetus may have no further implications for policy change.

One factor which I have found to affect the willingness of city councils, for example, to consider demands for policy change, has to do with whether or not the council employed a committee system (Downes, 1969; Downes and Greene, 1973). If a council uses a committee system, it can accept demands, refer them to committee, and then delay or cancel action on them. In this case, acceptance of demands is merely a symbolic response. Or, if a council is unsure whether it eventually wants to adopt the policy change, it may place the issue in committee until it has reached a decision. Referral to committee does not commit a council to action—it may simply be good public relations or serve to delay action on the issue until concern or controversy generated by the issue have subsided. The council can then act under less public scrutiny. One could hypothesize, on the basis of these findings, that councils which use committees (or their functional equivalents) will accept and consider demands from a wider range of individuals and groups in the community than councils which do not make use of committees. Furthermore, councils with committee systems may initially, at least, avoid conflict over demands presented to them.

In the *third stage,* efforts are made by local policy makers to come to terms with the issue. A number of phenomena (variables) determine whether in this stage the issue will be transformed into public policy or whether some intermediate or no action will be forthcoming. These include the level of demand and conflict, the availability of resources, the ability of policy makers and other participants to mobilize and coordinate political action to obstruct or modify the proposed policy change.

Some suburban municipalities are far more likely to become embroiled in conflict at this stage in the policy-making process, particularly if controversy developed in the first or initial demand articulation stage. For example, I have found (Downes, 1969) that suburban communities undergoing very high rates of population growth (over 500% from 1950-1960) not only had more issues coming before their councils which gave rise to issue disagreements but issue disagreements based on enduring council factions. Although medium growth rate communities, those whose populations increased from 16 to 100%, did have a great deal of issue conflict among city councilmen, such disagreements were seldom based on enduring factions. In these municipalities, conflict did not appear to be dysfunctional and impede council task performance as it did in very high growth rate suburbs.

Furthermore, issue disagreements, when they did occur among suburban councilmen, were far more easily resolved if councilmen interacted smoothly—in a cooperative rather than a noncooperative manner. In a small decision-making body like a suburban city council, affective relations among councilmen are extremely important and have a direct impact on the group's ability to perform its various tasks (Verba, 1961; Barber, 1966). For example, councils which interact smoothly—ones with high or positive affective relations—are more likely to be able to work out their disagreements over issues and, hence, vote unanimously on them at public meetings. On the other hand, issue disagreements within councils with low or negative affective relations are far more likely to spill over into the public arena. Split voting on issues, constant conflict or immobilism tend to characterize the proceedings of these councils. In addition, if community groups are regular participants in decisions, and the policy-making process is constantly engulfed in conflict, policy makers may be more reluctant to consider issues which are potentially conflictual. Hence, policy-making will be confined to relatively safe or noncontroversial issues.

The *fourth stage* envisages a number of longer-term outcomes having to do with the implementation of the policy change. Particularly important at this stage is the relationship between *public policy,* or the actions of local government (whatever they choose to do or not to do); *policy outputs,* or the service levels achieved by these actions; and *policy impacts,* or the effect the service has on a given problem or population (Sharkansky, 1968). If one examines closely the public policies which have been enacted to solve central city and suburban problems, it is quite evident that many have had little effect on upgrading the services which people actually receive, let alone done much to solve the problems they currently confront. Whether or not they do have their intended impact depends largely upon such phenomena as the level of demand and conflict, the availability of resources, and the response to policy change by policy implementors (administrators).

In this conceptualization of the political process, the outcomes of each stage in the quest for policy change are not viewed as the result of a unique set of phenomena but rather of different combinations of the same general variables, usually various components of a municipality's policy-making process. However, if any one factor could be singled out as a more important determinant of whether or not changes in municipal policy outcomes will occur, it would be the way in which political leaders respond to demands for political change, particularly the way in which they use their political authority and power. It is particularly important that local policy makers respond positively to demands for political change. If they are unwilling or unable to mobilize support for change, plan and commit resources, or undertake innovative action, change in policy outcomes are unlikely to take place in central cities and suburbs (Smelser, 1968: 192-280; Coleman, 1957).

Of course, most municipal problems are not responded to or resolved quite as sequentially or logically as the foregoing staging framework suggests. For instance, public policy changes are not necessarily the result of demand articulation and aggregation by groups of angry citizens. Instead, most appear to just happen, being largely the result of internal bargaining among local (state or federal) policy makers and implementors. Other policies are simply imposed on central cities and suburbs, either directly by state or indirectly by federal legislative, administrative, and judicial institutions.

It should be quite obvious, however, that there are numerous points at which citizens and their leaders can intervene in this conversion process and use their power to either obstruct or facilitate meaningful problem-solving by government institutions. The way in which they will use their power, of course, is largely dependent on whether they agree or disagree with the particular demand, issue, or policy. It should also be clear that politics involves more than simply making or not making public policy decisions. A critical stage in the process of changing who gets what, when, how, has to do with the implementation of the public policy. Whether and how public policies and programs are implemented has a rather direct bearing on both policy outcomes and impacts—on whether the problem which precipitated the demand for policy change will be satisfactorily solved. Or, in other words, on who does and does not benefit from a particular allocation of resources.

SUBURBAN PROBLEMS AND
THE CONFLICTS THEY CREATE

Populations in suburban municipalities have undergone a variety of changes since 1960 (Wirt et al., 1972). Some of these changes are new, while others represent a continuation of changes initiated in previous years. For example, populations in these municipalities are changing in their size; rate of growth; density; spatial distribution; and class, racial, ethnic, religious, and political composition. Although inner- and middle-tier suburbs have largely exhausted their supply of land and hence grew very little in the last decade, population growth and building continue unabated on vacant land in the outer tier—on the periphery of suburbia. While the population composition of middle- and high-growth suburbs became more homogeneous, that of low-growth (inner-tier) suburbs has become more heterogeneous.

Important political consequences can, of course, flow from changes such as these. Most important, from our perspective, is the fact that many of these changes can give rise to new or exacerbate old problems. If these changes and the problems they create touch upon important aspects of people's lives and also affect people differently the stage may be set for controversy and conflict (Coleman, 1957: 3-6). This would be particularly true if some people felt actions could be taken that would cope with or solve problems brought about by change. The development of controversy also depends on the type of problem and the kind of community in which it occurs. As I have already pointed out, of particular importance is the way in which residents and their political leaders respond to conflicting demands for policies designed to solve problems.

What types of problems create controversy and conflict in suburbia? As one might imagine, it is very difficult to discuss suburban problems since so little agreement exists as to just what conditions in suburbs constitute problems. What one person perceives as a problem in need of solution, others may merely view as an inconvenience. In addition, the problems of suburbs in larger, older metropolitan areas on the eastern seaboard, in the Midwest, and to a lesser extent on the West Coast, may be quite different, for example, from those found in suburbs in smaller newer metropolitan areas in the Southwest. Thus, as one moves from suburb to suburb, and from suburbs in one metropolitan area to those in another, the problems one encounters may differ not only in degree but also in kind. For example, most suburbs in smaller newer metropolitan areas are mainly confronting development problems. However, while suburbs on the outer periphery of larger, older metropolitan areas face similar problems, those in the inner tier adjacent to the central city confront redevelopment ones.

The upper class began migrating outward and spilling over central-city boundaries in the 1800s (Glaab and Brown, 1967; Kramer, 1972). As new modes of transportation developed, more people were able to leave the central city and its vast array of problems behind. The rush to suburbia accelerated drastically during the mid-1900s (1940-1960) as first the middle and then the lower-middle classes migrated to new and old municipalities beyond the central city's legal boundaries. Being in new communities, suburban policy makers initially faced physical or material problems —with the need to allocate resources for purposes of developing municipal facilities and services. As more and more people migrated to the suburbs, even greater investments in physical facilities became necessary. Sewage and water systems, streets, schools, city buildings, police and fire stations all had to be constructed. Thus, the great political conflicts in suburban municipalities have been historically fought out over fiscal and land-use problems—how much money should be spent and for what purpose, what tax rate should be levied, and how much land should be used for residential (single or multiple family dwelling units), commercial, or industrial purposes. Conflicts, of course, over similar problems still continue today. Although these problems may appear less important or serious than those currently confronting policy makers in central cities, how they are resolved has important consequences for the problems suburban municipalities will encounter in the future.

By 1970, the *suburbanization of the central city* was well advanced, and by the year 2000 it should be nearly complete. By then, only those who are economically unable to leave—the poor and aged—will remain in most of our older and larger central cities. Research indicates that the patchwork of suburban municipalities one finds in America's metropolitan areas resemble in their class, ethnic, religious, racial, and political characteristics, former central-city neighborhoods (Schnore, 1965; Wirt et al., 1972; Haar, 1972). Internally, most suburbs are quite homogeneous, yet, when compared with each other, they are quite distinct. Suburbia is as diverse and heterogeneous today as the central city has always been. The bulk of the white population of many central cities now resides in suburbia, as does its economic base, and chances of either returning are slim indeed.

Americans, as they have always done, continue to move outward in metropolitan areas, leaving their problems behind for the bulldozer. This appears to be part of an American tradition of using something until it has spoiled or begun to deteriorate and then moving on—*the spoiler tradition*. In the outer reaches of suburbia, growth and development continue, while in the inner suburbs—those closest to the central city—deterioration, blight, and decay become more apparent each day. It is quite clear that not only has the population and the economic base of America's larger and older central cities spilled over into suburbia, but also their problems and conflicts.

For example, one has only to drive or walk through the inner tier of suburbs surrounding our larger and older central cities to realize that many central-city problems are beginning to spill over into suburbia (Wilson, 1968). For example, we would observe pollution of the air, water, and land; decay and blight of the municipality's physical plant—housing is deteriorating, rundown, and dilapidated, commercial and industrial buildings are vacant or boarded up, streets are in need of repair, and so on. People are also segregated by class and by race. In suburbia, two societies are developing, one black and one white—separate and unequal.

If you stop and talk with officials, they will most likely stress another problem—the suburb's *fiscal plight* (Campbell and Sacks, 1967). The suburb's tax base is deteriorating due to the aging of its physical structures and the movement of business and industrial concerns further out into suburbia. The upper, middle, and much of the lower-middle class have also fled into the outer reaches of suburbia, leaving behind a population which is increasingly poor, black, old, and dependent on the suburb's institutions for survival; a population which demands that the city spend substantially for welfare, public housing, public health, and so on. As a result of these changes, the suburb is simply no longer able to generate (if it ever was to begin with) the resources necessary to support its increasingly dependent population. Hence, deterioration in both the quantity and quality of services these inner tier of suburbs are willing and able to provide has occurred.

The problem, on the other hand, foremost in the minds of lower-middle-class whites who remain in this inner tier of suburbs (and, of course, whites throughout suburbia) is *crime and violence* (Clark, 1970). They no longer feel safe walking the streets, for they fear, either rightly or wrongly, the spillover of crime and criminal elements into their neighborhoods.

Blacks living in these areas, however, see a different set of problems. They view as their primary concerns unemployment and underemployment; police practices; inadequate housing, education, and programs to overcome their poverty; lack of adequate black representation in governmental policy-making; the discriminatory administration of justice; poor recreational facilities and programs; racist and other disrespectful white attitudes; inadequate and poorly administered welfare programs; generally inadequate municipal services; and discriminatory consumer credit practices. The major problem, then, facing black Americans in suburbia (as elsewhere) is racism (Clark, 1965; Knowles and Prewitt, 1969).

> By racism we mean the predication of decisions and policies on considerations of race for the purpose of subordinating a racial group and maintaining control over that group. . . . Racism is both overt and covert. It takes two closely related forms: individual whites acting against individual blacks, and acts by the total white community against the total black community. We call these individual racism and institutional racism. The first consists of overt acts by individuals, which cause death, injury, or violent destruction of property. This type can be recorded by television cameras; it can frequently be observed in the process of commission. The second is less overt, far more subtle, less identifiable in terms of specific individuals committing the act. But it is no less destructive of human life. The second type originates in the operation of established and respected forces in this society, and thus receives far less public condemnation than the first type. . . . Institutional racism relies on the active and pervasive operation of anti-black attitudes and practices. A sense of superior disposition prevails. Whites are better than blacks; therefore blacks should be subordinated to whites. This is a racist attitude and it permeates this society on both the individual and institutional level, covertly and overtly [Carmichael and Hamilton, 1967: 3-5].

It should now be, if it was not before, apparent that the problems of central cities have implications for the types of problems suburban municipalities will confront. They are interrelated. Or perhaps it would be more accurate to argue that whether or not and in what way central-city problems are resolved has important consequences for the kinds of problems suburban policy makers will encounter. This is most apparent when one realizes that the vast migration to suburbia and the problems it has given rise to have been largely the result of deterioration in the quality of life in the central cities (Downs, 1969: 6-74). Correspondingly, the continuing migration of Americans to the outer reaches of suburbia has been accelerated in recent years by deterioration in the quality of life in the inner suburbs where the spillover of problems once confined to central cities is most apparent.

TYPES OF SUBURBAN PROBLEMS

Must we be content, however, to simply list suburban problems? Can any meaningful distinctions, for purposes of increasing our understanding, be made between the problems we have just discussed? Do they have any distinctive characteristics? For example, do they differ in any significant ways? And, if not, in what ways are they similar? Although, admittedly, a difficult task, let us examine some alternative criteria for distinguishing among types of suburban problems (Wilson, 1966).

First, and in very general terms, we can distinguish between physical or material problems and human problems. *Physical or material problems* refer to those having to do with (1) the conditions (quality) of the environment—how polluted are our suburb's water, air, and land, for example; and (2) the conditions (quality) of physical facilities—how blighted are a suburb's dwelling units, schools, commercial, and industrial buildings, and so on. We could also include the condition (quality) of a suburb's roads, street lighting, transportation system; and parks, recreational, health, sewage (waste disposal), water (purification), police, and fire protection facilities, as physical or material problems.

Human problems on the other hand, originate in individuals: (1) in their psychic or physical states, feelings, attitudes, beliefs, and values; or (2) in relationships taking place between individuals. In the former case, we are referring to, for example, problems such as mental retardation and illness; disease, hunger, poverty (lack of education and income); and prejudice, feelings of alienation, powerlessness, or relative deprivation. The latter includes problems such as discrimination (racism—both its overt and its more subtle covert institutional forms), crime, family instability, and so on. Subsumed here would also be problems arising from variations in the quantity and quality of services provided to residents of suburbs.

Most physical or material problems can be solved; that is, solutions to such problems are technically feasible even though they may not be politically acceptable. Such problems are created because large numbers of people are living in highly interdependent settlements. In new suburbs, physical or material problems require *development* decisions, as suburbs grow and age and their environment or physical facilities deteriorate, *redevelopment* decisions become necessary. In either case, solutions to these problems require policy makers who are both willing and able to enact and then implement public policies designed to improve the condition (quality) of a suburb's environment and physical facilities. It is quite clear that, in the case of physical or material problems, our knowledge about these problems and hence our ability to solve them, far outdistances our willingness to do so. Just the opposite appears to be true of human problems—policy makers appear more willing to solve these problems but as yet do not have the knowledge and, hence, the ability to do so (Wilson, 1966: 37).

This leads us to several additional criteria which can be used to distinguish among types of suburban problems (Wilson, 1968). *First,* whether the problems are susceptible to rather precise formulation and study or not. *Second,* whether alternative ways of coping with the problems can be conceived and their effectiveness evaluated with a certain degree of rigor. *Third,* whether the problems have approached a crisis state or not—that critical turning point for better or worse. And *fourth,* whether the obstacles to solving problems are primarily political and, to a certain degree, economic or not.

THE SERIOUSNESS OF SUBURBAN PROBLEMS

How serious are the problems we have been discussing? Do any of these problems constitute a crisis? There is a great deal of disagreement over both these questions. Nevertheless, unless a problem is perceived as serious or as a crisis, it is doubtful whether any action will be taken to solve it. First, people disagree about the seriousness of the problems we have been discussing. A *serious problem* is one that is not easily solved; one which also has important or dangerous consequences. It is easy to imagine how people who live in different suburbs and who belong to different social classes—religious, political, racial, or ethnic groups, for example—might have very different perceptions of which problems are serious. For instance, the poor, unemployed black resident of a rapidly deteriorating inner suburb would most certainly view his constant hunger, inability to find adequate shelter, and propensity to become ill as serious problems. He is fighting to survive in what he views as a basically hostile world. On the other hand, the wealthy resident of an upper-class residential suburb views his most serious problems as keeping "undesirables" from moving into or committing crimes in his community. His basic needs for food and shelter have been satisfied, so he is concerned with other problems. To our inner-suburb resident, such problems have never occurred, and, even if they had, neither would be viewed as serious.

People also disagree about which, if any, suburban problems constitute crises. A *crisis* implies a bad or distressing situation brought about by a combination of causes; a critical (crucial) turning point (juncture) whose outcome will make a decisive difference for better or worse. Is a problem a crisis when it affects the essential welfare of individuals or the good health of a society? If people are dying from starvation and disease, or if the intellectual capabilities of their children are being stunted by malnutrition or lead paint poisoning, or if they have no place to live, or if they are constantly unemployed, or if they are constantly ill—do these constitute crises? Who has to experience these problems or how many have to die, starve, or be psychologically or physically maimed before a crisis will be perceived to exist?

There is a great deal of disagreement about whether or not the problems we have been discussing constitute crises. One recent presentation of the no-crisis perspective is found in Banfield (1968). Banfield (1968: 23) argues that much of what has happened in the typical central-city, suburb, or metropolitan area can be understood in terms of three phenomena or imperatives:

> The first is demographic: if the population of a city increases, the city must expand in one direction or another—up, down, or from the center outward. The second is technological: if it is feasible to transport large numbers of people outward (by train, bus, automobile) but not upward or downward (by elevator) the city must

expand outward. The third is economic: if the distribution of wealth and income is such that some can afford new housing and the time and the money to commute considerable distances to work while others cannot, the expanding periphery of the city must be occupied by the first group (the "well-off") while the older, inner parts of the city, where most of the jobs are, must be occupied by the second group (the "not well-off").

According to Banfield, these phenomena establish certain parameters which shape the process of metropolitan growth. As they have shaped the growth of central cities and metropolitan areas in the past, so they may be expected to shape that of metropolitan areas in the future. Despite some recentralizing tendencies, it is idle to talk of bringing large numbers of the well-off back into the central city. Decentralization of people and physical facilities will continue, with outward expansion checked only by the supply of vacant land.

But what of the role of local government in this process?

> Except that it may relax one or more of them and so change the logic of metropolitan growth, government—even the wisest and most powerful government—must work within the limits set by these imperatives. It may hasten the unfolding or the logic of growth and it may make such adaptation—very important ones sometimes—as are possible within it but given the premises, it cannot prevent the conclusions from following [Banfield, 1968: 41].

Banfield is arguing that governments are powerless to shape or cope with the consequences of population growth, technology of transportation, and the distribution of income—the *forces* shaping metropolitan growth. The die has been cast, the outcome inevitable, determined, and foreordained due to the inevitable logic or unfolding of these forces. "If towns are to grow into cities and cities into metropolises, old residential districts must decline and disappear" (Banfield, 1968: 42).

Banfield (1968: 257-258) also feels that the logic or metropolitan growth—economic growth, demographic changes, and the process of middle and upper classification—will ultimately solve the problems of lower-class people residing in our central cities and the inner tier of suburbs. He argues that government can and will do very little to solve these problems. "What stands in the way of dealing effectively with these problems (insofar as their nature admits of being dealt with) is mainly the virtue of the American political system and of the American character. It is because governmental power is widely distributed that organized interests are so often able to veto measures that would benefit large numbers of people" (Banfield, 1968: 256). A number of questions should be raised about this rather deterministic perspective. First, should we leave (as we have done historically) solution of these problems to the normal unplanned workings of the marketplace—to chance at best? Second, should we as citizens in a democratic polity stop trying to solve problems simply because our proposals for political change may be unacceptable to current elites or infeasible given current technology? Or do we have a responsibility to see that our political leaders enact the policies or programs and allocate the resources necessary to bring about the ultimate resolution of central-city and suburban problems. Are we as citizens, our political leaders, and societal institutions so tied to the present—the hallowed status quo—that we can no longer think about, plan for, and allocate the resources necessary to realize an alternative future? A future in which the quality of life in this country will be greatly improved? A future in which *all* Americans will have the opportunity to realize their fullest potential as human beings?

The arguments and perspectives developed in Banfield's (1968) *The Unheavenly City* are simply indicative of the malaise which exists in urban America. It is a feeling that, because nothing has solved central-city and suburban problems in the past, no policies or programs are likely to be developed which will bring about their resolution in the future. This feeling exists among both citizens and their political leaders. Along with such feelings has been the resurgence of the following myth: that government which governs best governs least, or, if we leave well enough alone, the logic of metropolitan growth, or whatever, will solve our problems. Could it be that central-city problems will simply move to suburbia? Or that institutional racism, that cancer which currently infects so many of our institutions, will not go away? Is this society, its citizenry, leaders, and institutions—so present-oriented, so tied to past and present domestic failures that it is no longer capable of innovative problem-solving action?

Our society's domestic priorities are not clear. Many of the public policies which decision makers have enacted to solve urban problems appear to be working at cross-purposes.

We do not know what we are trying to accomplish. . . . Do we seek to raise standards of living, maximize housing choice, revitalize the commercial centers of our cities, end suburban sprawl, eliminate discrimination, reduce traffic congestion, improve the quality of urban design, check crime and delinquency, strengthen effectiveness of local planning, increase citizen participation in local government? All these objectives sound attractive—in part because they are rather vague—but unfortunately they are in many cases incompatible (Wilson, 1966: 28-29).

Instead of solving problems, these policies and programs either accentuate existing ones or give rise to new ones. Obviously, there must be substantial agreement on priorities before any meaningful problem-solving will take place.

GOVERNMENT'S ROLE IN SOLVING
SUBURBAN PROBLEMS

The good or healthy society or community would be one that permits an individual's highest purposes to emerge by satisfying all his basic needs—by allowing that person to realize his fullest potential as a human being (Maslow, 1959: 202-222). On the other hand, if an individual is thwarted or unable to realize these needs, and becomes sick, either physically or mentally, such sickness can be ultimately traced to a sickness in a society and its institutions. This is because, in the final analysis, forces outside the individual are responsible for obstructing his realizations of these basic needs. Yet, as I have pointed out, the problems of our nation's suburbs impinge upon and affect, or we could say infect, people's lives differently. As a result, people disagree over which, if any, of these problems is serious. And, in turn, which, if any, constitutes crises. They also disagree over priorities and, hence, rank problems along these two dimensions quite differently.

Obviously, such disagreements impede problem-solving. But are there any other reasons why suburban problems are not being solved? Ever since people of the United States began congregating, initially in central cities and later in suburbs, they have encountered problems (McKelvey, 1963, 1968). Some of these problems—like poverty, illiteracy, lack of job skills, disease, and prejudice—people brought with them when they initially immigrated from abroad or later migrated from rural areas to central cities and from central cities to suburbs. Such problems were often accentuated as central cities grew during the 1800s and early 1900s, and their population increased in both density and heterogeneity.

As cities grew, they also confronted additional problems, ones which required development of and investment in municipal facilities and services. These problems revolved around the need for sewage and garbage disposal; pure water; education, law enforcement, health, and transportation facilities and personnel; fire protection; streets; and housing. All these problems arose quite early in America's central cities, for they are ones which face any central city or suburb no matter what the composition or problems of its population; sewage and garbage must be disposed of, water purified so it is safe for drinking, streets built, children educated, and so on.

Most central cities and suburbs in America, however, were not planned and instead grew in a rather piecemeal, uncoordinated manner. Local governments only enacted public policies designed to solve these problems either when they reached crisis proportions —people began to die from contaminated water—or when it became obvious that the private sector was unable or unwilling to undertake satisfactory problem-solving actions. This tendency of government to rely on the private sector for meaningful problem-solving has been largely the result of the following:

> Under the American tradition, the first purpose of the citizen is the private search for wealth; the goal of a city is to be a community for private money-makers. . . . From the first moment of bigness, from about the mid-19th century onward, the success and failure of American cities has depended upon *unplanned outcomes of the private sector*. . . . What the private market could do well, American cities have done well; what the private market did badly or neglected, our cities have been unable to overcome. . . . The 20th century failure of urban America to create a humane environment is thus the story of an enduring tradition of *privatism* in a changing world [Warner, 1968: x-xi].

The consequences of this tradition are particularly apparent when central cities and in recent years some suburbs, begin to age and their populations and economic base to change. An entirely new set of problems arises as the quality of life begins to deteriorate. To solve these problems or the problems arising from these contextual changes, requires substantial investment in redevelopment for purposes of rebuilding, revitalizing, and, in some cases, expanding municipal facilities, institutions, and services. However, despite the increased need for the investment of fiscal and human resources in redevelopment, particularly in the tier of inner suburbs surrounding our larger, older central cities, neither private nor public institutions have been willing and able to undertake such action.

It is quite clear that municipalities in transition, those which are aging and undergoing changes in the class and racial composition of their population and economic base, will not long remain viable unless they are able to meet the service needs of their changing population (Sternlieb and Beaton, 1972; Bradburn et al., 1971). For example, the upper and middle classes, and more recently the working class, have already made it abundantly clear they will not long remain in cities which do not provide good schools, good police protection and "good government." When the quality of life in suburbs begins to deteriorate, people will leave; although the more well-to-do will leave first the not so well-to-do are not far behind.

> Under such conditions of downward mobility the suburb has all the problems of the city without its assets. Limited capacity of many if not all homogenized suburbs to maintain themselves may be one of the most serious objections to them. In most of the tract subdivisions of suburbia there are neither the resources nor the commitment to pursue a policy of community conservation and stabilization. It is easier after a brief flurry to cut and run than to risk one's effort and capital in a struggle whose issue must seem dubious [Long, 1972: 49].

People disagree, then, over the seriousness of the problems we find in suburbia and which, if any, constitute crises. But they also disagree over just what the proper role of government should be in solving these problems. Should local (state and federal) governmental institutions play an active, neutral and passive, or negative role? Should they (Williams and Adrian, 1963):

(1) *Promote economic or residential growth*—a growth orientation—which usually results in local policy makers spending substantial amounts for planning, sponsoring an industrial development commission, devel-

oping an industrial park, selling or leasing city land to developers, or constructing and then leasing or selling buildings to developers;

(2) *Provide and secure greater life amenities*—a high service, high expenditure orientation—which results in local policy makers emphasizing increased expenditures not only for basic services but also for such things as parks, recreational facilities and programs, libraries, urban renewal and redevelopment, and public housing;

(3) *Arbitrate conflicting interests*—a conflict resolving orientation—encountered most often in heterogeneous communities (ones divided along class, racial, ethnic, or religious lines) and ones undergoing rapid population changes, in which local policy makers spend the bulk of their time attempting to handle and resolve conflicting demands generated by different perceptions of community problems and needs. Such an orientation usually results in incremental adjustments in public policy only, or immobilism;

(4) *Maintain existing social character*—in some ways similar to a caretaker orientation but in this case local policy makers (and residents) are not as concerned with keeping taxes and expenditures low as keeping "undesirables" (poor, black, Jewish people, for example) out of their municipality through the use of planning, zoning ordinances, and building codes;

(5) *Maintain existing services only*—a caretaker, low service, low tax orientation—which results in local policy makers not spending for any but basic services such as education, police and fire protection, and street maintenance.

We know that historically in this society the tendency has been for government to adopt a caretaker role and rely on the private sector for problem-solving. However, if we have learned anything from studying the growth, development, and decay of our larger, older central cities, it should be that reliance on the private sector has led to disaster! The private sector has been unwilling or unable to solve central-city problems. Why should we expect them to be any more willing or able to do so in suburbia? If the problems of our central cities and suburbs are to be solved, governments at all levels must establish problem priorities and then act positively and aggressively —they must enact and then implement policies designed to solve such problems. In effect, they must play an active rather than passive or negative role.

POLITICAL OBSTACLES TO MEANINGFUL
PROBLEM-SOLVING IN SUBURBIA

THE DISTRIBUTION AND USE OF POLITICAL POWER

Suburbs, unlike their counterpart neighborhoods in central cities, do have the authority to enact problem-solving policies. Whether most are both willing and able to do so, however, appears quite problematic. Many suburban governments (like most others, for that matter) simply refuse to view themselves as problem solvers. As a result, suburban policy makers seldom anticipate problems or define the ones they have very accurately. They are also seldom willing to engage in meaningful cooperation with other municipalities in an effort to solve problems which "spill over" municipal boundaries. One reason for this is their fear that an active rather than passive or negative problem-solving role will generate controversy and conflict. Another has to do with the "second face" of power (Bachrach and Baratz, 1970: 6)—that is, with how "power may be, and often is, exercised to confine the scope of decision making to relatively 'safe' [and, hence, noncontroversial] issues." For example,

> political systems and sub-systems develop a 'mobilization of bias,' a set of predominant values, beliefs, rituals, and institutional procedures [formal and informal] ('rules of the game') that operate systematically and consistently to the benefit of certain persons and groups at the expense of others. Those who benefit are placed in a preferred position, to defend and promote their vested interests. More often than not, the 'status quo defenders' are a minority or elite group within the population in question [Bachrach and Baratz, 1970: 43-44].

The primary method for sustaining a given mobilization of bias is what Bachrach and Baratz (1970: 44) referred to as non-decision-making. A nondecision is simply a decision to use power to suppress or thwart any challenge to the values or interest of policy makers or implementors. This means that demands for change in the existing distribution of benefits and privileges in a community—for changes in policy outcomes—will be obstructed and blocked by those who have power before they are voiced; or kept covert; or killed before they gain access to the relevant authorities; or, failing all these things, maimed or destroyed in the decision-making or implementing stages of the policy-change process.

Those who have resources and, hence, power—elites in the community—may use any of various means to obstruct change. At each stage, they may use constraints—for example, harassment, imprisonment, beatings, and even murder, as whites have done to blacks for years, particularly in the South; or inducements—for example, individuals and groups demanding policy changes may be coopted, bought off, or given token and symbolic reassurances by power holders—or persuasion. Elites may also "invoke an existing bias of the political system—a norm, precedent, rule or procedure—to squelch a threatening demand or incipient issue" (Bachrach and Baratz, 1968: 45);

> For example, a demand for change [and hence the individual or group making it] may be denied legitimacy by branding it as socialist, unpatriotic, or in violation of an established rule or procedure. Challengers can also be deflected by referring the demand or issue to a committee for a detailed and prolonged study or by steering them through time-consuming ritualistic routines that are built into the political system. Tactics such as these ... are particularly effective when employed against impermanently or weakly organized groups (e.g., students, the poor) which have difficulty withstanding delays.

Elites may also *change*—reshape or strengthen—an *existing bias* in order to block demands for political change. For instance, additional rules and procedures for processing demands may be established. Finally, B confronted by A who has superior power resources may simply not make a demand on A for fear of the consequences that might follow. In this instance, B anticipates a negative reaction from A and hence makes no demand.

Like the covert institutional forms racism takes in this society, the way in which power is used to obstruct political change may sometimes be quite subtle and difficult to study. However, one begins to learn about these more subtle uses of power by first studying a community's policy-making process, particularly the way in which political authorities respond to demands for political change. Attention should be directed toward investigating a set of concrete decisions, particularly ones involving key issues. A key issue "is one that involved a genuine challenge to the resources or authority of those currently dominating the process by which policy outputs [outcomes] in the system are determined" (Bachrach and Baratz, 1968: 47-48). A key issue of current importance involves citizen demands for greater participation in making decisions that affect our lives. This, of course, can only be accomplished by opening up governmental institutions and redistributing political power and authority. Furthermore, if, in studying policy-making, in a community one learns that citizens have either overt grievances (those that have already been expressed and become issues) or covert ones (those which are still outside the political system), this may indicate that power is being used to thwart or prevent serious consideration of demands for political change. In addition (Bachrach and Baratz, 1968: 49-50):

> in the absence of conflict—a power struggle, if you will—there is no
> way to accurately judge whether the thrust of a decision really is to
> thwart or prevent serious consideration of a demand for change.

Nevertheless, just because there are no overt or covert grievances or political conflict does not mean that power is being used for anything other than to maintain the status quo. For example, many suburban communities have low levels of political conflict, or perhaps none, because policy-making is limited by elites to relatively noncontroversial matters (Downes, 1969). Policy-making is limited to relatively "safe" issues. This may simply be the result of a great deal of agreement among participants or it may be that certain groups in these communities and, hence, problems, demands, and issues are being excluded from the policy-making process. In effect, the system may be closed to certain individuals and groups. Furthermore, because controversial issues are being kept out of the policy-making arena, in low-conflict municipalities city councils may simply not be discussing or attempting to solve important communitywide problems.

If power can be used to facilitate as well as obstruct political change, then the question of "who governs" the community should be supplemented by investigation of who is especially disfavored under the existing distribution of benefits and privileges. We should be concerned not only with who benefits but with *who does not benefit and why* from the activities of elites. We should ask to what extent political power and authority are being used to maintain a political system that tends to perpetuate "unfair shares" in the allocation of benefits; and how, if at all, new sources of power and authority can be generated and used to alter the situation (Bachrach and Baratz, 1968: 50-51).

Obviously, it does make a difference who governs. Most importantly, it will have an important impact on the biases which are mobilized in a community—on the values and political myths, rituals, and institutional practices—which dominate policy-making and implementation. In addition, the attitudes, beliefs, values, and myths which those who govern bring to the political arena have important consequences for the way in which political power is used and, hence, for who does and does not benefit from governmental activity. It will have an obvious impact on whose grievances and demands power holders respond to and to whom they are accountable. It will also have an impact on the openness of policy-making and implementation processes, and on which public policies are enacted and then implemented, and, hence, on who does and does not benefit from governmental activity. Overall, then, those who govern have important effects on which, if any, problems get solved.

It is quite easy to conclude that most suburbs are largely governed by a series of *sometimes governments,* whose presence is rarely felt or rarely missed, but whose presence, I would argue, is sorely needed. The one thing many suburban governments have learned to do well is to be negative and do nothing. Their primary activities become indecision and inaction.

THE DISTRIBUTION AND USE OF POLITICAL AUTHORITY

Still another reason for the lack of meaningful problem-solving in suburban municipalities has to do with the fact that political authority is probably most fragmented at the local level. In every metropolitan area, there are tens, hundreds, and sometimes even thousands of governments—municipalities, special districts, counties, and so on—each of which has the authority to provide some portion of the services citizens require. Furthermore, within many of these local governments, political authority is further fragmented among rule-making, implementing, and adjudicating institutions.

Reformers have traditionally argued that the fragmentation of political authority in a metropolis has given rise or accentuated the following types of problems (Campbell, 1970: 61):

(1) the unequal distribution of fiscal resources;

(2) gross inequities in services provided;

(3) the absence of areawide authorities to cope with problems such as those listed above and pollution, transportation, housing, planning, education, land-use, health, and so on, which spill over municipal boundaries;

(4) wasteful duplication of service efforts and inefficient use of resources;

(5) inability of citizens to fix responsibility for government action or inaction; and

(6) political segregation of suburbanites and their political leaders from central-city problems.

However, most citizens and their political leaders, particularly those residing in suburbia, do not view these problems as serious and certainly not as crises (Campbell, 1970: 62-63). Hence, they oppose any attempt to overcome the severe fragmentation of authority which currently exists in all but a few metropolitan areas, through, for example, such means as metropolitan reorganization—the centralization of political authority in a single government. Furthermore, citizens view any attempt to redistribute political authority in such manner as posing a serious threat to their local autonomy.

> The advantages of submerging local sovereignties are less tempting than the psychic rewards of the status quo, at least, to the most highly involved actors who desire to remain big frogs in small puddles is backed by the folk conservatism of their constituents and a powerful set of local vested interests [Long, 1972: 79].

Thus, the major demands for metropolitan structural reform come from prominent, vocal, and articulate "good government" reformers who are capable of commanding attention. These are the intellectuals of local government; some central-city officials and federal administrators; central-city Chambers of Commerce; downtown businessmen, bankers, realtors, and other civic notables; and the metropolitan press. But, as we have pointed out, they lack widespread popular support.

Currently, then, suburban problems are most often handled on an ad hoc, piecemeal basis by the myriad of geographically and functionally fragmented governments one finds in most metropolitan areas. These are governments which either will not or cannot, because they lack the fiscal resources, authority, and organizational capability, provide adequate services and change policy outcomes. The traditional response to local-level problems has been a little rechanneling of tax money here and there, a dash of home rule for one place, and a jigger of regional coordination in another (Campbell, 1970: 114). However, today's problems may require a new response; one that incorporates comprehensive areawide planning and coordination, well-financed problem-solving strategies, and systematic evaluation of institutional performance.

Yet, there is, at present, very little cooperation among the local units of government and certainly no planned, coordinated effort to solve problems. The federal government has provided some fiscal incentives in an effort to encourage cooperative problem-solving activity at the local level, but this has not given rise to much meaningful cooperation (Campbell, 1970: 139-168). However, those with vested interests in maintaining the status quo oppose any comprehensive, planned approaches to solving central-city and suburban problems. They are particularly opposed to setting up federal and state departments of urban affairs or multipurpose metropolitan governments. Such changes, particularly the way they would redistribute (centralize) political authority, pose a threat to politicians and bureaucrats who have major influence over the internal content of basic services like education, public safety, the administration of justice, health care, transportation, and so on. These are elected and appointed officials who have over the years effectively insulated these and other services from public scrutiny behind the twin barriers of specialization and professionalism. Although federal, and to a lesser extent state, grants-in-aid do provide much-needed funds, thereby relieving some of the fiscal pressures on local government, they have the effect of reinforcing specialization and the maintenance of functional autocracies at the federal, state, and local levels.

> The proliferation of federal government grant programs has become something of a scandal. For some time, it had been the object of attack by the governors who found it eroding their independence of action and budgetary control. . . . It is a piecemeal and patchwork edifice without architect or design, representing the results of coalitions and pressure, adding this and that program to the structure. . . . What seemed and seems most amiss about the whole business is the lack of any overall order or concerted design. Like

Topsy, it has just grown. But, unlike Topsy, its consequences are more than comical. The grants program the states in ways they would frequently not choose. . . . It is not just the complex maze of federal grants, requiring special experts in Washington offices for states and cities as well as corporations which causes concern but the growing realization that this historic accumulation takes on a life of its own . . . once a program is in, its share of the governmental market gets reviewed, not its cost effectiveness and least of all its relatedness to a well-considered or even considered general program [Long, 1972: 58-59].

By using both the carrot, fiscal incentive, and the stick, enforced compliance, both federal and state governments could change the inequitable distribution of resources and also the lack of meaningful problem-solving by local governments that currently exist. But neither has chosen to do so. One positive development is that the courts may cause the fifty states or the federal government to act. Some state Supreme Courts have recently ruled that the use of the property tax to finance education gives rise to such variation of educational expenditures from school district to school district, as to represent a violation of the equal protection clause of state constitutions. If these rulings are upheld by the U.S. Supreme Court, gross inequities in expenditures for other services will also be open to attack—in effect, it will be doubtful whether the courts will be able to restrict their concern to education.

Since local governments are creatures of a state, primary responsibility for the way in which political authority is fragmented in America's metropolitan areas and the problems this has given rise to rests with the states (Campbell, 1970: 169-209). The fifty states have made it far too easy for communities to incorporate and yet difficult for these same communities to annex or consolidate—extend their boundaries—as they grow, their problems become more complex, and their need for resources expands. Forced reliance on the property tax as the primary means of raising revenue has reinforced gross inequities in available fiscal resources and, hence, inequities in the quantity and quality of services municipalities are able to provide, should they be willing. The federal government, unknowingly at first, has also contributed to the growing discrepancy between municipal needs and resources by making available low-interest, low down payment, federally insured F.H.A. and V.A. loans for the purchase of housing. This, as we now know, enabled the upper and middle classes to leave many central cities and the inner tier of suburbs surrounding them, an exodus which has seriously undermined the capacity of these cities and suburbs to serve the remaining population.

The states also have not required areawide planning in their metropolitan areas and, as a result, have exerted very little control over land use. State legislatures have also refused to set forth minimum service standards for local governments and appropriate penalties such as forced consolidation, to assure compliance. In effect, states have abrogated their constitutional responsibilities to their local units of governments who are, after all, legally "creatures of and utterly dependent upon" their state governments.

It is quite clear that local governments, particularly suburban ones, are most severely restricted constitutionally, fiscally, and organizationally in their authority and, hence, ability to solve problems. If the problems currently plaguing our nation's central cities and suburbs are to be solved, either responsibility for solving them must be assumed by the fifty states or the federal government, or central cities and suburbs must be provided the requisite fiscal resources, technical expertise, manpower, and authority to do so. Of course, the three levels of government and the various institutions at each level could undertake cooperative action to solve these problems. However, they will only be effective if existing functional autocracies or fiefdoms can be dismantled or their activities coordinated so cooperative problem-solving can take place. Only then will the duplication of effort, inefficiencies, and inequities in the present system of sharing be removed.

The evidence is overwhelming that, if those who have political authority continue to use it as they have in the past, central-city and suburban problems will not be solved. For example, many suburban municipalities continue to use their authority over zoning to exclude blacks and other minorities from suburbia, thus locking them into the decaying, problem-ridden central cities. These same communities have shown very little interest in cooperating with central cities and other suburbs in the quest for solutions to pressing areawide problems. Most local units of government jealously guard their autonomy, and perhaps rightly so, as long as their policy outcomes do not have negative consequences for their residents or for people outside their boundaries. Then, the state or some areawide authority must be able to hold such governments accountable for their actions. Any municipality, then, has a right to exist, so long as it does not become a "public nuisance."

We will not know whether they are a public nuisance unless we begin to evaluate the performance of municipal institutions. By so doing, we would hope to increase our understanding of the consequences of governmental activity (and inactivity), particularly whether public policy is having its intended effects. Very little research of this sort has been undertaken, however. One of the reasons for the lack of research evaluating the consequences of public policy is that such research is not only very difficult but also often resisted by officials responsible for the policy's implementation. Nevertheless, we must do a great deal more in this area. For example, in order to evaluate whether the consequences of policy actions are good or bad and for whom, we must raise and answer the following questions (Lineberry and Sharkansky, 1971: 196-200):

(1) *Cost:* how much does the particular policy or program cost? Who pays—who bears the tax burden and how is it distributed among a given municipality's population?

(2) *Service output:* what is the policy output or level of services attained for the particular expenditure of funds?

(3) *Service distribution:* who benefits and how are the benefits (services) distributed among neighborhoods, social classes, racial groups, and so on?

(4) *Policy impact:* what is the policy's impact on the target situation or group, on situations or groups other than the target (spillover effects), and on future as well as immediate conditions? Was the impact intended or unintended, anticipated or unanticipated, symbolic or tangible?

(5) *Feedback:* what is the nature of the feedback? Is it positive or negative? Are citizens, particularly target groups, satisfied or dissatisfied? Do demands increase or decrease? Is support for governmental institutions increased or decreased? Does learning by decision makers and policy adjustment or fundamental revision take place if feedback from citizens or evaluative research is negative?

The states also continue *not* to use their preeminent authority over local units of government to foster meaningful problem-solving by or in central cities and suburbs. Instead, they have abrogated their responsibility. Although the federal government, like governments in the fifty states, could encourage a comprehensive planned assault on central-city and suburban problems, it has not as yet done so. Instead, the federal government continues to maintain a myriad of functionally independent fiefdoms, each of which is attempting to solve some specific problem of the many which plague our central cities and suburbs. This disjointed, uncoordinated, underfunded, incremental approach to problem-solving simply has not worked in the past, and it will not work in the future. The federal government and, to a lesser extent, the fifty state governments, are, mainly because of their superior capacity to generate resources and assume functional responsibilities, in a better position to provide leadership in solving these problems. Whether they will, however, remains an unanswered question.

CONCLUSION

The current quest for political change in our society is based on the following assumption set forth in the Declaration of Independence:

> Governments are instituted among men for certain ends; that among these are life, liberty, and the pursuit of happiness; that whenever a government becomes destructive of those ends, it is the right of the people to alter or abolish it. The government is not synonymous with the people of the nation [or the community]; it is an artificial device, set up by the citizens for certain purposes; it is endowed with no sacred aura; rather it needs to be watched, scrutinized, opposed, changed, and even overthrown and replaced when necessary.

Much of the impetus to demands for fundamental changes in governmental institutions, processes, and outcomes at all levels results from the failure of such institutions to engage in meaningful problem-solving activities. This failure is most obvious in our larger, older central cities. However, as the problems of central cities increasingly spill over into suburbia, the negative consequences resulting from the unwillingness or inability of many suburban governments to solve problems will also become more apparent (Downes, 1974).

One result will be that demands for political change will increase. This will further exacerbate conflict over the following issues:

(1) Who is responsible for solving suburban problems?

(2) What level of government is both willing and able to do so?

(3) How can suburban municipalities which will not solve their problems be held accountable?

(4) How can suburban municipalities which cannot solve their problems best be assisted?

(5) How can the negative consequences which decentralization of political authority in suburbia has for solving problems which spill over municipal boundaries be overcome?

If, when, and how these conflicts are resolved will have important consequences for future generations of suburbanites.

REFERENCES

AGGER, R. E., D. GOLDRICH, and B. E. SWANSON (1964) The Rulers and the Ruled. New York: John Wiley.

ALTSHULER, A. A. (1970) Community Control: The Black Demand for Participation in Large American Cities. New York: Western.

BACHRACH, P. and M. S. BARATZ (1970) Power and Poverty: Theory and Practice. New York: Oxford Univ. Press.

BANFIELD, E. C. (1968) The Unheavenly City: The Nature and the Future of Our Urban Crisis. Boston: Little, Brown.

BARBER, J. D. (1966) Power in Committees: An Experiment in the Governmental Process. Chicago: Rand McNally.

BRADBURN, N. M., S. SUDMAN, and G. L. GOCKEL (1971) Side by Side: Integrated Neighborhoods in America. Chicago: Quadrangle.

CAMPBELL, A. K. [ed.] (1970) The States and the Urban Crisis. Englewood Cliffs, N.J.: Prentice-Hall.

––– and S. SACKS (1967) Metropolitan America: Fiscal Patterns and Governmental Systems. New York: Free Press.

CARMICHAEL, S. and C. V. HAMILTON (1967) Black Power: The Politics of Liberation in America. New York: Random House.

CLARK, K. B. (1965) Dark Ghetto: Dilemmas of Social Power. New York: Harper & Row.

CLARK, R. (1970) Crime in America: Observation on Its Nature, Causes, Prevention, and Control. New York: Simon & Schuster.

COLEMAN, J. S. (1957) Community Conflict. New York: Free Press.

DOWNES, B. T. (1974) The Politics of Change in Central Cities and Suburbs: An Inquiry into America's Urban Crisis. North Scituate: Duxbury.

––– [ed.] (1971) Citieis and Suburbs: Selected Readings in Local Politics and Public Policy. Belmont, Calif.: Wadsworth.

––– (1969) "Issue conflict, factionalism, and consensus in suburban city councils." Urban Affairs Q. 4 (June).

––– and K. R. GREENE (1973) "Conflict, compromise, and change: the politics of open housing." Amer. Politics Q. 1 (April).

DOWNS, A. (1969) Urban Problems and Prospects. Chicago: Markham.

DYE, T. R. (1972) Understanding Public Policy. Englewood Cliffs, N.J.: Prentice-Hall.

FANTINI, M., M. GITTELL, and R. MAGAT (1970) Community Control and the Urban School. New York: Praeger.

GLAAB, C. N. and A. T. BROWN (1967) A History of Urban America. New York: Macmillan.

HAAR, C. M. [ed.] (1972) The End of Innocence: A Suburban Reader. Glenview, Ill.: Scott, Foresman.

KNOWLES, L. and K. PREWITT [eds.] (1969) Institutional Racism in America. Englewood Cliffs, N.J.: Prentice-Hall.

KRAMER, J. [ed.] (1972) North American Suburbs: Politics, Diversity, and Change. Berkeley: Glendessary.

LASSWELL, H. (1958) Politics: Who Gets What, When, How. New York: World.

LINEBERRY, R. L. and I. SHARKANSKY (1971) Urban Politics and Public Policy. New York: Harper & Row.

LONG, N. (1972) The Unwalled City: Reconstituting the Urban Community. New York: Basic Books.

MASLOW, A. H. "A theory of human motivation," in L. Gorlow and W. Katkovsky [eds.] (1959) Readings in the Psychology of Adjustment. New York: McGraw-Hill.

McKELVEY, B. (1968) The Emergence of Metropolitan America, 1915-1966. New Brunswick, N.J.: Rutgers Univ. Press.

––– (1963) The Urbanization of America, 1860-1915. New Brunswick, N.J.: Rutgers Univ. Press.

SCHNORE, L. F. (1965) The Urban Scene: Human Ecology and Demography. New York: Free Press.

SHARKANSKY, I. (1968) "Environment, policy, output, and impact." Presented at the American Political Science Association Annual Meeting, Washington, D.C., September.

SMELSER, N. J. (1968) "Toward a general theory of social change," in Essays in Social Explanation. Englewood Cliffs, N.J.: Prentice-Hall.

STERNLIEB, G. S. and W. P. BEATON (1972) The Zone of Emergence: A Case Study of Plainfield, New Jersey. New Brunswick, N.J.: Trans-action Books.

VERBA, S. (1961) Small Groups and Political Behavior: A Study of Leadership. Princeton: Princeton Univ. Press.

WARNER, S. B., Jr. (1968) The Private City: Philadelphia in Three Periods of Its Growth. Philadelphia: Univ. of Pennsylvania Press.

WILLIAMS, O. P. and C. R. ADRIAN (1963) Four Cities: A Study in Public Policy Making. Philadelphia: Univ. of Pennsylvania Press.

WILSON, J. Q. [ed.] (1968) The Metropolitan Enigma: Inquiries into the Nature and Dimensions of America's 'Urban Crisis.' Cambridge, Mass.: Harvard Univ. Press.

––– (1966) "The war on cities." Public Interest 3 (Spring).

WIRT, F. M., B. WALTER, F. F. RABINOVITS, and D. R. HENSLER (1972) On the City's Rim: Politics and Policy in Suburbia. Lexington, Mass.: D. C. Heath.

12

Exclusionary Zoning: A Code Phrase for a Notable Legal Struggle

RICHARD F. BABCOCK

☐ HOWEVER ONE CHOOSES TO VIEW SUBURBIA—as a homogeneous mass of bigots or as a diverse and harassed society to which no generalization may be fairly applied—whatever conclusions the amateur sociologist draws, it is a fact that the courts today on an unprecedented scale are being asked to make value judgments about the ethical bases of suburban attitudes. Whether these judicial conclusions are based upon inadequate or conflicting data is, in one sense, irrelevant; those judgments become the law, and they can and do operate to affect the institutional systems and the social characteristics of many parts of our suburban areas. No cluck-clucking by commentators is likely to change that.

By judicial value judgments, I refer to the surge of lawsuits in the last few years attacking suburban zoning and subdivision laws as illegal because they operate to exclude the poor, particularly racial minorities, by forcing up the cost of housing.

I do not intend to argue the merits of this accusation; nor to describe the ingenious local rules that allegedly are employed to achieve this end; nor to parade the depressing statistics on the growing distance between job opportunities and low-cost housing —those roads are being heavily traveled. Instead, I propose to comment on what is taking place in the law and to offer a few clouded predictions on where this tide may wash.

First, background to set the scene.

Control by municipalities over land-use policy (primarily through zoning and subdivision ordinances) has become such an accepted part of our system this last half-century that it is easy to forget that this municipal authority is delegated by the state and that only fifty years ago it was a close thing whether municipalities, as agents of the state, would be allowed to get away with telling a landowner what he could and could not do with his land. The derivative nature of local control is central to any forecast of what is likely to take place during the next decade, and the early struggle to establish the validity of local land-use regulations is a predicate to an understanding of the strategies behind the current legal conflict.

Zoning law passed through two stages since 1916 before it reached its present phase. There were the early years in the 1920s when zoning swept the country. Outraged landowners, told they no longer could build apartments or sell their land for a gas station, marched into court to protest this unconstitutional taking of what they thought were inalienable property rights. The municipalities, after some perilous contests in court, finally won that round—at least, the legitimacy of comprehensive zoning was settled—in principal, that is (for a detailed discussion of this early history as it affected housing, see Babcock and Bosselman, 1963). What really happened was that the courts said, "Yeah, but . . .": zoning may be valid as a general concept, but the courts reserve the right to determine in each case, first, whether the application of that principle to a particular parcel of property is reasonable (lawful), and, second, whether a new control technique adopted in the name of zoning is valid. During the twenties and thirties, more often than not, the municipalities encountered "yeahbut" decisions that invalidated zoning as applied to particular parcels of land or outlawed attempts to stretch zoning beyond simple regulation of uses and bulk. By the end of World War II, zoning could be said to be recognized as a legitimate tool, but it was by no means a foregone conclusion that in any given case the municipality would win.

Then came the explosion in our metropolitan areas and the birth of the suburbia that has become the darling of the social commentators. Rising expectations of the migrants to suburbia were matched by rising costs; social judgments of these *nouveau arrives* combined with fiscal crises to make the suburbs more determined than ever to hold the line against further change. Suburbs tightened their zoning ordinances. They invented new regulatory techniques, and, far more frequently than before World War II, they told the landowner or prospective developer that he damn well would have to take the village to court before it would change the rules. In the 1950s, when the town was taken to court, it emerged, not so surprisingly, victorious. The courts were increasingly sympathetic to municipal efforts to direct growth in the manner and at the rate desired by the community. (If the predominant migration was outward from the cities, it was to be expected that judges would be among the migrants.) If one dares generalize, by the late fifties and early sixties the judicial scales seemed to have tipped in favor of municipalities. With only slight exaggeration, "general welfare" seemed to the courts to mean whatever a municipality wanted it to mean. The judiciary became increasingly receptive to efforts to stretch the embrace of the police power to include such practices as architectural control, the uncompensated elimination of nonconforming uses, "contract zoning," and other regulatory devices, sophisticated or silly, that, in their complexity, were to the early ordinances of the 1920s as the IRS Code is to a Dick and Jane primer.

By 1965, suburban municipalities had every reason to feel smug. The air was filled with the anguished howls of homebuilders who wanted to build something other than single-family detached houses on quarter-acre lots but now faced courts that no longer regarded property rights as more sacred than "the character of a community."

Then something happened. Perhaps it was Watts and its progeny; perhaps it was the realization that Brown v. the Board of Education did not mean as much as had been hoped as long as there remained de facto segregation due to housing patterns; it may have been the slow dawning upon the leaders of the building industry that if they would just stop talking about property rights and start talking about human rights they might enlist some new, if strange, allies in their battles with the suburbs; and builders might then touch a sensitive nerve in the courts. While the U.S. Supreme Court, for more than four decades, had shown a marked unwillingness to hear zoning cases (since Euclid and Nectow) there had emerged in the Warren Court what the commentators referred to as the "new equal protection clause." This was an interpretation of the Fourteenth Amendment, at least in cases involving the franchise and criminal procedure, which seemed to say that laws neutral on their face could be unconstitutional if they affected sensitive constitutional areas and did in fact penalize the poor (see Kurland, 1968).

What could be more "neutral" than zoning laws? Both the rich and the poor could move into Sleepy Hollow if they had the price of a $50,000 house. The higher hurdle was whether housing occupied as lofty a niche in our constitutional pantheon as did the right to vote and the right to a fair trial. As yet no one has the answer to that question, but beyond question the complacency of the suburbs has been shattered. All is by no means lost, but the early skirmishes in this latest joust suggest that the anti-suburban coalition of civil libertarians and homebuilders have chosen a new battleground where the old municipal tactics are no longer irresistible.

The tempo of recent lawsuits attacking "exclusionary" suburban zoning is so swift that any attempt to catalogue the happenings is dated before it reaches print. Names of lawsuits such as Sasso (Southern Alameda Spanish Speaking Organization v. City of Union City, 1970), Lackawanna (Kennedy Park Homes Association v. City of Lackawanna, 1970, 1971), and Lawton (Dailey et al. v. City of Lawton, Oklahoma, 1969), Kit-Mar (In re: Appeal of Kit-Mar Builders, 1971), and Girsh (Appeal of Girsh, 1970) are no sooner part of the zoning rubric than they are followed by Mt. Laurel (Southern Burlington County NAACP et al. v. Township of Mt. Laurel et al., 1972), Madison Township (Oakwood at Madison, Inc. v. Township of Madison and the State of New Jersey, 1971), and Blackjack (United States of America v. City of Blackjack, Missouri, n.d.).[1] The Index of Legal Periodicals is now and will increasingly be citing analyses of these decisions and of others which are working their way through the courts. The reader can find whatever he seeks in those critical sources. My purpose is more limited: to suggest some trends and to point out some oddities in these judicial exercises.

At the start, it is necessary to distinguish between the litigation in the federal and in the state courts. The few federal exclusionary zoning cases have two features in common: the zoning rules are alleged to operate against residents of the municipalities. In Lackawanna and Lawton, it was blacks who were found to be denied housing because of the zoning practices; in Sasso, it was Americans of Mexican descent. Unlike cases involving the franchise and criminal procedure, federal courts have not yet in housing cases declared that the reach of the equal protection clause extended to economic classes. Indeed, the U.S. Supreme Court, in James v. Valtierra (1971), held that a provision of the California constitution that requires a referendum before a municipality can authorize public housing (even though other housing did not require such an act) did not violate the Fourteenth Amendment. The discrimination, if any, said the Court, was by economic, not racial, class.

In a pending zoning case in the U.S. District Court in Chicago, where exclusion by race was alleged to be the consequence of local zoning policy, the defendant, the city of Evanston, a large suburb with a black population substantially greater than is found in most suburbs, moved to dismiss the complaint on the grounds it did not state a cause of action. Evanston cited Valtierra. The court, refusing to dismiss the complaint, had the following to say about the applicability of Valtierra to a zoning case where race was alleged to be at issue (Sisters of Providence of St. Mary of the Woods et al. v. City of Evanston et al., 1971):

> We believe that for the purposes of this action the cases involving referendums, though helpful in terms of standing, jurisdiction, and judicial attitudes towards discrimination in housing, are a category unto themselves. James v. Valtierra, 402 U.S. 137 (1971) and Ranjel v. City of Lansing, 417 F.2d 321, heavily relied on by defendant are thus not dispositive of the issues. The voting rights involved in those cases inject quite a different constitutional ingredient, one not present in pure zoning cases, and courts are quite understandably more willing to uphold referendum procedures rather than second guess or interfere with the fundamental suffrage right.

Whether Valtierra effectively blocks attacks in the federal courts based solely on the allegation that zoning operates to exclude the poor remains open. At issue in Valtierra was the franchise—a referendum—a notable constitutional favorite, and California, unlike many states, has long employed the referendum in a variety of civil affairs. At the least, Valtierra casts a chill over the use of the federal courts in exclusionary cases where a racial impact cannot be proven.

The second apparently limiting factor in the federal cases relates to the geographic location of the persons who are allegedly injured by the municipal practice. In the federal cases to date, the complainants have been residents of the offending municipality, and the thrust of the court opinions is that a duty is owed to the residents. In Sasso (1970: 295; italics added), the court said:

> Appellants' equal protection contentions, however, reach beyond purpose. They assert that the effect of the referendum is to deny

decent housing and an integrated environment to *low-income residents of Union City*. If, apart from voter motive, the result of this zoning by referendum is discriminatory in this fashion, in our view a substantial constitutional question is presented.

In Lackawanna and Lawton, the plaintiffs were residents of the respective municipalities. If there emerges from these decisions a doctrine that residency is a condition precedent to standing to protest the consequences of municipal rules which affect housing supply, the consequences are apparent: those suburbs that have few if any residents who belong to racial minorities try to maintain that posture in order to avoid any duty; a dubious prospect, indeed, for any regional approach to decent housing reasonably near to jobs for low- and moderate-income persons who are not already residents of the offending suburb.

Meanwhile, back in the states, the exclusionary zoning dispute was taking a different focus in the courts. (Again, accept as a given that the cases are too few for any but the most cautious evaluations.) Neither race nor residency has been required for standing in this clutch of decisions. Nor is it always clear just what constitutional imperative has been reached. The remarkable point about some of these cases is the sensitivity of the opinions to the regional consequences of suburban housing policy. One has the distinct feeling upon reading these state cases that the courts were not so much concerned with a clear violation of the equal protection clause or any other constitutional restriction on state action as they were convinced that zoning was being used for a purpose just never intended by the state legislatures; in short, that the exclusionary zoning was *ultra vires*—beyond the power granted by the state legislatures, whatever the federal constitutional infirmities.

Two Pennsylvania cases have received the most attention, partly because they were first and partly because their anti-municipal language was so sweeping. In one, In re Kit-Mar Builders, the Supreme Court of Pennsylvania struck down a 2- and 3-acre minimum lot size requirement. The majority opinion by Justice Roberts (In re: Appeal of Kit-Mar Builders, 1971: 768-769) said:

> It is not for any given township to say who may or may not live within its confines, while disregarding the interests of the entire area. If Concord Township is successful in unnaturally limiting its population growth through the use of exclusive zoning regulations,

the people who would normally live there will inevitably have to live in another community, and the requirement that they do so is not a decision that Concord Township should alone be able to make.

Zoning abounds with little ironies, and Kit-Mar has its share. For one, in this case that has been hailed as a breakthrough by those who seek to eliminate suburban legal barriers to low-income housing, all the plaintiff, Kit-Mar, wanted was the right to build houses on *1-acre* lots instead of the 2- and 3-acre minimum the township demanded. Some breakthrough! But what counts, so they say, is the language of the opinion, which insinuates itself into the legal lore and later emerges as dogma long after the facts of the case are forgotten.

The other twist in Kit-Mar is that the plaintiff would never have won, the legal journals would have been bereft of excited notes, were it not for one judge who could not accept Justice Roberts' passion for the regional obligations of suburban communities but did regard it as an outrage that the township would stand in the way of the landowner making a buck on his property. Justice Bell provided the necessary fourth vote for the majority that invalidated the ordinance in its application to the property, but he was compelled to write a separate concurring opinion. He said (In re: Appeal of Kit-Mar Builders, 1971: 772):

> I believe that this zoning ordinance, which has no substantial relationship to health or safety or morals, is an unconstitutional restriction upon an owner's basic right of the ownership and use of his property, and it cannot be sustained under the theory or principle of "general welfare."

The companion case to Kit-Mar in the Supreme Court of Pennsylvania in 1970 was In re Appeal of Girsh. Girsh proposed to build apartment units and the ordinance made no provision for apartments in the township. Justice Roberts, as in Kit-Mar, was able to marshall a bare majority (with Justice Bell's concurrence in the result) to hold that a suburban town could not totally exclude apartments. What this means is hard to say. Perhaps only that Pennsylvania suburbs with ordinances that forbid apartments will have to zone the town dump for apartments; or that they will simply reclassify for apartment use a parcel that has on it nonconforming apartments built before the alleged exclusionary ordinance was adopted. Using either ploy, a town may claim it has complied with the Girsh mandate. Not very persuasive, but the point is that *in each such case* another costly lawsuit would be required to show that compliance with Girsh was illusory.[2]

By that time, Justice Roberts might have lost his precarious majority.

Time may, however, be on Justice Roberts' side. Other state courts are being asked to deal with the issue of exclusionary zoning. If enough opinions in other states cite Kit-Mar and Girsh favorably, the Pennsylvania court could be more reluctant to repudiate its position as leader in a dramatic shift in judicial attitudes toward suburban land-use policy.

Time, I said, *may* be on the side of the critics of suburban practices. The accuracy of that observation depends in part on what the New Jersey Supreme Court does with two zoning cases that have been decided by trial courts in that state.

These cases are Oakwood at Madison, Inc. v. Township of Madison (1971) and Southern Burlington County NAACP v. Township of Mt. Laurel (1972). By the time this book is published, the Supreme Court of New Jersey probably will have spoken on each of these decisions by two New Jersey trial courts. In these two cases, each in an area distant from the other, two trial judges struck down municipal zoning ordinances that appeared to offend their respective senses of equity, if not some unarticulated provisions of the constitutions of New Jersey or of the United States.

In the Madison Township case, the plaintiffs are two developers and six individuals who live outside Madison Township and allege they and others of the class they represent cannot find housing in the township. (Note the alliance of homebuilders and the poor.)

The court described the circumstances in the township as follows:

> About 55% of the land area of the township is zoned R40 or R80. The R80 zone is new, the R40 zone expanded. Minimum lot size is one acre in R40 and two acres in R80. Minimum floor space is 1500 square feet in R40 and 1600 square feet in R80. According to the former township engineer, 80% of R40 (or about 5500 acres) and 30% of R80 (or about 2500 acres) is vacant and developable. Minimal acreage is vacant and developable in the R7, R10 and R20 zones. Since the 1930s there has not been a development on two-acre lots within the township. Since 1964 only one subdivision plan for one-acre lots has been proposed. Land and construction

costs are such that the minimum purchase price in R40 would be $45,000 and in R80 $50,000. Only those with incomes in the top 10% of the nation and county could finance new housing in R40; an even smaller percentage in R80.

The court concluded that the zoning scheme was exclusionary and invalid. The court did not base its decision on any provision of the U.S. or the New Jersey constitutions. What it did say was:

> In pursuing the valid zoning purpose of a balanced community, a municipality must not ignore housing needs, that is, its fair proportion of the obligation to meet the housing needs of its own population and of the region. Housing needs are encompassed within the general welfare. The general welfare does not stop at each municipal boundary. Large areas of vacant and developable land should not be zoned, as Madison Township has, into such minimum lot sizes and with such other restrictions that regional as well as local housing needs are shunted aside. Vickers v. Tp. Com., of Gloucester Tp., 37 N.J. 232 (1962), upholding a prohibition against trailer camps anywhere within a municipality, is not to the contrary.

The court (Oakwood at Madison, Inc. v. Township of Madison and the State of New Jersey, 1971: 358-359) concluded: "For all the foregoing reasons the Madison Township zoning ordinance . . . is held to be invalid in its entirety."

In Mt. Laurel, the court apparently decided to act the part of a referee in bankruptcy, to propose a way that the offending municipality could clear itself of sin.

The plaintiffs in this case, in addition to NAACP, included homebuilding corporations, poor residents of the community, and poor residents of areas outside Mt. Laurel.

The Mt. Laurel zoning ordinance effectively excluded multiple-family residents. Minutes of township committee meetings suggested the officials were going to see to it that no low-income housing was built, even though there were many residents living in substandard housing. All in all, a rather dismal record for the municipal attorney to defend.

After distinguishing Valtierra (referendums commonplace in California), quoting Justice Hall's dissent in Vickers, and questioning the rationale in Euclid (and quoting the lower court in that case), the trial judge concluded (on page 20 of the unpublished copy of the opinion):

The patterns and practices clearly indicate that defendant municipality through its zoning ordinances has exhibited economic discrimination in that the poor have been deprived of adequate housing and the opportunity to secure the construction of subsidized housing; and has used Federal, State, County and local finances and resources solely for the betterment of middle and upper-income persons, and the zoning ordinance is, therefore, declared invalid.

Similar exclusionary zoning cases are sprouting all over the Garden State's boroughs. New Jersey, as a part of the New York megalopolis, is a natural point of attack by the builders and civil libertarians, but this does not explain why more such cases are moving in that state rather than in Connecticut or Westchester County, New York. Possibly the singling out of New Jersey is due to the frightful condition of New Jersey housing and tax policy, and the absence of any evidence that the state itself is trying to solve housing crises in a manner similar to that of the New York Urban Development Corporation. Perhaps the presence in New Jersey of an office of Suburban Action, Inc., a nonprofit organization with its principal office in White Plains, New York, explains the action in New Jersey. Suburban Action is a determined organization dedicated to opening up the suburbs, and it believes the New Jersey suburbs and exurbs are ripe. A final explanation may be that lawyers sense that the New Jersey Supreme Court may be of a mind to reappraise all the nonsense that has encrusted local autonomy on housing and land-use policy. If so, this would be fitting. It was the lonely voice of New Jersey Supreme Court Justice Hall, dissenting in Vickers v. Township Committee (1962), who cast the first small shadow on the suburban horizon. Ten years ago, Justice Hall could not persuade a majority of his colleagues that it was not acceptable for a township to exclude totally all trailer parks ("mobile home" parks, Justice Hall labeled them), but he did write a dissent that will be remembered long after the majority opinion in that case is collecting dust (Vickers v. Township Committee, 1962: 145):

Certainly general welfare does not automatically mean whatever the municipality says it does, regardless of who is hurt and how much. . . . The . . . general welfare transcends the artificial limits of political subdivisions and cannot embrace merely narrow local desires.

Justice Hall's dissent in Vickers would be little more than a rallying cry for the critics of suburbia in New Jersey were it not for the decision of the same court eight years later isn DeSimone v. Greater Englewood Housing Corp. No. 1 (1970: 38-39), in which Justice Hall found himself writing the opinion of the court that held that it was legitimate for the city of Englewood to grant a variation for subsidized housing in a middle-class sector of the city because

> we specifically hold, as a matter of law in the light of public policy and the law of the land, that public or, as here, semi-public housing accommodations to provide safe, sanitary and decent housing, to relieve and replace substandard living conditions or to furnish housing for minority or underprivileged segments of the population outside of ghetto areas is a special reason adequate to meet that requirement of N.J.S.A. 40:55-39(d) and to ground a use variance.

So students of this new phase of the zoning game will watch New Jersey and keep an eye on names of communities such as Madison, Mt. Laurel, and Mahwah that will appear in the reported cases.

New Jersey and Pennsylvania courts are not the only state forums to share the task of resolving the numerous attacks on the traditional preeminence of municipalities on land-use policy. I know of cases that are under way in state courts in California, Illinois, and New York. It should not be forgotten that the U.S. Supreme Court did not decide Euclid until zoning had first been upheld in a number of state courts.

In zoning cases, as in other areas of the law, one must be on guard for decisions that at first seem to have only peripheral relation to your interest, but on further examination may suggest new openings, favorable or otherwise. For those concerned with suburban zoning practices, such a case is Golden v. Town of Ramapo (1972).

The town of Ramapo is located in Rockland County, New York, an area still largely undeveloped but in the path of the outward population migration. Concerned over the prospect of being over-whelmed by growth and without the financial wherewithal to provide services, the town prepared a master plan and a new zoning ordinance. That is not unusual. But then the town adopted a capital budget, setting forth public improvements for the next six years, and followed that with adoption of a capital program specifying additional capital improvements for the twelve years following the six-year life of the capital budget. What it then did was crucial. These detailed plans for sewer, water, roads, parks, and other capital improvements were accompanied by provisions in the zoning ordinance that effectively prohibited any subdivision of land until the necessary public improvements had been installed—next year or eighteen years from now—however the schedule worked out. In short, subdivision plans were prohibited unless and until the necessary public improvements were available for the acreage, either by public or private effort. A majority of the intermediate judges on the intermediate appellate court invalidated that "staging" ordinance on the ground that it was a taking—it forbade any use or development of property with compensation.[3] A majority of the New York Court of Appeals reversed and upheld the ordinance. The court was unmistakably concerned lest its approval be taken as sanctioning exclusionary practices under the guise of thoughtful planning. The court's awareness of the dilemma, its sensitivity to the need to balance competing equities, was apparent:[4]

> What we will not countenance, then, under any guise, is community efforts at immunization or exclusion. But, far from being exclusionary, the present amendments merely seek, by the implementation of sequential development and timed growth, to provide a balanced cohesive community dedicated to the efficient utilization of land. The restrictions conform to the community's considered land use policies as expressed in its comprehensive plan and represent a bona fide effort to maximize population density consistent with orderly growth. True other alternatives, such as requiring off-site improvements as a prerequisite to subdivision, may be available, but the choice as how best to proceed, in view of the difficulties attending such exactions (see Heyman & Gilhool, The Constitutionality of Imposing Increased Community Costs on New Suburban Residents through Subdivision Exactions, 73 Yale L. J. 5.1119; see, also, ALI, A Model Land Development Code, §3-104, subd. [6]), cannot be faulted.
>
> Perhaps even more importantly, timed growth, unlike the minimum lot requirements recently struck down by the Pennsylvania Supreme Court as exclusionary, does not impose permanent restrictions upon land use [citing cases]. Its obvious purpose is to prevent premature subdivision absent essential municipal facilities and to insure continuous development commensurate with the Town's obligation to provide such facilities. They seek, not to freeze population at present levels but to maximize growth by the efficient use of land,

and in so doing testify to this community's continuing role in population assimilation. In sum, Ramapo asks not that it be left alone, but only that it be allowed to prevent the kind of deterioration that has transformed well-ordered and thriving residential communities into blighted ghettos with attendant hazards to health, security and social stability—a danger not without substantial basis in fact.

We only require that communities confront the challenge of population growth with open doors. Where in grappling with that problem, the community undertakes, by imposing temporary restrictions upon development, to provide required municipal services in a rational manner, courts are rightfully reluctant to strike down such schemes. The timing controls challenged here parallel recent proposals put forth by various study groups and have their genesis in certain of the pronouncements of this and the courts of sister States [citing]. While these controls are typically proposed as an adjunct of regional planning (see Proposed Land Use and Development Planning Law, arts. 3, 4; ALI A model Land Development Code, art. 7), the preeminent protection against their abuse resides in the mandatory on-going planning and development requirement, present here, which attends their implementation and use.

We may assume, therefore, that the present amendments are the product of foresighted planning calculated to promote the welfare of the township. The Town has imposed temporary restrictions upon land use in residential areas while committing itself to a program of development. It has utilized its comprehensive plan to implement its timing controls and has coupled with these restrictions provisions for low and moderate income housing on a large scale. Considered as a whole, it represents both in its inception and implementation a reasonable attempt to provide for the sequential, orderly development of land in conjunction with the needs of the community, as well as individual parcels of land, while simultaneously obviating the blighted aftermath which the initial failure to provide needed facilities so often brings.

The proposed amendments have the effect of restricting development for onwards to 18 years in certain areas. Whether the subject parcels will be so restricted for the full term is not clear, for it is equally probable that the proposed facilities will be brought into these areas well before that time. Assuming, however, that the restrictions will remain outstanding for the life of the program, they still fall short of a confiscation within the meaning of the constitution.

Ramapo's record on housing policy left little room to charge it with a sophisticated scheme to keep out low- and moderate-income housing. The town had created a public housing authority and had fought off in court efforts by local citizens to halt the construction of subsidized housing (Fletcher v. Romney, 1971; Greenwald v. Town of Ramapo, 1970; Farrelly v. Town of Ramapo, 1970).

The rub is that Ramapo may be a model for other communities with a less-noble record and less-virtuous objectives. Justice Breitel, dissenting, put it this way:[5]

> But there is more involved in these cases than the arrogation of undelegated powers. Raised [by these cases] are vital constitutional issues, and, most important, policy issues trenching on grave domestic problems of our time, without the benefit of a legislative determination which would reflect the interests of the entire state. The policy issues relate to needed housing, planned land development under government control, and the exclusion in effect or by motive, of walled-in urban populations of the middle class and the poor. The issues are raised by a town ordinance, which, as one of the Appellate Division Justices noted below, reflect a parochial stance without regard to its impact on the region or the State, especially if it becomes a valid model for many other towns similarly situated.

Does Ramapo provide a cover for communities that seek more sophisticated ways to build walls? Let a thousand Ramapos bloom?

We are at a watershed in the legal rules governing land-use development. The old ways are at the barricades in both federal and state courts. It is not enough to protest that court-administered reform is a sloppy way to administer what has been rightly called a "quiet revolution" (Bosselman and Callies, 1971). It is in the best American tradition that the courts act before the legislatures, federal and state, move with their usual glacial speed. There is a pervading restlessness with the old parochial system of public regulation of land development. Unlike most other revolutions, this protest is supported both by money makers and do-gooders. To bring such odd allies together must mean something is really coming apart.

What are the prospects that the stir in the courts will lead to some reform in the legislatures to restructure the decision-making process?

I believe the chances are good if one does not ask that reform be by midnight coup. I have before suggested that "the long sleep of the states is over," and wherever one looks—to New Jersey, Massachusetts, Florida, Vermont, New York, California, Ohio—there is evidence that the states are beginning to take back some of the control over land-use policy they had for so long left to their creatures, the municipalities (Bosselman and Callies, 1971; Babcock, 1972). Admittedly, the motives for this state resurgence are varied and not necessarily consistent with a concern for more adequate housing. In some jurisdictions, the state has moved in to regulate development out of fear of too much growth. In many states, the renaissance has more to do with the current ecological kick, a reflection of a greater concern for the quality of habitats for bald eagles than of humans. No difference. The point is that an area of public regulation—land development—long vested in the municipalities is coming under greater state control, and whether it goes by the name of "environmental protection" or "zoning," the consequence is the same: a diminishing municipal voice in major issues of land development policy.

NOTES

1. The complaint alleges that residents of an unincorporated area near St. Louis formed a municipal corporation and enacted a restrictive zoning ordinance when it was learned that FHA 236 housing was going to be constructed. The suit is about to go to trial.

2. At the time this chapter was written, thirty months after the decision of the Pennsylvania Supreme Court in Girsh, the plaintiff had still not received his building permit and had been forced to go back to the courts for further relief. His latest effort is still pending before the Pennsylvania Supreme Court.

3. 324 N.Y.S. 2d 178, 1971.

4. 285 N.E. 2d 291 at 302.

5. 285 N.E. 2d 291 at 305.

CASES

DAILEY ET AL. v. CITY OF LAWTON, OKLAHOMA (1969) 296 F. Supp. 266 (U.S. Dist. Ct. W.D. Oklahoma).

FARRELLY v. TOWN OF RAMAPO (1970) 317 N.Y.S. 2d 837; App. Div. 2d 957.

FLETCHER v. ROMNEY (1971) 323 F. Supp. 189 (S.D. N.Y.).

Appeal of GIRSH (1970) 263 A. 2d 395 (Pennsylvania).

GREENWALD v. TOWN OF RAMAPO (1970) 317 N.Y.S. 2d 839; 35 App. Div. 2d 958.

JAMES v. VALTIERRA (1971) 402 U.S. 137.

KENNEDY PARK HOMES ASSOCIATION ET AL. v. CITY OF LACKAWANNA (1970, 1971) 436 F. 2d 108; cet. den. 91 S. Ct. 1256.

In re: Appeal of KIT-MAR BUILDERS (1971) 268 A. 2d 765 (Pennsylvania).

OAKWOOD AT MADISON, INC. v. TOWNSHIP OF MADISON AND THE STATE OF NEW JERSEY ET AL. (1972) 117 N.J. Super. 11, 283 A. 2d 353.

SISTERS OF PROVIDENCE OF ST. MARY OF THE WOODS ET AL. v. CITY OF EVANSTON ET AL. (1971) Memorandum opinion motion to dismiss complaint (U.S. Dist. Ct. N.D. Illinois, October 28).

SOUTHERN ALAMEDA SPANISH SPEAKING ORGANIZATION v. CITY OF UNION CITY (1970) 424 F. 2d 291 (U.S.C.A. 9th Circ.).

SOUTHERN BURLINGTON COUNTY NAACP ET AL. v. TOWNSHIP OF MT. LAUREL ET AL. (1972) Sup. Ct. N.J., Burlington Cty. L-25741-70-PW, May 1.

UNITED STATES OF AMERICA v. CITY OF BLACKJACK, MISSOURI (n.d.) U.S. Dist. Ct. E.D. Missouri 71C372.

REFERENCES

BABCOCK, R. (1972) "Let's stop romancing regionalism." Planning (July).

––– and F. B. BOSSELMAN (1963) "Suburban zoning and the apartment boom." Univ. of Pennsylvania Law Rev. 111: 1040.

BOSSELMAN, F. B. and D. L. CALLIES (1971) The Quiet Revolution in Land Use Control. Washington, D.C.: Council on Environmental Quality.

KURLAND, P. (1968) "Equal educational opportunity: the limits of constitutional jurisprudence undefined." Univ. of Chicago Law Rev. 35: 583.

13

A Question of Choice: Access of the Poor and the Black to Suburban Housing

LEONARD S. RUBINOWITZ

☐ PUBLIC POLICIES AND SUBSIDIES have provided a basic choice for the white and the relatively affluent in the past several decades, the choice of where to live within our metropolitan areas. Increasingly, these groups have opted to move to the suburbs. This steady stream from the central cities has not been the result merely of massive, coincidental private market decisions. Public funds built the roads which facilitated this trek to the suburbs, since commuting to central-city jobs was often necessary. Mortgage insurance, through the FHA, helped to finance the house in the suburbs. Suburban municipalities' zoning laws and practices helped to ensure that only those who could pay the price of admission had the choice.

Now the question is whether this country will provide the same locational choice to the poor and the Black. Will these groups have the *option*, the opportunity, for access to suburban land that has been granted to the relatively privileged and the absolutely white?

Through a variety of mechanisms, government has made housing available to low- and moderate-income families. The form of the program and the nature and depth of the subsidies involved have changed over time and will probably change in the future. It is likely, however, that some form of public subsidy will continue to enable some (but not all) low- and moderate-income families to have "access" to decent housing.

Whatever the subsidy mechanism, the question remains of whether low- and moderate-income families will be limited in their locational choice, as in the past, or whether the range of options available to these families will begin to approach equality with that of the rest of society. To move toward this state of equal opportunity, an obstacle course has to be crossed. Many of the barriers to low- and moderate-income housing in the suburbs are local in character. In addition, impediments have often been built into the subsidy programs themselves.

The last few years have witnessed the development and implementation of a variety of strategies, both public and private, for overcoming the suburban hurdles. These activities, most of which are at an embryonic stage, indicate a possibility that the question may be answered, ultimately, in favor of increased options. More strategies for "opening the suburbs" are being tried, in more places, each year. Many of the "victories" to date have been symbolic. Other efforts have met with defeat, at least temporarily. The resistance by the suburbs shows no signs of fading away. However, the objective of housing choice is increasingly a shared one, with different interest groups having their own reasons for favoring increased options. In sum, the ultimate outcome remains very much in doubt.

HOUSING PROGRAMS

This country has experienced major governmental intervention in the housing market since the 1930s. These programs have significantly influenced the housing options of the affluent and the poor, the white and the Black, alike. The mortgage insurance programs administered by the Federal Housing Administration, or FHA (now part of the Department of Housing and Urban Development, or HUD), stimulated and assisted the post-World-War-II exodus from the central cities. FHA mortgage insurance was available essentially on a whites-only, suburbs-only basis. Although not a direct subsidy, this insurance enabled many families who had not been homeowners to purchase housing which was being built in suburbia.

The other long-standing federal housing program, public housing, has provided housing for the poor, and increasingly, for the Black, almost entirely in the central cities of metropolitan areas. Sites have generally been located in the most segregated and deteriorated neighborhoods, and occupancy has also been on a racially segregated basis.

In sum, the historical pattern of a dual housing market, one for whites and one for Blacks, in metropolitan areas, has long been supported and buttressed by public programs.

In the last ten years, Congress has created an array of additional direct subsidy programs for low- and moderate-income people. These programs fill the gap between the public-housing program for the poor and the mortgage insurance programs for middle-income people. By 1960, moderate-income people had been priced out of the new housing market; subsidies were necessary to make new housing accessible to this income group. The decade of the sixties was a period of new numbers, letters, and words in the housing lexicon. In 1961, came the 221(d) (3) program, referred to by the section of the congressional statute creating it.[1] Under this program, new apartments were constructed for moderate-income families, and their rents were subsidized. This was followed by the rent supplement program in 1965.[2] This program has deeper subsidies and serves low-income families (much like public housing) in accommodations which are privately developed and owned.

The current digital soup of housing production programs was completed in the omnibus Housing Act of 1968. The section 235 program provides for homeownership for moderate-income families and the section 236 program (successor to 221[d] [3]) subsidizes rental housing virtually for the same income group.[3] Again, the numbers come from the sections of the law creating the program.

A U.S. Commission on Civil Rights (1971b) report indicates that the past patterns of metropolitan racial segregation perpetuated by HUD and its predecessors have continued in the administration of the section 235 homeownership program. New homes in the suburbs provide access to the suburbs for moderate-income families, but on virtually a whites-only basis. The central-city homes under this program are almost exclusively in the ghettos and Black-occupied.

A related development which is likely to become increasingly important over time is the "housing allowance." This subsidy is related to the family rather than to the housing unit. Families who cannot afford decent housing receive subsidies which enable them to secure standard housing of their own choice in the private market. The difference between the market cost of the housing and the amount the family can afford to pay is subsidized by the government.

Kansas City undertook the first housing allowance experiment, using funds received from HUD through the Model Cities program. The two-year experiment was sufficiently promising that HUD, with congressional authorization, embarked on a larger housing allowance experiment. A number of cities have been contacted, with at least one refusing to participate in the program.

The concept of providing the resources for people to make their own housing choices is an increasingly popular one. Recent interest in this approach is a result, in part, of the administrative failures of the production programs, ranging from problems of construction quality to scandals related to fraud and corruption within the bureaucracy. The upshot may be an increasing emphasis by Congress on enabling low- and moderate-income families to secure totally private housing and a phasing out of programs designed to produce housing specifically for lower-income groups. Under these or other alternative subsidy schemes, many of the obstacles to the movement of the poor to the suburbs remain.

THE SUBURBAN OBSTACLE COURSE

Exlusionary devices at the local level are numerous and varied. Traditionally, the substance of zoning ordinances has posed the major roadblocks to lower-income housing in suburbs. In some cases, ordinances completely exclude certain residential uses, such as apartments or mobile homes. Others impose density requirements which limit development to very low densities, thus pricing out low- and moderate-income housing. Still other ordinances require unreasonable minimum house size (usually in terms of floor space) or limit apartments to one bedroom, thus excluding large families. Somewhat more subtle, and equally effective, are the administrative practices in the local zoning process, which exclude low- and moderate-income housing, even where the ordinances may not do so directly. Apartments may be permitted with administrative approval, but such approvals are never forthcoming. Undesirable uses may be provided for specifically, but in undesirable places, where development will not take place. An apartment zone may be squeezed in between the town dump and the local glue factory.

Other devices available to suburbs for exclusionary purposes include the power to issue building permits, approval of tap-in to the sewer system and other administrative prerogatives which are seemingly neutral.

The U.S. Congress is a frequent and not unconscious ally of the exclusionary suburbs. The federal subsidy programs come equipped with constraints which, directly or indirectly, permit the local community to prohibit the use of the program within its borders. Of the current array of four major subsidy programs, the local governing body has a direct veto power in two and indirect "assistance" in this endeavor from the Congress in the other two.

When the rent supplement program was created by Congress in 1965, HUD considered using it as a means of providing integrated housing opportunities in the suburbs. When Congress got wind of these plans, the program was almost aborted. Finally, a small appropriation was approved, but Congress placed a condition on the use of these funds.[4] None of the funds could be used in any community without the express approval of the local governing body. Inaction by the community is thus sufficient to keep out rent supplement units.

Doing nothing also serves to keep low-rent public housing out of a community. Doing nothing may take the form of not creating a "local housing authority," the agency which administers the public housing program at the local level. Even if a housing authority exists which has jurisdiction over a suburb (e.g., in some states, housing authorities have countywide jurisdiction), the housing authority cannot operate within the boundaries of any municipality without the consent of the local governing body. Generally, that approval takes the form of a "cooperation agreement" between the local government and the independent housing authority.[5] The locality agrees to provide municipal services and to accept ten percent of shelter rents as a payment in lieu of taxes.

In sum, Congress in its wisdom has clearly made the low-income housing programs a matter of local option. A similar result is achieved in the moderate-income programs through the cost ceilings written into the law. Section 235 and 236 units may not exceed approximately $24,000 in their total development cost.[6] Through the use of any number of exclusionary devices described earlier, a suburb can merely price this housing out of town.

In 1972, Congress considered a bill granting local communities a direct veto power over the use of all federally subsidized housing.

Had it not been for some strong lobbying and an election year rush to adjourn, the bill might well have passed. A similar bill will almost surely appear in 1973.

The suburbs have an additional ally, in the form of inertia. Housing for the poor gets built in the suburbs only if sponsors and developers take the initiative and attempt to overcome the obstacles outlined above, as well as to endure the red tape involved in the federal subsidy programs. The few developers who take this trip risk coming out with a clear conscience and an empty wallet. Not surprisingly, most developers of subsidized housing prefer to take the path of least resistance and build where they are welcomed with open arms.

STRATEGIES

Given these many and varied obstacles, the road to the suburbs for the poor is likely to be a difficult one. Some help seems to be forthcoming, however. In addition to the emergence of exclusionary zoning lawsuits, an increasing number of strategies is being developed and tested around the country to make the suburbs accessible to low- and moderate-income housing. The volume of activity is growing in both the public and the private sectors.

Examples of public efforts to "open the suburbs" can be found at the county, regional, state, and federal levels. These approaches run the gamut from almost total dependence on voluntary cooperation to high degrees of coercion. All attempt to escalate a part of this crucial land-use decision above the local level, where it has been lodged for the past half-century.

Since exclusionary zoning cases remain the primary tool of private individuals and organizations for opening the suburbs, the main focus here will be on public and "quasi-public" strategies. Some attention will also be given to private efforts to increase the responsiveness of higher levels of government to the problem of suburban exclusion.

Most of the approaches, whether public or private, voluntary or coercive, local or federal, are recent in origin. None has been completely successful. Many are in the infant stage, and it is too early to assess whether they will mature, and, if so, what will be their impact. I will only attempt therefore to make guesstimates as to where these various approaches may lead and which ones appear to have potential viability, in terms of providing low- and moderate-income families the option of living in the suburbs. In theory, equal opportunity in housing is protected by the federal law. The purpose here is to consider how the reality might begin to approach the theory, for the metropolitan poor. Starting, then, at the county level, at least one county has attempted to affirmatively require housing developers to include low- and moderate-income housing in their development plans. Fairfax County, Virginia, is directly in the path of suburban development in the Washington area, due to the continued influx of primarily middle-income whites moving out of the city. As a result, the county is in the throes of a housing construction boom. Permits for 7,000 to 8,000 housing units were issued in 1969 and again in 1970. In 1971, building permits were issued at more than double the rate of the previous year.

The county board of supervisors decided to use its zoning power to regulate this growth in precisely the opposite way from that in which suburban localities ordinarily react—i.e., requiring the inclusion of low- and moderate-income housing, rather than excluding it through zoning requirements.[7]

The Fairfax County ordinance, passed in August 1971, requires that for any residential development of fifty or more units:

> an applicant for RPC (Residential Planned Community) zoning or for an amended development thereunder . . . shall provide or cause others to provide, under the development plan, low-income units which shall not be less (and may be more) than six per cent (6%) of the total number of units (other than detached single family dwelling units) specified in the development plan. The applicant shall also provide, or cause others to provide, the number of moderate-income dwelling units which, when added to the number of low-income dwelling units, shall not be less (and may be more) than fifteen per cent (15%) of the total number of dwelling units (other than detached single family dwelling units) specified in the development plan.

Under this scheme, all substantial housing projects were to have an economic mix and presumably a racial mix (given the demographic patterns of the Washington area) of occupants.

However, shortly after the passage of the ordinance, and before any real opportunity for testing its workability, the county zoning ordinance was challenged in the state courts, in De Groff Enterprises, Inc. v. Board of County Supervisors of Fairfax County, Virginia et al. (1971). The bubble burst, as the trial court held that the ordinance was beyond the powers which the State Zoning Enabling Act gave to the county board. The county petitioned for an appeal to the Virginia Supreme Court. It is worth noting that the trial court did not decide that the county ordinance was unconstitutional, but merely that it was not provided for under state law. Even if that decision is upheld in Virginia, the result might be different in a state with a more expansive zoning enabling act (or state courts that interpret the state law more liberally).

In that regard, the Montgomery County, Maryland, Council has under consideration proposed ordinances similar to the Fairfax County ordinance.[8] Montgomery County is the affluent, predominantly white suburban county immediately adjacent to the northwest section of Washington. Jobs are moving into Montgomery County in droves. HEW has built a major office complex housing thousands of employees in Rockville. Several major corporate employers have moved out to the open spaces of Montgomery County in recent years. Many of the employees, both public and private, cannot afford the private housing available in Montgomery County and must choose between a long, expensive commute from the inner city and a search for a new job in the city. The result has been serious problems for both employers and employees. High turnover and vacancy rates have given corporate leaders second thoughts. Some have told county officials that if they knew then what they know now about the housing situation in the county and the consequent problems, they would not have moved their operations to Montgomery County (Garrott, 1971: 73). Because of the importance of the industrial influx to the county's fiscal health, these complaints were undoubtedly an impetus for the submission of proposed ordinances requiring subsidized housing.

Whether or not the pro-housing forces prevail in Montgomery County, the coalition of interests supporting the proposals may have potential elsewhere. Suburban counties and municipalities need the revenue of the employers. The employers need employees. Employees need housing they can afford, accessible to their jobs. The interests of these diverse groups may come together in a way that brings the necessary pressure to bear to push through Fairfax-type ordinances or other steps toward "opening up" the suburbs.

Beyond actions at the county level, the next institutional response has been at the metropolitan or regional level. As the courts are ruling in the exclusionary zoning cases, the housing problem is regional in nature and scope. Population movements do not stop at political boundary lines. Increasingly, metropolitan and regional planning bodies recognize this fact of life and are developing regional housing plans.

Many regional agencies were created in the last twenty years, in response to the passage by Congress of the Section 701 planning assistance program.[9] Congress makes funds available for comprehensive regional planning under this program, on a matching basis with local funds. (There is, of course, no faster way to bring about the creation of public bodies than to make available federal funds to be applied for and spent by such agencies.) Shortly after the federal planning funds began to flow, the countryside was dotted with "Areawide Planning Organizations," APO in the federal jargon. These agencies then proceeded to produce massive volumes which did little more, in most cases, than provide small children with easier access to the dinner table and keep a number of federal bureaucrats ungainfully employed reviewing these plans for dotted "i's" and crossed "t's." Although these bodies may take many forms—e.g., Regional Planning Agencies, Councils of Governments (COGS)—they rarely have any significant degree of governmental authority. They are instead advisory, and, all too often, that adjective turns out to be a euphemism for impotent.

In 1968, Congress took a step toward specificity in its requirements for the expenditure of the planning funds. Areawide planning organizations were to include as part of their work program a "housing element."[10] Although this provision remained vague, it at least made clear that housing was to be a part of the planning agency's agenda.

Since 1968, a gradual change has taken place in the activities of planning agencies in the housing area. In at least one case, the planning agency latched onto the "housing element" requirement to justify its housing involvement (which it wanted to do anyway) to its constituency. In other cases, the federal requirement may have stimulated new activity. In still another instance, HUD used this requirement to condition planning funds on the development of an action-oriented housing plan (more on that when we reach the federal role).

The primary thrust of the planner's housing efforts has been the development of "allocation" or "distribution" plans for low- and moderate-income housing. These plans identify where, within a metropolitan area or region, the needed housing should be located. They do not usually designate specific sites for the housing. Instead, "subareas" or planning areas are identified within which specific quantities of housing should be placed or which should receive a particular level of priority for subsidized housing.

The first distribution plan to be developed, and the farthest down the road on implementation, is the Miami Valley Regional Planning Commission's *Housing Plan for the Miami Valley Region.* MVRPC is the Areawide Planning Organization, the recipient of federal planning funds, for a multicounty area centering on Dayton, Ohio. (The exact number of counties which are members of the commission varies from time to time. Because of the controversial nature of the housing plan, two of the rural counties that were part of the original five-county region periodically threaten to withdraw from the region. Of the approximately 900,000 people in the five-county region, 250,000 live in Dayton.)

The Miami Valley effort has been from the outset based on persuasion, cooperation, and education, rather than on coercion (for a detailed discussion, see Bertsch and Shafor, 1971b). The commission has thirty municipalities, in addition to the member counties. There are approximately forty commissioners, almost all of whom are elected officials, with each municipality having one representative and each county having two. The commissioners were persuaded to the staff to sanction and support the development of an analysis of the region's need for low- and moderate-income housing and a plan to distribute the needed housing throughout the region. Once having drawn up the plan, the staff presented it to the commission and to any public forum which was available in the five-county area. By the time the plan came up for formal adoption, commission members were virtually unanimous in support, in spite of vigorous opposition in some of the communities these officials represented.

Before looking at the record on the difficult trek of the plan from paper to reality, it is appropriate to take a quick look at the "Dayton Plan" itself. First, families who were considered "in need" of subsidized housing were defined in terms of rather arbitrary flat income figures, a $10,000 per year maximum in the urbanized counties and $7,500 in the rural counties. These annual income figures were intentionally pegged at a high level to be inclusive of large segments of the region's population, as a way of building political support for the plan. The unmet need, the number of additional housing units necessary for every family to live in a decent house within its means, was then determined to be in excess of 14,000 low- and moderate-income units. (This figure did not consider the availability of subsidy funds or the local production capability, but looked only at the "social need." For this reason, it was understood that it would be impossible to literally meet this "need" in the five-year period specified in the plan.)

Finally, the plan allocated the needed housing over some 53 planning units based on criteria related to the ability of a particular area to absorb new housing units—e.g., school capacity, tax base. A planning unit's allocation was also to serve as a ceiling, to avoid the concentration of subsidized housing in any one part of the region. The "ceiling" feature was responsible for much of the support the plan received in the region.

The plan was adopted in July 1970. The ensuing implementation period largely coincides with a time of vastly accelerated production of federally subsidized housing nationwide. This coincidence makes it difficult to draw conclusions from the before and after (the plan) statistics and determine the impact of the plan. It is impossible to know precisely what would have happened in the absence of the plan, at the initiative of private developers and other sponsors under the federal programs.

One thing is "perfectly clear," however. The planning commission did not use the traditional planners' cop-out—that their role is technical only—that they deliver the document to the politicians' desks, and their function ceases. On the contrary, the planning commission has taken the lead role in the implementation of this plan, a role which is of necessity heavily political. The planning commission has developed de facto authority well beyond its extremely limited legal authority. Any potential sponsor of subsidized housing in the Dayton area cannot get funding from HUD without doing business with MVRPC. Top HUD officials argued for years that lower-income housing problems can only be solved on a metropolitan basis. With MVRPC taking the first steps among planning agencies, toward a serious effort in this direction, HUD committed itself to supporting the Dayton plan and assisting in its implementation. This put MVRPC in an enviable bargaining position vis-à-vis developers.

Since the adoption of the plan, the trend in the development of lower-income housing has been away from the central city of Dayton and toward the suburbs of the region, as indicated below:

Low and Moderate-Income Units in the Miami Valley Area

	1969 (before the plan)	1972 (cumulative)	Proposed Units
In Dayton	95% (2979)	81% (4914)	22% (726)
Outside Dayton	5% (158)	19% (973)	78% (2605)

Based on this contrast, the likelihood is that the Miami Valley Regional Planning Commission has had a significant impact in bringing about the development of low- and moderate-income housing in the Dayton suburbs. Much of this housing would not have been built, or the federal assistance sought, had it not been for the pushing and pulling, the prodding and politicking of MVRPC.

A couple of qualifications. The suburban housing is virtually all moderate-income housing, since the local approvals necessary for the development of low-income housing have been as elusive there as in other metropolitan areas. Second, much of the housing has been built for, and occupied by, people who were already living in the suburbs. To date, inner-city Blacks have received few of the benefits, in terms of suburban access. (For political purposes, the plan was sold as an approach to meeting the community's own needs, an important weakness in terms of the suburban access objective. The suburban communities and the housing developers have taken this aspect of the plan quite literally.) Finally, a significant portion of the production has been for elderly occupancy. This approach may have been thought of as a way to get the suburban door ajar, but all too often the door has been slammed shut in the faces of low-income families after all the suburbanites' mothers or grandmothers have been let in. Suburban grandmothers are an unlikely opening wedge for inner-city Blacks, particularly in large families.

The progeny of the Miami Valley Plan are far from legion, but their numbers are increasing. Agencies which have adopted plans which seek to influence the location of low- and moderate-income housing within a metropolitan area include: Washington, D.C., Council of Governments (COG); Metropolitan Council of the Twin Cities, Minnesota (Metro Council); San Bernardino County, California; Southeastern Wisconsin Regional Planning Commission; Dade County, Florida; and Sacramento Regional Area Planning Commission.

Given the extremely limited formal authority of regional planning bodies, their effectiveness in increasing suburban housing opportunities will depend on their willingness to jump into an extremely controversial area and on their ability, once involved, to keep from drowning. The Miami Valley Commission has come out from behind the planners' apolitical guise and has had considerable success in the face of very long odds. Astute political planners may be able to duplicate this approach elsewhere. Even the preliminary returns are not yet in.

Like the regional planning agencies, the states are becoming involved in opening their suburban "creatures" to the poor and minorities. A number of approaches have been tried, ranging from creating state housing agencies which provide low- and moderate-income housing themselves, to involvement in the zoning process, thus restricting local prerogatives, to taking the exclusionary suburbs to court. In addition, state legislatures have considered, without passing, a variety of measures which would facilitate the introduction of low- and moderate-income housing into suburbia.

The number of state legislatures and administrations which have ventured into these politically hazardous waters is increasing, but remains small. Several models are now available, but other states seem reluctant to imitate these approaches.

One of the best-known "models" at the state level is the Massachusetts Anti-Snob Zoning Law.[11] This law establishes a quota for low- and moderate-income housing for each town in the state—i.e., a maximum number of units which the city or town must accept. Once 10% of the locality's total dwelling units are low- and moderate-income housing, the community may deny any subsequent request to permit development of such housing. The community has the same discretion if, at any point, low and moderate housing occupies 1.5% of the land zoned for residential, commercial or industrial uses. Finally, the law places annual limits on the units a locality must accept until the overall limits are reached.

In addition to placing an obligation on localities, the law establishes an appeals process at the state level, in case of an adverse decision at the local level on a proposed subsidized housing development. The State Housing Appeals Committee consists of five members and is lodged in the State Department of Community Affairs. If a town rejects an application for subsidized housing or imposes conditions unacceptable to the developer, an appeal is available to the Housing Appeals Committee. When it finds that the decision of the local zoning board is "inconsistent with local needs," the committee may direct the issuance of a comprehensive permit. This "one-step" permit eliminates the need to obtain all the separate local approvals, such as building permits, zoning and subdivision approvals, and health certificates. As a result, a reversal by the state committee prevents the locality from using regulatory powers other than zoning to exclude subsidized housing.

The administrative appeal process does not preclude additional review by the courts. Instead, the law adds a level of decision-making between the local zoning board and the state judiciary.

The principle has knocked heads, however, with some difficult realities, some involving the language of the statute itself and others related to the politics of suburban land development.

Substantial problems have been created by the vagueness of the criterion for denial of permits, inconsistency with "local needs." A project proposed for the town of Newton ran into hearings before the State Housing Appeals Committee which extended over a year, as the opposing counsel attempted to demonstrate that denial was consistent with this vague standard.

In addition to the vagueness problem the drafters created, they also left a major loophole or two. Communities do not have separate obligations for family and elderly housing, just one overall responsibility. As a result, some communities are rushing to meet the letter of the law by developing enough of the politically safe, even desirable, housing for the elderly, to meet their "quota." The state has indicated administratively that it will cut off the flow of funds for elderly housing unless a community is also attempting to meet its needs for low- and moderate-income family housing. This solution is not likely to be as satisfactory as a requirement in the law of an elderly-family balance.

Had the statute been free of drafting deficiencies, it would still have had a hard road to hoe. The law is not self-executing; housing does not appear automatically.

As a practical matter, the anti-exclusionary provisions operate on a case-by-case basis, as developers seek the necessary local approval to build subsidized housing in a particular community. Developers have not come rushing into the fray, probably because attempting to take advantage of the law and enter into an unwilling locality may have tremendous costs.

If an appeal becomes necessary, the hearings may be lengthy and expensive in themselves. This delay, like any other, is costly for a developer and may mean the loss of his option on the land or his financing. His decision to fight the battle in the first place may make him very unpopular with local officials in the area. As a result, he may find all manner of obstacles in his path on other developments he undertakes, subsidized or unsubsidized, in that community or others. His reputation as a troublemaker may be much like that of the developer who sues a community because of its exclusionary zoning. The good will which the developer needs to weave his way through the local administrative maze may all but disappear.

Several towns in Massachusetts have taken additional steps to block implementation of the law (Wall Street Journal, 1972). Two towns have challenged the legality of the law itself in court. These suits place a cloud over the law, making developers more reluctant to make use of the provisions.

The town of Chelmsford used its eminent domain power to condemn a piece of land which had been identified as a potential site for low- and moderate-income housing. The town then designated the site for "conservation" and denied the permit for housing. Finally, some towns have raised questions about the appropriateness of particular sites which have been proposed for low- and moderate-income housing.

In the face of these obstacles, it appears that the developer most willing and able to test the utility of the state law is the state itself, through the Massachusetts Housing Finance Agency. This agency finances the development of mixed-income housing—i.e., the law requires that all of their developments contain both low- and moderate-income components.[12] MHFA has made use of the state appeals process when it has been unable to secure the necessary zoning changes at the local level.

Favorable decisions from the appeals committee do not necessarily end the matter, however. At this point, the spectre of the loss of local control begins to become a reality for local residents, and the issue may be joined again in court. Suburbanites are likely to have the resources to exhaust judicial remedies and their incentive is to do so, since any delay is likely to be to their benefit.

In sum, the state legislature in Massachusetts has made an important statement of principle, that local communities are not sovereign and have a responsibility to provide opportunities for people of all incomes to live within their boundaries. As is often the case, however, the proponents of the status quo have a varied repertoire of devices to slow the pace of change. They have not yet exhausted their bag of tricks.

Another piece of alphabet soup, UDC, represents the state of New York's unique approach to providing low- and moderate-income housing. The state Urban Development Corporation was created in 1968.[13] UDC was given a wide range of powers to assist in the redevelopment process in the state, as well as the development of subsidized housing, using state and federal programs. The agency is authorized to operate in any community in the state, to purchase, lease, or condemn property and redevelop the property with industrial, commercial, or civic buildings or low-, moderate-, or middle-income housing.

Uniquely, UDC was authorized to override "the requirements of local laws, ordinances, codes, charters, or regulations applicable to such construction, reconstruction, rehabilitation, alteration or improvement . . . when, in the discretion of the corporation, such compliance is not feasible or practicable." In theory, after meeting its duty to consider local plans and seek the cooperation of local officials, UDC could build a subsidized apartment project in an affluent suburban community with minimum lot sizes of two acres.

It is no surprise that UDC has not proceeded in this manner. For the first couple of years of its existence, UDC responded to invitations from local officials to assist in the development of needed low- and moderate-income housing. In spite of this unspoken nonagression pact, the very existence of the "override" provision was sufficiently threatening that the legislature passed a bill rescinding the power in 1971. Only a veto by Governor Rockefeller preserved the power for UDC.

At that point, UDC had used its zoning override only in one or two unusual circumstances. Having survived the legislature's attack on its powers, UDC announced on June 20, 1972, that it was going to develop a series of subsidized housing projects in nine relatively rural communities in Westchester County, adjacent to New York City (Logue, 1972). Construction was to begin in 1972. The "Nine Towns Program" was based on a "fair share" principle, similar to that used by Miami Valley and other regional planning agencies. One hundred units of low- and moderate-income housing—town houses and garden apartments—would be built in each of the towns: Bedford, Cortlandt, Greenburgh, Harrison, Lewisboro, New Castle, North Castle, Somers, and Yorktown. In each case, seventy apartments would be for moderate-income families, twenty for low-income families, and ten for the elderly. Twenty percent of the units would be made available for minority-group families.

UDC announced that it would seek the cooperation of the towns involved, but that it would build the projects for low- and moderate-income families and the elderly, regardless of the receipt of this cooperation. The "override" authority was to be the "stick" designed to induce local public officials to negotiate about the proposed sites or provide substitute sites. Major zoning changes were clearly going to be necessary to implement this plan. Approximately 99% of the residentially zoned land in the county permitted only single family homes—no apartments. All but one of the sites proposed by UDC (the exception was in Cortlandt) were zoned, at that time, for single family homes, with minimum lot sizes as high as four acres.

Opposition mounted quickly, after the announcement of the Nine Towns Program. The "United Towns for Home Rule" established as its purpose, to oppose "the absolute and unfettered power of a state superagency to come uninvited into a community to override its zoning and to unilaterally impose its will upon that community" (New York Times, 1972a). Actions contemplated included demanding consultation with UDC, making opposition to UDC a campaign issue in the upcoming election, and seeking repeal of the agency's override power.

Whatever the precise process of exerting pressure, within two months UDC announced that it was suspending implementation in five of the towns—Cortlandt, New Castle, North Castle, Somers, and Yorktown—pending further local consultations to attempt to identify sites acceptable to both the local communities and the state (New York Times, 1972b).

At the same time, the town of Harrison went to court and obtained a temporary restraining order against UDC, based on the claim that the corporation did not go through the correct filing procedures to obtain rights to the 24-acre site there.

Some rays of hope appeared, however. Bedford invited the UDC to build an additional one hundred housing units, beyond what was originally planned. Greenburgh has voiced no strenuous objections to the housing plan, and in Lewisboro the UDC was looking for an alternative site to the one proposed initially.

The "box score" on the Nine Towns Program is likely to change almost from day to day. The UDC has embarked on a politically ambitious undertaking, regardless of the small number of units involved, and it is important that the effort bear some fruit, in terms of future suburban approaches by UDC, as well as other states. UDC must walk a difficult tightrope, trying to achieve its objectives while avoiding stirring up so much opposition that the legislature would clip its wings and strip the agency of the override and other powers.

Meanwhile, in neighboring New Jersey, Governor Cahill has also recognized the problems caused by suburban exclusionary zoning practices. Using as a stick the recent state court trend against exclusionary zoning, the governor's message to the legislature, "New Horizons in Housing," of March 27, 1972, proposed legislation providing for a voluntary "fair share" approach.

The governor argued that

> change is in the offing. It will probably be mandated judicially, if it is not developed through a coordinated plan controlled at the local level with the aid and assistance of the county and State.
>
> In essence, what we must achieve is a balance in housing; low, moderate and expensive, single and multi-family. I am not suggesting that the balance should be equal or that every community should be the same. . . . It should be emphasized that this is a purely voluntary plan; no municipality is mandated or forced to comply. It is however, an orderly approach and a reasonable alternative to probable judicial action that may well require construction in areas not especially suited under county and municipal planning for this kind of development.

The governor's alternative, the "Voluntary Balanced Housing Plan Act," contemplates an analysis of the housing need in the state every two years, with the overall need allocated by the state among its constituent counties. The counties would then have six months to suballocate their allocation among municipalities, based on availability of sites, access to jobs, tax base, number of low- and moderate-income households, and county and state plans. If counties did not do the necessary planning, the state would step in. Once municipal allocations were arrived at, communities could, on a voluntary basis, either designate areas for low- and moderate-income housing or develop policies that would assure that suitable land would be made available.

At some point, this bill may receive serious consideration in the New Jersey Legislature, assuming the state courts continue to place limits on exclusionary practices of New Jersey suburbs. The incentive for localities to support this bill would be that they would be participating in defining the limits of their responsibilities, rather than face the possibility that a state judge would perform this task and preempt the local decision-making process. Local participation might prevent the lawsuits or at least limit the relief ordered by the courts.

Nirvana not having been attained in Massachusetts, New York, or New Jersey, we take the next step westward, to Pennsylvania. There, the commonwealth, through the state attorney general, has filed suit (along with other plaintiffs, Suburban Action Institute and Bucks County Legal Aid) against one of its constituent political subdivisions (Commonwealth of Pennsylvania and Bucks County Interfaith Housing Corporation et al. v. County of Bucks et al., 1972). The defendant in the case is Bucks County (and its municipalities) which is suburban to Philadelphia and which, allegedly, indulges in practices that have the effect of excluding low- and moderate-income people from the county. The complaint was dismissed by the trial court, without any trial on the merits, but the state and the other plaintiffs have appealed the dismissal.

In tandem with the lawsuit, a second strategy has been attempted. The Bucks County Legal Aid Society sought to have the planning agency for the area, the Delaware Valley Regional Planning Commission (DVRPC) develop an allocation plan for low- and moderate-income housing, to include Bucks County.[14] This was not a frontal assault on the exclusionary ordinances and practices, but an approach to providing a framework within which to develop the housing and attack the practices, as necessary. To put pressure on the planning commissions, Legal Aid requested that HUD, the agency which provides DVRPC with federal planning funds, require the planning commission to develop a housing plan. The combined pressure appears to have been sufficient to stimulate the development of a housing plan. The real test will, of course, come in the implementation of the plan.

Finally, a number of states have established state housing finance agencies. The number seems to double every couple of years, as the states carve out an increasing role for themselves in middle-, moderate-, and low-income housing. None of the twenty-plus agencies was created specifically to develop subsidized housing in suburbia. Rather, they were intended to make available a ready source of financing for new construction, usually through the issuance of tax-free bonds. A lead time of a couple of years is necessary before a new agency produces any housing at all because the constitutionality of the law must be established before the bonds are marketable. Only about a half-dozen of these agencies have been in existence long enough to have established a "record," including agencies in Massachusetts, New York, New Jersey, Michigan, and Illinois. Of these, the Massachusetts Housing Finance Agency seems to have as a major objective the opening up of opportunities for low- and moderate-income families in locations where such housing had not previously been available. Housing developments are economically integrated, in conformity with the state law under which the agency operates, and many are racially integrated.

The record of the other financing agencies which have produced significant quantities of housing is mixed, in terms of location. Projects are built in central cities and small towns, with no apparent emphasis on using their programs as entering wedges into hitherto inaccessible suburbs. As with their cousins at the federal level, "production" is often the name of the game.

Perhaps the most promising thing that can be said about the role of the states in furthering housing opportunities for the poor in suburbia is that this activity has increased several hundred percent in the last five years. Virtually all the efforts described have gotten under way since the mid-1960s. Awareness of the problem and the possibilities of remedial action are increasing. State efforts to date have borne little fruit in terms of actual housing for the poor in suburbia, particularly for minorities. But the seeds have been sown so recently that it is too early to tell whether there will be a harvest.

Finally, on to the "feds." For opening up the suburbs, the federal government has the most diverse public tool kit of laws, executive orders, regulations, and good, old-fashioned administrative discretion. As at the state level, it is difficult enough to get the right rules of the game on the books. Making sure that the game is actually played that way may be many times more difficult.

The federal agencies most responsible for interpreting, applying, and enforcing the complex federal rule book are the Department of Housing and Urban Development and the Justice Department. In 1970, differences began to surface within the administration on the role the government should play in increasing housing opportunities for the poor and the Black. HUD was advocating a relatively more aggressive stance, while the Justice Department and the White House preferred a more passive posture. Discussions touched on a number of issues, including the HUD efforts to ensure equal housing opportunities in Warren, Michigan, HUD's proposed project selection criteria for subsidized housing programs, and a possible Justice

Department discrimination lawsuit involving a subsidized housing project in suburban Black Jack, Missouri.

Policy development, or "clarification," required many more months than was first thought necessary. In the interim, much of the rhetoric from the administration related to what the government was not going and was not authorized to do. Foremost among these "nonpolicies" was: there was not going to be any "forced integration" of the suburbs. (It was never clear who was not going to force whom not to do what.)

Ultimately, the resolution took the form of a statement by the President on federal policies relative to equal housing opportunity.[15] The statement purported to make no new policy, but rather to "define and explain" the existing policy, as codified in various laws, executive orders, and other administrative pronouncements. The statement included a number of clues on the nature and scope of future federal actions, as well as areas of inaction.

PROJECT SELECTION CRITERIA FOR SUBSIDIZED HOUSING
PROGRAMS AND AFFIRMATIVE MARKETING REQUIREMENTS

At the time of the President's statement, HUD had had drafts of so-called "project selection criteria" for subsidized housing ready for months, awaiting a green light from the White House to make them public. Prior to that time, the subsidized housing programs were administered essentially on a first-come-first-served basis. Applications for federal assistance have usually far exceeded congressional appropriations, with a resultant "pipeline." Of the projects which met statutory and administrative requirements, those which were received and entered the pipeline first were funded first. Projects were not compared. National objectives were not considered. The project selection criteria were developed to enable HUD to select from among "eligible" applications those which will go the farthest toward meeting equal opportunity and other goals. These criteria were to be applied to all of the subsidized housing programs.

The President's statement picked up on the theme of the obligation to administer housing programs so as to further equal opportunity:

Based on a careful review of the legislative history of the 1964 and 1968 Civil Rights Acts, and also of the program context within which the law has developed, I interpret the "affirmative action" mandate of the 1968 act to mean that the administrator of a housing program should include, among the various criteria by which applications for assistance are judged, the extent to which a proposed project ... will in fact open up new, nonsegregated housing opportunities that will contribute to decreasing the effects of past housing discrimination ... in choosing among the various applications for Federal aid, consideration should be given to their impact on patterns of racial concentration.

In furtherance of this policy, not only the Department of Housing and Urban Development but also the other departments and agencies administering housing programs—the Veterans Administration, the Farmers Home Administration and the Department of Defense—will administer their programs in a way which will advance equal housing opportunity for people of all income levels on a metropolitan areawide basis.

Based on these remarks, HUD began the process of formalizing project selection criteria and affirmative marketing requirements. Drafts were published and comments invited from interested parties. In January 1972, HUD formally published these regulations, to become effective the following month.[16] These criteria place great weight on the location of the proposed project, but consider other factors in selecting among applications for assistance. Of the eight criteria which are applied to applications by housing sponsors, three are directly related to increasing locational choice, although none is specifically related to the suburbs. Applications are rated under each of the criteria as superior, adequate, or poor.

(a) Criterion 2, Minority Housing Opportunities: The stated objectives here are: "to provide minority families with opportunities for housing in a wide range of locations" and "to open up nonsegregated housing opportunities that will contribute to decreasing the effects of past housing discrimination." A "superior" rating would be given if, for example, the project was to be located "so that, within the housing market area, it will provide opportunities for minorities for housing outside existing areas of minority concentration and outside areas which are already substantially racially mixed."

(b) Criterion 3, Improved Location for Low(er) Income Families: Objectives include: "to avoid concentrating subsidized housing in any one section of a metropolitan area or town" and "to locate subsidized housing in areas reasonably accessible to job opportunities."

(c) Criterion 4, Relationship to Orderly Growth and Development: Objectives include: "to develop housing consistent with officially approved State or multijurisdictional plans" and "to encourage formulation of area-wide plans which include a housing element relative to needs and goals for low- and moderate-income housing as well as balanced production throughout a metropolitan area." This criterion clearly envisions the use of allocation plans, a la Miami Valley, to encourage housing choice, as well as the use of the criterion itself as an inducement to develop allocation plans.

Other selection criteria, not directly related to the choice question, include the "need" for the housing in the area, environmental factors, ability of the sponsor to perform, potential to provide job and business opportunities for minorities, and provision for sound management.

On the whole, the thrust of the criteria is toward the suburbs. The criteria do not deal directly with the local exclusionary practices or the impediments built into the federal programs. However, the criteria may impact the inertia of the developers by giving them first shot at the limited subsidy funds if they can get into the suburbs.

These criteria are, however, subject to widely varying interpretation by the several dozen HUD area offices which administer them. The criteria were designed to be general and flexible, but the cost of pursuing that course may be that all of the projects in some metropolitan areas will be in Black areas of the central city and, in other metropolitan areas, all projects may be in the suburbs and none in the Black community. With general criteria and no overall plan or direction necessarily available (projects are evaluated on a case-by-case basis, generally), the outcome is unpredictable. The data, to date, are fragmentary. It is too early to determine the locational impact of the criteria.

Working in tandem with the project selection criteria are the "affirmative fair housing marketing regulations," which apply to both unsubsidized and subsidized housing assisted by FHA. Procedurally, applicants must submit an affirmative fair housing marketing plan for approval by HUD. The plan is to outline the means by which the applicant intends to attract buyers or tenants of all minority (as well as majority) groups to the housing for initial sale or rental. For multifamily projects, the plan is to remain in effect throughout the life of the project. The regulations state that:

such a program shall typically involve publicizing to minority persons the availability of housing opportunities through the type of media customarily utilized by the applicant, including minority publications or other minority outlets which are available in the housing market area. All advertising shall include either the Department-approved Equal Housing Opportunity logo or slogan or statement and all advertising depicting persons shall depict persons of majority and minority groups.

Applicants who do not comply with these regulations are subject to sanctions, including denial of further participation in HUD programs and referral of the matter to the Justice Department. Literal compliance with these marketing requirements may not, however, provide real opportunities for Blacks. The test is not "effect." All-white suburban projects may be the result of the project selection criteria and the affirmative marketing requirements, operating together. A sponsor who has met the minimum requirements is not penalized for such a result, and, in fact, he may receive funding for additional projects with segregated occupancy. Again, it is too early to determine the practical impact of these regulations which, in theory, will increase the housing choice of minorities.

COMPREHENSIVE PLANNING

In regard to the Section 701 planning program, the President's statement indicated that HUD "must follow statutory mandates":

Where comprehensive planning is supported by a federal grant under the 1954 Act, as amended in 1968, the plan must include a "housing element" to insure that "the housing needs of both the region and the local communities studied in the planning will be adequately covered in terms of existing and prospective in-migrant population growth." This provision has broad application, since such planning grants are often used to prepare the areawide plans which are a prerequisite for Federal financial assistance under the water and sewer, open space and new communities programs.

The statement correctly identifies a point of potential federal leverage. Without federal funds, regional planning agencies probably could not exist, much less develop comprehensive and functional plans. Without these plans, communities in the region may be rendered ineligible for important federal "hardware" grants. HUD is thus in a position to apply the "carrot and stick approach," by conditioning planning funds (and, implicitly, subsequent categorical grants) on the development of an action-oriented "housing element" for low- and moderate-income families. For the most part, however, HUD has been content to fund long-term-housing-needs studies and projections.

A major exception to this trend, which provides a potentially useful model, is HUD's actions involving the Southeastern Wisconsin Regional Planning Commission (SEWRPC). In early 1971, HUD informed SEWRPC that it was not permitted, under departmental regulations, to execute a contract under the federal planning program until an adequate housing element was provided. Consequently, the planning funds were held up.

HUD then specified that "our primary concern with the work program as it exists is its lack of a short term action plan or strategy," and proceeded to "offer an example of what might be done to meet that need" (Levin, 1971). The first major step proposed by HUD was the development of an allocation plan for distributing low- and moderate-income housing among areas of the seven-county region. HUD suggested that SEWRPC's first objective be the "identification of specific areas where low- and moderate-income housing should be constructed."

SEWRPC accepted these conditions, albeit reluctantly, and revised its work program to include an outline of a short-term action program for low- and moderate-income housing. In the summer of 1972, SEWRPC completed what amounted to an allocation plan for the region, for 2,000 units of low- and moderate-income housing, and embarked on implementation efforts.[17]

When planning agencies have taken the initiative themselves, HUD has used the carrot instead of the stick. This has included rhetorical and financial support for the Miami Valley Region, as well as the Washington, D.C., area (see Romney, 1971):

> The Department would be happy to reward Metropolitan Washington with a bonus of housing units beyond what the area would normally receive as a means of encouraging the Council of Governments' effort to establish a fair share plan on a 'real city' basis.

In the planning area, as in others, HUD has been more willing to use the carrot than the stick. HUD has scarcely scratched the surface of its potential to move planning agencies toward providing a regional framework for introducing low- and moderate-income housing into suburbia.

WORKABLE PROGRAM

A third tool in the HUD kit was described in the President's statement:

> Where the "workable program" requirement—imposed on local communities by the Housing Act of 1949, as amended in 1954, in connection with urban renewal and related programs—is a condition of eligibility, HUD may not make a grant in the absence of a HUD-certified workable program for community improvement. The program must make reasonable provision for low- and moderate-income housing, which must of course be available on a nondiscriminatory basis.

As a practical matter, the workable program is a very limited tool for opening up the suburbs. Urban renewal and the other programs for which the workable program is a prerequisite to funding are basically central-city and small-town programs. Although suburbs come in all shapes, sizes, and physical conditions, relatively few have reached the requisite condition for participation in the urban renewal program. Even fewer are willing to admit to this state of affairs.

The city of Warren, Michigan, represents a dramatic exception to this general pattern. Warren is a virtually all-white "suburb" of almost 200,000 people adjacent to Detroit. It was an early entrant in the Neighborhood Development Program (NDP), a "streamlined" form of urban renewal introduced at the end of the 1960s. The older portions of Warren, closest to the city of Detroit, consisted of extremely modest houses, many of which were in need of rehabilitation. In 1969, the city received approximately two million dollars in HUD assistance for the first year of its NDP. When it came time to renew the grant for the second year, the city's workable program had expired. HUD requested that the city develop, as part of its workable program application, a plan to assure the availability of housing, particularly for low- and moderate-income families, on a desegregated basis.

After extended negotiations, the city seemed ready to agree to a limited program, when three successive front-page stories in the Detroit newspapers announced that HUD had selected Warren as a "target" for its efforts to "force integration" in the suburbs. After HUD Secretary Romney assured the city officials that the administration was not seeking to "force integration," a further compromise was reached.

By the fall of 1970, the issue had generated sufficient political heat that the continuation of the Neighborhood Development Program (to which the "strings" were perceived as attached) was placed before the voters in a referendum.

After the federal program was soundly rejected by the citizens of Warren, the referendum was challenged in court, as being racially motivated (Pollak et al. v. City of Warren, HUD et al. n.d.).

The problems generated from HUD's experience in Warren were undoubtedly among the factors leading to the President's statement of 1971. Notwithstanding the reiteration in that statement of the housing requirements of the workable program, HUD seems to have used this tool somewhat less aggressively to achieve housing goals, since mid-1971. In suburbia, where few communities apply for workable program certification in the first place, the workable program appears to lack significant potential, in the absence of major revitalization by the administration.

TITLE VIII OF THE 1968 CIVIL RIGHTS ACT

HUD and the Justice Department share the responsibility for enforcement of the federal fair housing law. The President's statement outlined the statutory role of the Justice Department:

> Under the terms of Title VIII of the 1968 Civil Rights Act, the Attorney General is empowered to bring suits in Federal Court where he finds that racial discrimination in housing constitutes a "pattern or practice," or where housing discrimination cases raise issues of general public importance.

Three days later, the Justice Department filed its first suit under this act, which was specifically attacking the exclusionary practices of a suburb. In U.S. v. City of Black Jack, Missouri (n.d.), the Justice

Department alleged that this St. Louis suburb's conduct in blocking the construction of a subsidized housing project had "the purpose and effect of perpetuating racial segregation in Housing," in violation of Title VIII.

In December 1969, Park View Heights Corporation, a nonprofit organization, had contracted to buy a piece of land for a subsidized housing development. The site was in an unincorporated area of St. Louis County. County zoning permitted multifamily use, including the type proposed by Park View Heights. As the plans for a racially integrated housing development received public attention, opposition arose in the area of the site. Local citizens created the municipality of Black Jack, and the city council adopted a zoning ordinance, which effectively prohibited multifamily construction in the city. The Justice Department alleged that this ordinance, which prevented construction of Park View Heights, violated Title VIII, and requested the court to order the city to take the necessary steps to permit prompt construction of the development.

The Black Jack case does not appear to have set a pattern for Justice Department enforcement of Title VIII. Since initiating the Black Jack case (which is still pending), Justice Department actions have largely related to sellers, landlords or realtors who allegedly discriminated against minorities of any income in city, suburb, or small town.

In addition to the techniques outlined in the President's statement, the federal government has available other potentially useful devices for increasing access of the poor to suburbia.

NEW COMMUNITIES

The Housing Act of 1968 authorized HUD to guarantee loans to developers of large-scale new communities.[18] The program was limited to private developers, and a ceiling of $250 million in total loan guarantees was established. The purpose of the guarantee was to enable developers to secure the long-term, very substantial financing necessary for a new community. The lender was guaranteed against loss, as in the FHA single family mortgage insurance program.

In 1970, Congress doubled the ceiling on loan guarantees and extended the coverage of the law in a number of ways, including making public agencies eligible for assistance.[19]

The statute requires as a condition of assistance that the new community development program "makes substantial provision for housing within the means of persons of low and moderate income and that such housing will constitute an appropriate proportion of the community's housing supply."[20]

As of August 1972, HUD had approved ten new communities for loan guarantees exceeding $226 million. Many of the approved new communities are located in the metropolitan areas of major central cities.

The "project agreements" concluded, to date, between HUD and the new community developers indicate that these communities may provide a significant proportion of low- and moderate-income housing. However, only scant evidence is in yet on actual construction, and the indications are that developers who fall behind on this aspect of their commitment are receiving, at most, a slap on the wrist from HUD. Congress has established requirements which are extremely general, but sound in principle. It remains to be seen whether HUD will take this opportunity to increase the housing opportunities of the poor.

A-95 REVIEW PROCESS

Circular A-95 is an instructional memorandum developed by the federal Office of Management and Budget.[21] The circular authorizes certain state and regional agencies to review and comment on applications for federal assistance, under a whole array of categorical programs. A-95 is a device designed for coordination, to ensure that (1) applications from localities are consistent with regional and state needs and (2) federal funds are programmed in a coordinated, mutually supportive way, consistent with overall community needs.

Although A-95 is not mandated directly by any law, its statutory underpinnings are in the Inter-Governmental Cooperation Act of 1968 and the National Environmental Policy Act of 1969. Based on these statutes, the A-95 process opens up the review of applications to a wide range of considerations related to the environment, including housing and civil rights matters.

Procedurally, applicants for federal grants notify state and regional "clearinghouses" of their applications. The application is then reviewed and commented on by the clearinghouses and other interested agencies, including designated civil rights agencies in the area. The comments are then passed on to the federal funding agency, which can attach as much or as little weight to the comments as it desires. Although the comments have no legal force, they can amount to a veto in practice, if the federal agency chooses to take the process seriously.

In Connecticut, the focus has been on the federal water and sewer grant program. In at least two instances, the A-95 process has been used to block sewer grants to "exclusionary communities" until pledges to revise local zoning were made (Sikorsky, 1972).

Few areas of the country have begun to tap the leverage inherent in the A-95 process, but the potential here for opening up the suburbs may well be significant.

GENERAL SERVICES ADMINISTRATION

GSA is the landlord for federal agencies. It controls over 10,000 facilities across the country. Many of these are federally owned and tax exempt. Others are leased and represent valuable tax assets to their communities. Even the tax-exempt facilities provide very substantial indirect financial benefits to the locality. As a result, GSA has the potential for important leverage over localities competing for federal facilities.

In fact, the federal government has followed the trend of private industry to suburbia. In so doing, it has located or relocated its installations in areas to which low- and moderate-income and minority employees have extremely limited access. GSA has used almost none of its leverage to secure low- and moderate-income housing for the employees. This is in spite of the policy adopted by GSA in 1969 that it would avoid locations which did not have an adequate supply of low- and moderate income housing accessible to them.

In 1970, Executive Order 11512 was issued, requiring that, in selecting sites for federal facilities, GSA should study the availability of housing in a community, along with about twenty other factors. As a follow-up, HUD and GSA signed a memorandum of agreement, under which HUD would assist in assuring the needed housing supply, accessible to federal facilities.[22] A study was to be made of the housing supply in the area. If the necessary housing was not available, an "affirmative action plan" was to be developed. At the end of 1972, no such plans had been developed, anywhere in the country.

To prod GSA to use its leverage and meet its obligations, a lawsuit was brought concerning a new $25 million Internal Revenue center in Brookhaven, New York. The center was to house 4,000 employees. The federal court, in Brookhaven Housing Coalition v. Kunzig (1972) found that there was not adequate housing in the community for these employees, as well as the entire population of the area. Because of this lack of housing, the court held that GSA had violated the executive order. Instead of holding up construction, which was well advanced by April 1972, the preliminary injunction required GSA to use 220 units of surplus housing at a nearby air base for low-income residents of the county. The plaintiffs further intended to request the judge to impose a complete low-income housing plan on the town of Brookhaven.

The Brookhaven lawsuit may be the first of many such cases. Judge Judd established a precedent of a broad interpretation of the executive order, requiring housing availability for employees, as well as nonemployee residents. This is a test which can rarely be met without substantial "affirmative action plans." If the federal government does not take the initiative and use its leverage to increase housing opportunities, the courts will probably be resorted to repeatedly to assure that GSA meets its legal obligations.

EQUAL EMPLOYMENT OPPORTUNITY COMMISSION

The EEOC is another federal agency with some degree of leverage in suburbia, which it has been reluctant to use. Its influence is limited, but expanding, since it can not go to court in cases of employment discrimination, rather than just plead and cajole.

The EEOC is the enforcement agency created pursuant to Title VII of the Civil Rights Act of 1964.[23] Title VII is the equal employment section of the law.

Although it has been argued to the EEOC, both from outside and by its own staff, that corporate relocation to suburbia may constitute employment discrimination, the commission has been unwilling to adopt this position. In July 1971, a memorandum prepared by the EEOC general counsel's office suggested that suburban corporate locations may constitute employment discrimination:

> The transfer of an employer's facilities constitutes a *prima facie* violation of Title VII if (1) the community from which an employer moves has a higher percentage of minority workers than the community to which he moves, or (2) the transfer affects the employment situation of the employer's minority workers more adversely than it affects his remaining workers, and (3) the employer fails to take measures to correct such disparate effects.[24]

Suburban Action Institute (SAI), a "nonprofit institute for research and action in the suburbs," has also argued to the EEOC, as well as the corporations themselves, that corporate movements to exclusionary suburbs constitute employment discrimination. SAI called on General Telephone and Electronics, early in 1971, not to move from New York City to Stamford, Connecticut, until there was an adequate supply of housing available for lower-income workers there. Of the 5,700 acres of residentially zoned vacant land in he city, 99.5% is zoned for one-acre development or larger.

Similarly, in March 1971, Suburban Action objected to RCA's proposed move from New York City and Camden, New Jersey, to New Canaan, Connecticut. A complaint was filed with EEOC against the move to the restrictively zoned suburb (Davidoff and Gold, 1971). Although the EEOC took no formal action, RCA decided to reconsider its planned relocation.

In response to these and other complaints, the EEOC has taken no position. With its power of publicity and its newly acquired authority to go to court, EEOC has some potential for influencing corporate relocations, perhaps delaying them until adequate housing is provided. The potential remains untapped.

The federal laws, executive orders, and policies on the books thus constitute a potentially potent array of tools for increasing locational choice for the poor. Among HUD, the Justice Department, GSA, EEOC, and other federal agencies, there is substantial leverage available. The problem is actualizing that potential, in the face of the political reality of the power wielded by suburbia.

Because of the perceived inability of federal agencies to meet their equal housing responsibilities, private efforts are directed increasingly toward pressing for responsiveness by federal agencies. The Brookhaven lawsuit against GSA and the complaints to the Equal Employment Opportunity Commission reflect the trend of activist groups to pursue other paths in addition to exclusionary zoning cases.

As the agency with the most direct responsibilities for providing housing opportunities, HUD is increasingly becoming a target of activist groups. In the Shannon case in Philadelphia, a federal appeals court held that HUD must consider the racial impact of the housing developments it subsidizes (Shannon v. United States Department of Housing and Urban Development, 1970). This decision provided much of the impetus for the project selection criteria promulgated by HUD in early 1972.

Further impetus for these regulations came from the Gautreaux (v. Chicago Housing Authority, 1969) case in Chicago.[25] The court of appeals found that HUD's approval of sites and funding of segregated public housing was sufficient to implicate the department in this discriminatory pattern of site and tenant selection. After finding that HUD had violated the Constitutional rights of low-income Blacks in Chicago, the court of appeals sent the case back to the district judge to determine what HUD should be ordered to do to remedy the discriminatory effects of past practices.

The plaintiffs, represented by the American Civil Liberties Union, proposed that the court order that low-income public housing be developed throughout the suburban areas of Chicago.[26] The proposed order envisioned ignoring the local approval requirement built into the federal law, if necessary to ensure relief for the plaintiffs.

In the Gautreaux case, the original discrimination complained of was segregation practiced within the city of Chicago. The court held that the plaintiffs were entitled to desegregated housing opportunities. Based on the 1970 census, the ACLU argued that, to achieve the goal of lasting desegregation, it was necessary to provide the plaintiffs access to the suburbs. Otherwise, the city neighborhoods would turn increasingly Black, and the possibilities for desegregated housing would diminish over time. The judge took the proposed order under advisement.

Meanwhile, in Cleveland, public housing tenants sued in a frontal assault on the local approval requirement of federal law (Mahaley v. Cuyahoga Metropolitan Housing Authority, n.d.). These tenants challenged the constitutionality of the cooperation agreement provision because of its effect in keeping public housing out of the suburbs. Under Ohio law, local housing authorities, which administer public housing, have what amounts to countywide jurisdiction. However, these housing authorities cannot develop public housing in any municipality without a cooperation agreement with the local governing body. Consequently, almost all of the public housing in

Cuyahoga County is in the city of Cleveland. (The exception is East Cleveland, a heavily Black suburb adjacent to the Black east side of the central city.)

The Mahaley case, pending in the federal courts, seeks to have this local approval barrier removed, to facilitate the development of public housing in the Cleveland suburbs.

A third case aimed at developing public housing in the suburbs is Crow v. Brown (1972). There, the Atlanta Housing Authority, which has jurisdiction in parts of suburban Fulton County, was stymied by the county in its attempts to develop suburban public housing. The court ordered the issuance of the necessary building permits, as well as the development of a plan for suburbanizing public housing. Although HUD was named as a defendant in this case, the court found that HUD had sought to provide desegregated housing opportunities and was not liable in that situation.

Increasingly, then, private organizations are using the courts to move HUD toward developing public housing in suburbia. Such groups are also using the courts in related areas, which may have significant impact on the access of the poor to suburbia. For example, two federal appeals courts have ruled on metropolitan school desegregation plans, affirming one in principle in the Detroit area and reversing another in Richmond, Virginia (Bradley v. Milliken, 1972; Bradley v. Richmond, 1972). If the courts ultimately support metropolitanwide school desegregation plans, one of the incentives for excluding low- and moderate-income housing from suburbia will be eliminated. Similarly, the Serrano (v. Priest, 1971) case raises questions about the legitimacy of widely varying resources being made available among school districts within a state. The school finance cases could lead to major changes in the financing of local governments. The property tax may diminish in importance and, with it, fiscal zoning. An additional reason for excluding subsidized housing would be removed.

Finally, private strategies to increase access to the suburbs are not limited to the courts. Many of the approaches are political, ranging from corporate pressure to regional coalitions. Corporations are beginning to see the need for low- and moderate-income housing, to accompany their relocation in suburbia. In their own self-interest, corporations may lobby increasingly against exclusionary zoning or even form housing development corporations to do the job themselves.

The regional coalition approach is represented by the Metropolitan Housing Coalition, in the Chicago area. This coalition grew out of a series of meetings of a small group of suburban mayors, who saw the need for facing up to the problem of low- and moderate income housing in suburbia. The motives of the mayors varied, but the underlying theme was the retention of local control (as opposed to control by the courts) and a fair division of responsibility. In mid-1972, the coalition secured substantial funding to (1) build a political constituency for low- and moderate-income housing and (2) develop an allocation plan to ensure that communities would receive an equitable share of the needed housing, with no place having a major concentration of public housing.

CONCLUSIONS

For the future, we must assume that federal subsidy programs will continue to exist, in some form. This is not a safe assumption, but it is a necessary one if housing opportunities for the poor are to be expanded. The private market is building for a decreasing proportion of the market, as housing costs climb faster than income.

The form of federal subsidies in the near future may evolve from production subsidies toward housing allowances or vouchers. Remaining production programs may contain local approval requirements which do not exist for privately developed housing.

The obstacles to providing locational choice for the poor are likely to continue also and to take on new forms. At the same time, increasingly sophisticated strategies are being developed to overcome these hurdles. In the last half-decade, local exclusionary practices have been under attack from the Massachusetts Anti-Snob Zoning Law, the New York Urban Development Corporation, the Commonwealth of Pennsylvania, users of the A-95 process, "enforcers" of the GSA executive order, as well as in an array of exclusionary zoning cases.

At the same time, attempts have been made to chip away at the impediments in the federal programs themselves, in the Mahaley case, the Gautreaux case, and, to some extent, through the project selection criteria.

Finally, efforts have been made to induce housing sponsors and developers to move into the suburbs, through allocation plans, project selection criteria, as well as the Fairfax County ordinance.

In sum, many of the models for action exist. There is still a need for new models and refinement of ones already developed. The race is too close to predict, based on early, scattered returns.

NOTES

1. Housing Act of 1961, 12 U.S.C. §1715, 1964.

2. Housing and Urban Development Act of 1965, 12 U.S.C. §1701 s (Supp. I) 1965.

3. Housing and Urban Development Act of 1968, 12 U.S.C. §§1715z, 1715z-1, 1968.

4. The administration's requested appropriation of $30 million was cut to $12 million and the rider was attached. Each subsequent congressional appropriation for the rent supplement program has contained the rider.

5. At 42 U.S.C. §1415(7)(a)(i) 1964. The several-decade history of the Cooperation Agreement has been that local governing bodies have generally refused to sign such agreements for projects in nonslum areas.

6. Housing and Urban Development Act of 1968, 12 U.S.C. 1715z(i)(3)(B), 1715z-1 (i)(3) 1968.

7. Zoning Amendment 156, Fairfax County Code, Fairfax County, Virginia.

8. In addition to Fairfax-type requirements, these proposals include density bonuses, and a planned development zone proposal requiring twenty percent of the units to be moderate-priced, stated by the Montgomery County (Maryland) Planning Board in October of 1971.

9. Comprehensive Planning Assistance Program, Section 701, Housing Act of 1954, as amended 40 U.S.C. §461.

10. At 40 U.S.C. §461(a) 1968.

11. At 40 B. Mass. Gen. Laws Ann. §§20-23.

12. At 23A. Mass. Gen. Laws Ann. §§1-1, 1-17.

13. McKinney's Cons. Laws of New York, ch. 24 §§6254-6266.

14. The Bucks County Legal Aid Society and other organizations filed a petition with the Delaware Valley Regional Planning Commission "for the purpose of obtaining the enactment and enforcement of a housing element that provides for the construction of low-income and moderate-income housing in the Delaware Valley Urban Area."

15. From a statement by the President on federal policies relative to equal housing opportunity, June 11, 1971.

16. The Project Selection Criteria were published on January 7, 1972 in Volume 37, number 4, of the Federal Register. The Affirmative Fair Housing Marketing Regulations were published on January 5, 1972 in Volume 37, number 2, of the Federal Register.

17. From a Short-Range Action Housing Program for Southeastern Wisconsin—1972 and 1973.

18. At 42 U.S.C. §3901 et seq., 1968.

19. The Urban Growth and New Community Development Act of 1970, P.L. 91-609, 84 Stat. 1770, was approved December 31, 1970; 42 U.S.C. §3906, 1970.

20. Section 712(a)(7) of the Urban Growth and New Community Development Act of 1970; 42 U.S.C. §3903, 1970.

21. This circular was originally issued in July 1969.

22. From the Memorandum Between the Department of Housing and Urban Development and the General Services Administration Concerning Low- and Moderate-Income Housing of June 12, 1971.

23. At 42 U.S.C. 2000e et seq.

24. This document, dated July 7, 1971, reached the press without public announcement by the EEOC.

25. The original finding of discrimination was made against the Chicago Housing Authority (Gautreaux v. Chicago Housing Authority, 1969). On July 1, 1969, the district court ordered that future public housing would be built primarily in white areas to remedy past segregation (304 F. Supp. 736, N.D. Illinois). In 1971, the court of appeals held that HUD's cooperation constituted a violation of the Constitution and the 1964 Civil Rights Act (448 F. 2d 731, 7th Cir.).

26. The proposed order was submitted by the plaintiffs to the district court in September 1972.

CASES

BRADLEY v. MILLIKEN (1972) Nos. 72-1809-72-1814 (6th Cir.).

BRADLEY v. RICHMOND (1972) 462 F. 2d 1058 (4th Cir).

BROOKHAVEN HOUSING COALITION v. KUNZIG (1972).

COMMONWEALTH OF PENNSYLVANIA AND BUCKS COUNTY INTERFAITH HOUSING CORPORATION ET AL. v. COUNTY OF BUCKS ET AL. (1972) No. 27 T.D.

CROW v. BROWN (1972) 457 F. 2d 790 (5th Cir.).

DE GROFF ENTERPRISES, INC. v. BOARD OF COUNTY SUPERVISORS OF FAIRFAX COUNTY, VIRGINIA ET AL. (1971) November 11.

GAUTREAUX v. CHICAGO HOUSING AUTHORITY (1969).

MAHALEY v. CUYAHOGA METROPOLITAN HOUSING AUTHORITY (n.d.) No. C-71-251 (N.D. Ohio).

POLLAK ET AL. v. CITY OF WARREN, HUD ET AL. (n.d.) Civil no. 35843, U.S. Dist. Ct. for Eastern District of Michigan.

SERRANO v. PRIEST (1971) 96 Cal. Rptr. 601.

SHANNON v. UNITED STATES DEPARTMENT OF HOUSING AND URBAN DEVELOPMENT (1970) 426 F. 2d 809 (3rd Cir.).

UNITED STATES v. CITY OF BLACK JACK, MISSOURI (n.d.) No. 71 C372(1), U.S. Dist. Ct. for Eastern District of Missouri.

REFERENCES

ALEXANDER, R. (1972) "Fifteen state housing agencies in review." J. of Housing (January): 9-17.

BERTSCH, D. and A. SHAFOR (1971a) "A regional planning commission experience." Planners Notebook (April): 1-8.

——— (1971b) "A regional housing plan: the Miami Valley Regional Planning Commission experience." Planners Notebook (April).

BROOKS, M. (1972) Lower Income Housing: The Planners' Response. Chicago: American Society of Planning Officials.

CASSIDY, R. (1971) "G.S.A. plays the suburban game on a grand scale." City (Fall): 12-14, 72.

CUNNINGHAM, C. and J. SCOTT (1970) Land Development and Racism in Fairfax County. Fairfax, Va.: Washington Suburban Institute.

DAVIDOFF, P. and N. N. GOLD (1971) Letter to William H. Brown III, Chairman of the U.S. Equal Opportunity Commission, March 3.

Department of Housing and Urban Development (1971) Equal Opportunity in Housing. Englewood Cliffs, N.J.: Prentice-Hall.

ELLICKSON, R. (1967) "Government housing assistance to the poor." Yale Law J. (January): 508-544.

GARROTT, I. (1971) Statement before the U.S. Commission on Civil Rights, Washington, D.C., June 14-17.

JACKSON, S. (1972) "New communities." HUD Challenge (August): 1-4.

LEVIN, E. M., Jr. (1971) Letter to George C. Berteau, Chairman of the Southeastern Wisconsin Regional Planning Commission, March.

LOGUE, E. J. (1972) Remarks announcing the Westchester Nine Towns Program, June 20.

National Housing and Development Law Project (1970) "Handbook on housing law." Berkeley, California.

New Jersey, State of (1970) "The housing crisis in New Jersey."

New York State Urban Development Corporation (1972) "Fair share."

New York Times (1972a) "Suburbs fighting state agency's plan to override local zoning." (June 17).

--- (1972b) "Westchester towns win a moritorium on U.D.C. housing." (August 7).

REILLY, W. and S. SCHULMAN (1969) "The state urban development corporation: New York's innovation." Urban Lawyer: 129-146.

ROMNEY, G. (1971) Statement, October 27.

ROSENBERG, W. (1970) "New York State's urban development corporation is an action agency." J. of Housing (December).

SIKORSKY, I. (1972) "A-95: deterrent to discriminatory zoning." Civil Rights Digest 5 (August): 17-19.

U.S. Commission on Civil Rights (1971a) "Hearing before the United States Commission on Civil Rights, held in Washington, D.C., June 14-17, 1971."

--- (1971b) Home Ownership for Lower-Income Families: A Report on the Impact of the Section 235 Program.

--- (1970) "Federal installations and equal housing opportunity." Washington, D.C.

U.S. House of Representatives, Committee of the Judiciary (1972) "Hearings before the Civil Rights Oversight Subcommittee of the Committee of the Judiciary, House of Representatives, on federal government's role in the achievement of equal opportunity in housing."

Wall Street Journal (1972) "Suburban stall: housing for the poor blocked despite curb on snob zoning." (October 17).

14

Suburban Police Departments:
Too Many and Too Small?

ELINOR OSTROM
ROGER B. PARKS

□ IN A BACKGROUND PAPER on "Crime in the Suburbs," prepared for the President's Task Force on Suburban Problems, John L. McCausland (1972: 63) stated that the "distinctive suburban problem here, as in so many areas, is fragmentation." McCausland cites the fact that there are 85 police agencies in the Detroit metropolitan area as obvious evidence that fragmentation is the most distinctive problem. He goes on to state:

> Most suburban police forces are too small to handle major disturbances or criminal investigations, or problems that require special expertise such as organized crime, narcotics, and vice. They are too small to support adequate training programs, or to be equipped with the latest electronic communications and computer equipment. Such suburban crime crosses jurisdictional lines, but few police departments can [McCausland, 1972: 63].

AUTHORS' NOTE: *The authors are appreciative of the major financial support provided by the Center for Studies of Metropolitan Problems of the National Institute of Mental Health in the form of Grant 5 RO1 MH 19911-02. Earlier support related to several of the studies reported herein was provided by the National Science Foundation, Grant GS-27383 and the Urban Affairs Center of Indiana University. We appreciate comments made by James McDavid, Nancy*

McCausland's analysis squares with the predominant view held by most current scholars. Daniel L. Skoler and June M. Hetler, in an article on criminal administration in metropolitan areas, stressed the grave nature of the growing crisis of law enforcement. The crisis was largely attributed to "fractionalization, fiscal impotence, duplication and lack of coordination" (Skoler and Hetler, 1970: 55). The President's Commission on Law Enforcement and the Administration of Justice (1967: 301) described the "machinery of law enforcement in this country" as "fragmented, complicated and frequently overlapping."

> America is essentially a nation of small police forces, each operating independently within the limits of its jurisdiction. The boundaries which define and limit police operations do not hinder the movement of criminals, of course. They can and do take advantage of ancient political and geographic boundaries, which give them sanctuary from effective police activity. Nevertheless, coordination of activity among police agencies, even when the areas they work in are contiguous or overlapping, tends to be sporadic and informal, to the extent it exists at all. This serious obstacle to law enforcement is most apparent in the rapidly developing urban areas of the country, where the vast majority of the Nation's population is located and where most crimes occur.

As a result of its analysis, the commission (1967: 308) recommended that:

> Each metropolitan area and each county should take action directed toward the pooling, or consolidation, of police services through the particular technique that will provide the most satisfactory law enforcement service and protection at lowest cost.

One year later, the President's Task Force on Suburban Problems echoed the same diagnosis in asserting that the low density of population in many small suburbs as well as "confusing political

Neubert, Vincent Ostrom, Dennis Smith, and Gordon Whitaker on an early draft of this paper. Paula Baker and Judy Keim provided skillful research assistance on this project. Patrick Bova at the National Opinion Research Center and Richard Block at Loyola University provided considerable help in getting us started in the use of NORC data.

boundaries make law enforcement especially difficult and create special needs for the expansion and rehabilitation of police services of all kinds." The commission further argued (President's Task Force on Suburban Problems, 1968: 17-18):

> If the metropolitan areas are to be made more efficient, their vitality released by the reduction of economic distress and racial tension, their mobility freed from the bottlenecks of inadequate and ill-coordinated transportation, it is essential to apply policies that are comprehensive, that treat the metropolitan area realistically as an organic whole. Disorder, crime, pollution, and overcrowding, housing shortages, the noise of aircraft and the congestion of roads—these and the host of other problems are no respecters of antique jurisdictional frontiers. They must be dealt with comprehensively.

The consistency with which scholars, presidential commissions and task forces, and others repeatedly assert the inadequacies of the "too many and too small" suburban police departments (see also MacNamara, 1950; Committee for Economic Development, 1972) leads one to suppose that extensive and confirming empirical evidence underlies these assertions. However, a search of these publications for citations to specific studies establishing a negative relationship between the size and multiplicity of suburban police departments with service levels reveals a paucity of references.

The most frequent form of "evidence" is a "list" of the number of police departments serving some particular metropolitan area. McCausland (1972: 63) lists the number of police departments serving the Detroit area. Skoler and Hetler (1970: 59, 63) list the number serving the Chicago and St. Louis metropolitan areas and also those serving Southern Illinois. The President's Commission on Law Enforcement and the Administration of Justice (1967: 119) listed the number of police departments serving U.S. metropolitan areas (313 county forces and 4,144 municipal forces). However, a simple listing of the number of jurisdictions serving a particular area is hardly adequate evidence that the multiplicity of jurisdictions is associated with poor police service or that suburban police departments are too small. A list is merely a census of the number of police departments. *It establishes no relationship to performance.*

Evidence that the costs of suburban police departments exceed those of center cities or that police service levels are lower in smaller jurisdictions is conspicuous by its absence in the many reports and articles recommending the elimination of suburban police departments through consolidation.[1] Occasional glowing descriptions are made of the performance of the few metropolitanwide police forces which have been established in recent years. Skoler and Hetler (1970: 61), for example, state unequivocally that the merger of Jacksonville-Duval County "appears to have significantly increased the effectiveness with which law enforcement services the Jacksonville-Duval County public." Their reference for such an assertion is a brief article written by a local promoter in *The Police Chief* after a mere four months of experience with consolidation. They also assert that police consolidation is "operative and substantially complete in Nashville-Davidson County," without discussing any effects of that consolidation whatsoever. One immediate effect of the consolidation was to *increase* all costs of local services dramatically,[2] and costs of law enforcement have remained at higher levels ever since. In 1965, two years after consolidation, a sample of fringe residents was asked how local services rendered by Nashville-Davidson County compared with those before the adoption of the merger. Fifty-eight percent responded that services were about the same (McArthur, 1971: 20). Eight percent responded that they were worse. Of local services, fringe residents were the most dissatisfied with the police services they were receiving (McArthur, 1971: 19). Sixty percent felt that their local taxes were too high (McArthur, 1971: 22). And yet the simple fact that police services in Nashville-Davidson County has been consolidated is seriously presented as presumptive evidence that "police consolidation may lead to more effective police service and fiscal economy" (Skoler and Hetler, 1970: 66).

Based on the general diagnosis that suburban police departments provide inadequate police services, one would also expect to find suburban residents dissatisfied with the police services they receive and in support of the frequent proposals made for merging their police departments with others in a metropolitan area. However, Campbell and Schuman (1972: 109) report that respondents living in the suburbs of Cleveland and Detroit are "particularly well satisfied with their police protection." Hawley and Zimmer (1970: 79) have reported that in large- and medium-sized metropolitan areas where most suburban residents were served by their own independent police forces, a majority of suburban residents evaluated their police services to be equal to or better than the police services offered by the center city of the area. Only in fringes around small metropolitan areas served predominantly by county or state police did less than a majority rank their services as less adequate than those provided in the center city. Even here, 45% ranked police services as equal or better.

Many attempts to merge suburban police departments with larger departments have been defeated at the polls. A proposal to make the Erie County (Pennsylvania) Police Department the single law enforcement agency in the county was defeated at the polls by the vote of suburban residents. A majority in 24 of the 25 towns and in all 16 of the villages voted against the measure (Skoler and Hetler, 1970: 60). Suburban voters in Dade County, Florida, have recently voted against further consolidation of the police departments serving that area. Suburban residents in Marion County, Indiana, are strongly opposed to the possibility of merging their smaller police forces with the Indianapolis Police Department (E. Ostrom et al., 1971: 106).

Proponents of consolidation have difficulty explaining why suburban residents oppose their proposals. The "political and social pressures linked to the desire for local self-government," a Task Force on Police (1967: 109) for the President's Commission on Law Enforcement and Administration of Justice argued, "offer the most significant barrier to the coordination and consolidation of police services." Others have accused suburban residents of abandoning the center cities and transferring their allegiance to more homogeneous communities surrounding metropolitan areas, where they can "keep out" the poor and the black (Wingo, 1972: 3; Haar, 1972: 51).

An alternative explanation of suburban voting patterns may exist. Suburban residents may not be convinced that merger of their own independent police forces into one large department will improve service levels or reduce costs as promised by proponents. Given the paucity of evidence presented in support of these promises, it would not be surprising to find some skepticism concerning the claims for more effective service by larger departments at lower costs.

Suburban residents in many metropolitan areas will be faced with new proposals during the next few years to eliminate their police departments and create one large unit serving an entire area. Given this insistent call for the creation of large, integrated police jurisdictions, it is important to ascertain what evidence is available to support the claims for consolidation. Consequently, the central question to be examined in this chapter is: *What evidence do we have that consolidation of police forces in metropolitan areas will lead to an increase in service levels or to a reduction in the costs of service without a reduction in service levels?* If significant evidence can be marshaled in support of the positive effects of consolidation, then resistance to it by suburban residents would indeed have to be viewed as primarily irrational, uninformed, or exclusionary. However, if evidence does not support the claims for consolidation, then the charge that suburban residents are uninformed or irrational in their resistance to consolidation is also left without support. The charge that suburbanites resist consolidation primarily because they are shirking their responsibilities to the poor and underprivileged rather than seeking the best possible service is also weakened if evidence cannot be marshaled to support the existence of better service or lower costs flowing from the creation of larger jurisdictions.

THE WORKING HYPOTHESES

The propositions underlying the positive claims for consolidation can be formalized in the following four working hypotheses:

H_1: An increase in the size of the jurisdiction providing police services to citizens will be positively associated with higher service levels.

H_2: For constant service levels, an increase in the size of the jurisdiction providing police services will be positively associated with lower per capita expenditures on police services.

H_3: A decrease in the number of police jurisdictions serving the citizens within a metropolitan area will be positively associated with higher service levels.

H_4: For constant service levels, a decrease in the number of police jurisdictions providing police services within a metropolitan area will be positively associated with lower per capita expenditures for police services.[3]

We shall examine these working hypotheses in the remainder of this paper.

PREVIOUS RESEARCH FINDINGS
RELATING SIZE OF CITIES
AND PER CAPITA COSTS

The question of the relative economies or diseconomies of scale has been of considerable interest to scholars in the field of local public finance. British scholars, shortly after the turn of the century, began to examine the statistical relationship between municipal expenditures and city size. C. Ashmore Baker (1910) found the lowest per capita expenditures for a group of 72 English cities tended to occur around 90,000 population. The per capita costs of all services except gas and electricity rose consistently in cities above that size. Additional British studies produced findings consistent with that of Baker: major "diseconomies of scale" for all urban services, including police, were found in larger cities, particularly those above 250,000 (London County Council, 1915; Oxford University, 1938; Phillips, 1942).

One of the earliest U.S. studies found a consistent rise in per capita costs of municipal expenditures for all American cities above 275,000 (National Resources Committee, 1939). Otis Dudley Duncan (1951) examined the municipal finance data for 1942 and found lowest per capita costs for municipal services in cities ranging in population from 50,000 to 100,000 (see also Ogburn, 1937). Following these early studies, the number of studies examining the "determinants" of municipal expenditures has burgeoned (for a review, see Bahl, 1968). "Nearly all empirical studies of city expenditure have shown a positive, though not always significant, relationship between the city's population and its per capita expenditures" (Bahl, 1969: 39).[4] Many studies have shown a positive relationship between city size and per capita expenditures for police (see Bahl, 1969; Brazer, 1959; Gabler, 1969, 1971; Fowler and Lineberry, 1972). Other studies have found *no* relationship between city size and per capita expenditures on police (Hirsch, 1959; Weicher, 1970). *No one has reported a negative relationship between city size and per capita expenditures on police.* While some authors have concluded that repeated findings of a positive relationship between city size and per capita expenditures on police are evidence of diseconomies of scale, particularly in the very large cities, interpretation of these findings is somewhat difficult. Bahl (1969:

39) has pointed to three conclusions one might derive from these findings:

(a) municipal costs vary directly with population because of diseconomies of size; or

(b) economies of size exist, but other factors associated with larger populations result in a higher per unit cost in providing the same quality of service; or

(c) the quality of public services is better in large than in small cities, and, therefore, it is impossible to isolate the pure effects of population size.

Some authors have been willing to make the assumption that service levels are equal in all cities under study and have consequently concluded from their empirical findings that *dis*economies of scale exist (Gabler, 1971: 138). Others have cast doubt on the commonly held assumption that economies of scale exist, but, given the lack of some measure of output other than the amount of money being expended, have been hesitant to go further in interpreting their data (Campbell and Sacks, 1967: 179).

The findings from this long tradition of empirical work are *not* supportive of H_2. The conclusion derived from these studies is that per capita costs of police for similar levels of service are likely to be higher in larger cities than in smaller cities. However, independent measures of output are needed for a more definitive examination of this working hypothesis. Research findings presented in the later sections of this paper will utilize independent measures of the quality of police services.

PREVIOUS RESEARCH FINDINGS RELATING MULTIPLICITY OF JURISDICTIONS AND PER CAPITA COSTS OF MUNICIPAL SERVICES

Several scholars interested in determinants of municipal expenditure patterns have examined the relationship between the number of governments in a metropolitan area and expenditure patterns. Bahl (1969) included such a measure in a seventeen-variable multiple-regression model utilizing 1960 expenditure figures for 198 American cities. Bahl (1969: 62) found a significant negative relationship (beta of -0.24) between the number of governmental jurisdictions serving an area and per capita operating expenditures. The relationship was again negative, but not significant for common functions including police (beta of -0.05; Bahl, 1969: 64). The relationship was positive for per capita police expenditures but not significant (beta of 0.08; Bahl, 1969: 68). In a stepwise regression, the "fragmentation" variable consistently contributed very little to explained variance for any of the expenditure patterns examined. Bahl dropped this variable from further analysis when he adopted a nine-variable model which could explain almost as much variance as his seventeen-variable model. Campbell and Sacks (1967: 179) did not find any significant relationships between their measures of fragmentation (population per government and area per government) and expenditure levels.

Brett W. Hawkins and Thomas R. Dye have recently investigated both the environmental correlates of fragmentation and its public policy consequences for 212 metropolitan areas based on 1962 data. When fragmentation was measured by the number of governmental units in a metropolitan area, the "environmental variable" with the highest correlation coefficient was the size of the metropolitan area (.71; Hawkins and Dye, 1970: 19). Age and income levels in the area were also associated with their absolute measure of fragmentation. However, Hawkins and Dye were surprised to find that the racial composition of the center city was not associated with their absolute measure of fragmentation. Nor were city-suburban differences in socioeconomic levels related. Hawkins and Dye (1970: 20) had expected to find relationships between both of these variables and fragmentation due to the "belief that middle class whites seeking to escape from the problems of less affluent central city populations would develop and incorporate independent suburbs, and thereby contribute to fragmentation." Hawkins and Dye (1970: 20) concluded from their analysis that "this does not appear to be an explanation of fragmentation." When they measured fragmentation in relative terms (governments per 100,000 population) most coefficients were generally weakened, and many changed signs. Thus, a relative measure of fragmentation was *negatively* associated with the size of the metropolitan area, its age, the percentage nonwhite in the central city, and city-suburban SES differences.

Prior to an examination of the public policy consequences of

TABLE 1
CRIME RATES, OFFENSES KNOWN TO THE POLICE IN 1971 BY CITY SIZE

City Size (Number)	Homicide	Forcible Rape	Robbery	Aggravated Assault	Burglary	Larceny $50 and Over	Auto Theft
Cities over 1 million (6)	21	42	862	388	2,070	1,199	1,177
500,000 to 1 million (21)	20	52	525	357	1,997	1,313	1,143
250,000 to 500,000 (30)	14	35	359	277	1,950	1,224	901
100,000 to 250,000 (98)	11	27	226	240	1,790	1,350	739
50,000 to 100,000 (260)	6	17	126	150	1,243	1,180	499
25,000 to 50,000 (509)	5	12	95	131	1,042	1,117	397
10,000 to 25,000 (1,224)	4	10	51	122	880	924	251
Cities under 10,000 (2,810)	4	8	30	128	722	763	173

SOURCE: FBI Uniform Crime Report, 1971, Table 9, pp. 100-101.
NOTE: Rate equals number of crimes per 100,000 inhabitants.

fragmentation, Hawkins and Dye first examined the strength of relationships between environmental variables and expenditures. They found a strong relationship (.57) between the size of the metropolitan area and expenditures for police. They also found expenditures for police related to income levels in the area (.51), age of the area (.39), and the occupational patterns of residents (.30; Hawkins and Dye, 1970: 22). When they compared the influence of environmental variables on expenditure patterns with that of fragmentation, "the influence of fragmentation appears very weak." In regard to police, the number of municipalities in an SMSA is positively associated with expenditures (.35), while the number of governments per 100,000 population is negatively associated (−.29) with expenditures (Hawkins and Dye, 1970: 23). Hawkins and Dye (1970: 23) conclude that the "dollar consequences of metropolitan governmental fragmentation have probably been overemphasized in the reform literature."

Given the lack of an independent measure of the service levels provided, it is difficult to interpret the meaning of the previous research findings concerning the relationship between the multiplicity of governmental units and per capita costs of police services. However, the findings from previous research efforts do cast doubt on the warrantability of the fourth working hypothesis that multiplicity of governmental jurisdictions are positively associated with higher per capita costs.

INDEPENDENT MEASURES OF POLICE
SERVICE LEVELS

As Bahl noted, a measure of the quality of a public service is needed before one can "isolate the pure effects of population size." Assessment of the effect of multiple jurisdictions also requires measurement of service levels. Little agreement exists about the appropriate methods for measuring police services. The FBI crime index has been utilized by some analysts as an inverse indicator of police service. That is, police are said to be doing a better job where the FBI index is lower. As shown on Table 1, a positive relationship exists between city size and the reported crime rates for each of the seven offenses included in the FBI index. If this were utilized as a measure of police service levels, one would have to conclude that service levels decline markedly as city size increases. Such a conclusion directly contradicts the relationship posited in the first working hypothesis.

However, the FBI crime index is widely regarded as unreliable and inappropriate for measuring police service levels (Biderman, 1966; E. Ostrom, 1973b; Seidman and Couzens, 1972). None of the other records routinely maintained by police departments offers any better indicators of service levels. Internal police records consist primarily of workload data subject to considerable inaccuracy due to internal pressures to "improve productivity" (Skolnick, 1967). Moreover, police data are entirely production-oriented. The evaluation by consumers of the services they are receiving is not recorded at any point in routine police records. To gain an adequate measure of service levels, one needs to include the evaluation by those receiving the services (Hatry, 1970; Clark, 1972).

Consequently, we will utilize as an independent measure of police service levels sample surveys of citizen-reported experiences with the police and citizen evaluations of police services in their communities. We will discuss findings from both a national survey and two comparative studies within single metropolitan areas. The national sample is from a survey undertaken by the National Opinion Research Center (NORC) for the President's Commission on Law Enforcement and Administration of Justice in 1966 (Ennis, 1967). For each independently incorporated municipality included within the NORC sample frame, we have coded additional information, including city size and police budget. For municipalities located in SMSAs, we have also recorded the number of municipalities and counties located within the SMSA as an estimate of the number of police departments in the metropolitan area. For this analysis, we have examined the responses of individuals living in separately incorporated municipalities above 10,000 in population. The total number of respondents included in the analysis is approximately 2,000 of whom 1,300 live in center cities, 500 live in suburban municipalities, and the remainder live in independent cities outside SMSAs. The sample frame included 102 cities and 42 SMSAs. We did not include suburban respondents living in unincorporated county areas since they would normally be provided police services by a large-scale county sheriff's department. Unfortunately, this distinction between separately incorporated suburbs and suburban areas located in unincorporated county territory is rarely made in most analyses of the differences in service levels between center city and suburbs (see, for example, Hawley and Zimmer, 1970). Without such a distinction, one cannot derive any conclusion regarding the effect of the size of police departments upon performance.

Second, we will present findings from two comparative studies of police service levels in similar neighborhoods within metropolitan areas. Indianapolis and Grand Rapids are the two metropolitan areas studied.[5] In each case, three separately incorporated suburbs, located immediately adjacent to the center city of the metropolitan area were included in the sample frame. Three adjacent neighborhoods within the center city—which were similar to the incorporated suburbs in terms of racial composition, housing patterns, and the income levels and occupation of residents—were selected for comparison. Given the homogeneity of the neighborhoods, the only major difference affecting respondents' evaluations of their police departments related to the size of the police department and the form of community organization.

The findings from these two types of studies complement each other. The Indianapolis and Grand Rapids studies are relatively limited in scope but were specifically designed to examine the relationship between the size of a police jurisdiction and police service levels. The research design controlled for factors, other than the size of the jurisdiction, which affect citizen's evaluation of the police serving their area. Thus, we would expect to find the strongest patterns of association in these two small-scale studies.[6] The national survey is general in scope but the sample frame was *not* specifically designed to investigate hypotheses such as those posed above. Consequently, the jurisdictions included vary along a number of dimensions other than size and the multiplicity of police departments in a metropolitan area. Thus, one would expect relationships of service level with size and multiplicity of jurisdictions to be relatively weak. However, if the findings from both types of studies are similar in direction, they provide reinforcing evidence for each other. The weaknesses of each type of study are compensated for by a strength in the other.

POLICE SERVICE LEVELS IN THE
NATIONWIDE SAMPLE

When utilizing survey data regarding citizen satisfaction with public services, more confidence is gained in the validity of empirical relationships when responses to different questions are analyzed individually rather than relying upon a single response or a scale aggregated in an ad hoc manner from several questions. One can ascertain whether there is a consistent pattern of responses to a variety of different questions. Consequently, in reanalyzing the responses to the NORC victimization survey, we have utilized the responses to ten different questions from the survey instrument. These ten responses have been grouped into three types of evaluations: (1) general evaluations of police performance, (2) feelings of safety, and (3) confidence in the police. In regard to general evaluations, the responses to a series of five separate questions were utilized.[7] These questions probed for evaluations of the local police in five aspects of service: enforcing the law, being prompt to answer calls for service, being respectful, paying attention to citizen complaints, and giving protection to people in a neighborhood. Responses to three questions were used to measure feelings of

TABLE 2
RESPONSE VARIATIONS AND SAMPLE SIZE FOR OUTPUT MEASURES

General Evaluation	Total Sample[a]	Central-City	Suburbs
1. Enforcing the law			
excellent	21.7	18.9	28.9
good	45.2	43.3	49.3
fair	24.4	27.3	17.0
poor	8.6	10.5	4.8
n	1947	1266	501
2. Promptness			
very good	40.3	37.6	48.2
pretty good	31.5	33.8	26.2
not so good	12.2	14.5	6.3
no opinion	15.9	14.1	19.3
n	1987	1294	512
3. Respectfulness			
very good	57.7	55.4	65.2
pretty good	27.1	28.8	19.9
not so good	7.3	8.4	5.5
no opinion	7.8	7.4	9.4
n	1983	1291	512

TABLE 2 (Continued)

General Evaluation	Total Sample[a]	Central-City	Suburbs
4. Attention to complaints			
very good	42.6	41.1	48.2
pretty good	30.3	31.9	26.0
not so good	12.4	13.9	8.0
no opinion	14.6	12.9	17.8
n	1986	1293	512
5. Protection to neighborhood people			
very good	41.3	38.2	52.3
pretty good	35.0	37.0	29.2
not so good	10.8	12.6	6.3
no opinion	12.7	12.1	12.3
n	1987	1294	511
6. Safe at night			
very safe	34.4	28.0	42.7
somewhat safe	23.0	23.2	24.1
somewhat unsafe	21.9	23.5	20.7
very unsafe	20.6	25.3	12.5
n	1983	1291	511
7. Concerned about break-in			
very concerned	20.0	22.5	17.0
somewhat concerned	38.1	37.9	39.3
don't worry at all	41.9	39.5	43.8
n	1989	1295	512
8. Likeliness of being attacked			
very likely	12.0	15.3	6.2
somewhat likely	23.7	28.4	16.2
somewhat unlikely	33.0	33.3	34.1
very unlikely	31.3	23.0	43.5
n	1963	1281	501
9. Confidence in police honesty			
almost all honest	50.4	42.5	63.9
mostly honest, few corrupt	36.5	42.0	27.5
almost all corrupt	4.1	5.1	1.6
don't know	9.0	10.4	7.0
n	1987	1293	512
10. Any unreported crime			
yes	5.3	6.2	3.5
no	94.5	93.5	96.5
don't know	0.2	0.3	0.0
n	1960	1275	509

a. Total sample includes respondents living in independent cities over 10,000 population as well as central-city and suburban respondents.

safety. These related to feeling safe walking alone after dark, fearing home burglaries, and feeling likely to be held up or attacked while walking at night. For indicators of confidence in the police, we utilized, first, the citizens' evaluation of the honesty of their police. The second indicator is whether anything of a criminal nature had occurred to the respondent's family which had not been reported to the police. Nonreporting of criminal victimization is attributed to a lack of confidence in the police. Table 2 presents response variations and size of sample for each question used as an output measure.

THE EFFECT OF SIZE IN DIVERSE
URBAN AREAS

The most consistent theme in the literature recommending consolidation of the police forces serving metropolitan areas is that large-scale police departments are necessary to provide high levels of police service. The logic behind this assertion seems to be that larger departments can be more specialized, more professional, more efficient, and that these characteristics will naturally lead to higher levels of service. While it is not possible here to test all the linkages posited, we have summarized the thrust of the argument in hypotheses H_1 and H_2. The first formalizes the larger size-higher quality argument, while the second allows us to test the efficiency argument. In general, the relationships found in the NORC data are not strong. Given the nature of social science data and the ordinal measures of police service levels utilized, this is not surprising. The relatively high correlations and multiple R^2s reported in many studies of police and other municipal services result from the exclusive use of aggregate data, with all the attendant methodological problems (Robinson, 1950). Yet significant relationships are found in the data. The consistent direction of these relationships over a variety of measures casts serious doubt upon the hypotheses which characterize the arguments for metropolitan police consolidation.

Table 3 presents the relationship (Kendall's tau) found between our measures of police services and the size of the jurisdiction as measured by population in 1965. *Every significant relationship runs counter to that predicted by the first hypothesis.* General evaluations of the quality of police service are lower as size of jurisdiction

TABLE 3

RELATIONSHIPS BETWEEN MEASURES OF POLICE SERVICES AND
SIZE OF JURISDICTION[a]

Measures of Police Services	All Respondents	Central City Respondents	Suburban Respondents
General evaluation			
enforcing the law	−.10*	−.05	—
promptness	−.11*	−.08*	−.06
respectfulness	−.06*	−.04	—
attention to complaints	−.06*	−.04	−.07
protection to neighborhood people	−.10*	−.08*	—
Feelings of safety			
safety while walking at night	−.16*	−.07*	−.05
fear of home break-in	.10*	.10*	—
likelihood of being attacked	.25*	.16*	.09
Confidence in the police			
rating of police honesty	−.16*	−.05	−.10*
unreported crime	.04	—	—

a. Size here is measured by 1965 population. All relationships (Kendall's tau) reported are
significant at the .05 level. Those with asterisks are significant at the .001 level.

increases. Feelings of safety decrease with size, fear of break-in and
attack increase with size, confidence in the police as measured by
rating of police honesty decreases with size, while unreported crime
increases slightly as the size of jurisdiction increases.

Since the relationships shown in Table 3 are generally quite weak,
one might conclude that the size of police jurisdictions in which
respondents lived did not affect their evaluations of service levels
received. Given the consistent pattern of significant relationships,
one might even conclude that increasing size of police departments is
associated with poorer service. Certainly, the data provide no support
for H_1, which predicts opposite signs for all indicators.

However, the respondents to the NORC survey whose answers
were included in this analysis lived in jurisdictions ranging in size
from approximately 10,000 to over 8 million in population. As a
measure of association, Kendall's tau reports the strength of a *linear*
relationship between, in this case, citizen evaluations and size of
jurisdiction. Linear measure of association may be inappropriate over
such a wide size range. Our finding that, in the main, increasing size
of jurisdiction is weakly but negatively associated with the quality of
police services as evaluated by citizens does not mean that a negative
relationship would hold over all size ranges.

In order to further investigate the effects of size of jurisdiction, we
have examined the relationships between this variable and citizen

TABLE 4

RELATIONSHIPS BETWEEN MEASURES OF POLICE SERVICES AND
SIZE OF JURISDICTION—OVER RESTRICTED SIZE RANGES[a]

| | Size of Jurisdiction | | | |
| | Suburban Respondents | | Central-City Respondents | |
Measures of Police Services	10-20,000	Over 20,000	10-100,000	Over 100,000
General evaluation				
enforcing the law	—	—	.20*	—.06
promptness	.13	—.10	.21	—.07*
respectfulness	.14	—	.12	—.05
attention to complaints	.22*	—.07	.28*	—.06
protection to neighborhood people	.19	—	.18	—.09*
Feelings of safety				
safety while walking at night	—.12	—	.20*	—.04
fear of home break-in	—	—	—	.09*
likelihood of being attacked	—	.09	—	.15*
Confidence in police				
rating of police honesty	—	—.10	.11	—.05
unreported crime	—	—	—	.04

a. Size here is measured by 1965 population. All relationships (Kendall's tau) reported are significant at the .05 level. Those with asterisks are significant at the .001 level.

TABLE 5

RELATIONSHIPS BETWEEN MEASURES OF POLICE SERVICES AND
NUMBER OF POLICE PER 1,000 POPULATION[a]

Measures of Police Services	All Respondents	Central-City Respondents	Suburban Respondents
General evaluation			
enforcing the law	—.11*	—.11*	—.05
promptness	—.10*	—.09*	—.11*
respectfulness	—.09*	—.09*	—
attention to complaints	—.10*	—.11*	—.06
protection to neighborhood people	—.12*	—.12*	—.09
Feelings of safety			
safety while walking at night	—.11*	—.09*	—
fear of home break-in	.09*	.12*	—
likelihood of attack at night	.21*	.19*	.07
Confidence in the police			
rating of police honesty	—.16*	—.12*	—.20*
unreported crime	.04	.05	—

a. All relationships (Kendall's tau) reported are significant at the .05 level. Those with asterisks are significant at the .001 level.

evaluations of police service over more restricted size ranges for both suburban and central-city respondents. These relationships are presented in Table 4. Due to the nature of the NORC survey, these data must be considered merely suggestive. As a national survey, the number of respondents living in the larger jurisdictions of each class is much larger than would be appropriate for our purposes. However, the relationships do indicate that significant nonlinearities may be involved. In the *smaller size ranges* for both suburban cities and center cities, service levels may increase as city size increases. Thus, evaluation by suburban residents living in cities below 20,000 increases as city size increases. The same is true for residents of center cities under 100,000. However, the direction of the relationship *reverses* for suburban residents living in cities above 20,000 in population and center-city residents living in cities above 100,000. Economies of scale may accrue up to some size range in the neighborhood of 20,000 population for suburban police departments and 100,000 for center-city police departments, but diseconomies of scale may occur in larger sizes.

Clearly, the problem is a complex one and requires an explanation based on more than a single variable. Those who advocate consolidation as *the* answer for all size ranges are definitely *not* supported by the data presented here. Yet, some consolidation at the lower end of the size scale may be appropriate. Research designed to investigate the effects of a multiplicity of interacting variables may well provide some clearer answers to these questions.[8]

As an alternative measure of the "size of a police department" we have examined the relationships between the number of policemen per 1,000 population and the same set of citizen evaluations. The relationships, shown in Table 5, tend to be slightly stronger than those using population as an indicator of size. Again, the relationships run consistently counter to H_1. Thus, *neither* an increase in the population size of a jurisdiction *nor* an increase in the police/citizen ratio is positively associated with higher service quality.

While for suburban residents living in cities *below* 20,000 population we found a positive, though weak, relationship between city size and evaluation levels, this relationship is reversed when policemen per 1,000 population is used as the indicator of size. As shown on Table 6, any increasing returns to scale in this range *cannot* be attributed to being able to employ more policemen in proportion to population. Rather, the data indicate that *increasing* numbers of police per 1,000 population is *strongly but negatively* associated with police service evaluations in this range.

TABLE 6

RELATIONSHIPS BETWEEN MEASURES OF POLICE SERVICES
AND TWO "SIZE" MEASURES—*SUBURBAN RESPONDENTS*
(10-20,000 population) ONLY[a]

	Size of Police Force	
Measures of Police Services	As Measured by Population Served	As Measured by Police per 1000 Population
General evaluation		
enforcing the law	—	−.22*
promptness	.13	−.37*
respectfulness	.14	−.23*
attention to complaints	.22*	−.31*
protection to neighborhood people	.19	−.24*
Feelings of safety		
safety while walking at night	−.12	.11
fear of home break-in	—	—
likelihood of being attacked	—	—
Confidence in police		
rating of police honesty	—	−.17
unreported crime	—	—

a. All relationships (Kendall's tau) reported are significant at the .05 level. Those with asterisks are significant at the .001 level.

We now turn to the efficiency claims of the advocates of police consolidation as posed in hypothesis H_2. To test the efficiency hypothesis, we examine the relationship between size of jurisdiction and per capita expenditures for police services, controlling for quality of police service. Our quality control for this analysis is the percentage of respondents in each city giving the highest rating to their police on each of the questions utilized above. Table 7 presents the Pearson product-moment correlations between per capita expenditures for police services and population of the city in which

TABLE 7

RELATIONSHIP BETWEEN PER CAPITA EXPENDITURES FOR POLICE
SERVICES AND SIZE OF JURISDICTION (population in 1965)[a]

	All Cities	Central Cities	Suburban Cities
Without controls	.06	.52	.28
Controlling for general evaluation measures	.06	.53	.23
Controlling for feelings of safety measures	.07	.45	.25
Controlling for confidence in the police measures	.07	.52	.26
Controlling for all quality measures	.07	.50	.22

a. Pearson product-moment correlation.

respondents are living for the total sample, for central cities, and for suburban cities; and the partial correlation coefficients between these two variables when quality is controlled. If differences in quality of service were the explanation for the positive relationship between size and cost, we would expect this relationship to disappear or, indeed, to change sign when quality is controlled. However, this does not occur. There is virtually no change in the coefficients when quality is controlled. Consequently, the efficiency argument is refuted. Hypothesis H_2 should be disconfirmed. For constant service levels, an increase in the size of the jurisdiction providing police services is *positively, not negatively,* associated with an increase in per capita expenditures on police services.

THE EFFECT OF FRAGMENTATION IN DIVERSE METROPOLITAN AREAS

Advocates of metropolitan consolidation have vociferously denounced the "crazy-quilt pattern" of jurisdictions serving citizens in metropolitan areas. Critics have found the multiplicity of jurisdictions to be everything from an "esthetic defect" (Bollens, 1961: 191-192) to a condition which makes it difficult for citizens to know where to call in an emergency or where to file a complaint. They argue, as hypothesis H_3 states, that a decrease in the number of jurisdictions is associated with higher service quality.

Table 8 presents data bearing upon this hypothesis. Here the association (again measured by Kendall's tau) between citizen evaluations and the estimated number of police departments serving a metropolitan area is shown. The data provide no support (and some contrary trends) for the hypothesis among those most directly affected by fragmentation—the suburban respondents living in Standard Metropolitan Statistical Areas (SMSAs).

Suburbanites, who are served by the multiplicity of smaller departments, gave responses indicating very weak but higher evaluations in regard to promptness, respectfulness, attention to complaints, and ratings of police honesty in metropolitan areas served by larger numbers of jurisdictions. No significant relationships were found in citizens' feelings of safety or unreported crime.

Surprisingly, central-city respondents, who are presumably only

TABLE 8

RELATIONSHIPS BETWEEN MEASURES OF POLICE SERVICES AND
NUMBER OF POLICE JURISDICTIONS IN METROPOLITAN AREAS[a]

Measures of Police Services	Central-City Respondents	Suburban Respondents
General evaluations		
enforcing the law	−.06*	.10*
promptness	−.06*	.07
respectfulness	−	.06
attention to complaints	−.03	.06
protection to neighborhood people	−.06*	−
Feelings of safety		
safety while walking at night	−.06*	−
fear of home break-in	.07*	−
likelihood of being attacked	.14*	−
Confidence in the police		
rating of police honesty	−.04	.13*
unreported crime	−	−

a. All relationships reported are significant at the .05 level. Those with asterisks are significant at the .001 level. Number of police jurisdictions estimated as equivalent to number of municipalities and counties in an SMSA.

TABLE 9

RELATIONSHIPS BETWEEN MEASURES OF POLICE SERVICES AND
ESTIMATED NUMBER OF POLICE JURISDICTIONS PER
100,000 POPULATION[a]

Measures of Police Services	Central-City Respondents	Suburban Respondents
General evaluations		
enforcing the law	−	−
promptness	−	−
respectfulness	.04	−
attention to complaints	−	−
protection to neighborhood people	−	−
Feelings of safety		
safety while walking at night	−	−
fear of home break-in	−.03	−
likelihood of being attacked	−.05	−.07
Confidence in the police		
rating of police honesty	−	.09
unreported crime	−	−.09

a. All relationships (Kendall's tau) reported are significant at the .05 level.

indirectly affected by the fragmented suburban jurisdictions, are the ones who provide responses weakly consistent with H_3. As the number of police jurisdictions within a metropolitan area increases, residents of the center city tend to rate their own service levels as falling slightly. This puzzling result, generally consistent with the "suburban exploitation" thesis of Hawley and others, cannot be fully interpreted in light of our analysis to date. We intend to investigate this relationship further and report our findings at a later date.

Looking at a relative measure of fragmentation of multiplicity, the estimated number of police jurisdictions per 100,000 population living in an SMSA, produces additional results which cast doubt upon H_3. Here, as shown in Table 9, all significant relationships run counter to those predicted. While the relationships which are found are weak, both center-city and suburban respondents feel safer, and suburban respondents exhibit greater confidence in their police, as relative fragmentation increases.

Efficiency advocates have argued that the existence of a multiplicity of jurisdictions in a metropolitan area is inefficient in economic terms, each jurisdiction's command and overhead structure being seen as essentially a duplication of effort. Our study of police services in the Indianapolis area indicated that such structures may well consume a larger share of resources in large jurisdictions than in smaller ones serving similar neighborhoods, thereby causing us to question the efficiency argument (E. Ostrom et al., 1972). Here we utilize the data from the SMSA portion of the nationwide sample to test hypothesis H_4, which summarizes this argument. Table 10

TABLE 10
RELATIONSHIPS BETWEEN PER CAPITA EXPENDITURES FOR POLICE SERVICES AND ESTIMATED NUMBER OF JURISDICTIONS CONTROLLING FOR QUALITY OF SERVICE

Uncontrolled Product Moment Correlation	0.34		
Partial Correlations Controlling for Quality			
Ratings by:	All Respondents	Central-City Respondents	Suburban Respondents
general evaluation rating	0.35	0.44	0.18
feelings of safety rating	0.31	0.19	0.22
confidence in police rating	0.27	0.33	0.18
all ratings combined	0.28	0.26	0.16

presents the product-moment correlation between per capita expenditures for police services in the SMSA's and the estimated number of police jurisdictions in these areas, as well as the partial correlation coefficients when quality of service ratings by both central city and suburban respondents are controlled.

Looking at the multiplicity of jurisdictions in absolute terms—that is, as the estimated number of police jurisdictions—the data provide some support for H_4. A larger number of police jurisdictions is positively associated with higher per capita expenditures for police services, although the relationship is greatly weakened when quality ratings of suburban respondents are controlled. The uncontrolled relationship is very similar to that reported by Hawkins and Dye (1970: 23) utilizing 1962 data.

However, when multiplicity is measured in relative terms, as the estimated number of police jurisdictions per 100,000 population in the SMSAs, the relationships are *directly and strongly contrary* to those predicted by H_4. Controlling for suburban respondents' quality ratings strengthens this negative relationship between per capita expenditures and the relative multiplicity of jurisdictions. Thus, the argument cannot be made that higher expenditures merely reflect higher service levels. For constant service levels, a decrease in the relative number of police departments in a metropolitan area is associated with higher costs per capita.

In sum, the data presented here provide very little support for the hypotheses relating to the adverse effects of fragmentation upon police services in metropolitan areas and a good deal of contrary evidence.

TABLE 11

RELATIONSHIPS BETWEEN PER CAPITA EXPENDITURES FOR POLICE SERVICES AND ESTIMATED NUMBER OF JURISDICTIONS PER 100,000 POPULATION CONTROLLING FOR QUALITY OF SERVICE

Uncontrolled Product Moment Correlation		−0.47	
Partial Correlations Controlling for Quality			
Ratings by:	**All Respondents**	**Central-City Respondents**	**Suburban Respondents**
general evaluation rating	−0.47	−0.42	−0.57
feelings of safety rating	−0.47	−0.46	−0.48
confidence in police rating	−0.47	−0.47	−0.48
all ratings combined	−0.49	−0.43	−0.54

THE EFFECTS OF SIZE IN
MATCHED NEIGHBORHOODS
WITHIN METROPOLITAN AREAS

In this section, we discuss the findings from two comparative studies of citizen experience with and evaluation of police departments serving their neighborhoods. The findings from these studies are related to the first and second working hypotheses concerning the effect of size on service levels and on per capita costs.

The Indianapolis study was undertaken in the spring of 1970 (E. Ostrom et al., 1972). The city-county consolidation of the city of Indianapolis with Marion County ("Unigov") did not affect the organization of police forces in Marion County. Three small, separately incorporated communities—Beech Grove, Lawrence, and Speedway—are located immediately adjacent to the boundaries of the Indianapolis Police District and within Marion County. Each has its own police force. The independent communities vary in size from approximately 13,500 to 16,500 population. The police departments serving these communities range from 18 to 26 full-time officers. The Indianapolis Police Department has 1,100 sworn officers. Three adjoining Indianapolis neighborhoods were also selected as sample areas. Each was matched as closely as possible to the adjoining independent community in terms of size, population, and socioeconomic characteristics. The population in the six neighborhoods is composed of predominantly white, middle-income families with similar educational backgrounds, employed in a similar range of occupations and living in owner-occupied residences. All six neighborhoods are located about equidistant from the center of the city of Indianapolis.

The Grand Rapids study was undertaken in the spring of 1971 by Samir Ishak (1972). The independently incorporated cities of East Grand Rapids, Kentwood, and Walker are located immediately adjacent to the city of Grand Rapids. Three neighborhoods within the city of Grand Rapids, immediately adjacent to the independent communities, were selected for comparative study. The population of Grand Rapids is approximately 200,000, while that of the independent communities ranges from approximately 12,000 to 20,000. The police departments serving these communities vary from 9 to 17 sworn officers. The police department serving the city of

Grand Rapids employs 313 sworn officers. The six neighborhoods chosen in the Grand Rapids area are very similar to each other and to the six Indianapolis neighborhoods, as shown in Table 12. A similar questionnaire was administered to a sample of 722 respondents in the six Indianapolis area neighborhoods and 1,143 respondents in the six Grand Rapids area neighborhoods.

Respondents in the small independent suburbs of both metropolitan areas consistently rated the police serving their neighborhoods higher than did respondents in the six neighborhoods located in the large center cities. In particular, respondents in the independent suburbs rated their police better with regard to responding rapidly, police-citizen relationships, the likelihood of police accepting bribes, and a general evaluation of the job being done. With regard to their specific experiences, respondents in the six independent suburbs were less likely to have been the victim of a crime, more likely to

TABLE 12
CHARACTERISTICS OF RESPONDENTS IN THE AREAS STUDIED
(in percentages)

	Indianapolis Area		Grand Rapids Area	
	Three Suburbs	Three City Neighborhoods	Three Suburbs	Three City Neighborhoods
Ownership of Housing				
buying home	85	89	95	88
renting home	8	9	2	4
renting apartment	8	2	1	4
n =	(369)	(347)	(557)	(553)
Age of Respondent				
16-30	27	31	25	31
31-40	24	34	32	31
41-50	21	21	19	20
over 50	28	14	23	18
n =	(339)	(321)	(565)	(576)
Length of Residence				
less than 2 years	21	21	16	27
2 to 5 years	23	25	24	30
6 to 10 years	19	36	22	24
more than 10 years	36	20	38	19
n =	(373)	(349)	(565)	(570)
Sex				
male	42	49	40	42
female	58	51	60	58
n =	(373)	(349)	(564)	(574)

have reported a victimization, and more likely to have received some form of assistance from the police. When the police were called, respondents in the six independent suburbs received faster service than their neighbors within the larger city. A smaller proportion of the respondents living in the independent suburbs thought that crime was increasing in their neighborhood than did respondents in the larger cities. A summary of the rank orders of each of the neighborhoods in each metropolitan area is presented in Table 13. Given the consistency of the pattern of responses in two different metropolitan areas, it would appear that respondents living in the smaller, independently incorporated suburban communities are receiving higher levels of police service than their neighbors living in similar surroundings but served by a large center-city police force. This evidence is directly counter to what would be predicted from the first working hypothesis and substantially challenges its warrantability. Given the similarity of these findings to those from a national sample, the likelihood that there is a positive association between city size and service levels is very small.

The costs of providing police services in the twelve neighborhoods included in the Indianapolis and Grand Rapids neighborhoods were also examined. At the time of the studies, the per capita expenditure for police within the city of Indianapolis was $21.33, while the per capita expenditure for police within the three incorporated suburbs of Indianapolis ranged from $11.80 to $14.47. The per capita expenditure of the city of Grand Rapids was $20.00, and the per capita expenditures for police within the three incorporated suburbs of Grand Rapids ranged from $5.81 to $16.13. However, the per capita expenditure for either of the large cities is *not* an appropriate measure of input for the neighborhoods studied. In order to utilize this figure, one would have to assume that expenditures were being devoted to activities which were equally distributed throughout these cities. In order to estimate expenditures for the subareas studied, several formulas were examined. The most appropriate estimate of expenditures involves an examination of the actual patrol practices of the department and the number of offenses reported in the particular neighborhoods under study.[9] Expenditures for criminal investigation were allocated to each neighborhood in the proportion of the number of offenses reported in the neighborhood to the number of offenses in the city as a whole. Other supportive services were allocated according to the ratio of the patrol forces assigned to the neighborhood to the patrol forces serving the entire city. Using

TABLE 13

RANK ORDER OF NEIGHBORHOODS WITHIN EACH METROPOLITAN AREA ON POLICE PERFORMANCE LEVELS[a]

	Citizen Experiences						Citizen Evaluations			Cumulative Rank
	Victimization	Willingness to Report	Level of Follow-Up	Willingness to Call for Assistance	Response Time	Promptness	Neighborhood Crime Trend	Police-Citizen Relations	General Evaluation	
Indianapolis Metropolitan Area										
Incorporated suburbs										
Beech Grove	4	4	3	3	3	3	2	3	3	3
Lawrence	3	1.5	2	2	1	2	1	2	1	2
Speedway	1	3	1	1	1	1	4	1	2	1
Indianapolis neighborhoods										
near Beech Grove	2	6	6	4	5	5	6	5	4	5
near Lawrence	6	5	5	5.5	4	6	3	4	6	6
near Speedway	5	1.5	4	5.5	6	4	5	6	5	4
Grand Rapids Metropolitan Area										
Incorporated suburbs										
Walker	2.5	1	5	1	2	3	1	4.5	2.5	2
East Grand Rapids	1	3	1	3	1	1	2	1	1	1
Kentwood	2.5	2	6	4.5	3	2	4	6	5	3
Grand Rapids neighborhoods										
near Walker	5.5	4	4	2	4	5	3	4.5	4	4
near East Grand Rapids	4	6	3	6	5	6	5	2	2.5	5
near Kentwood	5.5	5	2	4.5	6	4	6	3	6	6

a. Indicates highest percentage of citizens responding to highest rating category for each questionnaire item, ties indicate equal percentage.

TABLE 14
ESTIMATED PER CAPITA COSTS OF POLICE SERVICES PROVIDED
TO STUDY AREAS

Area	Estimated Per Capita Cost	
Indianapolis Area		
Incorporated suburbs		
Beech Grove	$12.04	
Lawrence	11.80	
Speedway	14.47	
Average for Incorporated Suburbs		12.76
Indianapolis neighborhoods		
near Beech Grove	12.94	
near Lawrence	9.87	
near Speedway	9.36	
Average for Indianapolis Neighborhoods		10.72
Grand Rapids Area		
Incorporated suburbs		
Walker	10.08	
East Grand Rapids	16.13	
Kentwood	5.81	
Average for Incorporated Suburbs		10.67
Grand Rapids neighborhoods		
near Walker	18.37	
near East Grand Rapids	10.61	
near Kentwood	37.02	
Average for Grand Rapids Neighborhoods		16.88

this method of allocation, the costs of providing services within the Indianapolis neighborhoods under study averaged $10.72 and, within the Grand Rapids neighborhoods under study averaged $16.88. A comparison of the estimated costs of providing police services in the twelve areas is shown in Table 14. In the Indianapolis area, the estimated costs of providing services to the Indianapolis neighborhoods are slightly lower than the costs of providing services to the incorporated suburbs. In the Grand Rapids area, the estimated costs of providing services to the neighborhoods in the large city are higher than the costs to the incorporated suburbs, while service levels are lower. Cumulatively, these findings do not lend support to H_2. The evidence from the Grand Rapids area directly challenges this hypothesis, since higher levels of services are being provided by the smaller police departments at lower costs. In Indianapolis, the smaller suburban police departments are providing higher levels of service for a somewhat higher level of expenditure.

CONCLUSION

We have attempted to examine the empirical warrantability of four working hypotheses derived from the arguments for consolidation of police within metropolitan areas and the consequent elimination of most suburban police forces. We have discussed the findings from previous studies which utilized aggregate statistical data. We have reanalyzed the responses to a nationwide sample survey of citizens living in a variety of urban locations ranging from small, suburban municipalities to very large center cities. We have also presented findings from two comparative studies of similar neighborhoods within single metropolitan areas served by small and large police departments. Two of the working hypotheses underlying recommendations for consolidation—those relating to size of jurisdiction—are *not* supported by the evidence. Thus, the following two hypotheses would appear unwarranted on the basis of our findings:

H_1: An increase in the size of the jurisdiction providing police services to citizens will be positively associated with higher service levels.

H_2: For constant service levels, an increase in the size of the jurisdiction providing police services will be positively associated with lower per capita expenditures on police services.

With regard to H_1, respondents living in larger units in both the nationwide study and in the two smaller, comparative, metropolitan area studies were less satisfied with the services which they received than were respondents living in smaller units. Regarding H_2, data presented from previous aggregate studies, from the NORC nationwide survey, and from one of the comparative metropolitan area studies indicate increased per capita expenditures in larger jurisdictions. The quality of services was held constant in the nationwide sample, and the quality of service was higher in the smaller jurisdictions of the metropolitan area sample.

Turning to the hypotheses relating to fragmentation, the evidence is somewhat mixed. Examining the first of these:

H_3: A decrease in the number of police jurisdictions serving the citizens within a metropolitan area will be positively associated with higher service levels,

we found that the responses of suburban residents in the nationwide sample run counter to the hypothesis whether fragmentation is measured in absolute or in relative terms. The responses among center-city respondents are also contrary to those predicted when relative fragmentation is considered, but tend toward agreement when fragmentation is measured absolutely.

The second fragmentation hypothesis:

H_4: For constant service levels, a decrease in the number of jurisdictions providing police services within a metropolitan area will be positively associated with lower per capita expenditures for police services,

is supported by the nationwide sample data when fragmentation is measured by the number of police departments serving populations within SMSAs. Even here, the relationship is seriously weakened when quality ratings of suburban respondents are controlled. When fragmentation is measured as the number of police departments per 100,000 population, the evidence is strongly against H_4.

The major conclusion which can be derived from the analysis presented above is that proposals for the elimination of suburban police departments by consolidation are *not* based on firm empirical evidence. The level of police services now provided to residents in independently incorporated suburban communities might well be *reduced* if smaller police forces were to be replaced by one large police force serving an entire metropolitan area. The costs of such police services would be likely to increase. However, it is also obvious that size and multiplicity of jurisdictions are only two of many factors affecting the performance of urban police forces. The types of services provided by police vary greatly from neighborhood-level patrol services to criminal investigation requiring modern laboratory and communication facilities.

The most appropriately sized unit for providing one type of service may not be the most appropriate for other types of services. Bigger is not always better, nor is smaller always better. Reform proposals based upon theoretical inferences which are contrary to relevant evidence are likely to produce counter-productive results. Such reforms may exacerbate rather than alleviate the social pathologies associated with the contemporary crisis of law and order (V. Ostrom, 1973). Both suburban *and* center-city residents might be better served if larger-scale police agencies were created to provide specialized services to existing suburban forces, and if smaller-scale police agencies were created *within* large center cities to provide neighborhood patrol services. In such a system of overlapping large- and small-scale police agencies, both suburban and center-city residents could gain the advantages of the appropriate scale of production for the types of services provided (see V. Ostrom et al., 1961; V. Ostrom, 1971).

It is somewhat difficult to explain the persistence of the strongly held beliefs favoring consolidation of police services without evidence to support these beliefs. Perhaps the difficulty arises from the absence of any alternative explanatory theory that would lead social scientists to pose alternative questions and entertain alternative hypotheses. It is paradoxical that the NORC study which we utilize as part of our data base was originally prepared for the President's Commission on Law Enforcement and the Administration of Justice. While the primary purpose of the study was to measure the amount of criminal victimization in the United States, a significant effort went into the construction of questions designed to measure attitudes toward the police, law enforcement, and feelings of personal security. Typical SES variables were selected to explain these attitudes, but *no* effort was made to include such variables as the type of municipality in which respondents lived, its size or the multiplicity of police jurisdictions serving different metropolitan areas. As noted above, the President's Commission recommended consolidation of police departments serving metropolitan areas as a means of improving performance. It would appear from its report that the Task Force on Police for the commission automatically assumed that fractionalization of police services was positively associated with lower service levels (see Task Force on Police, 1967: 68-70; also see Smith, 1971). Given the predominance of this view in the literature, the lack of questioning is understandable. However, given our findings presented above, it is essential that further theoretical and empirical efforts be devoted to the question of whether there are too many and too small suburban police departments. The answer can no longer be stated as a categorical yes.

NOTES

1. The assumption that evidence has been gathered in the past to establish that consolidation will lead to higher levels of police services and efficiency is clearly shown in this recent statement by Bernard L. Garmire (1972: 9), Chief of Police of the Miami Police Force: "An outstanding example of such a problem—one of the most persistent and

frustrating problems confronting our municipalities and states—is the Balkanization of the police forces. There are approximately 40,000 police agencies ranging in size from one man to 30,000 men. It is not necessary to elaborate on the disadvantage of such a fractured system; they are too well known. Perhaps only cities of 50,000 or more should be allowed their own police agencies; the state should police the smaller cities and the rural areas. In the metropolitan areas there should be consolidation of the smaller police agencies. If this were done, the number of police agencies could be reduced from 40,000 agencies to roughly 400-plus agencies—with clear gains in terms of effectiveness and efficiency."

2. An increase in the per capita costs of municipal services also occurred after the Ontario legislature created Toronto Metro (Kaplan, 1967). Some of the increases in costs can be attributed in the early years to an increase in capital equipment such as patrol cars and the like. However, the long-run increase in per capita costs raises serious questions about the "economy" argument in the police consolidation literature.

3. These working hypotheses are very similar to four of the questions utilized by Hawley and Zimmer (1970: 97) to operationalize the metropolitan orientation of respondents living in six different metropolitan areas. Their four questions and the answer which demonstrated the "metropolitan orientation" of the respondent were:

Query	Answer
The existence of a number of governmental units in a metropolitan area is wasteful.	Agree
Conditions have reached a point at which suburbs require increased amounts of public services.	Agree
Public services, if provided on an areawide basis, would be —	Better
If public services were provided on an areawide basis, taxes would—	Decrease or remain the same

A fifth question was also included in their set—the agreement to the statement that "suburbs should share the costs for services and conveniences provided by the central city." This particular aspect of the metropolitan reform ideology is not amenable to the same kind of empirical examination as are the other four statements.

4. Masotti and Bowen (1971), however, did not find such a relationship for eighteen Ohio cities.

5. See E. Ostrom et al. (1973b) for a detailed description of the Indianapolis study; see also Ostrom and Whitaker (1973) and E. Ostrom et al. (1973a); see Ishak (1972) for a detailed description of the Grand Rapids study.

6. Factors which may affect police service and citizen evaluation of police service include neighborhood wealth, density, racial composition, proximity to dissimilar neighborhoods and median age of residents. The effect of these factors was controlled by selecting neighborhoods which are very similar in terms of these characteristics.

7. Where necessary, responses have been recoded so that a higher numerical score represents a higher evaluation.

8. We have initiated a large-scale study in the St. Louis metropolitan area which will enable us to examine the effects of interacting variables on police performance.

9. See E. Ostrom et al. (1973a) for a description of the methods used to estimate costs; see Ishak (1972) for derivation of cost data for Grand Rapids area.

REFERENCES

BAHL, R. W. (1969) Metropolitan City Expenditures. A Comparative Analysis. Lexington: Univ. of Kentucky Press.

––– (1968) "The determinants of public expenditures: a review," pp. 198-220 in S. Mushkin et al. (eds.) Functional Federalism: Grants-in-aid and PPB Systems. Washington, D.C.: George Washington University.

BAKER, C. A. (1910) "Population and costs in relation to city management." J. of Royal Statistics Society 73 (December): 73-79.

BIDERMAN, A. D. (1966) "Social indicators and goals," pp. 68-153 in R. A. Bauer (ed.) Social Indicators. Cambridge, Mass.: MIT Press.

BOLLENS, J. C. (1961) Exploring the Metropolitan Community. Berkeley: Univ. of California Press.

BRAZER, H. (1959) City Expenditures in the United States. New York: National Bureau of Economic Research.

CAMPBELL, A. K. and S. SACKS (1967) Metropolitan America: Fiscal Patterns and Governmental Systems. New York: Free Press.

CAMPBELL, A. and H. SCHUMAN (1972) "A comparison of black and white attitudes and experiences in the city," pp. 97-110 in C. H. Haar (ed.) The End of Innocence: A Suburban Reader. Glenview, Ill.: Scott, Foresman.

CLARK, T. N. (1972) "Community social indicators and models." Presented to the Conference on Social Indicator Models, Russell Sage Foundation, New York, July 12-15.

Committee for Economic Development (1972) Reducing Crime and Assuring Justice. New York.

DUNCAN, O. D. (1951) "The optimum size of cities," pp. 632-645 in P. K. Hatt and A. H. Reiss (eds.) Reader in Urban Sociology. New York: Free Press.

ENNIS, P. H. (1967) Criminal Victimization in the United States: A Report of a National Survey. Submitted to the President's Commission on Law Enforcement and Administration of Justice. Washington, D.C.: Government Printing Office.

FOWLER, E. P. and R. L. LINEBERRY (1972) "The comparative analysis of urban policy: Canada and the United States," pp. 345-368 in H. Hahn (ed.) People and Politics in Urban Society. Beverly Hills: Sage Pubns.

GABLER, L. R. (1971) "Population size as a determinant of city expenditures and employment—some further evidence." Land Economics 47 (May): 130-138.

––– (1969) "Economies and diseconomies of scale in urban public sectors." Land Economics 45 (November): 425-434.

GARMIRE, B. L. (1972) "The police role in an urban society," pp. 1-11 in R. F. Steadman (ed.) The Police and the Community. Baltimore: Johns Hopkins Press.

HAAR, C. H. (1972) "The service community. the immediate environment," pp. 49-53 in C. H. Haar (ed.) The End of Innocence: A Suburban Reader. Glenview, Ill.: Scott, Foresman.

HATRY, H. P. (1970) "Measuring the effectiveness of nondefense public programs." Operations Research 18 (September/October): 772-784.

HAWKINS, B. W. and T. R. DYE (1970) "Metropolitan 'fragmentation': a research note." Midwest Rev. of Public Administration 4 (February): 17-24.

HAWLEY, A. H. and B. G. ZIMMER (1970) The Metropolitan Community. Its People and Government. Beverly Hills: Sage Pubns.

HIRSCH, W. Z. (1959) "Expenditure implications of metropolitan growth and consolidation." Rev. of Economics and Statistics 41 (August): 232-241.

ISHAK, S. T. (1972) "Consumers' perception of police performance: consolidation vs.

deconcentration; the case of Grand Rapids, Michigan metropolitan area." Ph.D. dissertation. Indiana University.

KAPLAN, H. (1967) Urban Political Systems: A Functional Analysis of Metro Toronto. New York: Columbia Univ. Press.

London County Council (1915) Comparative Municipal Statistics, 1912-1913. London: London County Council, Local Government and Statistical Department.

McARTHUR, R. E. (1971) Impact of City-County Consolidation of the Rural-Urban Fringe: Nashville-Davidson County, Tennessee. Washington, D.C.: Government Printing Office.

McCAUSLAND, J. L. (1972) "Crime in the suburbs," pp. 61-64 in C. H. Haar (ed.) The End of Innocence. A Suburban Reader. Glenview, Ill.: Scott, Foresman.

MacNAMARA, D.E.J. (1950) "American police administration at mid-century." Public Administration Rev. 10 (Summer): 181-189.

MASOTTI, L. H. and D. R. BOWEN (1971) "Communities and budgets: the sociology of municipal expenditures," pp. 314-326 in C. H. Bonjean et al. (eds.) Community Politics: A Behavioral Approach. New York: Free Press.

National Resources Committee (1939) Urban Government. Washington, D.C.: Government Printing Office.

OGBURN, W. F. (1937) Social Characteristics of Cities. Chicago: International City Managers' Assn.

OSTROM, E. (1972) "Metropolitan reform: propositions derived from two traditions." Social Sci. Q. (December).

——— (1971) "Institutional arrangements and the measurement of policy consequences." Urban Affairs Q. 6 (June): 447-476.

——— and G. P. WHITAKER (1973) "Does local community control of police make a difference? Some preliminary findings." Midwest J. of Pol. Sci. (February).

OSTROM, E., R. B. PARKS, and G. P. WHITAKER (1973a) "Do we really want to consolidate urban police forces? A reappraisal of some old assumptions." Public Administration Rev. (forthcoming).

OSTROM, E., W. BAUGH, R. GUARASCI, R. PARKS, and G. P. WHITAKER (1973b) "Community organization and the provision of police services." Sage Professional Paper in Administrative and Policy Studies. Beverly Hills: Sage Pubns.

OSTROM, V. (1973) The Intellectual Crisis in American Public Administration. University: Univ. of Alabama Press.

——— (1971) The Political Theory of a Compound Republic. Blacksburg: Virginia Polytechnic Institute Center for the Study of Public Choice.

——— C. M. TIEBOUT, and R. WARREN (1961) "The organization of government in metropolitan areas: a theoretical inquiry." Amer. Pol. Sci. Rev. 55 (December): 831-842.

Oxford University (1938) A Survey of the Social Services in the Oxford District. Volume I, Economics and Government of a Changing Area. London: Oxford Univ. Press.

PHILLIPS, H. S. (1942) "Municipal efficiency and town size." J. of Town Planning Institute 28 (May/June): 139-148.

President's Commission on Law Enforcement and Administration of Justice (1967) The Challenge of Crime in a Free Society. New York: Avon.

President's Task Force on Suburban Problems (1968) "Final report excerpt," pp. 13-17 in C. M. Haar, The End of Innocence: A Suburban Reader. Glenview, Ill.: Soctt, Foresman.

ROBINSON, W. S. (1950) "Ecological correlation and the behavior of individuals." Amer. Soc. Rev. 15 (June): 351-357.

SCOTT, S. and E. L. FEDER (1957) Factors Associated with Variations in Municipal Expenditure Levels. Berkeley: University of California Bureau of Public Administration.

SEIDMAN, D. and M. COUZENS (1972) "Crime, crime statistics, and the great American anti-crime crusade: police misreporting of crime and political pressures." Presented at the annual meeting of the American Political Science Association, Washington, D.C., September.

SKOLER, D. L. and J. M. HETLER (1970) "Governmental restructuring and criminal administration: the challenge of consolidation," pp. 53-75 in Crisis in Urban Government: A Symposium: Restructuring Metropolitan Area Government. Silver Spring, Md.: Thomas Jefferson Publishing.

SKOLNICK, J. H. (1967) Justice Without Trial: Law Enforcement in Democratic Society. New York: John Wiley.

SMITH, D. C. (1971) "The constitution of police legitimacy." Bloomington, Ind.: Indiana University Department of Political Science Papers on Political Theory and Policy Analysis.

Task Force on Police of the President's Commission on Law Enforcement and Administration of Justice (1967) Task Force Report: The Police. Washington, D.C.: Government Printing Office.

WEICHER, J. C. (1970) "Determinants of central city expenditures: some overlooked factors and problems." National Tax J. 23 (December): 379-396.

WINGO, L. (1972) "Introduction," pp. 1-8 in L. Wingo (ed.) Minority Perspectives. Washington, D.C.: Resources for the Future.

Part V

THE ECONOMICS

OF SUBURBIA

THE ECONOMICS OF SUBURBIA

Both public sector and private sector economics play significant roles in suburban growth and development. Until recently, the viability of the suburbs was dependent on economic strength of the central city. That situation appears to be changing rapidly, as suburbia grows increasingly independent economically, primarily at the expense of the city. The two essays in this section provide a current assessment of the suburban economy and its relationship to the economic structure of the city and the metropolitan region.

In "A Preface to Suburban Economics," Thompson examines the development of "dual economies" in large metropolitan areas and the prospects for consolidation; he reassesses the physical form of suburbia in the context of municipal decision-making and speculates on the relationship between residential density and public service costs; and, third, he suggests a "capital-stock" analysis which will allow us to predict suburban futures by the regularities of aging and depreciation.

Berry and Cohen offer a detailed discussion of economic decentralization. The classic model of economic centralization based on positive externalities like scale, transit costs, and labor market is rejected as inappropriate. A new social dynamic, in which whites and developers of major suburban retail/residential projects avoid the central city, is replacing the economic dynamic of conventional wisdom. New technology makes this socially preferred decentralization economically feasible.

15

A Preface to Suburban Economics

WILBUR R. THOMPSON

□ SUBURBS HAVE GROWN TO A SIZE and importance that has not gone unnoticed. It is not necessary therefore to write on this subject matter as if nothing had gone before. Also, with well over one-third of the national population living in "suburbs" (37% living in metropolitan areas but outside central cities), it would be most surprising if these places did not span much of the full range of urban life. This modest effort can therefore do little more than pick and choose from among an almost unlimited variety of material.

Some of the most familiar subject matter has been passed over or, at least, not treated in a familiar way. In some cases the bypassed material has been well handled elsewhere, or, in any event, little could be added here. More often, this opportunity was taken to review the suburbs from some new perspective. For example, rather than go back over the well-worked ground of political fragmentation or the suburban-exploitation-of-the-central-city hypothesis, the opening section of this work reflects instead on the development of "dual economies" in our large metropolitan areas. What is the logic of a suburban manufacturing economy linked to but still significantly separate and distinct from the central-city service economy? And are there economic foundations of the prospects for political consolidation of the two?

Again, in the second section, rather than repeat the old arguments about urban sprawl, the physical form of suburbs is reassessed in a context of municipal decision-making, with an eye to the total social interest. The political arithmetic of the trade-off between total population, land area, and density is discussed, and some speculations are offered on the relationship between residential density and various public service costs. Land use and transportation are introduced by reflecting briefly on the problems of expressway location: are boundary locations necessarily best?

Finally, the suburbs (and central cities) are seen as stocks of capital, subject to long waves of depreciation and replacement. Not only does this approach afford deep insights into the nature of city-building and urban development in general, but the regularities of the depreciation process build predictive powers that are not to be lightly regarded. The likely consequences of the aging of suburbs on income patterns and transportation needs over the next few decades are briefly explored. But most of all, this capital-stock view of the city suggests that, even if "what will be, will be, the future *is* ours to see"—darkly, through the inexorable process of aging.

FEDERATED LOCAL ECONOMIES

E UNUM PLURIBUS

The rise of the suburb has reached the point where the metropolitan population outside of central cities outnumbers that within them. The combined population of suburbs, satellites, and fringe-area townships attains an advantage of 2 : 1 in many of the largest metropolitan areas, and this rises as high as 3 : 1 in Boston and Pittsburgh. We long ago divested ourselves of the simple-minded notion of the suburb as a bedroom community with some convenience shopping (the modern expression for the corner store). But the time has come to go well beyond that small step and reconsider an even more basic conception of the suburbs: the assumption that the suburban economy is a virtually indistinguishable integral part of a unified and virtually indivisible metropolitan area economy. The economist in particular has found it most convenient to assume that all the built-up space around the central city (the "urbanized area" of census definition) and the surrounding rural, nonfarm land beyond, out to about a one-hour commute from downtown, is a single economic entity; the local labor market.

But on even casual observation, the largest of our metropolitan areas have come to look more like overlapping submarkets for labor, housing, and shopping—indistinct enough to be consolidated in the collection and reporting of data but still distinct enough to deserve treatment as something more than political subdivisions or rich and poor residential districts. (This blurring of the identity of the local labor market arising out of centrifugal growth of the city is separate from and additive to the overlapping at the edges of urban areas through the formation of great linear strips of cities—megalopolis.)

Manufacturing began its long march to the suburbs well before World War II, but wartime shortages placed a moratorium on the construction of new plants and confined the wartime expansion of production to existing central-city plants. Then the postwar building boom acted to disperse plants as fast as housing; the suburban share of manufacturing employment rose from less than one-third the metropolitan area total to well over one-half in the postwar period, very closely paralleling the suburbanization of population (lagging slightly behind). We have much poorer data on services, but a rough estimate would be that this sector split, with retail trade dispersing with population, lagging at first but now catching up, and professional services still concentrated in the central city—perhaps two-thirds or more (Kain, 1970; Birch, 1970).

In broad terms, the largest metropolitan areas have been evolving rather rapidly into two easily identifiable, although heavily overlapping, subeconomies. The central city is more and more a place of business, professional, and governmental service—the workplace of lawyers, consultants, financiers, and officials of all kinds; the suburbs are more and more the heart of the manufacturing district. Retail trade and personal services are performed in both places, convenient to their customers. One of the main forces which integrate these "dual economies" is long-distance commuting to work; suburban white-collar workers heading downtown in the morning pass blue-collar workers from the central city heading for outlying plants and domestics on the way to suburban residences.

But even the ties of cross-commuting have been drawn thin as the money and time cost of the long journey to work has stimulated a sorting-out process that is matching home and workplace. Suburban manufacturing plants generate middle-skill work that fits well the residential preferences of *upper-middle*-income blue-collar workers.

(Manufacturing workers benefit from the combined power of unions and oligopolistic employers to derive above-average wages and incomes from average skills.) The locational preferences of manufacturing plants match the residential preferences of well-paid manufacturing workers, and this economic symbiosis is reinforced by the plants' dependence on trucks, the workers' dependence on cars, and the dependence of both on low-density land use—sprawl.

This drawing together of home and work can also be seen in central cities, more or less blurred by the long and difficult process of rebuilding the cores of old cities. Professional services, with heavy interpersonal interaction, the need for face-to-face contact, and speed in communication still cluster in or near the old downtown and serve as a magnet for selected residences. If professionals are, on the whole, more urbanized than, say, manufacturing workers, and if the former are more inclined than the latter to avoid long automobile journeys to work, there is at least a latent demand among professionals for inner-city living.

Low-skill personal service workers have little choice but to live in the older housing of the inner city near to the best of a bad lot of public transit. Still, there is convenience in the proximity of the offices and stores in which they work. (This is not to deny a reverse direction of causation: inner-city personal service workers may be low skilled because of lack of access to middle-skill manufacturing work.) A central location near the hub of the public transportation system is also an advantage in finding and getting to a new job, a major consideration for these easy-to-replace, low-seniority workers —"last hired, first fired." Again, personal service work, unconstrained by assembly lines and other production imperatives and less disciplined by labor-management formalities, provides a richer supply of the part-time work that is critical to the employment of female heads of households, second earners in the family, the elderly, and handicapped workers. And the preceding reads like a list of inner-city residents. Finally, public service employment, free from competitive pressures, could, with sophisticated planning, become an important source of part-time work, and these jobs are located more than proportionately in the inner city.

FACTORY OR STADIUM; OFFICE OR MUSEUM?

Our largest metropolitan areas may already be unified more for selected recreational and cultural purposes than for economic purposes, more as consumer markets than as local labor markets. There is considerable logic to a spatial pattern of trip-making in which less-frequently made trips to special events are, on the average, longer than the routine daily journey to work. During this first long cycle of urban development, from the center outward, professionals are most likely to live in new houses on the urban fringe and work in downtown offices, and factory workers are likely to live in older houses closer to the center and work in plants drifting outward. The second time through the long cycle, a more random distribution of housing by age and price is to be expected. Different qualities of building materials, different levels of home maintenance, and the cumulative disequilibrium processes that are generated in neighborhoods (both those that bring back a neighborhood and those that hasten its decay) all will work to leave small clusters of older houses spotted among the new houses of the second city. It should be more possible to live near one's work in the year 2000, and even earlier.

What may well become the major force that draws the metropolitan area together is that the great city is a mass market. Residents from all corners of the sprawling metropolitan area are quite willing to cross its full reaches to: (1) attend professional football, baseball, basketball, and hockey games; (2) participate in outdoor recreation, especially water sports; (3) attend universities, museums, and theatres; and (4) pursue vague objectives in specialized, comparative shopping expeditions. We define and delineate our large metropolitan areas as local labor markets and then fabricate "consolidated standard metropolitan areas" (e.g., New York-Northeastern New Jersey, Chicago-Gary-Hammond) when the geography gets away from us. Perhaps it is now time to review and redefine city regions ("metropolitan areas").

"FORM FOLLOWS FUNCTION": INDUSTRY-MIX AND POLITICAL CONSOLIDATION

A strong thread of economic determinism may run from the local industry mix through to the prospects for political consolidation of central city and suburbs. A first working hypothesis might be that specialization in manufacturing tends to produce a weak central city and a ring of strong suburbs. Suburbanites in this context would, it seems, be less interested in sharing fortunes with the central city and, if the latter is dominated by blacks, more sensitive to political power than to economic strength, and political consolidation would find few friends in either camp. By way of illustration, industrialized Cleveland seems to exemplify this situation: neither the central-city blacks nor the suburban whites supported consolidation in a recent referendum.

The contrasting case is that of the professional service city, with its much stronger central business district ("downtown") and the presence of high-income as well as low-income households in the central city. Consider Indianapolis, the state capital of Indiana. Contemporary comment on its recent political consolidation argued that the Republican suburbs forced the issue here to neutralize a growing democratic majority in the central city, employing a state legislative act (no referendum) to achieve political consolidation. (Black political power will also be contained if, as expected, this one-quarter of the total population captures only about one-eighth of the new legislative seats.) But suburban whites stand to lose political control of central cities everywhere. Why was the imminent loss of Indianapolis of grave concern and Cleveland's loss of so little concern? Indianapolis, with its state capitol and rapidly expanding professional-governmental service sector does contrast sharply with Cleveland's more economically depressed inner city.

Drawing loosely from memory from the small sample of attempted political consolidations that have been widely reported, other manufacturing cities that have rejected a central city-suburban merger range through: Chattanooga (with 41.1% of its 1970 total employment engaged in manufacturing), Augusta (34.3%) and St. Louis (30.5%), with Cleveland in the middle of the group (34.5% in manufacturing). Reinforcing the argument, merger referenda have won in less-industrialized places: Jacksonville (13.0% in manufacturing), Miami (15.3%), Baton Rouge (17.7%), Portland, Oregon (22.5%), Columbus, Georgia (27.4%), and Nashville (27.7%). (In fairness, two nonmanufacturing losers come to mind—Memphis, 21.9%, and San Diego, 17.4%.) But one could argue an asymmetry here: the deductive case for expecting weak-core manufacturing cities to reject consolidation seems stronger than the case for service cities supporting it. There is little or no *economic* rationale for the suburbs to join with the central city in manufacturing areas; there are mixed advantages and disadvantages for both parties in service-oriented economies—"regional cities."

PHYSICAL PLANNING MADE DIFFICULT

THE SIMPLE ARITHMETIC AND HARD CHOICES OF
PHYSICAL FORM

The physical form of a suburb is not easy to describe in words nor can it be precisely expressed in equations. It is not without good reason that urban planners draw pictures. Still, we can frame three very basic dimensions of any suburb in a simple mathematical identity:

$$\text{Total Population} = \text{Average Density} \times \text{Land Area}$$

No apology need be offered for presenting, so formally, so simple a formulation. There is considerable precedent in social science for the careful consideration of what appears on first blush to be a trivial equation: the demographer's expression of population change in terms of birth and death rates and gross flows of migration; the economist's "equation of exchange" ($MV = PQ$) in which total spending equals total value of goods purchased, by definition. Communities can make alternative sets of decisions that are internally consistent only within the bounds of this "iron law of space." But, more often than not, they act as if they can make independent decisions on all three variables, apparently innocent of the hard fact that, after the first two values have been set, the third has been fully determined and is beyond control. Because it is most uncommon for a new and growing suburb to make simultaneous and mutually consistent policy on target population, density, and land area, it is instructive to reflect on a few of the sequential paths of decision-making on physical form that are apparent on even casual observation.

A most common approach to community policy on physical form is to begin with that which is closest to the heart of the individual householder, especially if he is a homeowner: overall density in terms of number of houses per acre—average lot size. As much as any other single feature, the suburban movement was an escape from the crowding of the older, central city. Ever more prosperous and larger families, with at least one automobile, set the highest priority on the size of their backyard for sets of swing and gardening, sideyards set back from each other for privacy, and front yards set back for openness and status. Having determined the number of residences per acre, with variations ranging from one to the acre to five or six, depending on income levels and land prices, the community then turns to the lesser matters of boundaries and total population. Placing highest priority on the spacing between houses might appropriately, and with due seriousness, be designated the Territorial Imperative Model of Suburban Form.

A second, well-worn path to suburban design begins with the "natural" boundaries of the community: mountain slopes unsuited for building; expressways perceived, not always correctly, as divisive objects; and county lines. The Urban Topographic Model of Suburban Form has its own internal logic. One of the most complex and bedeviling features of the politically fragmented large metropolitan area is that of spatial spillovers—the lawyer's "neighborhood effects." These occur where one community's convenient incinerations become another community's smoke and smells; where one's shortcut to work becomes another's traffic problem. If "natural boundaries" do reduce spillovers—"good fences make good neighbors"—an early and binding decision on the land form would seem to be in the total social interest. Many suburbs have probably been driven to larger size than they would have preferred by their desire to reduce some adverse boundary effect—the risk of "bad planning" over the line.

A third and less-traveled path begins with the desired total population of the community, a criterion of the good life that may well rank higher in the future than it has in the past. More and more urban residents are seeking "human scale" in political and governmental bodies as well as in social groups. The small suburb has been defended in particular and in general ("political fragmentation") as a device for accommodating not only those who prefer small-scale association but also as a means of creating variety of life style—the "product differentiation" so appropriate to such a diverse, sophisticated and affluent market as that for human settlement (Ostrom et al., 1961).

Small population suburbs do perhaps have the greatest potential for social progress under large-scale urbanization and perhaps the greatest potential for ultimate social disorganization, too. But who among us knows how to use small subdivisions of the large and growing urban area as a constructive social device? The "road to hell is paved with good intentions"; besides, many of the intentions that lie behind the drive for small suburbs are, of course, not all that good. Because communities can be too small to achieve some reasonable degree of efficiency in the provision of public services and some reasonable degree of *im*personality in human relations, as well as too large to be responsive to human needs, placing the highest priority on total population is a variation on the theme of the endless search for the "Optimum Size 'City.' " What is the optimum size of a suburb?

HOW DENSE CAN WE BE?

One of the more puzzling questions about suburban physical form is the relationship between residential density and the cost of public services. The quest for the ideal house lot size consumes as much time at city council meetings and occupies their planning staff as often as almost any other matter. We may well conclude eventually that the total cost curve is flat—that is, the cost of providing water, sewers, streets, refuse removal, and fire and police protection is not much different per household across the full range from acre lots to apartments. The technical details of this matter lie well beyond the limits of this study, but some general impression of the nature of the question is worth forming.

To the economist used to well-behaved "U"-shaped cost curves, the individual components of this "total cost curve" present a bewildering mixture of odd shapes. Fire protection is apparently one of the more normal cost functions: it is expensive (per household) to protect very low-density areas because of the long runs from the fire station (and the private cost of damage, reflected in insurance premiums, is also greater because of the time delays in getting there). Thus, we have the beginnings of a decreasing cost function. Fire protection costs may turn upward at a very high residential density because of the need for special equipment for high-rise apartments, which are now blooming in the suburbs, and because of the risk of the fire spreading (more a private cost). But then U-shaped curves are familiar and comforting. If we pause here to add in a number of "utilities"—water, gas, and electricity—we have the core of the decreasing cost industries that provide the basis for the criticism of urban sprawl on efficiency grounds.

In clear contrast, sewage disposal generates a most unconventional cost curve and is as much responsible as any other service for creating our uncertainty about the effect of density on costs. The distinctive feature of sewage disposal is that intermediate densities of four or

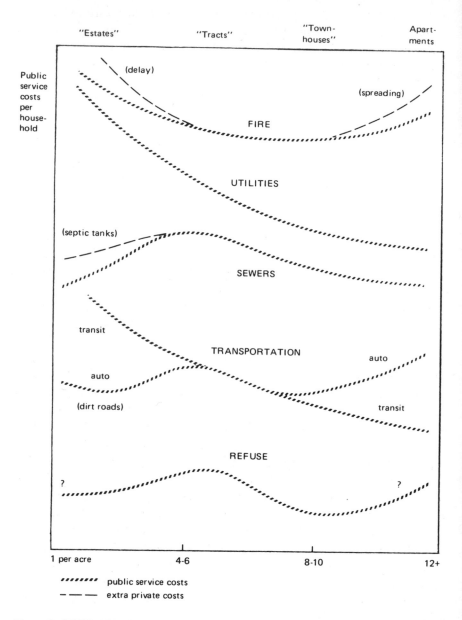

Figure 1: SOME GENERAL IMPRESSIONS OF THE NATURE OF PUBLIC SERVICE COSTS AS A FUNCTION OF RESIDENTIAL DENSITY

five to the acre—the typical detached, single family home—are probably the most expensive to service; that is, the cost curve is inverted. This follows from the fact that eight- or ten-to-the-acre "townhouses" and higher-density apartments require shorter runs of pipe per household, and, at the other end of the spectrum, one-acre lots usually can and do depend on septic tanks (soils permitting). In addition, storm sewers tend to be superfluous in one-to-the-acre subdivisions; roofs and paved surfaces account for a very small proportion of the ground area, and so absorption is high. Sewer service cannot, moreover, be lightly regarded in the cost accounting, because it is one of the most expensive demands of local government, especially in the early days, when mounting density forces the conversion from septic tanks to *underutilized* public sewers, and remembering that amortization periods on the bonds used to finance the sewer investment are shorter than the life of the sewers.

The provision of public access in the form of roads, streets, and sidewalks behaves a little like sewers, in that the changing mix of public and private costs complicates the cost function, but the uncertainty here is even greater. Reducing density from eight to four to the acre almost certainly increases access cost (per household) but further decreases to one-acre lots may not increase the cost of providing access because it is common to see a shift from paved to graded dirt roads and to go from sidewalks on both sides of the street to only one side or none at all. But five-acre lots would probably turn the cost curve upward. At the other extreme, higher densities increase automobile congestion, raising traffic control costs and causing time delays that add to many other costs, both public and private. But high densities offer the option of improved public transportation and cheaper access. That is to say, the urban transportation cost curve splits at medium high density, with the automobile access cost rising and the public transit access cost falling.

Refuse removal behaves a little like sewage disposal at very low densities: public service costs fall because households on acre-plus lots are more inclined to provide some or all of their own disposal, through burning or burying. High densities may offer some economies in collection because the households are closer together, but then the spillover effects of proximity—odors and unsightly accumulations—may make for more frequent pickups. All of which may add up to a (slightly) rising refuse removal cost per household with very high density, and then again it may not.

Perhaps the only thing that is clear from this tortuous discussion is that one should be very clear about which set of costs he is trying to minimize: private, public, or the sum of the two. Logically, the sum of private and public costs is most relevant to consumer welfare: fire department services plus private insurance premiums; city sewer and private septic tank costs synthesized. But suburban planning staffs, city councils, and homeowner-taxpayer associations are much more likely to center their attention on pure public service costs. This, too, could be logical and useful if the purpose of their discussions (analyses may be too strong a word) were to devise an appropriate set of user charges, or to move toward a true benefit theory property tax, that would exact a quid pro quo from each residential class. Set water, sewer, refuse collection, fire protection, and paving (or grading) charges as close to the full cost of providing those services as our knowledge permits and then caveat emptor. Or go one step farther and help new house buyers calculate the costs, both public and private, of different lot sizes (life styles). We could move away from the intuitive and implicit decisions that are reflected in our zoning regulations to the more explicit use of prices and *responsible* consumer choice.

Unfortunately, these partial equilibrium (engineering-type) cost analyses do not provide us with enough information to put together a balanced set of user charges, and tax rates, and tolls that will ensure the use of land in the public interest. It is not enough to calculate how much more paving or pipe will be required to cross a larger lot. Suppose large numbers of households should choose to live on full-acre lots or in high-density high-rise apartments—what would be the impact on public service costs? If anyone's lot is larger, everyone has to drive a little farther. If lot size doubles, on the average, the land area doubles and residences are, on the average, about one and one-half times as far from the pumping station or the fire station. No one has even begun to come to grips with the "externalities" of residential density decisions. On reflection, one gains at least a first impression that both very low and very high densities create heavy spillover costs that are not counted in the usual accounting. If so, there is a bias against intermediate densities—small lot sprawl—one that too easily reinforces our esthetic prejudices.

It is difficult to estimate the added costs borne by others already locked in place, such as central-city residents, when a new suburb zones for very low density, although it is clear that open space will be much farther away, and the next round of factories will also require much longer trips to work. It is more difficult still to estimate the added cost of travelling into the central city and getting water piped out from it for the *potential future* inhabitants of the next ring of suburbs that will be built *if* the metropolitan area continues to grow in numbers.

If a household building a new house on the edge of the urban area chooses an extra-large lot, pushing the built-up area well beyond the point it would have otherwise reached, that household spills out extra costs of overcoming space on those to come. Housing decisions are biased toward too big lots by too low prices (user charges and taxes) to sprawl. If, however, we were to extrapolate recent growth trends and levy charges on acreage that reflect expected future (spillover) costs and forced a reduction in residential lot size, and if that projected growth did not occur, we would be guilty of forcing *false economy* in the use of land. And if we were just to wait and see and then raise the land tax and user charges, *ex post,* if the region grows more than expected, serious inequities and misallocation of land results if the lot size decision is not reversible—if one cannot subdivide the land. (It is not inconceivable that we could overestimate the physical extension of our cities, especially if we continue to approach zero population growth.)

DO EXPRESSWAYS DIVIDE OR UNITE?

The cliche of "pluralism" in transportation planning must frequently confront a dilemma in the location of the route. The mode of public transportation that is most often the most appropriate for the low-density suburb is buses on main arterials, preferably limited-access highways (expressways), enjoying the use of "exclusive lanes" whenever possible. But that expressway may not be well placed to serve some future high-speed bus system because, typically, the communities in that vicinity must face and decide route location well ahead of a perceived public transportation crisis.

Generally, the proposed expressway is seen as a divisive force, and each community tries to push the route into a neighboring community, but not too far away for easy access. And the splitting of a political subdivision can create public service inefficiences as well as social problems. Infrequent bridges over the expressway does lengthen the routes traveled by city police cars, fire engines and sanitation trucks, and rush hour bottlenecks may increase delays and raise costs even more.[1]

But if nearby political subdivisions are alert, countervailing forces come into play, and the result of the many forces, after long delay, is usually an expressway route that wanders along the boundary lines of the many political subdivisions. If the route must be a little longer and the price of the land acquired a little higher, the political costs of the alternative are much higher. Apparently, citizen-motorists agree that boundary line expressways not only (1) leave communities intact and in touch, but (2) separate communities from each other in a way that helps build their individual identities.

The extra dimension that complicates this apparently simple matter is the new vision of expressways as the future roadbed of public transit. The Federal Department of Transportation and various metropolitan-area transportation authorities have been duly impressed by the great good fortune of Chicago that an early decision was made to leave an extra-wide median strip in the Kennedy Expressway, thereby keeping open the option of rapid transit at some later time. The problem is that public transit is better placed so that it runs through the middle of communities rather than around their edges. Transit stops tend to draw commercial facilities and office complexes that become town centers around which a sense of community can develop.

The flexible use of expressways seems destined to become an important public issue because we stand to make very great financial gains if we can develop a pluralistic transportation system over these very expensive facilities. One might indeed argue that, far from lagging in the construction of a rapid transit network, we have already completed a large part of that system and we need now "only" to master a political strategy for sharing the use of that huge investment in the total social interest. Recent experiments in the northern Virginia suburbs of Washington, D.C., with exclusive lanes for buses during rush hours, may mark the beginning of a new age in urban transportation.

The dilemma is clear: expressways designed for the exclusive use of private vehicles—automobiles and trucks—are divisive and probably do serve best as boundaries, but expressways designed to be used increasingly for public transportation probably, on net, unify and enhance the communities that they "split." But a flexible land-use strategy that will accommodate both the automobile and public transit dualism of the next stage in low-density movement, while preserving and even building a sense of community, is not nearly so evident.

TAKING STOCK OF THE FUTURE

WHEN SMALLER IS OLDER AND BIGGER IS YOUNGER

A flat statement is made, with only slight risk, that there is nothing worthy of note on the preferred rate of growth of a suburb—or the set of rates that apply under various conditions— despite the clear fact that rapid growth imposes severe strains on a community, especially its public sector. There is a small literature on the impact of the growth rate of the metropolitan area as a whole on the welfare of the full population. Deductively, a good case can be made for an overall local growth rate roughly equal to the natural rate of increase or the national average rate—fast enough to maintain full employment in the local economy but not so fast as to produce shortages and congestion in private and especially public goods and services. But a natural (= national average) rate of population increase of 13-14% per decade (the 1960-1970 U.S. change) becomes leveraged upward to 100% or more in the suburbs where the vacant land is concentrated. Suburbs may indeed triple or quadruple in a single decade (e.g., Scottsdale in the Phoenix metropolitan area and Southfield near Detroit). Are there no criteria at all with which to judge such astronomical rates of increase in population?

Because the suburb typically begins life as a bedroom community, its principal purpose is to supply new houses at the rate demanded by the expansion of the metropolitan area of which it is a specialized component. Thus it tends to be built up very rapidly and reaches early adolescence with almost all nearly new houses and almost all upper- and upper-middle-income households. Starting at the top, there is no place to go but downhill, albeit very slowly. But the perception of being an affluent community persists well beyond the fact; to the residents, time stands still.

The arithmetic of growth and status is inexorable: a community must grow ever faster to remain new and young, and, in general, it must remain new to remain affluent. Imagine a suburb with one-half its housing stock less than ten years old—say, 2,000 out of 4,000 units in the initial period (1950). This community must build 4,000 new houses between 1950 and 1960 to continue to have one-half its housing less than ten years old in 1960—that is, this place must double in size. Now with a stock of 8,000 houses aging, the community must again double the preceding construction rate to

8,000 new units between 1960 and 1970 to remain one-half new. Each decade, the community must double its size and double its rate of residential construction to retain its original age distribution of housing. Suburbs do grow this fast but not for long; land annexation is not all that easy or open-ended.

The moral of this simple story is that there is a hard trade-off between the local rate of growth, community size, and the age distribution of its housing. A community can slow its rate of aging by choosing to grow very fast and become very large—a mixed bag to most suburbanites. To elect slow growth and to try to stabilize population size is to elect to age faster than otherwise. (Chronological aging can occur with more or less physical deterioration, as Marlene Dietrich and Maurice Chevalier have taught us.) An individual can, of course, continue to live indefinitely in a constant-age community by moving regularly. And many (most?) do. But places are bound by the arithmetic of growth and capital depreciation—stocks and flows.

AND THE RICH WILL GET POORER, AND THE
POOR WILL GET RICHER

When only 10 to 20% of the metropolitan area population live in the suburbs, it is easily possible for them to have only the upper- and upper-middle-income households. But when their share mounts to over one-half and soon two-thirds to three-quarters of the populace, they must of necessity embrace almost all income classes and regress toward the median income. The principal instrument that will guide the suburbs toward a more normal distribution of income is their housing stock. Given the large amount of postwar construction in the inner suburbs, there will be a massive stock of 40- to 55-year-old housing there in the year 2000, and the occupants of these houses will probably fall into the middle- and low-middle-income classes. To this, we can add the smaller stock of prewar housing, probably then inhabited by low-income families. Surely it would be a relatively easy exercise to monitor the movement of the housing stock along that projected path to validate the hypothesis, correct the estimates, and prepare for the eventuality (Thompson, 1965: 130-132).

The central city would seem also to be headed toward a more normal distribution of income, carried along by the current of capital depreciation and normal replacement. There is both good reason and strong evidence to expect central cities to recycle, all in good time. Many of them have gone through two decades of sharp depopulation, for some a near prerequisite to repopulation. (The decline in population in the city of Detroit was 9.7% from 1950-1960 and 9.5% from 1960-1970; St. Louis was 12.5% and 17.0%, respectively; Boston, 13.0% and 8.1%.) The emptying out of the central city serves to speed the recycling of the inner city in at least two ways. First, it opens up large reaches of vacant land to accommodate the larger-scale building characteristic of today; New-Towns-in-Town will, for example, require very large tracts and therefore very substantial depopulation. Second, the emptying out of the central city acts also to pull the inner-city ghetto farther away from the frontier of rebuilding; urban renewal is fragile and vulnerable to the burden of massive poverty. This is, of course, not to say that low-income housing should not be included in the redevelopment, in proportionate amount.

This discussion of the trend in central-city income is not discursive. If the metropolitan area as a whole tends toward a normal distribution of income, as is characteristic of any large economy, then the case for the suburban economy evolving toward normality can be made either directly or indirectly. One can either demonstrate the ramifications of an aging suburban housing stock or the likely recycling of the central city. If the income of either of these major components of the total area approaches a normal distribution, the other must also. The argument here is that the "second coming" of the metropolitan area will produce similar "levels" of income in the inner circle and outer ring.

A slight refinement needs to be made in that projection for the sake of consistency with arguments above. The expectation here is that there will be an intermediate period in the second wave of city-building during which the suburban and central-city income "levels" will converge but out of substantially different distributions of income, neither of which will be normal. Following the arguments of the first section, above, the suburbs would have a preponderance of middle-income families (a leptokurtic distribution) and the central city more of the rich and poor (a bimodal distribution), perhaps on the way toward normal distributions and perhaps for as far ahead as we can see. But median family incomes should converge.

WHEN THE SUBURBS COME FILTERING DOWN

There is a large and largely unnoticed gap in the literature on a subtle relationship between the filtering down of the housing stock and recent (postwar) trends in the residential preferences of the upper-middle-income class. Rising incomes and a much sharper increase in automobile ownership led to the substantial reduction in residential densities that has drawn everyone's comment, while foreshadowing a future transportation problem that has gone virtually unheralded.

The new housing of the 1910-1920 period was readily passed down to the next lower-income class in the twenties, and the new housing of that latter decade also filtered down well, first to middle-income families in the thirties and forties and then to the better-off, lower-income families during the past two decades. Successive generations of housing were alike in that they were all built on city lots not much larger in size and therefore almost as accessible to public transportation as those that went before. But will the housing of the fifties and sixties filter down as well, first to the low-middle and then to the low-income households in the seventies, eighties, and nineties?

Certainly, our low-density postwar suburbs will be much more difficult to service with public transportation. As inadequate as central-city public transportation systems are to the needs of the day, perhaps as many as one-half the suburbanites of the country have virtually no means of movement except the automobile. This housing requires at least one automobile to survive and two or more to live a little. Suburban households in each of the three counties of the Detroit metropolitan area had reached the point six years ago where ninety percent of them had at least one automobile and half of them had two or more.

To date, heavy dependence on the automobile has not been regarded as all that great a personal hardship and is seen even less as a social concern by those who live in the suburbs and who have, for the most part, left the poor and elderly behind. They may ignore or devote themselves to the transportation needs of their children until the latter can drive the second car. But the young parents who moved to the suburbs in 1950 are now in their fifties and will in another two decades be either less mobile or growing public safety hazards as tremulous drivers. And the poor will arrive in the early postwar (nearest-in) suburbs at about the same time. The opening of the suburbs to blacks will tend to highlight the virtual absence of public transportation; in more than one-half the nonwhite families, the wife also works. What are the prospects for extending public transportation into the suburbs?

THICK SUBSIDIES FOR THIN TRANSIT SYSTEMS

For the metropolitan area as a whole, public transit ridership has declined from about three-fifths to one-fifth the total market for movement. What are our options? Headways between vehicles could be increased threefold from, say, five-minute to fifteen-minute waits. Or two-thirds of the transit routes could be abandoned, lengthening the walk to a transit stop from one to two blocks to four or five blocks. The size of the bus can be reduced by two-thirds ("jitneys"), but not the bus driver's wage. (Unless semi-professional drivers can be used, such as retired workers, housewives, or other part-time laborers.) Because frequent departures from all places to all other places are a highly indivisible service which cannot be contracted without serious deterioration in quality, those who remain dependent on public transportation as others shift to automobiles become not just relatively disadvantaged, but also absolutely worse off in movement.

There is perhaps a way out of the public transportation dilemma even for the low-density suburb, and one that does not require either large and unending subsidies for almost empty buses or a mass reconversion of automobile users into bus riders, neither of which seems very realistic at this time. With the appropriate public investment and land planning, we might offer the suburban dweller this choice: choose freely where you would like to live if you are prepared to provide your own transportation (automobile or taxi), or be free to choose good public transit by accepting limitations on where you can live.

The heart of the public transportation problem is that we have not posed squarely the hard trade-offs; we should not offer that which we cannot deliver. We have, in the past, led people to believe that they could live wherever they wished and still expect good public transportation at a low price, whether they used it or not. One is not free to choose to live in, say, northeastern Nevada and demand gas heat and public sewer service at typical prices, or to live in Manhattan and expect cheap parking or a garden patch. And when most people did choose to live on large lots and drive automobiles, we could not make good on our implicit promise to provide short-walk, short-wait, reasonably rapid public transportation to all corners of the urban area.

Do the majority who have turned away from public transit in favor of the private automobile have a moral obligation to help those remaining with transit? If motorists feel no debt to those who *choose* not to drive, what about those who have no option but transit—the poor, the elderly, the young, and some of the handicapped? One would be hard put to argue that the motorist-majority have an open-ended obligation to saturate the land surface of the whole urban area with public transit vehicles on five-minute headways over every mile of road. A reasonable compromise might be for motorists to accept financial responsibility for their fair share of the developmental period losses of a high-quality "thin system" of public transit, and for the users of that system to accept personal responsibility for relocating their residence, if need be, to sites within easy reach of that skeletal system.

Clearly, any household choosing to move in close to the officially designated, high-quality transit route would commit itself to a substantial private investment in money, effort, and social ties. Such an act would call for a high degree of confidence in the public commitment to maintain a high quality of service over that facility over a time period long enough to allow the household to amortize this heavy investment in the new location. A new fixed-rail facility—subway, elevated, or monorail—or even the less expensive conversion of an existing railroad, such as the new Lindenwohl line from downtown Philadelphia into the suburbs of southern New Jersey, constitutes such a commitment. A large fixed investment in public transit comes close to being its own performance bond. But this mode requires higher densities than are typical.

In contrast, a decision to run even the most attractive new buses on exclusive lanes of freeways may be competitive with rail in an immediate sense, but the hard fact that the new service could be quickly and easily discontinued would impart an aura of impermanence to that action that no words of public assurance could quite dispel. Perhaps some "hostage" other than a heavy fixed investment in immobile capital could be given, such as a performance bond that covers the relocation expense of households that moved to be near the transportation facility and would have to move again were it to be discontinued. Compensating payments, however, are cumbersome to administer.

Because land near the transit stops would probably rise in price, perhaps very sharply, careful provision would have to be made by the responsible government to ensure that there would be low-income housing near the new transit line. The affluent professional or executive who prefers transit to the automobile—the man who would rather read than drive—will, of course, be able to command housing near the new line by bidding land away from the elderly, handicapped, and other low-income transit users for whom the facility was primarily intended. The assumption is implicit that the subsidies required to house low-income persons on expensive land near transit stops would be less than the subsidies required to extend a comparable public transportation system to them here and there on scattered sites.

It is therefore disheartening today to see small clusters of townhouses and single apartment buildings being constructed on scattered sites throughout the suburbs, with no thought to their eventual connection to public transportation—this, when we know that most of them are being bought by persons in or near retirement who have (or should have) very near horizons as automobile drivers. Besides, we know that, as we live longer as individuals, we tend to grow older as a nation, especially if the current decline in our birth rate continues. Again, we have not, apparently, made any notable *net* progress in cleaning up the internal combustion engine of the private automobile at feasible cost—one kind of emission goes down and another goes up, and we do not know which to worry about most. Nor are we dispersing throughout the country, on net. Our land planning is obsolete due to clustering, aging, and traveling more. The one shortcoming that seems to be as true of the suburbs we are building today as it was of those of the early postwar period is that we still have not come to grips with the complexities and subtleties of movement in and access to the region in the social interest. Our suburbs still seem, in many ways, to be one of the most static parts of our dynamic society, and they seem destined for premature (technological) obsolescence as we continue into the age of ecology —natural and human. Not that we will be able to afford to discard them; our current suburbs will almost certainly be with us, in substantially their present form, for half a century or more to come, whether they fit the occasion or not.

NOTE

1. This follows only to the extent that each of the many local political subdivisions produces its own public services internally. If large outside suppliers are used (e.g., county refuse collection, central-city protection), the delivery of the service is no more hobbled by mid-community transportation arteries than by boundary-line ones, assuming that the outside supplier is serving neighboring ones, too. An areawide supplier will have to plan around the bottlenecks wherever they occur.

REFERENCES

BIRCH, D. L. (1970) The Economic Future of City and Suburb. New York: Committee for Economic Development.

CLAWSON, M. (1971) Suburban Land Conversion in the United States. Baltimore: Johns Hopkins Press.

GAFFNEY, M. (1964) "Containment policies for urban sprawl," in Approaches to the Study of Urbanization. Lawrence: University of Kansas Press.

KAIN, J. (1970) "The distribution and movement of jobs and industry," pp. 1-43 in J. Wilson (ed.) The Metropolitan Enigma. Garden City, N.Y.: Doubleday Anchor.

OSTROM, V., C. M. TIEBOUT, and R. WARREN (1961) "The organization of government in metropolitan areas." Amer. Pol. Sci. Rev. (December).

THOMPSON, W. R. (1965) A Preface to Urban Economics. Baltimore: Johns Hopkins Press.

16

Decentralization of Commerce and Industry: The Restructuring of Metropolitan America

BRIAN J.L. BERRY
YEHOSHUA S. COHEN

CONVENTIONAL WISDOM AND CLASSICAL MODELS

□ THE CONVENTIONAL WISDOM about the nature of urbanization processes that has been widely accepted throughout the social sciences as a guide for understanding the logic of metropolitan growth and change is, today, in need of substantial rethinking, reevaluation, and reshaping. This is the conclusion of this paper.

The classic idea was expressed no more straightforwardly than by Hope Tisdale when, in 1942, she wrote that

> urbanization is a process of population concentration. It proceeds in two ways: the multiplication of the points of concentration and the increasing in size of the individual concentrations. . . . Just as long as cities grow in size or multiply in number, urbanization is taking place. . . . Urbanization is a process of becoming. It implies a movement . . . from a state of less concentration to a state of more concentration.

From this widely accepted view, Louis Wirth derived a whole theory of the human consequences of urbanization, using as primary causal variables the size, density, and heterogeneity of the large city. And, consistent with the conventional wisdom, geographers and land economists in the interwar years produced a classic image of the North American city.

Much of this image involved models of commercial and industrial location within the city. Every city was thought to require at its heart a strong, viable, growing central business district, with adjacent central industrial and commercial zones. At strategic locations within the main sectors of the city, and in a hierarchy of successively smaller communities and neighborhoods, there ought, the model said, to be a hierarchy of business districts providing for the shopping needs and service requirements of the population. Major traffic arteries were seen to support ribbons of highway-oriented businesses, heavy commercial uses, and specialized functional areas such as the "automobile row." Heavy industry was found next to major transportation facilities—ports, and railroad lines and spurs. Light industry was seen as gravitating to the suburbs.

While a variety of factors were said to be operating in creating this spatial structure (the locational advantages of ports, strategic positions on transportation arteries, availability of fuels, and raw materials all figure prominently in the writings of the period) the principal force was said to be that of urbanization economies.

Urbanization economies involve externalities: the consequences of the close association of many kinds of activity in cities, producing mutual scale advantages that lower the costs and improve the competitive position of individual firms. Such cost reductions were demonstrated, in a variety of studies, to arise from several sources. The argument is as follows:

(1) *Transport costs.* Large cities have superior transport facilities offering significantly lower transport costs to regional and national markets. A city located at a focal point on transport networks is especially suited for easy assembly of raw materials and for ready distribution of products. There are thus market advantages for speedy and cheap distribution which may be accentuated by the advantages arising out of the size of the local consuming market. In the biggest cities, this local market will account for a very substantial share of service and residentiary activities and of the regionalized activities that need to minimize transport costs, since the local population may form a large

part of the total national market. Greater London, for example has one-fifth of the British consumer market, and the New York metropolitan area has been called "one-tenth of a nation." In both these cases, these markets account for a significantly larger proportion of the higher quality and fashion demands of the respective nations than this. Thus it may be that, because of its market dominance, the larger city is the ideal place in which to locate a plant even if it is serving a national market, because substantial demands are local and the balance of the output is more easily distributed. A New York location, for example, will obviate the necessity of providing extra facilities in New York, such as showrooms and warehouses, that would otherwise be needed if the firm were located outside New York.

(2) Labor costs. The large-city labor market is extensive, diverse, and dynamic. In this circumstance, the labor demands of single firms are only a small part of the total demands for labor. Thus, recruitment is easier. This becomes especially important where firms have seasonally varying labor needs. The big-city labor market also offers a whole range of skills; consequently, the advantages of the city over the small town are obvious, even though the wage rates of the large city may be somewhat higher than those of the small town. Facilities for labor that are readily available in the large city may often also have to be provided in the small town, which raises the real labor costs to the firm.

(3) Advantages of scale. The larger the city, the higher the scale of services it can supply. Thus, fire and police, gas, electricity, and water, waste disposal, education, housing, and roads are all in general better in the larger city than in the small town. Firms developed in otherwise unindustrialized areas generally have to spend much more upon social capital than those that have been developed in large cities. These kinds of scale advantages came into play particularly in the industrialization of the largest metropolitan areas in the period following World War I, although interrupted by World War II. Urbanization economies are those that, in particular, have a magnetic effect upon those kinds of industries in which coal was replaced as motive power by electricity, and in which rail and water were replaced as a means of transportation by trucking over the roads. This new industry is "light" industry, concerned with the manufacturing consumers' goods. These goods are often branded and serve sheltered markets because of imperfect competition deliberately created by advertising mechanisms. For these industries, transport costs are a low proportion of total costs, because the materials used tend to be semi-processed and because the products are compact and have high value in relation to weight. Labor requirements are often unskilled or semi-skilled, and the firms

therefore are more likely to be attracted by large cities, especially by the markets and the marketing facilities of these large cities. Thus, there is a "snowballing" which tends to be accumulative, for, as light industry is attracted to large cities, the magnetic power of large cities for this kind of industry is increased, creating conditions favorable to further growth. This kind of an effect has been called "circular and cumulative causation."

But if these kinds of beneficial externalities were seen to be the forces drawing economic activities into cities, negative externalities were invoked as a means of explaining the spatial distribution of these activities within the city. Traffic congestion, parking problems, and difficulties of loading and unloading were all seen as militating against the most crowded parts of the urban region. Increased competition for labor, combined with the higher costs of urban living (especially rent and transportation) were seen as factors driving up money wages and impairing the competitive capabilities of activities with high wage bills. And, most significantly, the high land values resulting from intense competition for centrally located space in the growing city were seen to be instrumental in "sorting out" activities in the urban area, both within and beyond the central business district.

Industries remaining in or close to the central business district (CBD) were observed to be those which find a great advantage in the accessibility of the CBD, the point of convergence for all city and regional transport routes, offering unusual convenience for buyers and for the assembly of workers. But because industry in central business districts was seen to face severe competition for sites, a hierarchy of land usage was said to result. The "highest" uses in the hierarchy were seen to be those which derive the greatest advantage from centrality and which could therefore pay most by turning the advantage into profit and rents. Housing was low on the list, except for high-class apartment houses. Most types of industry were also seen to be driven out because, for them, the advantages of centrality are relatively small, and the costs of production are lower on the periphery of the city. Many shopping functions likewise were seen to move out to lesser farmers districts central to local community and neighborhood markets.

Therefore, as the process of land use competition took place, the higher uses occupied the central area, and the only economic activities that remained localized within the boundaries of the central business districts were of the following kinds:

(1) The *financial district* of banks, insurance companies, lending institutions, brokers, and the like, relying upon speedy personal contact within a closely intertwined set of activities and relationships.

(2) *Specialized retail functions,* including department stores that require large population support at the point of convergence of population, firms that offer high-quality and rare goods which satisfy their need for great population support only at the point of maximum convergence of the population of the urban region.

(3) *Social and professional functions,* including the *headquarters and main office function.*

These activities tended to squeeze out housing and industry from the most accessible points. The manufacturing and service industry surviving in the inner city thus had very special features and tended to be found on the edges of the central area in the back streets because land values there were lower, in the classical view. For this kind of industry, central location was imperative; there was no alternative. The industry had to be in the inner city because of its marketing needs, which could not be separated from its manufacturing functions. A simple showroom in the central area would not do. Also, many central manufacturing industries were observed not to have moved because the inner areas supported relatively immobile skilled labor forces. Some also were said to be localized there because of the extreme subdivision of processes within the cluster of interdependent firms. Such firms use ground very extensively; their space requirements in relation to their level of output are very low. In this sense, they can use relatively obsolete multistory buildings in a variety of kinds of adapted space. This kind of space is, of course, available at the lowest rents until it is cleared and rebuilt for some other, more intensive, higher-paying kind of use. Industry in central areas thus tends to move from building to building, being displaced as those buildings are cleared for reuse and revenue. The scale of operation is generally small. Two good examples are printing and clothing.

For job printing in particular, the marketing factor is important, since rush jobs must be completed in a very short space of time. Firms are generally located on the margins of the professional and financial districts. Newspapers also tend to be produced in central locations, usually in very large, modern buildings near to the zone of highest land values. Here, centrality is related to the receipt of news. There has to be close contact among the editorial staff, printers, and the reporters who gather the news, and speedy distribution to the market is, of course, also imperative. On the other hand, periodicals, books, and standardized and regular kinds of job printing are not found in cities, but tend to decentralize into the suburban areas and rural and small-town locations.

Central-area clothing and related industries involve those branches where fashion and style are important—for example, ladies' dresses, gowns, furs, handbags, and fashion jewelry. Styles involves rapid changes in design, which means that output for stock is impossible; large varieties and small quantities are essential. Thus, small scale of activity is appropriate. There is no division of processes or line methods of production, and only a few persons work on a single garment. The work is highly skilled. Thus, in central London there are more than 2,000 firms making women's clothing. This kind of industry has no special space requirements. It can use converted obsolescent buildings, as the space required is small, but, on the other hand, skilled labor is essential. The trade is thus highly localized, with interdependence arising out of the provision of accessories by specialized and separate firms—embroidery, belts, buttons, and so on. Reliance on common specialists arises out of the fashion nature of the trade. The crucial point binding the industry to central locations is marketing, the essentiality of immediate displays to buyers. The industry has to be close to a place where buyers can cluster.

Manufacturing and selling are close together because the scale of firms is so small that they cannot afford to run separate premises and because selling and manufacturing are concerns of separate firms. Subcontracting is common, and people to whom the work is subcontracted need to be located close together. On the other hand, other branches of the clothing industry have quite different spatial distributions. Standardized products made by line methods, where there is subdivision of processes and where there are no requirements for a skilled labor force have decentralized.

The printing and clothing industries highlight the economies of urbanization and localization that held industry to the central areas of cities, surrounding the central business district. In the classic models, as the CBD grew and land values escalated, other activities would decentralize in a regular sorting-out process. Land values were considered the most important centrifugal force because they imposed increased costs upon industry. Moreover, in inner-city areas they resulted in cramped and restricted sites, in multistory buildings, in a shortage of storage space, in parking difficulties, in deficiency of light and air, and environmental pollution. Therefore, the pressure of space pushed out industry seeking space, light, and air to the periphery of the big city, to smaller towns, or to nonurban areas.

DECENTRALIZATION: A CONTINUOUS PROCESS

Steady decentralization has been observed in North American cities throughout the twentieth century, apparently attesting to the validity of the classical models. See, for example, the evidence in Figure 1, prepared using a table appearing in the recent work of urban economist Edwin S. Mills (1972: 99). In this diagram, the average exponential density gradient of population and several types of employment for a sample of cities is plotted against time on semi-logarithmic graph paper, to show whether decentralization was continuous, or evidenced marked changes over the year. For each item plotted (except manufacturing during the Great Depression), continuous decentralization is seen to have been taking place as cities have grown and transportation has improved.

Findings such as these lead some commentators—for example, Banfield (1968)—to argue that all that is happening to cities, even today, is very logical, a straightforward extension of past trends to which the classic models still apply:

> Much of what has happened—as well as of what is happening—in the typical city or metropolitan area can be understood in terms of three imperatives. The first is demographic: if the population of a city increases, the city must expand in one direction or another—up, down, or from the center outward. The second is technological: if it is feasible to transport large numbers of people outward (by train, bus, and automobile) but not upward or downward (by elevator), the city must expand outward. The third is economic: if the distribution of wealth and income is such that some can afford new housing and the time and money to commute considerable distances to work while others cannot, the expanding periphery of the city must be occupied by the first group (the "well-off") while the older, inner parts of the city, where most of the jobs are, must be occupied by the second group (the "not well-off"). The word "imperatives" is

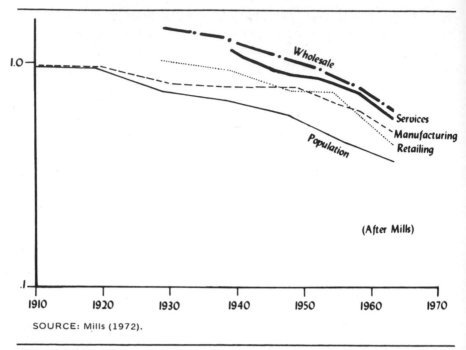

SOURCE: Mills (1972).

Figure 1: DECENTRALIZATION TRENDS SINCE 1910

used to emphasize the inexorable, constraining character of the three factors that together comprise the logic of metropolitan growth.

One indeed might conclude from inspection of Figure 1 that Banfield's argument is true—that since the last burst of inventions producing the upward and downward growth of the city (1880, the skyscraper; 1886, the subway; 1889, the elevator; Eberhard, 1966), an outward urge has prevailed, with the last major contributing change the effects of the onset of the automobile after 1920.

But also notice that there were increases in the rate of decentralization for both manufacturing and retailing in the years following World War II. It is these cases which are the focus of much current discussion, indicating to some that radical restructuring of metropolitan forms may be taking place, and it is to this issue that we now turn. The point we want to discuss was raised perceptively by the historian Oscar Handlin (and Burchard, 1963), who argued that

Some decades ago . . . a significant change appeared. The immediate local causes seemed to be the two wars, the depression, the new shifts in technology and population. However, these may be but manifestations of some larger turning point in the history of the society of which the modern city is part. The differences between city and country have been attenuated almost to the vanishing point. The movement of people, goods and messages has become so rapid and has extended over such a long period as to create a new situation. To put it bluntly, the urbanization of the whole society may be in the process of destroying the distinctive role of the modern city.

Are we indeed seeing such a radical restructuring of society, of which the changes in industrial and retail location are simply symptomatic?

SOME STATISTICS OF CENTRAL-CITY DECLINE

Certainly, many statistics appear to support this view. Between 1947 and 1958, the central cities of the New York region lost 6.0% of their manufacturing jobs, while the suburbs gained 37.2%. In other central cities of the old industrial heartland, comparable figures were: Chicago, −18.5 and +49.4%; Philadelphia, −10.4 and +16.4; Detroit, −42.9 and +41.5; Boston, −15.3 and +33.5; Pittsburgh, −25.3 and +18.1; St. Louis, −21.1 and +41.7; Cleveland, −22.4 and +98.4. In the period 1958-1963, the central cities of SMASs exceeding 250,000 population lost 338,000 manufacturing jobs, while 433,000 jobs were added to the suburbs. Between 1958 and 1967, net manufacturing employment changes for selected cities are reported in Table 1. Most of the older central cities in the nation's "heartland" continued to experience substantial decline. Where there was growth from sources other than annexation, it was in the "rimland" regions of California, Florida, and Texas, where the greater city areas kept much of the "action" within the city limits.

As one observer remarked, it is hard to keep a good record of those companies which have left or are planning to leave. A partial list of those who have moved or plan to move all or part of their facilities out of New York to the surrounding suburbs gives an indication of the dimensions of the problem:

TABLE 1

PERCENTAGE INCREASE IN MANUFACTURING EMPLOYMENT
1958-1967 (in thousands) SELECTED METROPOLITAN AREAS

Area	Central City	Suburbs
Washington, D.C.	8.5	141.8
Baltimore, Md.	−5.9	22.0
Boston, Mass.	−11.8	17.0
Newark, N.J.	−10.3	12.2
Paterson-Clifton-Passaic, N.J.	−1.3	33.8
Buffalo, N.Y.	−1.9	3.4
New York, N.Y.	−10.3	36.0
Rochester, N.Y.	18.3	55.9
Philadelphia, Pa.-N.J.	−11.6	30.0
Pittsburgh, Pa.	−13.8	3.7
Providence, R.I.	4.5	12.4
Chicago, Ill.	−4.0	51.6
Indianapolis, Ind.	23.1	36.3
Detroit, Mich.	−1.8	47.6
Minneapolis-St. Paul, Minn.	8.9	146.2
Kansas City, Mo.-Kans.	0	68.3
St. Louis, Mo.-Ill.	−14.9	41.4
Cincinnati, Ohio-Ky.-Ind.	10.6	2.7
Cleveland, Ohio	−5.3	42.6
Columbus, Ohio	18.1	0
Dayton, Ohio	12.9	99.5
Milwaukee, Wisconsin	−6.3	55.6
Miami, Fla.	5.7	116.0
Tampa-St. Petersburg, Fla.	3.8	280.0
Atlanta, Ga.	8.9	86.0
Louisville, Ky.-Ind.	16.7	48.4
New Orleans, La.	12.3	29.2
Dallas, Tex.	41.9	131.0
Houston, Tex.	41.6	12.6
San Antonio, Tex.	28.5	62.5
Los Angeles-Long Beach, Cal.	10.3	23.0
San Diego, Cal.	−24.5	27.5
San Francisco-Oakland, Cal.	−23.5	29.3
San Bernardino-Riverside-Ontario, Cal.	57.3	60.0
Denver, Colo.	6.9	112.7
Portland, Ore.-Wash.	16.5	68.0
Seattle-Everett, Wash.	−25.7	244.7

To Connecticut: American Can Co.; Lone Star Cement Corp.; Avco Corp.; Bangor Punta Corp.; Howmet Corp.; U.S. Tobacco Corp.; Olin Corp. (chemicals division); Hooker Chemical; Chesebrough-Pond's, Inc.; Technicolor, Inc.; Christian Dior Perfumes Corp.; General Telephone and Electronics Corp.; Consolidated Oil; and Stauffer Chemical.

To Westchester County: IBM, Inc.; Pepsico, Inc.; Dictaphone Corp.; General Foods Corp.; Flintkote; and AMF.

To northern New Jersey: CPC International; Union Camp; and the
American Division of BASF. Even the Fantus Company, the relocation
firm that helped plan many of these moves, has taken up new offices in
Englewood Cliffs, New Jersey (population 5,810)—along with CPC;
Thomas Lipton; Scholastic Magazine; and Volkswagen.

Elsewhere, the story is the same. In 1970, St. Louis lost 43
companies to the suburbs. In two recent years, Boston lost about 75.
In Cleveland, recent losses include the frozen foods division of
Stouffer Foods; National Screw and Manufacturing; National Copper
and Smelting; Fisher-Fazio-Costa (a major food chain); and the
headquarters of B. F. Goodrich Chemical Co. Likewise in Detroit,
from which S. S. Kresge, the retail chain, plans to move its corporate
offices, along with the Michigan Automobile Club; Delta and Pan
American Airlines; and R. L. Polk, publishers; not to mention Circus
World, the toy manufacturers. Things are so glum that former mayor
Jerome Cavanaugh sometimes refers to "Detroit's sister cities
Nagasaki and Pompeii." In Los Angeles, the financial district is being
deserted by many of the major banks, beginning with Crocker-
Citizens National in 1968 (Cassidy, 1972).

On the retail side, the data reveal trends of the same magnitude
(see Tables 2 and 3). Deflated retail sales declined in many central
cities between 1958 and 1967, and there were distinct patterns by
city-size class. Central business districts in particular lost their retail
function, especially in the smaller metropolitan areas.

The story can be repeated for many personal services. In Chicago,
for example, in 1950, 62% of all the metropolitan area's medical
specialists had their offices in the Loop, another 23% were to be
found elsewhere in the central city, and only 15% were in the
suburbs. By 1970, the Loop's share had dropped to 25%, the city
had another 25%, and fully 50% were to be found in the suburbs.
Altogether, Chicago lost 400 firms and 70,000 jobs between 1950
and 1965 in the inner third of the city. This exodus continues with
little abatement in the dingy, crime-ridden, and decaying West Side
neighborhoods. Forty-five factories employing 3,875 persons left
Chicago for the suburbs in 1969. In 1970, 23 firms employing 1,455
persons. On the West Side, loss of 7,512 industrial jobs meant that
210 retail businesses folded, 4,550 persons employed in nonmanufacturing businesses lost their jobs, and retail sales there declined
$23.1 million (Young, 1972).

TABLE 2

PERCENTAGE INCREASE IN RETAIL SALES DEFLATED BY GENERAL PRICE INCREASE 1958-1967, FOR SELECTED METROPOLITAN AREAS

Area	Central City	Suburbs
Washington, D.C.	10.5	134.8
Baltimore, Md.	4.9	128.2
Boston, Mass.	−1.4	79.2
Newark, N.J.	−14.1	37.1
Paterson-Clifton-Passaic, N.J.	0.9	74.5
Buffalo, N.Y.	−9.9	54.7
New York, N.Y.	9.7	60.2
Rochester, N.Y.	18.1	91.3
Philadelphia, Pa.-N.J.	6.2	65.4
Pittsburgh, Pa.	7.8	28.7
Providence, R.I.	−36.3	73.1
Chicago, Ill.	5.3	86.6
Indianapolis, Ind.	20.0	160.8
Detroit, Mich.	0.7	86.4
Minneapolis-St. Paul, Minn.	7.9	149.7
Kansas City, Mo.-Kans.	55.2	64.3
St. Louis, Mo.-Ill.	−7.6	76.2
Cincinnati, Ohio-Ky.-Ind.	4.6	129.4
Cleveland, Ohio	−15.2	269.1
Columbus, Ohio	22.8	141.9
Dayton, Ohio	3.6	125.5
Milwaukee, Wis.	7.5	108.3
Miami, Fla.	−2.5	98.2
Tampa-St. Petersburg, Fla.	30.9	108.9
Atlanta, Ga.	37.7	153.9
Louisville, Ky.-Ind.	14.0	101.8
New Orleans, La.	21.0	141.9
Dallas, Tex.	36.6	119.2
Houston, Tex.	55.9	63.3
San Antonio, Tex.	36.4	79.9
Los Angeles-Long Beach, Calif.	22.2	75.4
San Diego, Calif.	25.6	91.8
San Francisco-Oakland, Calif.	16.3	81.6
Denver, Colo.	11.1	132.4
Portland, Ore.-Wash.	28.1	180.3
Seattle, Wash.	18.0	152.5

Table 4 provides yet another insight into the changes. It shows that a very large share of new private nonresidential construction in the period 1954-1967 was taking place in the suburbs—for example, over 80% of new industrial building in Boston, or 98% of all new community facilities in Dayton, Ohio. And if the statistics are adjusted to control for annexation, the story is apparently even worse. For example, the mean annual changes in employment between 1948

TABLE 3
RETAIL CHANGE, BY CITY SIZE CLASS 1954-1967

	Population of Metropolitan Area (in thousands)			
	3,000+	**1,000-3,000**	**500-1,000**	**250-500**
(a) Percentage Change in Sales: Current Dollars				
CBD	12.1	8.3	−3.7	−6.7
Central city	34.3	26.8	58.4	61.4
Suburbs	132.2	175.0	209.0	193.1
(b) Percentage Change in Numbers of Establishments				
CBD	−26.0	−26.9	−38.2	−37.6
Central city	−26.3	−23.7	−8.4	−8.4
Suburbs	29.9	30.3	51.3	48.0

and 1958 in the central cities of metropolitan areas were 2.74 for manufacturing, not controlling for annexation, and −0.6 with controls. For wholesaling, the figures were 1.92 and 0.7 and, for retailing, 1.00 and −0.4. The statistics suggest a field day for the prophets of doom.

PROPHECIES OF DISASTER

To the reader of the daily press and of the news magazines, loss of jobs in manufacturing and the decline of the retail function of the central business district presages disaster. Consider, for example, the following random clippings from stories appearing in 1972:

> the loss of the Giants was an almost comic sidelight to a more serious problem: the exodus of dozens of businesses, including many of the "Fortune 500," from New York to the surrounding suburbs or other cities. And since conditions in Gotham often forecast what may happen elsewhere, America's older cities are finding that they, too, are loosing [sic] some of their long-time corporate occupants to the burgeoning suburbs. The growing industrialization of the overwhelmingly white suburbs, coupled with the entrapment of minorities in the central cities, paints a picture with the word CRISIS splashed across it in boldface. . . . If business [and suburban] communities continue in their present mindlessly selfish way we will Los Angelesize our land, Balkanize our region's finances, and South Africanize our economy [Cassidy, 1972].

TABLE 4

PERCENTAGES OF PRIVATE NONRESIDENTIAL CONSTRUCTION IN THE SUBURBS, 1960-1965 AND 1954-1965

Percentage of valuation of permits authorized for new nonresidential building in—

	Atlanta	Boston	Chicago	Cleveland	Dayton	Detroit	Indianapolis	Los Angeles	New Orleans	New York	Philadelphia	St. Louis	San Francisco
1960-1965													
All types	47	64	65	56	62	69	41	59	42	38	65	41	60
Business	44	68	64	60	66	69	49	60	49	39	70	39	63
Industrial	71	81	77	61	56	70	52	85	58	61	75	67	84
Stores and other mercantile buildings	44	74	67	74	78	80	55	63	66	64	75	75	72
Office buildings	25	52	58	38	53	55	21	41	10	21	52	32	38
Gasoline and service stations	63	91	54	57	98	58	54	60	60	51	66	55	72
Community	50	61	64	44	49	71	33	61	37	31	60	37	58
Educational	59	63	64	51	28	68	24	72	35	29	67	67	57
Hospital and institutional	59	38	56	15	56	61	14	69	44	25	38	35	52
Religious	69	92	73	84	56	81	56		35	55	77	86	62
Amusement	31	59	80	60	99	86	58	35	41	19	59	85	74
1954-1965													
All types	43	68	63	58	a	71	44	62	a	44	67	a	63
Business	41	70	61	59	a	73	50	63	a	44	69	a	64
Industrial	66	82	73	60	a	75	61	86	a	75	76	a	84
Stores and other mercantile buildings	40	74	67	73	a	77	52	66	a	71	72	a	72
Office buildings	21	51	39	37	a	58	21	41	a	18	51	a	37
Gasoline and service stations	60	82	59	62	a	65	56	62	a	65	73	a	73
Community	48	67	66	44	a	70	40	63	a	38	68	a	64
Educational	57	72	69	61	a	79	46	59	a	34	72	a	53
Hospital and institutional	32	41	58	33	a	62	10	70	a	32	43	a	53
Religious	59	86	68	81	a	74	59	70	a	61	80	a	65
Amusement	30	64	75	57	a	43	52	50	a	33	72	a	55

SOURCE: Newman (1967).
a. Not available.

The second great migration to the suburbs—the exodus of families and offices—has profoundly altered the economic life of cities, just as the great migration of middle-class families altered the social patterns. Four out of five new jobs created in major metropolitan areas in the past ten years have been outside the cities. This disproportion has been a major cause of the "urban crisis," damaging the ability of cities to pay their way and to provide employment for the people who live in them [Gooding, 1972].

While concern has mounted in the last decade over the plight of the inner cities in vast metropolitan areas, the aging downtowns of hundreds of smaller cities have been quietly deteriorating with little national notice . . . losing businesses and office-users just as surely as their metropolitan cousins. . . . "In the old days, the farmers all came to town on Saturday night," . . . a long-time resident recalled sadly. "Not any more, it's just dead." And a teenage girl driving what is locally known as "the Circuit" said blithely: "This town is for nothing but old fogeys. It's a bummer" [Kneeland, 1972].

REASONS FOR THE EXODUS

The press is also full of speculations about the reasons for the exodus of manufacturing and retailing, although there is far from any consistency in journalistic perceptions. Some see the shift from the pull of low land values to the push of urban problems:

Business is moving out for a variety of reasons. The explanation immediately after World War II was the attraction of cheap land in the suburbs, permitting single-story factories that were convenient for truck loading. In recent years, the motive is more push than pull. Executives complain about the abominable phone service in many cities, horrendous commuting conditions, rapidly rising crime. Even bomb threats are mentioned, when GT&E actually had a bomb go off in the building, the bosses lost no time in making the final decision to get the hell out. Then there are problems with the work force. Many young women seem to be avoiding the big cities, while young execs no longer consider a move to the New York office a promotion; indeed, they demand differential pay to cover the increased cost of living. There is also the desire to get away from it all, which was one of the big reasons why Xerox moved its top men from Rochester to pastoral Connecticut: the company president felt that they would get a better perspective of the whole company from the new, more isolated locale. But the biggest appeal of the suburbs, of course, is that much of the population, housing and development is there, or headed there [Cassidy, 1972].

Others see the lure of new kinds of prestige locations, closer to the preferred places of residence of the businessmen and decision makers:

Around the perimeter of O'Hare International Airport are any number of businesses that have O'Hare stuck in their name. The O'Hare area, which had no office space 10 years ago, now accounts for nearly 30 per cent of all suburban office space in the Chicago area. "Using O'Hare in the company name has the psychological aspect of giving the firm prestige," says Loren Trimble, an expert on the area's economic development. "O'Hare carries the connotation of being near the world's busiest airport, of being modern and in the jet age." Proximity to the airport explains part of the phenomenal growth of the northwest suburbs in the last decade. O'Hare has had a very direct effect on the building of fancy hotels, convention facilities and office buildings in the area.

But it is only one of many reasons for the rapid industrial and commercial growth and, most experts agree, not the main one. Most maintain that things would be booming in the area without the world's busiest jetport.

"I say that the principal reason we have 500 new companies in Elk Grove Village is that it is close to where the boss lives," said Marshall Bennett, a partner in Bennett & Kahnweiler, industrial real estate brokers. His firm developed Centex Industrial Park in Elk Grove Village, the largest of its kind in the nation. The northwest suburbs are now Chicago's biggest competitor for the dollar of the industrialist, conventioner and nightclubber. The area is one of the four fastest growing areas in the country. Why this once rural economy grew so rapidly to an industrial and commercial giant is laid mostly to transportation. First came the people. They started the trek to the suburbs in the early 1950's. Later, the bosses and decision-makers in industry started bringing their factories closer to home. Old and dreary saw-toother-roof factories were abandoned in favor of horizontal layouts in spacious suburban buildings. They were mostly light industries that produced little noise, smoke, odor or other irritants to suburban residents. Many were warehouses. However, employes in these factories and warehouses have not been as keen as their bosses at being around O'Hare. Many of the workers, particularly the blue collar ones, still live in Chicago, while many of the northwest suburbanites still work in the city. This situation has led to considerable reverse commuting. For example, about 75 per cent of the 25,000 employes at Centex Industrial Park in Elk Grove commute from Chicago, while the majority of Elk Grove Village residents work in Chicago [Young, 1972].

Yet others look at identical situations and see only polarization centering on race:

A series of downtown events have highlighted the problem of the Chicago downtown area and raised fears that the ghostlike character of downtowns in most major U.S. cities will come to pass in Chicago as well. The abandoned appearance of downtowns throughout the country are a product of two factors: race and market and the interplay between them.

The society that is in the majority judges an area's decline or renaissance in this country (whether it is a downtown area or a residential community) by the behavior of white residents and consumers—i.e., by an area's attraction and appeal to white people. This is attributable in part to our social and economic stratification and most particularly to the income distribution in our population.

Therefore, since whites possess the wealth, the range of choices available to them, and the ways in which they act on such choices, tend to predestine an area's future. When whites decide to leave an area, a self-destructive process is perceived in white eyes to have been set in motion. This process feeds on itself and results in an inevitable reduction in the money necessary to support commercial facilities, pay for housing and building maintenance and make possible new construction. This in turn further accelerates the exodus of white populations and increases black populations, vacancies and abandonment [Meltzer, 1972].

SOME ALTERNATIVE STATISTICS

What actually is happening and why? Before an answer can be provided, it must be said that all is not told by the statistics in the previous section. Indeed, there have been some dramatic changes of a directly contrary kind, and the counterpoint they play to the loss of manufacturing jobs and retail sales must be orchestrated into a

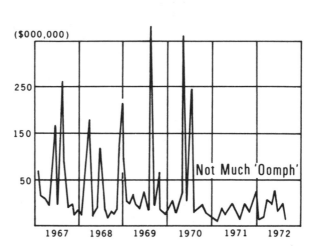

INDUSTRIAL DEVELOPMENT INVESTMENTS
Metropolitan Chicago

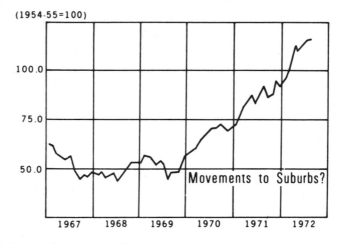

INDEX OF VACANT INDUSTRIAL BUILDINGS
City of Chicago

SOURCE: Chicago Association of Commerce and Industry.

Figure 2: RECENT INDUSTRIAL STATISTICS FOR THE CHICAGO REGION

balanced interpretation of contemporary urban dynamics. For example, who would acknowledge, reading the interpretations and data on Chicago already presented, that between 1953 and 1964 $750 millions were invested in construction and improvement of manufacturing plants in Chicago (Hartnett, 1971)?

The problem is that whenever a negative statistic is observed in the city today, someone hastens to cry wolf. Thus, when the usual annual summertime peak in new industrial investment commitments vanished in metropolitan Chicago in 1971 and 1972, and industrial vacancies in the central city started to mount, the Chicago Association of Commerce and Industry produced illustrations asserting that flight to the suburbs was the cause, as Figure 2 reveals. But is this a valid interpretation of the statistics? All the numbers so far have represented *net* change. This may be the source of major errors in interpretation, masking the fact that net change, plus or minus, is made up of many elements.

Consider the data on the sources of employment change in various parts of New York in the years 1967-1969 reported in Table 5 (Leone, 1972). Net change in each area is seen to represent the resolution of interacting birth, death, immigration, emigration, and local growth processes. Clearly, given such complexity, any simple explanation is likely to be not only suspect, but almost undoubtedly faulty. Although Manhattan's CBD lost 10,625 jobs due to move-outs and 17,890 jobs due to businesses closing, 24,602 new jobs were created in new businesses and local nonmovers expanded their employment by 38,804! Much of this was associated with the office boom, to which we refer later. The decline in jobs in Brooklyn was due to an excess of deaths over births, because that area experienced net immigration. In Queens, too, there was net immigration, and the decline in that case was due to local nonmovers reducing their employment by 10,702.

The dynamic was different in each case. No wonder, then, the difficulty of many observers to square the gloom-and-doom perceptions with the most massive downtown office boom in the nation's history that has transformed central business districts of the nation's major metropoli. Gross floor space in private office buildings in the Manhattan CBD increased from 126 million square feet in 1936 to only 128 million in 1950. But by 1960, the total was 160 million, by 1963, 184 (with an additional 18.8 in public buildings), and by 1970, 226 (with 20.8 in public buildings). In the Dallas CBD, comparable fugures were 1936: 4.7 millions; 1950: 6.0; 1960: 15.6;

TABLE 5
EMPLOYMENT CHANGE IN NEW YORK, 1967-1969

	Manhattan CBD	Brooklyn	Queens	New York SMSA
Births	+24,602	+5,882	+3,713	+45,146
Deaths	−17,890	−8,083	−3,901	−40,574
Immigrants	+5,851	+5,777	+6,892	+9,018
Emigrants	−10,625	−4,960	−3,078	−5,579
Local relocators	+8,136	+1,098	+961	+14,184
Nonmovers	+38,804	−1,338	−10,702	+11,696
Net change	+48,878	−1,624	−6,115	+33,891

SOURCE: Leone (1972), using data from the Dun & Bradstreet Corporation's "Dun's Market Identifier" files, prepared for analysis at the National Bureau of Economic Research.

1963: 17.3; and 1970: 22.5 (Armstrong, 1972). In Chicago, 8.2 million square feet of office space was added to the Loop in the period 1967-1972. In the ten years 1960-1969, the valuation of new office space authorized for construction in the nation's major office centers was, in millions of dollars: New York SMSA, 1,659; Los Angeles SMSA, 1,220; Washington SMSA, 812; Chicago SMSA 710; San Francisco SMSA, 601; Boston SMSA, 393; Detroit SMSA, 327; Atlanta SMSA, 279; Philadelphia SMSA, 250; Cleveland SMSA, 172; Seattle SMSA, 169; Milwaukee SMSA, 101. This investment boosted private and public office space by 44% in these SMSAs, which, along with Pittsburgh and St. Louis, account for over 70% of the headquarters and headquarters employment of the nation's top 500 industrials, and over 60% of all of the nation's central administrative office employment (another 7% is in Minneapolis-St. Paul, Houston, Dallas, Cincinnati and Kansas City; Armstrong, 1972).

INTERPRETATION

These apparently contradictory trends, which evidently have defied the abilities of most popular commentators to comprehend their basic importance, are in fact a product of a revolutionary transformation of American metropolitan areas that renders conventional wisdom inoperable.

The initial development of road transport and highway improvements did indeed breed a continuous and gradual decentralization of people and jobs in the core-oriented metropolis through the first half of the twentieth century. But then came a series of technological changes that significantly modified the forces operating and produced the change in rates of decentralization observable in Figure 1.

On the manufacturing side, superhighways, large-scale trucks, and piggyback combination of road and rail transport reversed the pull of rail terminals and docks in central areas on industrial location. Previously, extremely efficient interregional transportation conspired with slow and costly local road transport to the industry to central terminal locations. Suddenly, this constraint was removed, and peripheral locations became as (or even more) accessible to the metropolitan region than was the traditional city center. With the removal of traditional constraints, the negative externalities of the central city loomed large in choice, both perceptually and in fact. The massive decentralization of industry that followed was dictated more in its locational choices by social factors and prestige locations than by traditional dollars and cents. Low-status locations, poor, polluted, and black neighborhoods were left and avoided by new concerns.

For retailing, the significant development was the invention of the planned shopping center, a dramatic and drastic technological change. Although the first planned centers were built in the 1920s, it was not until after 1950 that the rapid diffusion of the new technology transformed retail relationships. In the early years, developers *followed* the growing suburban market with a hierarchy of neighborhood, community, and regional center. The first centers were oriented to the more affluent sectors of the biggest cities and diffused from them into smaller places and lower-income suburbs. It is in the smaller cities that the change has been most devastating.

As one commentator (Kneeland, 1972) remarked, Portland, Maine

had a population of 75,000 in 1960, a figure that dropped to 63,000 in 1970 as residents left the aging seaport for the rapidly developing suburbs.

In 1958, the year before the first shopping center opened in the area, gross sales of downtown retail merchants were $140-million. Now there are 10 shopping centers in Greater Portland, including the recently opened Maine Mall, an enclosed center with about 50 shops and two huge department stores. Downtown retail sales have fallen to about $40-million annually.

Like a number of other cities, Portland seems to be somewhat resigned to downtown's loss of retail dominance to the shopping center.

In the 1960s in the bigger metropolitan areas, a fundamental change in shopping center developers' concepts became apparent. A recent article in *Fortune* magazine (Breckenfeld, 1972) captured some of the flavor of the change:

The shopping center has become the piazza of America. In big metropolitan areas and smaller cities alike . . . indoor piazzas are reshaping much of American life. Giant regional shopping centers have risen by the hundreds across the nation and are still going up by the score. To an amazing degree, they are seizing the role once held by the central business district, not only in retailing but as the social, cultural, and recreational focal point of the entire community . . . grim, formal unwelcoming old cities scarcely fit the more relaxed life style of the newly affluent middle class with considerable leisure . . . in many cities middle-class white shoppers are beginning to abandon downtown to the poor and blacks except from nine to five, Monday through Friday (office working hours).

Again, the suggestion of the social dynamic is there replacing much of the economic dynamic of conventional wisdom. This social dynamic involves not only the avoidance of the central city by the majority white residents, but also the creative, style-setting leadership envisaged by major shopping center developers who are busy building on the metropolitan periphery ahead of residential growth. By leading rather than following, the new developers seek to establish the direction of metropolitan growth, to set the style and tone for the suburbs to follow, and to capitalize on the new opportunities thereby created.

While part of the social dynamic involves this creative development of new opportunities on the expanding periphery, this is only possible because of the eagerness of most Americans to find a safe haven in the new forms of suburbia thereby created. As President Johnson's Commission on Crimes of Violence reported:

If present trends are not positively redirected by creative new action, we can expect further social fragmentation of the urban environment, formation of excessively parochial communities, ~~greater~~ segregation of different racial groups and economic classes ... and polarization of attitudes on a variety of issues. It is logical to expect the establishment of the 'defensive city,' the modern counterpart of the fortified medieval city, consisting of an economically declining central business district in the inner city protected by people shopping or working in buildings during daylight hours and 'sealed off' by police during nighttime hours. Highrise apartment buildings and residential 'compounds' will be fortified 'cells' for upper-, middle-, and high-income populations living at prime locations in the inner city. Suburban neighborhoods, geographically removed from the central city, will be 'safe areas,' protected mainly by racial and economic homogeneity and by distance from population groups with the highest propensities to commit crime. Many parts of central cities will witness frequent and widespread crime, perhaps out of police control.

What all of this means, for example, is that the black resident of the metropolis finds himself in a central-city ghetto abandoned by both whites and, increasingly, employment. The flight of white city dwellers into the expanding peripheries of metropolitan regions is an accelerating phenomenon as minorities move toward majority status in the city center. The exurban fringes of many of the nation's urban regions have now pushed one hundred miles and more from the traditional city centers. More important, the core orientation implied in use of the term "central city" and "central business district" is fast on the wane. No longer is it necessary to have a single, viable, growing heart. Today's urban systems appear to be multinodal, multiconnected social systems in action, in which the traditional centralization of the population into metropolitan areas has been counterbalanced by a multifaceted reverse thrust of decentralization. The situation is very different from the period at the end of the nineteenth century from which we derive the concept of urbanization. Decentralization and an outward urge have replaced centralization and core orientation; differentiation and segregation substitute for the integrative role of the melting pot.

The essence of the new urban system is its linkages and interactions, as changed by changing modes of communication. Both places of residence and places of work are responding to social rather than to traditional economic dynamics. At the same time, new communications media, notably television, have contributed to cognitive changes by providing the universal perception of decaying central cities, the new home of the former residents of the now-emptied periphery; the immediate on-the-spot experience of their riots; the careful documentation of their frustrations; and acute awareness of emerging separatist feelings. It is no accident that the suburbanization of white city dwellers has increased, supported by rising real incomes and increased leisure time. Similarly, decentralizing commerce and industry avoid the lower-status central-city location, and the new office complex seals itself in by defensive techniques. Gradients of distance-accretion are now beginning to replace those of core-centered distance-decay within the larger megalopolitan complexes as persons of greater wealth and leisure seek home and work among the more remote environments of hills, water, and forest, while most aspire to such settings as an ideal. In consequence, core-dominated concentration is on the wane; the multinode, multiconnection system is the rule, with the traditional multifunctional core simply a specialized one among many. It is the spontaneous creation of new communities, the flows that respond to new transportation arteries, the waves emanating from new industrial and retail growth centers, the mutually repulsive interactions of antagonistic social groups, the reverse commuting resulting as employment decentralizes, and a variety of other facets of social dynamics that today combine to constitute the new urban systems in America.

These trends and changes should, of course, be put back into the context of changing technology. Concentrated industrial metropoli *only* developed because proximity meant lower transportation and communication costs for those interdependent specialists who had to interact with each other frequently or intensively and could only do so on a face-to-face basis. But shortened distances also meant higher densities and costs of congestion, high rent, loss of privacy, and the like. As soon as technological change permitted, the metropolis was transformed to minimize these negative externalities. The decline of downtown retailing and of central-city manufacturing, alongside the office boom within the major metropolitan areas, are but manifestations of this fundamental transformation of American urbanism, for which a new definitive theory to match the conventional wisdom of the first half of the twentieth century has yet to be written.

REFERENCES

ARMSTRONG, R. B. (1972) The Office Industry. New York: Regional Plan Assn.

BANFIELD, E. C. (1968) The Unheavenly City. Boston: Little, Brown.

BERRY, B.J.L. (1970) "The geography of the United States in the year 2000." Transactions of Institute of British Geographers 51.

BRECKENFELD, G. (1972) "Downtown has fled to the suburbs." Fortune (October).

CASSIDY, R. (1972) "Moving to the suburbs." New Republic (January 22).

COHEN, Y. S. (1972) "Diffusion of an innovation in an urban system." University of Chicago Department of Geography Research Paper 140.

Commission on Population Growth and the American Future (1972) Population and the American Future. Washington, D.C.: Government Printing Office.

EBERHARD, J. (1966), "Technology for the city." International Sci. and Technology 57: 18-31.

GOODING, J. (1972) "Roadblocks ahead for the great corporate move-out." Fortune (June).

HANDLIN, O. and J. BURCHARD [eds.] (1963) The Historian and the City. Cambridge, Mass.: MIT Press.

HARTNETT, H. D. (1971) "A locational analysis of those manufactoring firms that have located and relocated within the city of Chicago, 1955-1968." Champaign-Urbana, Illinois.

KNEELAND, D. E. (1972) "Quiet decay erodes downtown areas of small cities." New York Times (February 8).

LEONE, R. A. (1972) "The role of data availability in intrametropolitan workplace location studies." Annals of Economic and Social Measurement 1: 171-182.

MELTZER, J. (1972) "Loop betrays deeper ills of city." Chicago Sun-Times Viewpoint (October 22).

MILLS, E. S. (1972) Urban Economics. Chicago: Scott, Foresman.

NEWMAN, D. K. (1967) "Decentralization of jobs." Monthly Labor Rev. 90: 7-13.

YOUNG, D. (1972) "Industry flees decaying city." Chicago Tribune (May 7).

17

Financial and Desegregation Reform in Suburbia

FREDERICK M. WIRT

☐ INTERACTION AMONG LEVELS OF AMERICAN GOVERN-MENT has provided the historical generator for the politics of federalism. As scholars have frequently noted and policymakers just as frequently repeated, no level of government is truly autonomous. However, there is much rhetoric about "local control" and "states' rights," and, varying with time and issue, much local political muscle to back the rhetoric. Equally true, however, with many policies we can also see the centralization of function and resource which drains that rhetoric of any but emotional meaning (for an important conceptualization of this process, see Shields, 1952). This article is a study in the contemporary forces of centralization working upon American suburbs in that most expensive, extensive, and vital policy—education—and of the research needs thereby engendered.

The central thrust of what follows is that, despite suburbanites' dream of finding more personal autonomy in the curving drives and sweeping lawns out past the city's rim, the dream is failing. Simply put, forces both external and internal to these communities are reducing these citizens' control over their schools. This pull of centralization may well be affecting other local policies as well, as some of the forces noted below are not special to school policy (see Reagan, 1972), but our policy focus here is more narrow.

THE CENTRALIZING FORCE OF
PROFESSIONALISM

CONTOURS OF THE NATIONAL FORCE

Schools began in America as extraordinarily private and local affairs. They were most often, and designedly, instruments of moral instruction, intimately allied with the church against evil. Early Massachusetts law providing for such protective indoctrination was popularly termed the "Old Satan Deluder Act." An early, popular textbook had a title which illustrates the connection and also illuminates the vapidity of contemporary titles: *Spiritual Milk for Babes Drawn from Both Breasts of the New and Old Testaments* (Meyer, 1967: ch. 2). From this simple, private, and moral orientation, it is many years and conflicts to the nonreligious, specialized, diversified, and bureaucratic system of today. However, what is important about that change for our analysis is the driving force of professionalism and its consequences for popular notions of "local control."

In the last part of the nineteenth century, there occurred what Cremin (1964) has termed "the transformation of the school" in American society. Across a wide range of American communities, a middle-class elite of businessmen and professionals imposed upon the schools new concepts of curriculum, personnel, and administration. Where schools were once highly decentralized,[1] they became centralized under a single board elected citywide and at-large, outside prevailing political party processes (Salisbury, 1967). Where once teachers and administrators were hired as simply another part of the patronage basket,[2] now appointment stemmed from qualifications met in university schools of education and not in the boss's saloon. Where once budgeting, procurement, and financing for schools were irradiated with graft and favoritism,[3] now these functions moved into the hands of professionally trained specialists motivated by universal norms of rationality, honesty, and efficiency. All this is what was meant by the effort to "get the schools out of politics," but the results were also to transfer power from one civic group to another, itself a political process. For example, in St. Louis for the period 1897-1927, the proportion of upper-class professionals on the centralized school board jumped from 4.8 to 58.3% and businessmen from 9.0 to 25.0%, but wage earners also fell from 28.6 to none (see Gersman, 1970; for a Chicago study, see Counts, 1927; for Philadelphia, see Issel, 1970).

THE CONSEQUENCE FOR LOCAL CONTROL

Note that the practices precipitating reform were primarily locally controlled; as with other policies, local control is no guarantee of purity. Rather, the reform movement was a state control syndrome. Reformers shifted the decisional arena from the community to the state legislature and froze into state statute and constitution their values and implementation programs. We have long known that occupational licensing statutes represent a political victory for one group contending with others over the allocation of resources. In this case, imposed upon local districts were state requirements for: the training and certification of teachers and administrators; approved curriculum and in some cases textbooks; technical standards of health and sanitation, safety, building construction, administrative reporting, and nutrition; the raising of taxes, and the selection and duties of the local school boards. A graphic indication of this "transformation" of local control is provided in Table 1, showing the frequency of state standards for numerous local school activities by the 1920s.

What local control has come to mean by the 1970s has never been specified empirically for all the states, but a regional variation is reported. In the Southeast, state control of local schools has been extensive and long-standing, partly a historical deposit from the Civil War's evisceration of local resources and functions. However, in the Northeast, decentralization has been an equally characteristic pattern, influenced by historical objections to royal centralization (Bailey et al., 1962). But even in these polar cases, countervailing tendencies operate—in the Southeast, local options do exist, and, in the Northeast, the state does mandate some practices. There is also variation in state money entering local districts, with its potential for influence. State contributions to local school costs averaged 40.8% nationally in 1970, but the states varied from Hawaii's 87.0 to New Hampshire's 8.5% (National Education Association, 1970).

But a curious omission in the scholarship of school politics is knowledge of the extent and intensity to which local control can be empirically verified for aggregates of districts and states. We cannot

TABLE 1
MINIMUM DISTRICT STANDARDS REQUIRED AMONG
THIRTY STATES, 1925

Item	Number of States	Item	Number of States
1. Site		Library—care in selection	15
Area of playgrounds	15	Globe—kind, size	18
Ornamentation	15	Musical instruments—pho-	
2. Buildings		nograph, organ, or piano	15
Exterior structure		6. Teacher	
Condition of repair	21	Academic and professional	
Interior structure		qualifications	28
Walls—finish, color scheme	17	7. Organization	
Floors—material, finish,		Average daily attendance	15
color scheme	20	Length of term	15
3. Lighting		Daily program—posted,	
Placement of windows	22	followed	16
Glass area	25	State course of study to be	
Shades	20	followed	16
4. Heating and Ventilation		8. Hygiene of the school plant	
Jacketed stove, standard		Seating of pupils	
heating and ventilating		With reference to size of	
plant, or furnace	20	seats	20
5. Equipment		With reference to light	20
Outdoor		Individual drinking cups or	
Flag	17	sanitary bubblers	15
Playground apparatus	20	Toilets	
Indoor		Number—two	16
Pupils' desks—material,		Type	20
kind	26	Condition	17
Blackboard—amount of		9. Community	
space	20	Frequency of community	
Dictionary—kind	15	meetings	18
Maps—kind, number	17	Community spirit	18

SOURCE: George J. Collins, "Constitutional and Legal Basis for State Action," in **Education in the States: Nationwide Development since 1900,** Edgar Fuller and Jim B. Pearson, eds., (Washington: National Education Assn., 1969), 14.

automatically assume that state control follows the state dollar down into the district. For it is arguable that the reverse may operate, as district agreement to take on new state programs can bring state funds upon compliance.[4] But there are no doubt complaints from local schoolmen that much of what the district does operates within constraints and requirements set by the state (for a recent review of these constraints on the school board, see Bendiner, 1969). Thus, roughly 80% of the local budget is frozen into personnel salaries, the schedules for which stem from state law and escalate increasingly under pressures from organized teachers (James et al., 1966). It is a curious kind of local control, therefore, in which options over spending 4 out of 5 dollars are not locally controlled.

Similarly, the definition of a teacher, principal, or superintendent basically rests in the dictates of state law. That definition no longer is a matter for local school officials to create, while their firing is increasingly proscribed. Even the effort to evaluate them in terms of their own professional objectives is bitterly fought by staff in both ghetto and suburb alike. It is a curious kind of local control which cannot control its personnel.

On matters of curriculum, there is more local influence, it is true, but even here outside constraints operate. What constitutes an acceptable curriculum has been mandated in most states, at least in minimum terms. Thus, the elements of a "college prep" track or "commercial" or "vocational" education have long been rooted in state requirements and implementing administrative codes. The quality of these courses may vary, of course, from those of the existing handful of rural, one-room schoolhouses to the rich curricula of affluent suburbs. But the bulk of the question—what should be taught?—is answered from outside the district. And even inside it, detailed aspects of the answer come usually from the professional whose "expertise" often overwhelms the laymen.[5] It is a curious kind of local control which cannot shape basically what will be taught, ostensibly the primary purpose of local control.

The preceding merely sketches a complex and detailed phenomenon which operated among all school districts in America for a century. At the end of it, the authoritative agent of local control—the school board—appears as an inept and incapacitated product. In the summary words of a recent Carnegie Corporation report on this agency:

> It must be plain from all that has gone before that in three major aspects, all vital to public education integration, teacher militancy, and finances, the American school board has reached a point where what was mere inadequacy has come close to total helplessness, where decline and fall are no longer easily distinguished [Bendiner, 1969: 165; more fully elaborated in Wirt, n.d.].

Professionalism, then, has become a powerful agent of centralization, a private development cast in governmental clothes. That is, like all "public" policy, this transformation of the school stemmed from private group needs and values made to impinge upon government in order to achieve an "authoritative allocation of values" (for an extended application of the Easton framework of analysis, see Wirt and Kirst, 1972). Note that school reformers did not escape the political quality of policy-making when they escaped party politics. Rather, their success transformed local schoolmen into agents sensitive and responsive to national professional norms. Those they transplanted had been equally responsive, but to different norms, those of gain or enhancement for ethnic or party groups competing with other local groups. For almost a century, then, the enemy of local control has not been the federal government but the professional in the bosom of the community.

PROFESSIONALISM AND SUBURBIA

In this development, suburbs shared even more. When the professional reform commenced, there were relatively fewer suburbanites than now, of course, In 1900, only one-third of Americans lived in a metropolis, whether center city or suburb; by 1970, that proportion lived in suburbia alone, and, by 1985, there will be more suburbanites than the entire U.S. population of 1920 (Wirt et al., 1972: ch. 2). That shift must be seen in company with another taking place—the drive toward school consolidation. Between the early 1930s and mid-1960s, the number of school districts in the country fell from about 126,000 to 21,000; by mid-1972, they were about 17,000.[6] What mostly disappeared here were the multitude of small rural districts, only some of which were suburban. The result has been that, in a very reduced set of districts, the proportion of suburban districts has risen along with the share of national population. In short, the kind of district in which professionals can inject their values and exercise their influence has become increasingly suburban.

But the influence of professionalism in reducing suburban autonomy has not been uniform. Status considerations mediate that influence, for the character of the population of the district seems to make a difference in the professional's influence. This has been demonstrated both in cross-sectional and longitudinal analyses of sets of communities. There is an extensive literature showing the sharing of values among schoolmen and middle- (particularly upper-middle-) class constituents, so that the latter accepted school policy directions more frequently; this could be measured by the success rate of school levies, which characteristically have been approved more frequently (and with higher percentages) the greater the status level of the community (most extensively in support of these findings, see Carter, 1963; Carter and Sutthoff, 1960). Minar (1966) has shown, among 48 Chicago suburbs, that the status dimension also differentiated these districts in local conflict over school policies and in acceptance of professional schoolmen's "conflict management" skills. That is, working-class suburbs were more likely to have more turnout on school referenda, more dissent on such votes, and other indications of severe questioning of professionals' decisions; the reverse qualities characterized school politics in more affluent suburbs.

Similar differentiation appears when longitudinal analysis is applied. Thus, the work of Iannaccone and others, using Southern California Districts, shows a developmental pattern in districts where the population either grows numerically or changes in its status characteristics (see Iannaccone and Lutz, 1970; the dissertations they cite; Walden 1967). When suburbs mushroom or alter, the old community values consonant with the school values, which the professionals administered, are sharply challenged by newcomers bringing in different values. The school boards of such suburbs suddenly become a forum for agitation for change; schoolmen resist such new demands; high-turnout elections create turnover in board membership; new superintendents are appointed, reflecting the new board's values, to replace administrators and faculty, until finally a new equilibrium is reached between community and professional values.

While all this may suggest the ultimate authority latent in any community to control its schools, that conclusion may be overdrawn. After all, what transpires in such conflict is the replacement of one set of professionals with another, all within a set of state-imposed constraints which limit community options. More bluntly, what may well often be the case is that new suburbanites transmit into the local school system demands for what seems new and desirable which have actually been received earlier from professionals themselves through popular media coverage. It looks like much the same interaction between a *Reader's Digest* article on a new wonder drug and the subsequent and prompt demand of patients upon their doctors for that drug.

Among other suburban policies, a similar process may be seen of new values and demands clashing with the old—new land uses which bring zoning ordinances, new expectations about service from public utilities which bring new municipal operations, new evaluations of acceptable tax loads which bring new or expanded levies (for a conceptual and empirical treatment of these suburban policy conflicts, see Wirt et al., 1972: part 4; Downes, 1968). With these policies, as with education, oldtimers may exclaim against the "newfangled" and "extravagant"; in terms of social theory they are protesting challenge to established regime norms and resource allocations. This mixture of the symbolic and the material seeking authoritative action constitutes the essential political quality of educational decision-making in suburbs, as in other sites.

In such a contest, the civil bureaucracy—of which professional schoolmen constitute a legion—increasingly forms part of a new combination of power in big-city politics (Salisbury, 1964). Their influence is even greater in the smaller suburb where the potential group base for community politics is more limited than in the core city. That influence is magnified even more by other salient political qualities of suburbia: the prevalent nonpartisanship of elected suburban government, which means the absence of parties as competitors with bureaucrats for the stakes of local power; the strong preference of suburbanites for governmental services administered on the basis of technical proficiency and efficiency rather than on the grounds of patronage and special favor; and, in general, the low level of political conflict in these communities. There may well be "episodic events" in such conflict, as Martin characterized suburban school politics a decade ago, but in the normal course of events one sees much more evident a characteristic "routines of politics" (Sharkansky, 1970). Professionals are trained to be "community relations" specialists, which upon inspection means manipulating the public to support professional decisions while generating an aura of citizen control. For most decisions in suburban school affairs, the authority may lie with the community, but the effective influence is operated by professionals, an arrangement which seems pretty much to satisfy most suburbanites, judging from the rarity of local widespread conflict over school issues (for an elaboration on this thesis of the "dry arroyos" of school-community relationships, see Wirt and Kirst, 1972: ch. 4).

In short, suburban school authorities are far removed in time and power from the early days of the public school, when boards of education set policy, examined teachers—and abruptly hired and fired them—stoked the wood-burning stove, scanned textbooks and teachers for signs of moral deviation, and negotiated for—and sometimes even erected—a new building. As with so many other institutions of contemporary American society, the ideology of another and purportedly more golden era hovers over a set of relationships whose reality actually contradicts the myth. What "free enterprise" is to the modern economy, "local control" is to the modern community school system. Both speak to values of decentralized and fragmented power, but the reality works the other way.

Professionalism and consolidation are private *external* forces that arose outside school districts from decisions about universal norms set in professional schools which became cloaked in public authority by state action. Another set of private forces, *internal* to the district, however, is stimulating other, external forces that in turn will centralize decision-making for school policy. In this category are those private fears which collectively oppose contemporary school policies. Specifically, these are the status and racial fears which create barriers to status and racial desegregation and the financial fears which increasingly create opposition to local school levies. The latter will concern us first, as more is known about it.

THE FEAR OF PEOPLE

The enormous suburban trek stemmed from both push and pull motivations. The attractions outward had to do with newly realizable ambitions to live in better surroundings, "better" meaning housing, schools, and public services. But this migration also resulted from the push of reaction against core-city conditions, one of which was the arrival of Negroes fleeing the South. Whether from push, pull, or both, by the opening of the 1970s core cities, particularly in the Northeast and North Central, were increasingly black and the suburbs white. Despite the movement of millions of blacks to the suburbs during the 1960s, nationally they constituted less than five percent of suburbanites. As Figure 1 shows, that proportion has been remarkably constant since 1900 in the North, although it represents a new low in the South (for further aspects of the black suburb, see Wirt et al., 1972: ch. 4).

This metropolitan color sorting rests partly upon economic grounds; higher property values, confronting a black group charac-

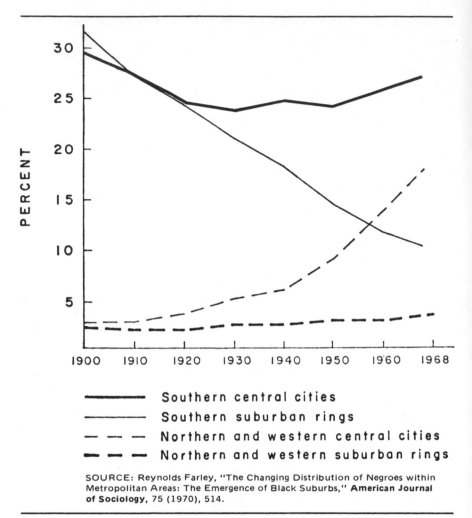

SOURCE: Reynolds Farley, "The Changing Distribution of Negroes within Metropolitan Areas: The Emergence of Black Suburbs," **American Journal of Sociology**, 75 (1970), 514.

Figure 1: CENTRAL CITY AND SUBURBAN BLACK POPULATION, 1900-1968

teristically possessing more limited resources than whites, would serve to keep the former out of suburbia even if whites were free of all racial prejudice. Curiously, there is the as yet unexplained social phenomenon that, until the 1970 census, suburban blacks had *lower* income, education, and occupations than core-city blacks; the comparison is reversed for whites in most metropolises (Wirt et al., 1972; U.S. Bureau of the Census, 1971). But another sorting mechanism in these suburbs has clearly been prejudice, whether rooted in status or racial antagonisms. There is a history of these, from the once-prevalent racial covenants to contemporary suburbs' refusal to permit public housing within its borders. And on occasions without number, violence and intimidation have accompanied white suburbanites' more formal resistance.

But we should not fall into the stereotype that the whites of suburbia are so different from the whites of core cities, on this or many other issues. Thus in Table 2 is seen Hensler's secondary analysis of a University of Michigan national survey sample response, with locale and education controlled. Her regression analysis tested the independent additive effects of individual attributes and of locale on racial attitudes. She discovered for the education query in Table 2 that status variables had little effect on those attitudes, and that, controlling for these individual attributes, suburbanites were only four percent more likely to oppose federal intervention in education than were white core-city dwellers (Wirt et al., 1972: 110-111).[7] After an extended analysis of this, as well as of special surveys in the Boston and Los Angeles areas, Hensler (Wirt et al., 1972: 129-130) concludes that

> the common view that people in city and suburb do diverge in their orientations toward community, polity, and current issues does receive some support. But suburban attitudes upon closer analysis

TABLE 2

PERCENTAGE OPPOSING EXPANSION OF THE RIGHTS OF BLACKS, WHITE METROPOLITAN POPULATION OUTSIDE THE SOUTH, 1968 BY LOCALE AND EDUCATION

Issue	Less than H.S. Grad		H.S. Grad		Some College +		Total	
	city	suburb	city	suburb	city	suburb	city	suburb
Federal gov't should *not* guarantee fair treatment in jobs to blacks	46.0	57.6	50.9	54.1	42.1	43.5	46.3	51.1
Federal gov't should *not* guarantee school integration	40.4	50.6	50.0	50.9	35.2	45.9	42.1	48.9
Whites have right to exclude blacks from neighborhood	18.5	23.0	10.8	18.1	10.5	17.4	13.1	19.2
Favor some degree of segregation	68.3	72.2	60.6	65.0	42.1	42.1	57.5	59.1
Total n	63	90	66	120	57	115	186	325

SOURCE: Frederick M. Wirt, Benjamin Walter, Francine Rabinovitz, and Deborah Hensler, **On the City's Rim: Politics and Policy in Suburbia** (Lexington, Mass.: Heath, 1972), 110.

TABLE 3

PER PUPIL AND PER CAPITA[a] EDUCATIONAL AID, 1962
(school district basis)

	$ Per Pupil		$ Per Capita	
	CC	OCC	CC	OCC
East				
Baltimore, Md.	106	162	20	32
Boston, Mass.	50	42	6	8
Newark, N.J.	50	66	16	12
Buffalo, N.Y.	194	288	25	60
New York, N.Y.	234	297	30	66
Rochester, N.Y.	186	305	24	67
Philadelphia, Pa.-N.J.	140	132	18	25
Pittsburgh, Pa.	93	184	11	34
Midwest				
Chicago, Ill.	111	103	15	20
Indianapolis, Ind.	97	126	21	28
Detroit, Mich.	137	163	24	39
Minneapolis, Minn.	133	218	19	47
St. Paul, Minn.	129	218	18	47
Kansas City, Mo.-Kans.	130	89	21	31
St. Louis, Mo.-Ill.	129	141	18	25
Omaha, Neb.-Iowa	33	40	5	11
Cincinnati, Ohio-Ky.-Ind.	51	145	8	23
Cleveland, Ohio	50	66	7	13
Columbus, Ohio	50	114	9	28
Toledo, Ohio	52	202	8	48
Milwaukee, Wis.	93	55	13	12
South				
Birmingham, Ala.	154	152	32	37
Atlanta, Ga.	102	152	21	49
Louisville, Ky.-Ind.	95	137	18	28
New Orleans, La.	194	200	29	39
Oklahoma City, Okla.	96	61	23	14
Memphis, Tenn.-Ark.	107	119	22	33
Dallas, Tex.	140	172	27	38
Houston, Tex.	143	201	32	52
Norfolk, Va.	101	117	17	28
West				
Long Beach, Calif.	114	316	36	59
Los Angeles, Calif.	172	316	51	59
San Diego, Calif.	195	282	37	74
San Francisco, Calif.	189	254	24	59
Denver, Colo.	74	122	14	35
Portland, Ore.-Wash.	123	217	21	54
Seattle, Wash.	234	289	42	79

SOURCE: Seymour Sacks, **City Schools/Suburban Schools: A History of Fiscal Conflict** (Syracuse: Syracuse University Press, 1972), 89.
a. Based on 1960 population.

are revealed as correlates primarily of race and social class and not characteristic of the suburbanite *qua* suburbanite. . . . Suburbanites' attitudes are different from those of their urban counterparts because of the kind of people they are, *not* because of where they live.

Whatever the origin and context of these suburban attitudes, it is clear that they provide an extensive emotional barrier to any desegregation of their heavily white schools. In the short run, they offer no optimism about their willingness to accommodate deseg- regation. Accordingly, those seeking such change have turned to external forces, specifically court action, to work the change. We will return to this possibility shortly.

THE FEAR OF TAXES

If fear of blacks is one internal private force shaping future educational policy in suburbia, fear of increased taxes is another. But in this case, these citizens not only do not oppose solutions from the outside; they may often earnestly seek them. The analysis of school finances is incredibly complex because of a myriad of state and local practices across the spectrum of American federalism. Yet, in that confusion, it is certain that suburbs have a disproportionate share of per capita state and national aid to education compared to core cities (for an extensive analysis of this urban phenomenon, see Sacks, 1972: esp. ch. 3, on the state aid statement; on the national aid evidence, see Berke et al., 1971: 57; for an important introduction to school finance policy in the nation, see Benson, 1968; for an extensive bibliography, see Sacks, 1972: 179-194). Sacks (1972: 174-175) summarizes a broad set of factors which might potentially measure the suburb's dominant position.[8]

> The relationship between income and expenditure indicates that for any given level of income the per-pupil expenditure in the suburbs would be larger than in the city, and, since the average income level is higher in the suburbs, this increases expenditures. Similarly, the effect of a dollar of state aid is higher in the suburbs than in the large cities.

The results by 1960 may be seen in Table 3, which shows not merely the interstate variation in state aid but also the prevalence of suburbs getting relatively greater aid. Of the 37 SMSAs featured there, only a fraction showed larger city than suburban aid—seven on per-pupil and only three on per capita measures. As Berke has shown, federal aid under the Elementary and Secondary Act of 1965 did not alter this suburban advantage. Table 4 shows his analysis for six major states in a special Ford study.

But again, let us not fall prey to the suburban stereotype. The reason state aid can have such an effect upon suburban school expenditures lies in the variety of economies and school systems which constitutes the spectrum of suburbia. The point of this variety is presented concisely in Table 5 with 1962 data. It is this very variety which accounts for the greater influence of state aid upon suburbs than upon core cities. Where, within a state resources are more homogeneous, as among the core cities, state aid will show less variance in its influence than where resources range widely, as among suburbs (Sacks and Ranney, 1966: 107). One of the influences of state aid has been the subsequent stimulation of local districts to take on greater tax burdens.

TABLE 4

FEDERAL AID AND TOTAL REVENUE BY CENTRAL CITY, OUTSIDE CENTRAL CITY, AND NONMETROPOLITAN AREAS, 1967

State	Federal Aid	Total Revenue	% Federal Aid
California			
central city	$39	$ 684	5.8
outside central city	40	817	4.8
nonmetro	54	641	8.4
New York			
central city	68	876	7.7
outside central city	31	1,037	3.0
nonmetro	31	923	3.4
Texas			
central city	38	479	7.9
outside central city	36	485	7.4
nonmetro	63	535	11.8
Michigan			
central city	29	683	4.2
outside central city	17	666	2.5
nonmetro	30	629	4.8
Massachusetts			
central city	69	675	10.2
outside central city	38	779	4.8
nonmetro	n.a.	n.a.	n.a.

SOURCE: Joel S. Berke, et al., **Federal Aid to Public Education: Who Benefits?** (Syracuse: Syracuse University Research Corporation, 1971), 57.

TABLE 5

**MEANS OF SOCIOECONOMIC AND FISCAL CHARACTERISTICS OF
SUBURBAN AND CENTRAL CITY SCHOOL SYSTEMS IN THIRTY-FIVE
STANDARD METROPOLITAN STATISTICAL AREAS[a] (in dollars)**

	Outside Central City Areas ("suburbs")[b]	Central Cities	Difference
Socioeconomic Independent Variables			
Per capita personal income	2,182	2,068	+114
Public school enrollment as a percentage of population	22.6%	17.2%	+5.2%[c]
Fiscal Independent Variables			
Per capita education state aid	37.66	20.73	16.87[c]
Per student education state aid	165.54	124.92	+40.62[c]
Per capita noneducation expenditures	126.44	161.70	−34.76[c]
Current educational expenditures as a percentage of all local current expenditures	55.0%	32.4%	+22.6%[c]
Dependent Variables			
Per student current educational expenditures	439.11	376.33	+63.78[c]
Per capita total educational expenditures	127.24	67.96	+62.78[c]
Local school tax proxy	83.80	47.23	+37.57[c]
Exhibit: U.S. Totals			
Median family income[d]	7,114	5,943	†1,141
Median family income	6,707	5,943	+764

SOURCE: Seymour Sacks and David C. Ranney, "Suburban Education: A Fiscal Analysis," **Urban Affairs Quarterly**, 1 (1966), 107.
a. Since two areas have two central cities there are actually thirty-seven observations.
b. Includes all communities outside the central city but within the metropolitan area.
c. The difference of means is significant at the .01 level of confidence.
d. Urban fringe is used here in place of outside central city.

Against this pattern of suburban benefit, it would be useful to know the degree to which suburbs joined the national opposition to the school property tax which began a decade after Sputnik had generated enormous financial effort. Unfortunately, full records which would enable us to distinguish the suburban reaction from that in core city or rural outland do not exist. No national source collects such records for local levies, and many states also fail to maintain central records.[9] But, in the course of research for this article, the impressions gathered from research directors in the National Education Association, American Association of School Administrators, and U.S. Office of Education are in agreement—there is no indirect evidence that suburbs have acted much differently from other locales.

Accordingly, it seems likely that suburbs shared in the national movement of tax referenda defeats, although, again, not all suburbs can be assumed to act in the same fashion. It is quite possible that, along the dimensions of suburban status and resource differences, tax defeat has been inversely related to those dimensions. As noted earlier, there is a history of voters with greater affluence and education supporting schools (including tax referenda) more strongly than those with lesser resources. The outcry in the late 1960s against the local property tax—chief support of local schools—should have been proportional to the incidence of its relative burden. Working- and middle-class suburbs contain citizens more likely feeling a relatively greater burden, more likely to challenge schoolmen's priorities for expenditure (to judge from Minar's work noted earlier), and more likely to be drawn from their usual nonparticipation into voting—and voting "no" (on the subjective dimension of preference accounting for intersuburban policy differences, see Williams et al., 1965; Sacks and Ranney, 1966; Wirt et al., 1972: chs. 9-12).

There is presently no literature of research on such propositions applied to suburbs, but Goettel's (1971) study of 261 school budget elections in New York in the spring of 1969 is an important suggestion on how such research may be designed. Given the variety of district status both in this sample and in suburbs, the findings for this sample may extend to suburbia also. The dependent variables here were voter participation and dissent, while the explanatory factors included status and wealth, as well as board decisions, whether controllable (e.g., percentage increases in school budget), or noncontrollable (e.g., percentage of school budget for debt service). Some of Goettel's findings are special to the politics of education in that state—particularly the upstate-New York City divisions—but others seem more widely applicable (Wirt, 1972). Thus, only about half the variation in participation and dissent could be accounted for by fiscal and economic factors or background variables (less than forty percent in the most sensitive analysis); ironically, decisions not controllable by the school board (especially the share of the budget to be raised from the local property tax) were "the most consistent stimuli of participation and dissent"; teacher salary increases, on the other hand, were no such stimuli; voter resistance in this issue seems unrelated to other local governments and their policies; and, as others have noted, there is a direct relationship between participation and dissent in these school conflicts. Because of the weakness of status and fiscal explanations of participation and dissent, Goettel

(1971: 23-27) suggested that a better explanation may lie in a "latent negativism that . . . may be stimulated as much by citizen perceptions of the effectiveness of the local schools as by costs of education." Such negativism could be activated by community conflict over a nonfiscal school issue, by organized opposition and contests over school board seats.

While these psychological dimensions in conflict over desegregation and finances broaden our understanding of the politics of education, research literature in such modes is skimpy, particularly on finances. The newness of the desegregation thrust into suburbs has worked against such research, while the attitudinal analysis of election contests over finances is rare, whether focused on city or suburb (for exceptions, see Carter in Carter and Sutthoff, 1960; Horton and Thompson, 1962; Willis, 1967-1968; Zimmer and Hawley, 1968). These internal private forces, then, reflective of personality needs which generate social demands, have been barely touched in analyses of school policy. But the results of such attitudes can be seen in legal and political pressures upon the political system, as we shall now note.

SCHOOL FINANCE REFORM AND SUBURBIA

THE CHALLENGE TO SUBURBAN ADVANTAGE

The decision of the California Supreme Court in August 1971, in Serrano v. Priest, may well be as historic in its impact upon American education as Brown v. Board of Education in 1954.[10] Because the Serrano decision crystallized much discontent with existing school financing, it precipitated widespread litigation and subsequent legislation to remove the discontent. A year later, similar litigation was under way in half the United States, and that which had reached the decision stage echoed the California decision.

The principle enunciated there was bluntly put, that school quality which children receive should not be "a function of the wealth of a pupil's parents and neighbors." A federal court in Minnesota amplified this shortly thereafter by claiming that "the level of spending for a child's education may not be a function of wealth other than the wealth of the state as a whole (Van Dusartz v.

Hatfield).[11] The specific target of this attack was the property tax, the main source for financing public education in this country; as befits a varied nation, this tax is widely variable in its rate, associated valuation, and subsequent production of school funds. Much is still unclear about implications of the Serrano principle, and only some aspects of it are to be treated here. Certainly the policy context is turbulent, for these calls to reform are inserted into a national scene characterized by

> widespread distrust of the efficacy of public schools, a public sentiment that school taxes are inequitable and too high, a heated concern over problems of race and religion, and a perception on the part of educators that they face a crisis in their ability to raise sufficient revenues to support quality education. Into this situation, the courts have introduced a new element, a command to reform systems of educational finance that discriminate against people living in poor districts [Berke and Callahan, 1972: 28; see this source for a full treatment of the fiscal and legal issues].

Some fiscal discontent may be seen in the variety of fiscal policies in suburbs. Table 6 shows that, among suburbs, the valuation per pupil in dollars differs from low to high property valuation categories by factors of 3, 11, 6, 12, and 4.5; the resultant revenue per pupil varies anywhere from 30 to 100% among low and high suburbs. Such large variation in school support cannot sit well with those suburbanites who see around them "better" schools. That discontent long ago led to state support designed to narrow the discrepancies of local resources. But the variation between states is as severe as that within states. For example, the range within a state may be measured by an index of high to low district per pupil expenditures. Here, North Carolina and West Virginia do well with an index of 1.4, but Wyoming and Texas display an enormous range of 23.6 and 20.2. Here every dollar spent for education in some districts is matched by $20 or more in other districts. More money means more quality, although that relationship is not linear by any means. At least it provides the personnel most professionally trained to educate. Thus, a recent San Antonio area study shows suburban districts with highest per pupil revenue are also highest in salaries, teachers with master's degrees, and professional personnel, while lowest in staff with emergency permits and pupils per counselor; districts with lowest revenue show obverse qualities (Berke and Callahan, 1972: 36).

TABLE 6
PER PUPIL SUBURBAN PROPERTY VALUE AND SCHOOL
REVENUES IN FIVE METROPOLITAN AREAS—1967 (in dollars)

Prop. Valuation Category	Boston Suburbs		Los Angeles Suburbs		New York Suburbs	
	Valn. (n)[a] Per Pup.	Rev. Per Pub.	Valn. (n) Per Pub.	Rev. Per Pub.	Valn. (n) Per Pub.	Rev. Per Pup.
High	44,767(3)	824	57,414(3)	958	60,842(5)	1,411
Moderate high	26,343(11)	780	17,176(16)	686	31,384(16)	1,172
Moderate low	20,554(9)	760	7,195(16)	630	18,413(17)	1,043
Low	15,481(5)	595	5,079(3)	663	10,997(3)	1,009

	Houston Suburbs		Detroit Suburbs	
	Valn. (n) Per Pup.	Rev. Per Pup.	Valn. (n) Per Pup.	Rev. Per Pup.
High	140,719(1)	928	27,138(4)	899
Moderate high	64,356(5)	571	14,750(12)	724
Moderate low	27,146(4)	466	9,282(11)	629
Low	12,494(2)	482	6.550(4)	599

SOURCE: Policy Institute, Syracuse University Research Corporation, Syracuse, New York. Based on official finance documents provided by state education and aufiting agencies. Cited in Joel S. Berke and John J. Callahan, "Serrano v. Priest: Milestone or Millstone for School Finance," Journal of Public Law (1972), 31.
a. Number of school systems.

If we look not at the range of suburban school resources and results, but rather at the mean of such ranges for suburbia and its core city, the latter invariably suffers severe disadvantages (for details on the following, see Berke and Callahan, 1972; Sacks provides collaboration). Core cities are hurt on their income side by containing larger proportions of low-income and smaller of high-income citizens. On the expenditure side, they are laden with "municipal overburden," the necessity to pay for more high-cost urban services. Nationally, central cities in the largest metropolitan areas average $600 per capita on total local expenditures for all services, while suburban rings average a third less—$419. In 1970, of the 72 largest areas, in only 2 did the suburban ring average higher per capita expenditures (Dallas and San Antonio), while two others were equal (Patterson-Clifton-Passaic and Houston); these were even fewer than in 1957, when, among suburbias, 5 were larger and 5 equal to their central cities.

But rarely does suburbia pay less for schooling; among these 72 areas in 1970, in only 1 (New Orleans) did the suburban ring fail to have a higher proportion of its total in education than did its central city. These locale disparities in expenditures reflect a similar disparity in taxes, with core tax rates averaging 40% higher than those in suburbia. In 1970, only 2 suburbias had higher rates (Buffalo and Houston), while 2 were equal (Patterson-Clifton-Passaic and San Diego); in 1957, the disparities were not so extensive, however, as 4 suburban ring rates had been larger and 7 equal.

This pattern of difference may not be simply a function of the cities having less valuable property to be taxed. When city and noncity districts in New York State were controlled for similar per pupil property valuation, cities still had lower school tax rates, state aid, and per pupil expenditures, and students showed less academic achievement and affluence. Despite the variety of suburbs, then, a careful comparison of them with central cities and ruralia shows the first has "the lowest degree of educational need, the best resource base to finance education, the least competing demands for fiscal resources, and ironically, proportionately greater intergovernmental support for their expenditures" (Berke and Callahan, 1972: 47).

REFORM ALTERNATIVES IN SUBURBIA

While all such measures of existing needs, resources, and allocations are relatively concrete and demonstrable, when one seeks to introduce an equalizing policy into this matrix the results are less known. Serrano and its judicial offspring have not dictated the means of achieving the end, preferring instead "fiscal neutrality," which leaves much legislative discretion to select appropriate implementations. Three possibilities have been discussed. States could assume the costs of local schooling, then distribute funds on different bases—"one scholar-one dollar," (equal per pupil expenditures), or equal dollars but increments for particularly costly educational activities (e.g., vocational) or for pupils suffering educational deficiencies. A second approach amplifies the long-standing—and ostensibly equalizing—practice of state aid programs; here special formulas for "power-equalizing" would enable every district to be equal by a combination of its local revenue and state assistance to fill the gap. Finally, redistricting might pool districts regionally so as to devise equal property valuation.[1][2]

The foregoing sets the background for a nationwide politics of redistribution of tax resources. Discussion of the first two above—full state assumption and power-equalizing formulas—clearly and publicly call for expanding state taxes to provide needed funds, usually by shifting to some combination of taxes other than those on property. Redistricting calls for redistribution of valuation-geographical boundaries so as to achieve new funds. And all will affect suburbia, now in a position of advantage—even accounting for less than affluent suburbs—as described earlier. But the pattern of effects arising under these redistributional formulas has been little examined, and it is too soon to have a study of the political response of suburban delegates in state legislatures.

That not all these proposals augur well for suburbanites is seen in a recent analysis of thirteen states to determine the relative benefits of full state assumption among core city, suburban, and rural districts (Berke and Callahan, 1972: 52-60). In the *cities* studied, taxes would increase without relative gains in expenditures; in suburbs, taxes would fluctuate mildly but expenditures would be less; in *rural* districts, tax increases would be moderate but expenditures would increase substantially. In political terms, then, full state assumption would seem to work against those suburban areas gaining most representation recently and against those areas with the largest weight in state legislatures—city and suburbs. If this sample analysis is valid, full state assumption may run into trouble in some states, for not even rural representatives are open to schemes which increase school taxes, even when the proportionate return would be greater than now enjoyed.

The political reaction will be shaped differently by varying state conditions, however. Even in the thirteen-state study noted above, central cities in seven would benefit by an increased per pupil expenditures (Boston, Newark, New York, Minneapolis, St. Louis, Milwaukee, Seattle), and four suburban areas would also benefit, three substantially (Newark, Denver, Minneapolis). In short, the reaction of suburbia in this as in any of many other policy matters is not uniform but differentiated by variations in resources, preferences, and needs.

The key question may be the *relative benefit* a locale enjoys in the redistributive policy. The redistributive effect could be positive if a district received more than it paid in taxes or if, even though both expenditure and tax declined, the latter's decline was greater than the former's. This analysis for the thirteen states noted above is shown in Table 7. Cities and suburbs gain very little in such a prospect, for in only two central cities and two suburbs were there such positive net results, while eleven of the rural areas enjoyed relative benefits. The fiscal effect of full state assumption seems to

TABLE 7

FULL STATE ASSUMPTION, PERCENTAGE CHANGE IN TAXES AND EXPENDITURES IN THIRTEEN STATES, 1969

	% Change in Taxes	% Change in Expenditures
Boston (C)	+29	+16
Norfolk County (S)	--	−9
Dukes County (R)	−45	−30
Newark (C)	+8	+21
Essex County (S)	+33	+11
Atlantic County (R)	−28	−13
New York City (C)	+11	+5
Westchester County (S)	+5	−1
Schoharie County (R)	+18	+28
Indianapolis (C)	+9	−6
Marion County (S)	−9	−19
Washington County (R)	−20	−2
Minneapolis (C)	+19	+9
Hennepin County (S)	+12	+8
Clearwater County (R)	+23	+58
St. Louis City (C)	+7	+14
St. Louis County (S)	+15	+1
Montgomery County (R)	+12	+34
Cleveland (C)	−24	−4
Cuyahoga County (S)	−10	−32
Champaign County (R)	−37	−31
Milwaukee (C)	+19	+12
Milwaukee County (C)	+13	−11
Marathon County (R)	+17	+11
Atlanta (C)	−17	−15
Fulton County (S)	−28	−33
Lanier County (R)	+5	+23
Louisville (C)	--	−2
Jefferson County (S)	−15	−4
Webster County (R)	−12	+28
Oakland (C)	+4	−7
Alameda County (S)	−2	−6
San Luis Opisbo (R)	+6	+6
Denver (C)	+19	−5
Jefferson County (S)	+32	+35
Dolores County (R)	+3	+33
Seattle (C)	+11	+2
King County (S)	−3	−1
Skagit County (R)	+18	+35

C = Central City Area
S = Suburban Area
R = Rural Area

SOURCE: Joel S. Berke and John J. Callahan, "Serrano v. Priest: Milestone or Millstone for School Finance," Journal of Public Law (1972), 61-2.

redistribute funds generally to the advantage of ruralia, although the specific pattern will vary, of course, from state to state. A different conclusion may be drawn by employing different measures of redistribution (for alternative measures, see Berke and Callahan, 1972: 60; on the significant importance of the measure of "wealth" of local units, see Riew, 1970), but voters are more likely to see, understand, and react politically to comparative percentages of tax and expenditure *increases* of the kind displayed in Table 7.

Other political elements need study in the reform, such as the role of political parties in state legislative decisions on such matters. In congressional voting, suburban representatives speak much less for their locale than for their party (for the data, see Wirt et al., 1972: ch. 14); the latter is far more salient than constituency on a wide range of issues. Little research deals with this party versus suburban effect in state legislative voting, and none analyzes state educational policy in this manner.

Yet another consideration for future research is the role of the courts in overriding constituency demands. We know that the judiciary has insisted upon its definition of Constitutional rights against adverse demands of electoral majorities in numerous matters, including desegregation. In school finance litigation, the courts of modern times are expanding the meaning of equality, although as yet with far less opposition known in racial litigation. A hint of judicial firmness was seen in a decision invalidating Texas's school finance system on Serrano grounds. Speaking at Christmas 1971, a three-judge federal district court gave the state legislature one session to correct the inequities of that system; if not, "the court . . . will take such further steps as may be necessary to implement both the purpose and spirit of this order (Rodriguez v. San Antonio Independent School District).[13] Such judicial firmness has been the definitive variable in the effective elimination of physical segregation in Southern schools, despite the wavering and opposition of presidents and congresses through almost two decades (for a systematic analysis of those decades, see Wirt and Kirst, 1972: ch. 9). The court has here been a parameter affecting policy outcome and overriding that community political and economic structure which has been so important an explanatory variable in other policies. Any research design into school policy which omits this black-robed variable may well miss the reality of the policy process.

One final consideration in fiscal reform concerns us before turning to another school issue—the degree to which this reform affects the centralization of school control in suburbia. Inspection of the three reform implementations—full state assumption, power-equalizing formulas, and redistricting—finds that each in its own way contributes to that historical shift. Full state assumption may well mean that many suburbs will suffer a negative redistributional effect, as noted. That is, suburbs would lose even more control over monies which will be redistributed elsewhere. Moreover, consider the variants of this proposal which would provide extra funds for special educational problems, whether vocational education or improving children with special handicaps. These really speak to suburbs at the lower end of the status scale, as well as to the publicized deficiencies of core-city schools—not especially characteristic of suburbs, however. Both kinds of program carry with them special centralizing qualities. Vocational education over a half-century has developed a parallel bureaucracy in American schools, not unlike a barony in the days of King John (for references among six major states to vocational education, see Wirt and Kirst, 1972). Local districts now find little control possible over these programs, nor is there little political possibility of increasing such control when new monies flow into the "Voc Ed" subsystem. Even those lower-income suburbs, which would receive help under formulas diverting more money into districts with large numbers of low-achievement students, will feel a centralizing effect if states demand evidence that local programs are effective in removing these deficiencies. Districts have complained strongly about this responsibility under ESEA federal funds, but the "accountability" concept may be gaining acceptance as some state departments of education improve their research and data collection functions.

Power-equalizing functions offer merely an updated and more heavily financed version of extant state assistance programs, none of which has prevented the undermining of local control in suburban or other districts. Redistricting, as noted earlier, seems most likely to compel more affluent suburbs to surrender some of their fiscal resources and independence. Indeed, all three of these alternatives necessarily work to funnel funds from more affluent suburbia. However, these suburbs have been as loud in complaint as less affluent suburbs and core cities against the burden of property taxes. But so early is the public discussion of this issue that little research evidence exists that suburbs are divided over any of these reform alternatives. The opportunities for social research in this policy field are enormous, as probably every state but Hawaii (where all financing is centralized) will shortly be immersed in political conflict. All now agree that the property tax is too heavy and alternatives must be adopted in accordance with the Serrano imperative. The conflict will begin with the realization that to change is not a simple matter of all agreeing to let the state do it.

METROPOLITAN DESEGREGATION

If school finance reform is so new that little economic and no political analysis exists on the topic, then the effort by courts to compel metropolitan desegregation, by its even greater recency, is a blank slate. This effort involves the court mandate in Richmond, Virginia, and Detroit for suburban school districts to accept nonwhite students in order to break down dual desegregation now existing in core and suburb (Bradley v. Richmond School Board).[14] The Constitutional argument is that only this method can remedy the lack of equal educational opportunity which local districts have created, in result if not in design. The alteration is said to require new school organization unlike the current pattern of small, self-contained containers clustered within metropolises and able to resist the encroachments of nearby districts. Rather, it is argued, there is no bar in principle against consolidating this multitude of districts into a single regional district; such consolidation has been at work all over the nation throughout this century, as noted earlier. Now it would be employed to mix students throughout a metropolis in order to redistribute school resources.

As the 1972-1973 school year opened, no metropolis in America was so organized. In a few, a handful of black children were being bused to nearby suburbs under cooperative arrangements with such core cities as Boston, New Haven, and New York. For the purpose of consolidating expensive audio-visual aids or vocational education, New York State had developed a set of regional cooperative bodies incorporating numerous local districts—Boards of Cooperative Education Services. There are suggestions that a metropolis might consider itself as a single taxing base and thereby level school income; at least one now returns forty percent of any tax roll increase attributable to a given suburb's new industry back to a metropolitan pool for redistribution to less fortunate suburbs (Metropolitan Council, 1971).

But the thrust of the Richmond and Detroit decisions goes far beyond these limited scopes and onto a subject touching racial and status fears and prejudices of enormous political impact. As of the fall of 1972, both district court orders had been stayed on appeals which could eventuate in a U.S. Supreme Court decision in 1973. But it is clear that the reform deeply antagonizes a majority of those touched by it. White suburbanites and their representatives in state and national legislatures have become increasingly touchy on any effort to desegregate their schools even on a small scale, so this call for a massive restructuring of districts finds few allies out beyond the city limits. There are blacks who believe the future course of the races should involve separate institutions in order to better build racial pride, and they resist also. What this change would entail in financing and in reallocation of resources has not been fully analyzed, although in the Richmond and Detroit cases the judges wrote lengthy expositions of the need for such change and some of its attendant costs.

All of this was further inflamed by arising in a presidential election year in which the incumbent was outspoken against any busing "to achieve a racial balance" and used federal resources to reinforce districts seeking to block busing mandates. His opponent was less firm in his opposition to busing, while candidates for lesser offices and across the conservative-liberal continuum were staunch in their "anti-busing" stand. It is all rather like the "anti-communism" panic of two decades earlier. Americans themselves, in frequent polls, firmly stood *for* desegregation and *against* busing, and Floridians in the 1972 primary gave large majorities on both sides of these issues when offered as state referenda. Against this rising thermometer of anti-desegregation attitudes and actions were suddenly laid lower court mandates that suburbia must go much farther than it ever had in assisting segregation problems of the core city.

At this period, the judiciary stands in the position it knew after the Brown decision in 1954—even less strong, as the Supreme Court has not supported the metropolitan desegregation policy. As before, there is no support in the other two branches of national government (for conditions under which the Supreme Court can successfully innovate, see Schubert, 1965, and the application to Southern desegregation in Wirt and Kirst, 1972: ch. 9). The new political weight of suburbia in state legislatures cannot be expected to support a move that inflames constituents, and news media give even busing—less this reform—little support as 1972 closed. Yet it must be remembered that the depth of opposition facing the judiciary in 1954 was just as great, although it may be much broader today, given the national thrust of this reform. Despite that earlier opposition to Southern desegregation, however, much has been accomplished in just a few recent years. In Table 8 appear two sets of data on regional concentration of blacks and their desegregation, which have important implications for the past and the future of this policy.

The four columns on the right side show remarkable shifts in desegregation in those areas where federal effort has worked hardest to eliminate de jure desegregation and the least where that effort has been weakest to eliminate de facto segregation.[15] In just two years, the percentage of Deep South schools with almost all black students had been cut in half, while the percentage of blacks in formerly all-white schools better than doubled, encompassing over one-third of all black students in that region (by 1972, the proportion was over 40%). The Border South showed a similar but smaller shift. But outside the South, especially in the big states, the reverse trend was at work. About half the Deep South rate of blacks was in majority-white districts in these big states, a particularly important result in light of the latter's large concentration of blacks nationally. At the very least, such figures suggest that the loneliness of the Supreme Court in the 1950s was not in vain (Wirt and Kirst, 1972: ch. 9).

TABLE 8

REGIONAL CONCENTRATION OF U.S. BLACK STUDENT POPULATION AND DESEGREGATION, 1970

		% Black Students in Majority-White Districts			
	% Black Students 1970	majority-white schools		schools with 90%+ black students	
		1968	1970	1968	1970
Deep South[a](7)	32.7	15.4	36.5	80.6	42.8
Border South[b](10)	22.1	19.6	23.4	68.9	60.8
Big Non-South States[c] (9)	40.5	19.6	18.1	55.4	56.6
Other States[d] (24)	4.7	30.8	34.0	41.6	38.7
Total U.S.	100.0	19.4	24.9	63.3	52.6

SOURCE: Abstracted from Department of Health, Education and Welfare news release (HEW-A66, June 18, 1971).
a. Alabama, Florida, Georgia, Louisiana, Mississippi, North Carolina, South Carolina.
b. Arkansas, Delaware, Washington, D.C., Kentucky, Maryland, Oklahoma, Tennessee, Texas, Virginia, West Virginia.
c. California, Illinois, Indiana, Michigan, Missouri, New Jersey, New York, Ohio, Pennsylvania.
d. Excludes Hawaii.

Table 8 also reveals in its first column an important fact in the seemingly nationwide debate over desegregation. About half the states of the Union contain about 95% of all blacks—an extraordinary concentration. Moreover, the sources for this table report that 90.7% of all Spanish-surname public school students were concentrated in 1970 in just 9 states.[16] Yet one easily gets the impression from the media that desegregation is a problem in 50—not 24—states. The role of these states with few minority students needs further study. In one model of national policy-making, they may serve as a moderating influence on the passions aroused in the other half of the Union, although party considerations and the nationalization of public attitudes may well override that role.

The consequences are enormous for suburban schools if this metropolitan desegregation policy were supported by higher courts. One can easily foresee an even larger-scale round of litigation than that which followed the Brown decision. For this reform would make starkly evident to many what this article has argued already operates very extensively—namely, the degree to which local control of schools has been diluted during this century. That confrontation will generate cognitive dissonance of a classic proportion. Private academies will emerge as one way out of the dilemma for some suburbanites, but their record in the South has been mostly one of little use and unconscionable expense.

Moreover, such a distributional policy opens up all the "push" factors which originally had moved white city dwellers out to the delights of suburbia. One aspect of that "push" had been the uncertain consequences of living with different life styles; metropolitan desegregation renews them once again. If that reform were reinforced by the courts, suburban America would have to reexamine its own fears in light of the realities that desegregation would bring. Southern whites have learned that the world did not stop when blacks entered their schools; the prevalent reports are that it was not as bad as anticipated, although, even there, second-generation problems of "resegregation" are at work.

In short, this suburban school reform is so new that speculation is all one can offer. Yet, were courts to mandate it, the state legislative response would provide a rich study in the diverse response of American federalism to common policy needs. But more than that, such a mandate would work even more to expose the unreality of the local control ideology.

In one aspect, however, both the finance and desegregation reforms which work to foster centralization have a common feature. They represent a new shape to federalism in which policy mandates are set by national agencies but are administered locally in response to the mediating influence of a diverse society (for a recent elaboration of this thesis, see Reagan, 1972). Both state assumption and power-equalizing strategies reflect this centralization of school financing in state sources, while metropolitan desegregation centralizes policy in federal hands. As such, both would provide new answers to the old question of how to develop effective policy in a federal system which reflects both national and local preferences, resources, and needs. We have not here argued whether any particular programmatic answer is good, although our preferences are strong for centralization in school finance and desegregation. Preference statements alone are insufficient, however, for even among common preferences there can be wide disagreement about the methods and organization for realizing policy preferences. In the case of finance reform, suburbanites are about ready to be caught up in that realization, for agreement on ends does not preclude disagreement on means. In the case of desegregation reform, they also agree in opposing both the end and the means. It is not all that certain yet, however, how their views in either case will affect the outcome. But it is certain that schools will be the central focus of policy controversy in suburbia in the decade ahead.

NOTES

1. Around 1900, a ward-based committee system of "community control" was widespread; Philadelphia had 43 district school boards of 559 members (see Tyack, 1969).

2. In San Francisco of the 1870s, a teacher had to provide a fee of $200 to the party boss to get the job. As one reformer lamented, "Poor girls, desiring to earn an honest living, have gone to this demon's groggery, around the entrance of which loitered gamblers and sports, to see if the coveted place could be bought at a cheaper rate" (see Callow, 1956: 269).

3. That same San Francisco boss's board chairman could be under fraud indictment and still be given the congressional nomination, while his school director could move to abolish the office of bookkeeper on the grounds of economy (see Callow, 1956: 275).

4. A recent study using rough measures for a handful of states, found little correlation between a state's share of local costs and the presence of state programs (see Levin et al., 1972: vol. 1, ch. 5).

5. Bendiner (1969) found this a common complaint of board members.

6. The figure shifts each year; for annual figures, see reports of the U.S. National Center for Educational Statistics of the U.S. Office of Education.

7. The wording of the question was: "Some people say that the government in Washington should see to it that white and Negro children are allowed to go to the same schools. Others claim this is not the government's business. Have you been concerned enough about this question to favor one side over the other? (If yes) How do you feel? Do you think the government in Washington should (1) see to it that white and Negro children go to the same schools; (2) other, depends; (3) Leave these matters to the states and local communities?" There was also a direct relationship between the size of the suburb and opposition to federal intervention in education (Wirt et al., 1972: 113).

8. Note, however, that his analysis fails to include property tax data because none with comparable meaning across jurisdictions was available.

9. One federal source does aggregate and publish totals, but only by states and only for bond elections (see National Center for Education Statistics, series).

10. Cal. 3d 584, 487 P 2d 1241, 96 Cal. Reptr. 601 (1971).

11. 40 U.S. Law Weekly 2228 (D. Minn. Oct. 12, 1971).

12. Full state assumption has been proposed for consideration in the New York and Maryland state legislatures (see New York Commission to Study the Quality, Costs, and Financing of Elementary and Secondary Education, 1972; Citizens Commission on Maryland Government, 1971). Both reflect the influence of Charles S. Benson and James W. Guthrie, of the University of California (Berkeley). District power-equalizing in its newest form has been urged by Coons et al. (1970: chs. 5-6). The reapportionment approach is seen in President's Commission on School Finance (1972: 68-70; see also Berke et al., 1972).

13. 40 U.S. Law Weekly 2399 (W. D. Tex. Dec. 23, 1971).

14. U.S. Sup. Ct., Docket No. 72550, cert. granted Feb. 15, 1973.

15. As the Richmond and Detroit decisions make clear, and as is being centrally argued in a Denver case to be decided by the Supreme Court in the 1972-1973 term, distinctions between these two are of questionable validity. When law could, but does not, act to change practice, law is a partner to that practice.

16. Arizona, California, Colorado, Florida, Illinois, New Jersey, New Mexico, New York, and Texas.

REFERENCES

BAILEY, S. K. et al. (1962) Schoolmen and Politics. Syracuse: Syracuse Univ. Press.

BENDINER, R. (1969) The Politics of Schools. New York: Harper & Row.

BENSON, C. S. (1968) The Economics of Public Education. Boston: Houghton Mifflin.

BERKE, J. S. and J. J. CALLAHAN (1972) "Serrano v. Priest: milestone or millstone for school finance?" J. of Public Law: 23-71.

BERKE, J. S., A. K. CAMPBELL, and R. J. GOETTEL [eds.] (1972) Financing Equality of Educational Opportunity: Alternatives for State Finance. Berkeley, Calif.: McCutchan.

BERKE, J. S. et al. (1971) Federal Aid to Public Education: Who Benefits? Syracuse: Syracuse Univ. Press.

CALLOW, A., Jr. (1956) "San Francisco's blind boss." Pacific Historical Rev. 25.

CARTER, R. F. (1963) Voters and Their Schools. Stanford, Calif.: Institute of Communication Research.

――― and J. SUTTHOFF (1960) Communities and Their Schools. Stanford, Calif.: Institute of Communication Research.

Citizens' Committee on Maryland Government (1971) "A responsible plan for the financing, governance and evaluation of Maryland's public schools." Baltimore, November.

COONS, J. E., W. H. CLUNE, and S. D. SUGARMAN (1970) Private Wealth and Public Education. Cambridge, Mass.: Harvard Univ. Press.

COUNTS, G. S. (1927) The Social Composition of Boards of Education. Chicago: Univ. of Chicago Press.

CREMIN, L. A. (1964) The Transformation of the School: Progressivism in American Education, 1876-1957. New York: Vintage.

DOWNES, B. T. (1968) "Suburban differentiation and municipal policy choices: a comparative analysis of suburban political systems," pp. 243-268 in T. N. Clark (ed.) Community Structure and Decision-Making: Comparative Analyses. San Francisco: Chandler.

GERSMAN, E. M. (1970) "Progressive reform of the St. Louis School Board." History of Education Q. 10: 8-15.

GOETTEL, R. J. (1971) "The relationship between selected fiscal and economic factors and voting behavior in school budget elections in New York State." Presented to the annual conference of the American Educational Research Association.

HORTON, J. W. and W. E. THOMPSON (1962) "Powerlessness and political negativism: a study of defeated local referendums." Amer. J. of Sociology 67: 485-493.

IANNACCONE, L. and F. W. LUTZ (1970) Politics, Power and Policy: The Governing of Local School Districts. Columbus: Charles E. Merrill.

ISSEL, W. H. (1970) "Modernization in Philadelphia school reform, 1882-1905." Pennsylvania Magazine of History and Biography 94: 381-382.

JAMES, H. T. et al. (1966) Determinants of Educational Expenditures in Large Cities. Stanford, Calif.: Stanford School of Education.

LEVIN, B. et al. (1972) Public School Finance: Present Disparities and Fiscal Alternatives. Washington, D.C.: Urban Institute.

Metropolitan Council (1971) The Impact of Fiscal Disparity on Metropolitan Municipalities and School Districts. Minneapolis.

MEYER, A. E. (1967) An Educational History of the American People. New York: McGraw-Hill.

MINAR, D. (1966) "Community basis of conflict in school system politics." Amer. Soc. Rev. 31: 822-835.

National Center for Education Statistics (series) "Bond sales for public school purposes." Annual Reports.

National Education Association (1970) Estimates of School Statistics, 1969-70 and Rankings of the States, 1970. Washington, D.C.

New York Commission To Study the Quality, Costs, and Financing of Elementary and Secondary Education (1972) Report.

REAGAN, M. D. (1972) The New Federalism. New York: Praeger.

RIEW, J. (1970) "State aids for public schools and metropolitan finance." Land Economics 46: 297-304.

SACKS, S. (1972) City Schools/Suburban Schools: A History of Fiscal Conflict. Syracuse: Syracuse Univ. Press.

––– and D. C. RANNEY (1966) "Suburban education: a fiscal analysis." Urban Affairs Q. 1.

SALISBURY, R. (1967) "Schools and politics in the big city." Harvard Education Rev. 37: 408-424.

––– (1964) "Urban politics: the new convergence of power." J. of Politics 26: 775-797.

SCHUBERT, G. (1965) Judicial Policy-Making. Chicago: Scott, Foresman.

SHARKANSKY, I. (1970) The Routines of Politics. New York: D. Van Nostrand.

SHIELDS, C. (1952) "The American tradition of empirical collectivism." Amer. Pol. Sci. Rev. 46: 104-120.

TYACK, D. B. (1969) "Needed: the reform of a reform," pp. 29-51 in National School Boards Association, New Dimensions of School Board Leadership. Evanston, Illinois.

U.S. Bureau of the Census (1971) Social and Economic Characteristics of the Population in Metropolitan and Nonmetropolitan Areas: 1970 and 1960. Washington, D.C.: Government Printing Office.

WALDEN, J. C. (1967) "School board changes and superintendent turnover." Administrator's Notebook 15.

WILLIAMS, O. et al. (1965) Suburban Difference and Metropolitan Policies: A Philadelphia Story. Philadelphia: Univ. of Pennsylvania Press.

WILLIS, C. L. (1967-1968) "Analysis of voter response to school financial purposes." Public Opinion Q. 31: 648-651.

WIRT, F. M. (n.d.) "The slender reed." (unpublished)

——— (1972) "The allocation politics of New York State aid to education," in J. S. Berke and M. W. Kirst (eds.) Federal Aid to Education: Who Benefits? Who Governs? Lexington, Mass.: D.C. Heath.

——— and M. W. KIRST (1972) The Political Web of American Schools. Boston: Little, Brown.

WIRT, F. M., B. WALTER, F. F. RABINOVITZ, and D. HENSLER (1972) On the City's Rim: Politics and Policy in Suburbia. Lexington, Mass.: D. C. Heath.

ZIMMER, B. and A. H. HAWLEY (1968) Metropolitan Area Schools: Resistance to District Reorganization. Beverly Hills: Sage Pubns.

Part VI

LOOKING AHEAD

LOOKING

AHEAD

It may be a bit premature to be concerned with the future of suburbia since we have hardly begun to comprehend the present. But the future is being shaped by present policies of both public and private sectors. Perhaps no set of policies is more significant than those related to land use and development. In this concluding section of the book is an article on satellite cities by Bernard Weissbourd and a short epilogue by the senior editor.

Weissbourd offers a proposal, based on the "new communities" provisions of the 1968 Housing Act, to provide for more orderly metropolitan growth, control air and water pollution, rationalize the transportation system, and reduce racial segregation. This would happen if the federal government, in cooperation with state and local agencies, would acquire vast acreage on the urban periphery and encourage the development of satellite cities to house part of the anticipated population growth during the remainder of this century.

In the Epilogue, Masotti discusses how the future of suburbia is being shaped, not by authorized public agencies and suburban government, but rather by the combined efforts of highway and utility planners, land speculators, financial investors, mortgage bankers, and residential and commercial developers with little or no accountability to the public. He concludes that suburbia, like the city before it, is destined for uncoordinated sprawl because the private marketplace, and not public purpose, guides its growth.

18

The Satellite Community as Suburb

BERNARD WEISSBOURD

☐ THE TERM "URBAN CRISIS" has different meanings to different people. Those most immediately affected by hunger, lack of health services, unemployment, poor education, and housing shortages are the black and the poor—and their neighbors. Problems of air and water pollution, traffic congestion, and municipal finance, on the other hand, clearly affect everyone in the metropolis—whether he lives in the suburb or in the city. But, to the poor, these problems are not as immediate as their poverty. The "urban crisis" has a significantly different impact on different segments of the population, living in different sections of the metropolis. The severity of the problems also varies widely between metropolitan areas. In general, it is directly proportional to the size of the population, the growth of the population, and the size of the ghetto.

What strategies can we employ to combat these pressures? Financial aid to the cities, either by means of revenue-sharing or

EDITORS' NOTE: *Mr. Weissbourd's chapter was originally published as a two-part essay entitled "Satellite Communities: Proposal for New Housing" in the February and March, 1972 issues of* Center Magazine. *It is reprinted here with the kind permission of the author and the Center for the Study of Democratic Institutions.*

through the federal government's bearing a larger share of expenditures for education and welfare, will certainly be essential. Unwieldy governmental machinery will have to be bypassed, since problems of this urgency will not wait for the reorganization of fragmented governmental jurisdictions.

The program proposed here focuses on the problems of migration, pollution, transportation, unemployment, and segregation. Specifically, it recommends that:

(1) The federal government, in cooperation with the states, should embark upon a land-acquisition program for new satellite communities. Acquiring enough land in the metropolitan areas, where the problems are the greatest, to accommodate half the population growth expected by the year 2000 would require the surprisingly small investment of less than three billion dollars. If the land were properly located, this investment could yield a profit, unlike other governmental housing programs, which are typically operated at a loss.

(2) Anti-discrimination laws should be affirmatively enforced through the lending policies of the FHA, VA, federally insured savings and loan associations, and through the denial of federal funds to builders of segregated developments.

(3) Existing federal subsidies for housing should be redirected toward families earning between six thousand and twelve thousand dollars a year, and income limits on persons eligible for subsidized housing in new satellite communities should be removed.

(4) An income-maintenance program should be established for lower-income people to assist them in buying or renting older housing.

(5) The federal government should provide funds to rebuild substandard central-city areas, perhaps as "new towns-in-town," after making a sufficient number of new homes available through the new satellite communities program. This would assure immediate accommodations for those displaced from the buildings to be demolished.

The heart of this program is the acquisition of land for new satellite communities which would house part of the population growth of selected metropolitan areas. The need for new communities is now widely acknowledged, although there is widespread confusion regarding their fundamental purpose, as well as their financing, location, and size.

For example, David Rockefeller has called for a ten-billion-dollar private corporation to build satellite new communities on land acquired through the aid of a federal land corporation. *The New City,* a book produced by an impressive group of senators, congressmen, and national organizations of cities, counties, and mayors, advocates:

> financial assistance ... from the federal government to enable the creation of one hundred new communities averaging one hundred thousand population each and ten new communities of at least one million in population.

Moreover, Congress has passed new legislation authorizing federal loan guarantees for financing new communities. This new legislation identifies four types of new communities as eligible for federal assistance: satellite communities (on the outskirts of existing metropolitan areas as an alternative to subdivision development), urban growth centers (additions to small towns and cities), new towns-in-town (within central-city areas), and independent new cities unrelated to metropolitan areas that are already in existence.

Yet, Congress, by enacting legislation providing funding for all four kinds of new communities, has avoided answering such basic questions as: what are the fundamental purposes of new communities? What kinds and sizes of new communities are most urgently needed? Where should they be located? The program proposed here is based on the conviction that new communities developed under current legislative policies will be located in the wrong places, will be too few, and only a small percentage of them will be desegregated.

I propose to show that, while we need new communities for many reasons, their most important function is as part of a desegregation strategy that deals realistically with race and class, that we need new satellite communities rather than isolated new towns, and that the federal government should begin now to acquire the land for these new satellite communities instead of continuing to guarantee loans for private land acquisition. Finally, I propose to identify those metropolitan areas with the greatest need—where the land should be acquired for new satellite community programs significant enough to have an impact on the ghetto.

DESEGREGATION–RACE AND CLASS

The 1970 census returns demonstrate all too clearly that the dire prophecies of black inner cities encircled by white suburbs are being fulfilled. At least four major American cities, Atlanta, Newark, Gary, and Washington, now have central cores that are over fifty percent black. In addition to these, seven other cities have black ghettos with more than forty percent of the central-city population. One reason is the flight of white populations from the central cities. In Chicago, for example, the white population declined by more than half a million between 1960 and 1970; in Detroit, there was a decline of 345,000 during the same period. The other factor, according to the Bureau of the Census, was that the "increase in the black population of the central cities proved to be both large and widespread, thus changing the racial mixture substantially." Though this situation is recognized as critical, most current proposals fail to deal realistically with political obstacles resulting from attitudes toward race and class.

Before addressing ourselves to this question, let us turn our attention toward another aspect of the problem. It has now become commonplace to recognize that the new suburbs are also the location of the new jobs, particularly the new industrial jobs, since industries find it more economical to build single-story plants on cheap suburban land than to build multistory plants on expensive city land. Only office jobs are increasing in the cities, and this trend is expected to continue. So we daily witness the remarkable phenomenon of inner-city blacks filling the expressways en route to their industrial jobs in the suburbs while, at the same time, suburban whites are fighting traffic in the opposite direction to get to their office jobs in the cities. This is a consequence of the separation of place of work from place of residence carried to absurdity.

The core of the problem is race. Black people need places to live near their work, yet white suburbia's level of anxiety at the prospect of an influx of black neighbors was expressed by President Nixon when he opposed Secretary Romney's proposal for using federal funds to open up the suburbs by calling it "forced integration," and there are voices within the black community advocating total separatism.

The strong resistance of white suburbia to the acceptance of black neighbors, in large part, is based upon a fear of inundation by an alien culture. Each suburban community fears that, if it becomes the first of its neighboring areas to welcome blacks, it will become the central focus of the entire black migration. Given this white resistance and the understandable reluctance of black families to move into an inhospitable environment, legal and legislative efforts can open up the suburbs to only a few highly motivated black families intent upon securing better housing, better jobs, and better education for their children. Large numbers of black families will not be affected, nor will the lower class or the unemployed.

Nevertheless, overcrowding in the ghetto and the resulting pressure on the white neighborhoods adjoining it must be relieved. Where housing is scarce and segregation prevails, the growing black population can only expand into the adjoining white neighborhoods. Block by block, as these white residents are squeezed out of their neighborhoods, they tend to reassemble in other ethnic or cultural enclaves farther out, which become that much more resistant to any black entry.

Land is scarce in the city, and urban renewal has destroyed more housing than it has created. It is very difficult to reproduce the high density of our ghettos except by building high-rise apartments—very high; ghetto residents now generally consider these to be undesirable places to raise their children. The only land available, then, for housing the growing black population of our cities—if they are not simply to displace their white neighbors—is in the areas next to be developed at the suburban fringes or on inner-city industrial land no longer suitable for industry. Once population pressures in the ghetto have been relieved through new construction in outlying areas, our inner cities can be rebuilt. London, for example, built new towns before the East End was cleared, so that the residents of the East End had someplace to go before their homes were destroyed.

There is still another aspect of the situation that should not be neglected. Besides simply relieving the overcrowding in the ghetto, it is also necessary to speak to the condition of despair in which most of its inhabitants find themselves. New satellite communities provide an opportunity for white America to convince black America that there is someplace in the United States where there can really be hope, where Americans put into practice the ideals we profess, where there are good jobs, good housing, and a good education for children. Many black Americans may prefer to stay where they are, rejecting the new communities and the racial mixture they offer, but they would still be aware that this opportunity exists if they choose it.

Black separatists argue that the growth of black majorities in our cities will increase both the political and the economic power of the black community. There is certainly strength in this position. The business communities which occupy the center of those American cities with populations over forty percent black will have to learn how to cope with this reality. On the other hand, it would not seem plausible that the separatist movement would deny that same choice of environment to blacks that the majority society has always enjoyed. In any case, white society now has the power to open the way for black occupancy of homes in the suburbs and in new communities, and, if it is used, blacks can choose whether or not to live there.

Historically, the ethnic neighborhood functioned effectively as a place where the immigrant could adjust to the New World while he preserved his own culture and roots. His children usually became part of the majority culture. It is glaringly obvious, however, that this process is not operating for black people and that the barriers to upward mobility are greater for them than they were for the Irish, Italians, Jews, and other groups. Limited progress does occur, of course. Opportunities for employment and education are expanding, but the increasing ghettoization of black people is an insurmountable barrier to their entrance into the larger society unless it can be somehow reversed.

We·will be confronted for many years to come with a situation in which a number of central cities will be dominated by black groups. This means that one route for the black community will be the development of their own economic and cultural institutions, depending on their own initiative and on negotiations with white people having economic power in the city. Here an analogy to the biculturalism of Quebec might be relevant, where the French majority faces an English minority holding economic dominance.

An alternative route for black advancement lies in desegregation and the accompanying improvement in education, employment opportunities, and housing that a desegregated society can provide. The term "desegregation," however, does not imply that every street or apartment building has black residents in precise proportion to the total population. Obviously, economic status, distance from work, arbitrary personal preferences (including those based on race), and many other factors will continue to affect where and how people choose to live. Desegregation does require, however, that black people have the same opportunities to make choices that white people have. We must continuously be reminded that the success with which we develop an environment that will close the unconscionably wide gap between the opportunities available to black and white people will not only determine the future of many of our cities, it will also have an enormous influence on our national character and on our international relationships.

Any desegregation strategy must begin by making a proper distinction between race and class. This distinction has long been blurred, perhaps deliberately. For years the executive branch of the government "sold" Congress on housing legislation without ever mentioning the word "black." It was considered wiser to talk about "poverty," and the fact that there are twice as many poor whites as there are poor blacks in the United States.

President Nixon's statement of policies relative to equal housing opportunities has been attacked because of its lack of commitment to racial and economic integration. However, the statement deserves to be read in its entirety because, for the first time, the executive branch of the federal government has made the important distinction between racial and economic integration and has recognized the difference between opposition to building low-income housing in wealthy suburban areas and the movement of relatively affluent black families into those same suburbs.

Anyone with any experience in leasing apartments or in selling homes knows that it is much more difficult to mix different economic classes than to mix different races in this country. The 1970 census shows, in fact, that we are getting some small mixture of middle-income black and white families in suburban areas, and there are numerous examples of integrated buildings and neighborhoods in the cities. However, comparable examples of economic integration of rich and poor are rare.

The President's statement also acknowledges for the first time the role that the Federal Home Loan Bank plays in the housing market through its regulation of savings and loan associations. Almost all present financing for new homes and apartments involves the federal government in one way or another. About half the outstanding mortgage debt on single-family homes is either FHA- or VA-insured, and most of the balance is financed by savings and loan associations with deposits insured by the federal government and investments regulated by federal law. The remaining factor in home and apartment mortgage financing is the life insurance companies, and they most certainly would cooperate with any Equal Opportunity

Program affecting the FHA, VA, and savings and loan associations. If these lenders or loan guarantors were to provide funds only for equal opportunity developments, almost all new construction of homes and apartments would be desegregated, just as equal opportunity employers have partially succeeded in desegregating the work force.

It will be necessary, however, to design separate programs for different income groups. Only families earning more than ten thousand dollars per year are able, at current market prices, to afford new homes or apartments outside the South. In the northern and western metropolitan areas approximately 32% of all black families earn more than ten thousand dollars a year, as compared to 52% of all white families. This means that in a metropolitan area with a population which is, say, 20% black (though most, of course, have a smaller percentage of black families), approximately 13% of the total families who could afford to move into suburban areas might be black.

However, because the white families are already there, and because black families might hesitate to move into a potentially hostile environment, a much smaller percentage of black families can be expected to purchase or rent new housing in already built-up areas or in traditional suburban subdivisions, even with a free housing market. Since the percentage is likely to be relatively small, any action taken by the federal government in relation to open housing should not arouse enormous fears. Experience suggests that such a mix would be stable. Moreover, the fact that practically all new developments would have to be equal opportunity developments under such a plan means that no single development would run the risk of becoming the focus of more extensive change. Builders would be able to prove compliance by demonstrating that they have, in fact, sold or rented to black families or show that they have unsuccessfully advertised for, and have not discriminated against, them. Thus a modest degree of racial desegregation would be achieved; it will not, however, produce any substantial amount of economic mixture, since only families earning more than ten thousand dollars a year would be involved.

RACIAL AND ECONOMIC BALANCE

To facilitate the housing of black industrial workers near suburban plants, other programs will be needed. This may be a more difficult goal because of the greater resistance to mixing economic classes than to mixing races. A better opportunity for achieving racial and economic heterogeneity is offered by new satellite communities. Desegregation here would signify real options in terms of where and how people want to live. Given a choice of locations in new communities, it is quite possible that people would sort themselves out: some into white neighborhoods, some into black, and some into mixed. It is also possible that some management people living in the new communities will isolate themselves from the factory workers, while others will prefer to mix with them. These options must always be kept open, because the principle of free choice is quite as important as any of the other concepts discussed here.

Housing middle-income people and industrial workers in new satellite communities will require some form of subsidy, since some of them may be earning only about six thousand dollars a year. This means that such a family can afford little more than $125 a month for rent. Of all black families living in the North and West, 65% earn at least six thousand dollars a year, as compared with 79% of all white families in those areas. This means that in a metropolitan area with a population, say, 20% black (again this figure is larger than average), approximately 17% of the total families who could move into such new communities would be black. Experience, again, indicates that such a population mix in new communities would be stable. However, even though black families would be welcome in such new communities, in contrast to the hostility of the existing suburbs, it is doubtful whether the percentage of blacks, at least in the initial phases of the program, would be as high as 17%.

Income groups earning less than six thousand dollars a year present an entirely different problem. It is possible that the best way to deal with poverty is with money. Income maintenance or some form of negative income tax might be more effective answers to the problems of low-income families than housing subsidies. People in this income bracket may prefer to select older housing, spending a larger proportion of their total income on other necessities. If we are interested in achieving some degree of economic heterogeneity, new communities should then be prepared to accept their share of the new housing for this group in proportion to their presence in the metropolitan area as a whole.

The results of this program are likely to be that a small percentage of relatively affluent black families will live in the same suburban neighborhoods as white families of a similar economic level, in new housing. A larger percentage of both working-class and affluent black people will live in desegregated new satellite communities with whites of the same economic status. Poor families, both black and white, will live largely in the older housing and apartments, with a small percentage in new homes and apartments and in new satellite communities.

High-rise public housing, on the other hand, should be reserved exclusively for adults—either the elderly or young single or married people without families. The buildings are adequately constructed and can be utilized in much the same way as luxury high-rise apartments, by adults. For large families, income maintenance might permit them to buy older housing, preferably in a neighborhood with good schools, and provide a much more suitable environment for the children.

This approach is the opposite of the one we are now pursuing. We have for some time been providing new housing for poor people through public housing projects and for the relatively affluent through government subsidies for single-family homes. Instead, we should be providing new housing for middle-income people earning between six thousand dollars and twelve thousand dollars a year and allow those with lower income to acquire older housing through income maintenance. People with incomes over twelve thousand dollars can afford their own housing, subsidized as at present through special income tax treatment of home ownership.

What has been proposed, then, is a four-point program:

(1) Affirmative action to enforce anti-discrimination laws through FHA, VA, and federal savings and loan association lending policies so that black families able to afford single-family homes can move to the suburbs.

(2) Housing subsidies for middle-income people earning between six thousand dollars and twelve thousand dollars a year to enable them to afford new housing in new satellite communities.

(3) An income maintenance program for lower-income people so that they may be able to purchase suitable older housing.

(4) A new satellite communities program offering housing of varying price range in a choice of black, white, or racially mixed neighborhoods.

THE HIDDEN COST OF
SUBSIDIES

The federal government pours at least $10 billion a year in direct or indirect subsidies into the housing market. The federal budget directly allocates approximately $4 to $5 billion a year to housing, including appropriations for urban renewal, public housing, mortgage loan interest, subsidy programs, and the like. This figure, however, does not include rent payments made for welfare recipients, which are indirectly financed by the federal government through matching funds to the states (in New York City alone, these costs amount to $8,000,000 a year). The federal budget figure also does not include the major benefit involved in the whole system of FHA and VA mortgage guaranties, nor does it include the funds supplied by the savings and loan associations, which have financed the large portion of so-called "private" housing development. By insuring deposits in savings and loans and requiring them to invest these funds in home mortgages, the government has made available a continuing supply of money to the housing market at a lower interest rate than could be obtained in a competitive money market. It would be difficult to calculate the extent of this benefit.

The outstanding mortgage debt, however, of the savings and loan associations exceeds $160 billion, and the outstanding FHA and VA guaranties represent an even larger amount. If the interest differential is one percent, the benefit involved amounts to more than $3 billion a year. The overall cost to the government of its borrowings may be increased by these loans and guaranties, but to a degree next to impossible to measure.

To these subsidies must be added the cost to the federal government of the loss of revenues deriving from its tax treatment of home ownership. The deductions allowed home owners for mortgage interest payments and real-estate taxes is another form of subsidy. It has been estimated that, in 1966, the deductions for mortgage interest reduced revenues by $1.6 billion, and those for property taxes caused a reduction of $1.4 billion. It is also estimated that the failure to tax the value of occupying a home (imputed income) reduced revenues by $4 billion. The total revenue lost to the federal government by special tax treatment of homeownership, therefore, amounts to some $7 billion a year, or more than the total amount in the federal budget earmarked for housing. Perhaps this analysis reflects a rather extreme point of view. However, if we were to follow the example of the Canadian government, which does not tax imputed income but does not allow the deduction of mortgage interest or of property taxes either, the gain in revenues would amount to $3 billion a year.

Perhaps an example will clarify what is meant by imputed income. If a homeowner has a house worth $30,000 on which he has a mortgage of $20,000, his equity investment is $10,000. If that investment had been invested instead at eight percent interest, it would be earning $800 a year on which a taxpayer in the forty percent bracket would pay $320 in income taxes. However, the homeowner has a return on his $10,000 investment of the privilege of occupying his own home, for which he pays no tax. This annual value is known as imputed income.

The inequities involved in this arrangement are manifold. For example, the more expensive the house, the larger the tax benefit of homeownership. If the homeowner, on the other hand, were to decide to rent his house to another family, he would have to pay income tax on the amount by which the rent received exceeded his mortgage payments, operating costs, and real estate taxes, except for the special tax treatment for depreciation of real estate, which in the early years will also make his $800 to $1,000 return on his investment tax free. Instances of such special tax treatment exist throughout the tax laws. The investment credit for new plants and equipment is a comparable example.

My aim here is not to reform the tax laws but rather to emphasize the size and extent of the subsidies available to homeowners and landlords in the form of tax benefits as compared with the subsidies for moderate-income housing, where a three percent interest subsidy on a $16,000 mortgage amounts to only about $480 a year. Its after-tax value to a taxpayer in the forty percent bracket, however, is only $192 a year, or considerably less than the subsidy to the homeowner or landlord owning more expensive property in the example above.

It will undoubtedly take a major effort to organize and rationalize the whole subsidy system in the housing market. I am not suggesting that we do away with income tax deductions for mortgage interest and real estate taxes or for depreciation. I am rather suggesting that we try to understand their impact on the housing market and make sure that comparable subsidies are made available to lower-income people as well.

Unfortunately, many people become quite moralistic where subsidies for the poor are concerned and tend to worry about the problem of cheating. The opportunity to cheat is built into many of our housing programs when they specify minimum-income levels. If a man's family is likely to be evicted from its apartment when his income increases, he is very likely either to avoid making more money or to be somewhat uncommunicative about the fact that he does. Such income limitations should be abolished in satellite communities. If a person earning more money still wishes to live among his lower-income neighbors, he should be encouraged to do so for the sake of economic heterogeneity. If he goes out and buys a home or rents a more expensive apartment, he is receiving another kind of subsidy, in any event. Furthermore, eliminating such income limitations will aid in overcoming racial prejudice. Experience has shown that the best way to encourage desegregation is to offer a bargain. Such integrated developments as Prairie Shores in Chicago and St. Francis Square in San Francisco have largely been successful because they offered well-designed apartments at a little less money. This policy creates a waiting list of potential tenants who are more concerned with getting in than they are with who lives next door.

Of the $10 billion that the federal government is spending in the housing market, a large portion has the effect of encouraging white homeownership in the suburbs and public housing for black people in the inner city. In other words—whether intentionally or not—it has resulted in increasing segregation. If we really were to decide to use these funds for the opposite goal, we could—within a decade—create many desegregated new communities, decrease overcrowding in the ghetto, reduce the pressure on its adjoining white neighborhoods, and encourage some black families to move into what are now all-white suburban areas.

Specifically, new satellite communities should receive a priority in the allocation of federal subsidies for low- and moderate-income housing in order to make it possible for everyone who works in the new community to live there. The amount of subsidy should vary with the income of the homeowner or renter, so that no more than 25% of his income is required for rent or mortgage payments.

NEW COMMUNITY OPTIONS

Why should we build new satellite communities? Why not consider a low- and moderate-income housing program in suburban locations instead? The fact is that the present level of resistance to the construction of such housing in already established suburban areas is so great that it is unlikely that very much low- and moderate-income housing will be built there. In bypassing these areas and creating new satellite communities on the vacant land beyond, the political opposition to desegregation is significantly reduced. Furthermore, building in such outlying areas where vacant land is still available requires a scale of development for roads, sewer and water, and other amenities, at least on the level of the planned unit development and approaching the scale of satellite communities, in any event.

Most of the advantages of new communities have been fully described elsewhere, so only a brief list will serve our purposes here. New communities offer options for living in black, white, or mixed neighborhoods; they can provide an integrated school system; they open up opportunities for employment in the new suburban industrial plants; they reduce automobile traffic by allowing people to live near their work, thus lessening air pollution and the drain on our diminishing oil resources.

Certainly, new satellite communities provide an organizing idea which permits the planning needed to rationalize our transportation systems as well as our land use. Furthermore, only in new satellite communities will it be possible to aggregate the housing market efficiently in order to make industrialized housing feasible, while their higher density would permit us to protect the environment by saving land and preserving open space. Definite financial benefits would result from federal land acquisition for new satellite communities, and these savings could be channeled into housing needed to accommodate population growth and replace substandard housing.

As mentioned earlier, influential proponents of new communities have called for ten new cities of one million people each and one hundred new towns of one hundred thousand each. It is thought that a population of at least one hundred thousand is required in order to sustain the services necessary to a self-sufficient community. New satellite communities, however, need not be self-sufficient, since they can depend on the central city for many of these necessary services. Consequently, they might be as small as eight to ten thousand people—i.e., neighborhood size, large enough to sustain an elementary school, community center, nursery, convenient shopping center, service station, recreational facilities, and other local amenities. Such neighborhoods exist in the new towns of Färsta and Vallingby, near Stockholm, and in Tapiola, outside Helsinki.

These are largely town-house and apartment communities, as opposed to single-family homes, which are more characteristic of new communities in the United States. If one of the purposes of new satellite communities is to house the workers employed by outlying industries, they will have to consist partly of town houses and garden apartments, since only persons earning at least ten thousand dollars a year can afford new single-family homes in the northern and western United States at today's prices. However, these town houses and apartments could be sold as condominiums rather than rental units, if ownership is the desired goal.

Though new satellite communities might in some instances, consist of as few as 8-10,000 people, in others larger populations are desirable, particularly if desegregation is our goal. Communities consisting of four such neighborhood units, or approximately 35,000 people, could offer a choice of white, black, or integrated housing, with an integrated school system as well as other community activities. Moreover, a new satellite community of 35,000 people could be linked to the central city and to other similar communities by mass transportation. This becomes even more feasible if the town houses and apartments are centrally located near the mass transit station or bus stop, with single-family homes surrounding this central core. Such concentration of population at mass transit origins and destinations is necessary if it is to be workable at all.

Even larger new satellite communities are possible, though it is generally true that the larger the community, the more difficult the land acquisition process. Evanston, Illinois, an old suburban community, for example, has a population of 80,000 and a school system which is integrated without excessive busing.

In metropolitan areas with very small black populations, like Minneapolis and Seattle, other means of desegregation are possible, so that new satellite communities there might be as small as 8,000 people. In those areas with large black populations where desegregation is a primary goal, new communities might better be built for 35,000 people or more.

Very little land is required for a new satellite community of 8-10,000 people. Three thousand dwelling units at a modest apartment density of twenty units to the acre could be built on only 150 acres of land. An additional 20 acres could easily accommodate a school, shopping center, and other community facilities. In most metropolitan areas, tracts under 200 acres are relatively easy to assemble. At mixed town-house and apartment density, such a community might require 400 acres; if four such neighborhoods are joined together to create a community of 35,000 people, 1,600 acres would be more than adequate. Another 400 acres could provide space for a high school, additional recreational facilities, and for industrial and commercial uses. However, in most metropolitan areas, tracts as large as 2,000 acres are usually quite difficult to assemble. In most places, the government's power of eminent domain will be necessary to assemble the land for new satellite communities.

A community of 40,000 people on 2,000 acres provides a density of 20 persons to the acre. The city of Chicago (including all of the land devoted to offices, industry, commerce, recreation, and other uses) has a density of more than 70 persons per acre, or more than 3.5 times as dense. Older suburbs of Chicago, like Berwyn, Cicero, and Evanston also have much higher densities—i.e., 30 persons to the acre, with mixed residences including single-family homes, duplexes, town houses, and apartments. Even some of the newer subdivisions, built since World War II and consisting entirely of single-family homes, have similar densities.

The lower density of 20 persons per acre has been chosen to demonstrate that, even at this low density, only 1,000,000 acres of land will be needed for a new satellite communities program accommodating half of the expected growth in selected metropolitan areas during the next thirty years. The cost of acquiring this land at $3,000 per acre (the land for both Reston and Columbia was purchased for about $1,500 an acre) is only $3 billion. So, if we build at higher densities, less land and less money will be required; if the land can be purchased for less than $3,000 an acre, a further saving can be obtained.

Half the expected population growth between 1970 and the year 2000 in 38 selected metropolitan areas amounts to 21,000,000 people. These selected metropolitan areas account for over 85% of the population, 87% of the black population, and 90% of the population growth in all metropolitan areas of over 500,000 people. A more modest but more concentrated program involving 24 selected metropolitan areas would account for 72% of the population, 83% of the black population, and 72% of the population growth in metropolitan areas of over 500,000 people. These, then, are the metropolitan areas with the greatest degree of segregation, the greatest pressure on the environment, and the greatest problems resulting from size.

The more modest program would require 800,000 acres in these selected areas for new satellite communities. This land could be acquired for less than $2.5 billion spent once, as opposed to the federal government's annual expenditure, both direct and indirect, of $10 billion for housing.

LAND ACQUISITION

I have suggested that the federal government, in cooperation with state agencies, acquire one million acres of land in certain metropolitan areas for satellite new communities. In addition to the value of early land acquisition to a new satellite communities program, the estimated $3 billion required for this investment could be expected to yield a profit to the government.

Private land acquisition for new community development can presently be financed through federal loan guaranties. It appears that a number of such projects will be undertaken if only as a result of the rather favorable terms offered. The major flaw in this approach is that the land best suited to the development of new communities is seldom available under a single ownership and cannot be assembled without invoking the government's power of eminent domain. Consequently, many of these new communities will not be located where they are most needed. Moreover, piecemeal new community development competes with ordinary subdivision development, which makes desegregation difficult.

A bolder strategy is needed. If the federal government were to immediately acquire the land required for new satellite communities, thus pinpointing publicly their intended locations, many of the worst fears would be allayed. As David Rockefeller notes, a federal agency with "powers for planning and obtaining sites for new towns" might "perhaps provide guidance in terms of national land use planning." Whether the federal agency involved is part of the executive branch of the government or is a separate entity altogether, such as the TVA (which I tend to prefer), the land should initially be acquired and *owned* by that agency. When land is privately owned, local zoning and building codes hamstring the development of new communities. This problem, however, tends to disappear when the new satellite community is part of a federal "enclave."

Local authorities, of course, should be involved in the process of planning the location of new communities. Perhaps the best agency for coordinating local with areawide planning is the metropolitan area planning authority, which exists in nearly every major metropolitan area. In the Chicago area, for example, most federal grants to municipalities in northern Illinois must first be approved by the Northeast Illinois Planning Commission in order to show compliance with the commission's guidelines.

The profits to be made from a land acquisition program can accrue to the public's benefit. Private real estate investors now reap the profits from increased land values created by public construction of subways, bridges, roads, airports, sewer plants, and other public utilities. Advance knowledge of highway interchanges creates instant fortunes in real estate. If land value profits created by public expenditures are recaptured for the public benefit, they can be reinvested in various aspects of new community development. In England, for example, land investments for new towns have nearly all been profitable for the government, although private investors in this country might expect a higher return.

The actual development of new communities, unlike land acquisition, should be accomplished by private developers rather than by the government. Urban-renewal procedures might serve as a model here, although we might experiment with other mixtures of government and private responsibility. Land for urban renewal is acquired by a public agency which also plans the major streets, public utilities, and land uses. Parcels of land are then sold to private developers. In this case, there is usually a write-down on the land, because these projects are generally undertaken in depressed areas, while in the case of new satellite communities, the land could be resold at a profit.

Of course, there is more to a new community program than buying the land. Roads and streets, sewers, water, and other utilities must be provided. While the land itself may cost less than three thousand dollars per acre, the infrastructure (roads, streets, utilities, and so on) for such a project might well cost as much as fifteen thousand dollars per acre. This expenditure, however, is not entirely made at the beginning of the project since only major roads, storm water interceptors, and part of the sewage treatment plant must be installed then. Perhaps five thousand dollars per acre would be a more realistic figure to represent the primary investment in the infrastructure. The balance can ordinarily be provided by land sales. The point is that these particular expenses will have to be met in any event if the land is to be developed at all. State and local governments ordinarily bear the burden of these expenditures, which are more costly for ordinary subdivision than for new community development. There is no reason why these expenses should not be shared by the private developer to whom the land is sold, perhaps with loans guaranteed under the New Communities Act.

Urban land values have increased more than 400% within the last twenty years. A single-family lot which sold for one thousand dollars in 1950 cost over five thousand dollars in 1970. This increase is greater by far than the increase in the cost of living in general and greater even than the increase in building costs themselves. Properly located land is a unique form of monopoly: the supply is limited, and increased demand sends prices soaring. If the land is acquired by a governmental agency, however, the profits—as explained above—will accrue to the public or be used for further new community development.

Numerous federal and state agencies presently are involved in the planning of roads and utilities. The federal government, moreover, is financing the acquisition of open lands. It would seem to be highly desirable to coordinate all these various efforts, so that planning for roads, mass transit systems, sewers, and water can coincide with land acquisition for new satellite communities and thereby guide the growth of our metropolitan areas.

A final and singularly important reason for suggesting that land be acquired now by the federal government is to establish the credibility of the new communities program. People need tangible evidence that we actually do have a national commitment to desegregated new communities. In the building industry, announcements of new projects are generally credible only when the builder owns the land.

I believe that the land for new satellite communities should be acquired in the areas next to be developed in our fastest growing metropolitan areas. This belief is based upon the judgment that these are the places where desegregated new communities are most necessary, and also that these are the areas where population growth will occur anyway. The Nixon Administration, however, has proposed a very different policy.

In July of 1970, the President's National Goals Research Staff issued a report entitled "Toward Balanced Growth—Quantity with Quality." In effect, it recommended that both the government and the private sector intervene in order to change the pattern of population settlement and prevent the further growth of mega-lopolis. The Nixon Administration has thus called for the concentration of resources to bolster urban growth centers by improving employment opportunities and providing professional services in regions of limited growth.

While federal assistance and intervention in the development of urban growth centers may be desirable, as a substitute for a national urban policy it is sheer disaster. Even if migration to the cities continues, the influx will progressively dwindle. Since massive migration is already coming to an end, the plan for urban growth centers actually comes too late. Moreover, it is simply not possible to keep the largest metropolitan areas from growing without overhauling the entire economic distribution system, and the cost of such an enterprise would far outweigh the benefits. Major metropolitan areas will continue to grow for the same reasons that they have grown in the past, and industries will keep on locating on the outskirts of such cities in order to serve their markets and attract labor for their plants and warehouses. The situation must be quite exceptional if an industry decides to locate in an urban growth center in Kentucky in order to serve a market in Chicago.

The large metropolitan areas will also continue to remain the centers of the boom in office buildings. Office jobs in the New York area are expected to almost double by the year 2000 and to increase 2.5 times throughout the country as a whole between 1965 and the end of the century. It is remarkable that, though the 24 largest metropolitan areas contain only 34% of the nation's total population, they provide 65% of all its office jobs—and this percentage is expected to decline only slightly, in spite of government intervention to promote urban growth centers. The number of white-collar jobs first began moving ahead of blue-collar jobs in the mid fifties, and it is expected that they will exceed all other forms of employment by 1980. Finally, metropolitan areas will continue to grow, even without significant migration, by the normal process of the birth of children to the residents. Serving these children and their families will be a motivational factor in the choice of location for many "service" industries.

If this is the real significance of metropolitan areas, then we must treat as commendable but, perhaps, futile the attempt by the administration to redistribute the population away from major metropolitan areas. Certainly, whenever possible, government offices, universities, defense contracts, and so on should be used to aid underdeveloped areas in the United States. This is simply a matter of not increasing congestion in built-up metropolitan areas. But to assume that any such effort can make a significant difference is a misconception. Whether we encourage new cities in the prairies or urban growth centers or further growth in metropolitan areas of less than 500,000 people will have very little impact on such metropolitan areas as New York, Los Angeles, and Chicago, which will all continue to grow.

Similar considerations preclude the building of many new cities in the prairies. It is possible to build a few, perhaps, and this effort might be encouraged, in locations where a new university or federal installation provides an economic base for such an independent new town as an experiment in new technology. However, in the main, no cost-benefit analysis can justify the federal government's intervention except where the action already is—i.e., in the areas to be developed during the next three decades on the outer fringes of the suburbs now encircling the major metropolitan areas. *After* these areas have been developed with new satellite communities, we can consider the creation of new towns-in-town on the vacated inner-city land which will become available to replace substandard housing no longer needed in the ghetto.

Some people give the term "urban growth center" a different meaning. They talk, instead, about adding to existing old satellite communities, of which Aurora, Elgin, Joliet, and Waukegan are examples in the Chicago area. This concept has some merit. There is no reason why these old satellite communities should not be enlarged, provided that a desegregation program is an integral part of the plan.

POPULATION GROWTH AND
HOUSING SUPPLY

It is argued that, even if some twenty million people lived in new towns, they would represent such a small percentage of the population as not to be significant. The 1970 census shows, however, that 20,000,000 people constitute more than half the total population growth expected between now and the year 2000 in the metropolitan areas with the worst problems. A program of properly located new satellite communities to house this anticipated growth in the metropolitan areas where it is expected to occur would certainly have a major impact.

It was once anticipated that the population would double between 1960 and the end of the century. The Nixon Administration, however, has been talking about a slower rate of growth, while Congress places the figure for population growth between 1970 and the year 2000 at only 75,000,000 people. Anthony Downs believes that the 1970 census data indicate that an increase of 50 to 60,000,000 is a more realistic figure, and estimates that most of this growth will occur during the earlier portion of this thirty-year period, with a point of zero population growth reached well before the turn of the century. My own calculations are based upon an increase of 70,000,000 people between 1970 and the year 2000.

The birth rate is still declining, as it has since 1957, with a consequent lowering of the estimates of future population growth. Nevertheless, large numbers of post-World-War-II babies are now producing children of their own and will probably cause the population to increase over the next two decades—even if they have proportionately fewer children than previous generations had. In addition, we gain some 400,000 persons each year through immigration. Whether the population will begin to stabilize after this latest wave of new families depends largely upon what people believe about how many children they should have. A stable population within the next thirty years could have enormous economic and social implications well beyond the scope of this analysis. At any rate, an increase of 70,000,000 people in a thirty-year period is a reasonably manageable figure, as compared to earlier estimates of 150 to 180,000,000.

Based on the larger projections of population growth, the Kaiser Commission established in 1968 a goal of 26,000,000 new housing units to be built over the next ten years. I believe that this goal is excessive. It was officially adopted during the Johnson Administration and has become the basis for the housing policy of the Nixon Administration. This goal is based upon estimates of the number of new household formations, housing units required to replace substandard units, and units scheduled for demolition. Though the administration's goal is based on too high an original estimate of future population growth, the error is partially offset by the fact that there are perhaps 10 to 12,000,000 households that we could consider inadequately housed, even though only 6,000,000 would be classified as "substandard" by the official definition of the U.S. Bureau of the Census. Moreover, if the rate of abandonments continues to increase as it has in recent years, we may need still more replacement units than heretofore anticipated. As a recent study of rental housing in New York City points out:

> The increase in the number of housing units withdrawn from use since 1965 has been startling. During the early nineteen-sixties, roughly fifteen thousand units were annually removed from the active housing stock for reasons other than their demolition to make way for new construction. In the period 1965 to 1967, the annual average rose to thirty-eight thousand units. These recent losses have not been confined to the worst part of the stock, as is usually the case.

It is obvious that there are numerous uncertainties inherent in such projections. For example, the number of new units required each year will depend partly on how fast we want to eliminate substandard housing. In addition, these projections must take into account the increased vacancy rate necessitated by the increasing mobility of the population. Since I believe that all these considerations are important, I would prefer to choose the higher rather than the lower goal, while recognizing that—if we conceive the timetable for eradicating the inadequate housing supply as fifteen rather than ten years—a lower rate of production would be adequate.

In spite of all these considerations, a goal of 2,600,000 new housing units each year seems excessive in relation to a population growth of 70,000,000 people in a period of thirty years. Even taking into account the replacement of inadequate housing, an increased vacancy rate, and the trend toward more abandonments, we should require no more than 40 to 45,000,000 new units in the next thirty years, or an average of at most 1,500,000 units a year. If we build 2,000,000 units a year for the next ten years to replace substandard housing, the rate might drop off to approximately 1,000,000 to 1,500,000 per year for the remainder of the period. If the building industry makes the attempt to gear up for the Kaiser Commission's 2,600,000 units per year, on the other hand, serious discontinuity will result later. A rate of 2,000,000 new units a year, therefore, seems a more appropriate goal.

The building industry is now producing 2,000,000 housing units a year, if we include mobile homes, so its present capacity is not the problem. The rapidly shrinking supply of construction labor, however, is serious enough to cause an eventual decline in production. Construction wages, as a result of the declining labor supply, are rising faster than those in almost any other sector of our generally inflationary economy, and the cost of housing has subsequently risen very rapidly. Though construction unions have been largely successful in minimizing minority participation, black labor, nevertheless, represents the only major labor pool left untapped; it is bound to become an indispensable component of any real solution to the problems of the industry. Given the government's "Operation Breakthrough" program, which was designed to encourage the industrialization of housing production, it is possible that the major thrust of black entry into the construction industry will be through the factory rather than in the field.

The most serious constraint, however, in the maintenance of adequate housing production has been the erratic fluctuations of "tight" and "loose" money. During periods of tight money, other industries tend to attract a larger share of available investment capital than does housing construction. The production of housing, therefore, falls off drastically until a period of loose money causes it to spurt suddenly ahead. While the federal government's policy of increasing the flow of money to the savings and loan associations and into subsidized housing programs does manage to smooth out these periodic fluctuations to some extent, a perceptive new approach for providing fresh capital to the housing industry will be necessary before these problems are entirely solved. Certainly, industry cannot be expected to invest substantial capital in industrialized housing until it is convinced that the demand for its product will not vary widely from year to year.

The excessively high projections of future population growth have not only led to erroneous estimates of the demand for housing but to the concept of "megalopolis" as well. This concept, while useful for some purposes, has had the negative effect of raising the false specter of an America running out of land. The megalopolis known as the Atlantic Region, which stretches from Massachusetts to Virginia, includes the cities of Boston, New York, Philadelphia, Baltimore, and Washington (or five of the fourteen metropolitan areas with populations over 2,000,000) and contains some 35,000,000 people. Other megalopoles are the Lower Great Lakes Region, the California Region, and the Florida Peninsula.

This situation sounds rather alarming unless two additional facts are also taken into consideration: there is no *other* identifiable urban region in this country that will exceed even 6,000,000 by the year 2000, and none of these so-called megalopoles is actually running out of open space. It is not even accurate to refer to "San-San," as the California Region is sometimes called (San Francisco to San Diego), since a large amount of open country still exists in that region between the mountains and the coast that could not possibly be filled up by the year 2000. Similarly, the concept of a "Lower Great Lakes Region" assumes a continuous urban belt between Detroit and Chicago. Though the towns of Kalamazoo, Jackson, and Battle Creek are certainly growing, an automobile trip on route I-94, connecting Chicago and Detroit, clearly shows that there are still miles and miles of undeveloped farmland along that stretch. Florida, of course, still has its share of Everglades, the palmetto scrub, and orange groves.

The idea that we are running out of land is patently not true. On the other hand, there are far more serious threats to urban life which require immediate attention. Both automobile ownership and consumption of electrical energy and fossil fuel is increasing at a much faster rate than the population as a whole. The concentration of people in megalopoles means a concentration of consumers whose capacity for consumption is potentially very much greater than its present level. If we are successful in raising the incomes of people now impoverished in inner-city ghettos and they, too, become consumers of automobiles, ranges, air conditioning equipment, and other accouterments promised them as part of the "good life," the problem will begin to take on very serious dimensions. Thus, while availability of land may not pose a real threat, air and water pollution, depletion of our oil resources, and the disposal of waste heat emerge as possibly the most ominous consequences of the growing concentration of people in megalopoles.

Similarly, there is an enormous pressure to develop ocean shores, lake frontage, woods, and other natural resources outside the cities which will intensify as the society becomes more affluent. The real danger highlighted by the concept of megalopolis is not that we will run out of land but that we might run out of clean air, water, and natural resources.

CRITICAL TARGET AREAS

The 1970 census data not only indicate that previous population growth estimates and required housing units have been greatly exaggerated, they also show that a large percentage of that growth as well as a large percentage of the black population is concentrated in a few metropolitan areas. If we rank metropolitan areas according to size, growth, and black populations, we can then begin to establish some criteria for deciding which of these areas requires the most urgent attention, for which new satellite communities offer solution. It will also enable us to suggest the kinds of satellite communities needed and to estimate the amount of land required for these new communities.

Writers from each of the American regions tend to project the problems of their own area onto the country as a whole. The particular problems besetting New York and Los Angeles, for example, are often described as though they were typical of the entire country. I suggest rather that New York and Los Angeles are uncommon. Cities like Chicago, Philadelphia, Detroit, and Baltimore require a different kind of approach. Smaller metropolitan areas with fewer black people and less population growth can utilize other strategies more appropriate to their needs.

The census defines a place as "urban" if it has a population of 50,000 or more people. A town of 50,000 does not usually have an "urban crisis," as the term is normally used, or share the problems of cities like Chicago and Cleveland. To use the term urban for such places sometimes misleads people into enlarging the scope of the urban problem. In fact, if we limit our attention to cities with populations in excess of 500,000, we will still find many metropolitan areas that neither manifest the problems nor require the solutions discussed here. Smaller metropolitan areas may need a strategy for desegregation or suffer from various other urban ills, but the scale of these problems is too small to warrant the construction of new satellite communities.

Table 1 lists those metropolitan areas in the United States with populations which exceeded 500,000 in 1970, ranked according to population size. Cities like New York, Newark, and other New Jersey cities have been grouped together in a single metropolitan area, as has Los Angeles, Anaheim, and so on, and Chicago, Gary, and East Chicago. These are actually unified metropolitan areas and are treated as such for purposes of this chart, though the census, for other reasons, lists them separately.

A look at the chart shows the concentration of urban population in a relatively few large metropolitan areas and an even denser concentration of the black population. For example, 69,000,000

Americans now live in the fourteen metropolitan areas with populations of approximately 2,000,000 or more; 82,000,000 live in cities over 1,000,000; more than 11,000,000 of these 82,000,000 are black, and 9,000,000 of that 11,000,000 live in central cities; 103,000,000 Americans live in metropolitan areas with populations over 500,000; 13,000,000 of these "urban" dwellers are black, and only 2,000,000 of the 13,000,000 live in the suburban areas surrounding the central cities.

Those cities most in need of desegregation can also be identified from the table. It indicates that, though a high proportion of the urban black population resides in a few large northern and western metropolitan areas, some smaller southern cities like Memphis, Birmingham, New Orleans, and Norfolk all have a very large proportion of black residents. These cities have their own distinct problems (and opportunities) and require some special attention.

The metropolitan areas listed on the chart may also be ranked according to size of growth between 1960 and 1970. It can be projected that perhaps 40 to 50,000,000 people will be added to these metropolitan areas during the next thirty years. Of course, it is one thing if an area of 1,000,000 people increases by thirty percent and becomes a city of 1,300,000; it is quite another if a metropolitan area like Los Angeles grows at a slower rate, but expands its population by 2,000,000 people. Even if a metropolitan area of 500,000 doubles in size, it will still not have the monumental problems of a city like New York, Los Angeles, or Chicago. The *increase* in population in Los Angeles during the last ten years was as large as the entire population of Cleveland or Baltimore or Houston. Yet only the city of Los Angeles could reach the present size of New York by the year 2000, only San Francisco and Philadelphia might grow to the size of Chicago, and only Washington might exceed the present population of Detroit within thirty years. Certainly, it is clear that no city of *less* than 2,000,000 people could possibly reach a population of 7,000,000 by the year 2000.

It is no longer appropriate to discuss metropolitan planning in broad, general terms. We need to consider each area separately and develop for each its own specific regional plan designed to protect the environment, provide for an adequate mass transportation system, and promote desegregation.

When all the 1970 census data have become available and we are able to evaluate the latest statistics on employment, education, welfare, housing, crime, transportation, and the fiscal condition of

TABLE 1

Metropolitan Area	Total Population 1970	Total Black Population	% Black Population	Total Population 1960	1970 Population Central City	1970 Black Central City	% Black
New York, Newark, Patterson, Clifton, Passaic, Jersey City	15,354,000	2,367,000	15.4	14,182,000	8,793,000	1,978,000	22.5
Los Angeles, Long Beach, San Bernardino, Riverside, Ontario, Anaheim, Santa Ana, Garden Grove	9,588,000	824,000	8.6	7,552,000	3,925,000	553,000	14.1
Chicago, Gary, Hammond, East Chicago	7,608,000	1,340,000	17.6	6,794,000	3,697,000	1,214,000	32.8
Philadelphia	4,816,000	844,000	17.5	4,343,000	1,949,000	654,000	33.6
Detroit	4,196,000	756,000	18.0	3,762,000	1,509,000	659,000	43.7
San Francisco, Oakland, San Jose	4,181,000	348,000	8.3	3,291,000	1,524,000	232,000	15.2
Washington	2,861,000	704,000	24.6	2,064,000	757,000	538,000	71.1
Boston	2,754,000	127,000	4.6	2,595,000	641,000	105,000	16.3
Pittsburgh	2,402,000	170,000	7.1	2,405,000	520,000	105,000	20.2
St. Louis	2,364,000	379,000	16.0	2,105,000	622,000	254,000	40.9
Baltimore	2,071,000	490,000	23.7	1,804,000	906,000	420,000	46.4
Cleveland	2,064,000	333,000	16.1	1,909,000	751,000	288,000	38.3
Houston	1,983,000	383,000	19.3	1,418,000	1,233,000	317,000	25.7
Miami, Ft. Lauderdale	1,888,000	267,000	14.1	1,269,000	582,000	100,000	17.2
Minneapolis, St. Paul	1,814,000	32,000	1.8	1,482,000	744,000	30,000	4.0
Dallas	1,556,000	249,000	16.0	1,119,000	844,000	210,000	24.9
Seattle, Everett	1,422,000	42,000	2.9	1,107,000	584,000	38,000	6.5
Milwaukee	1,404,000	107,000	7.6	1,279,000	717,000	105,000	14.7
Atlanta	1,390,000	311,000	22.3	1,017,000	497,000	255,000	51.3
Cincinnati	1,385,000	152,000	11.0	1,268,000	453,000	125,000	27.6
San Diego	1,358,000	62,000	4.6	1,033,000	697,000	53,000	7.6
Buffalo	1,349,000	109,000	8.1	1,307,000	463,000	94,000	20.4
Kansas City	1,254,000	151,000	12.1	1,093,000	507,000	112,000	22.1
Denver	1,228,000	50,000	4.1	929,000	515,000	47,000	9.1
Indianapolis	1,110,000	137,000	12.4	944,000	745,000	134,000	18.0
New Orleans	1,046,000	324,000	31.0	907,000	593,000	267,000	45.0
Tampa, St. Petersburg	1,013,000	109,000	10.8	772,000	494,000	87,000	17.6
Portland	1,009,000	23,000	2.3	822,000	383,000	22,000	5.6
SUBTOTAL	82,468,000	11,190,000	10.4	70,572,000	35,645,000	8,996,000	25.2

TABLE 1 (Continued)

Metropolitan Area	Total Population 1970	Total Black Population	% Black Population	Total Population 1960	1970 Population Central City	1970 Black Central City	% Black
Phoenix	968,000	33,000	3.4	664,000	582,000	28,000	4.8
Columbus	916,000	106,000	11.6	755,000	540,000	100,000	18.5
Providence, Pawtucket, Warwick	911,000	21,000	2.5	821,000	340,000	16,000	4.7
Rochester	883,000	58,000	6.5	733,000	296,000	50,000	16.8
San Antonio	864,000	60,000	6.9	716,000	654,000	50,000	7.6
Dayton	850,000	94,000	11.0	727,000	244,000	74,000	30.5
Louisville	827,000	101,000	12.3	725,000	361,000	86,000	23.8
Sacramento	801,000	38,000	4.7	626,000	254,000	27,000	10.7
Memphis	770,000	289,000	37.5	675,000	624,000	243,000	38.9
Fort Worth	762,000	83,000	10.9	573,000	393,000	78,000	19.9
Birmingham	739,000	218,000	29.5	721,000	301,000	126,000	42.0
Albany, Schnectady, Troy	721,000	24,000	3.3	658,000	256,000	20,000	7.8
Toledo	691,000	57,000	8.3	631,000	384,000	53,000	13.8
Norfolk, Portsmouth	681,000	168,000	24.7	579,000	419,000	131,000	31.3
Akron	679,000	54,000	8.0	605,000	275,000	48,000	17.5
Hartford	664,000	51,000	7.6	549,000	158,000	44,000	27.9
Oklahoma City	641,000	54,000	8.5	512,000	366,000	50,000	13.7
Syracuse	636,000	23,000	3.7	564,000	197,000	21,000	10.8
Honolulu	629,000	7,000	1.2	500,000	325,000	2,000	0.7
Greensboro, High Point, Winston-Salem	604,000	118,000	19.6	520,000	340,000	101,000	29.7
Salt Lake City	560,000	4,000	0.7	448,000	176,000	2,000	1.2
Allentown, Bethlehem, Easton	544,000	6,000	1.2	492,000	215,000	5,000	1.5
Nashville	541,000	96,000	17.8	464,000	448,000	88,000	19.6
Omaha	540,000	37,000	6.8	458,000	347,000	34,000	9.9
Grand Rapids	539,000	23,000	4.3	462,000	198,000	22,000	11.3
Youngstown, Warren	536,000	51,000	9.4	509,000	203,000	44,000	21.7
Springfield, Chicopee, Holyoke	530,000	24,000	4.6	494,000	281,000	23,000	8.5
Jacksonville	529,000	118,000	22.3	455,000	529,000	118,000	22.3
Richmond	518,000	130,000	25.1	436,000	250,000	105,000	42.0
Wilmington	499,000	61,000	12.2	415,000	80,000	35,000	43.6
TOTAL	103,041,000	13,397,000	13.0	88,059,000	45,681,000	10,820,000	23.7

the cities, perhaps some useful comparisons can be made in order to determine the effect, if any, the size of the metropolitan area, the growth rate of the population, and the size of the ghetto have on these problems. In the meantime, perhaps we can draw some tentative conclusions:

(1) We should concentrate on the metropolitan areas with the largest populations, because that is where the people are, and size itself creates problems of transportation, pollution, and race relations.

(2) We should concentrate on the metropolitan areas with the largest growth, because that is where the opportunities are, and that is where the environment will be destroyed if we do not act quickly.

(3) We should concentrate on the metropolitan areas with the largest black populations, because that is where desegregation is most crucial.

VARYING LAND REQUIREMENTS

Applying these criteria to the metropolitan areas with populations over 1,000,000 it is possible to establish certain categories and to estimate the amount of land required for both a minimum and maximum new satellite community program.

The first category contains only New York and Los Angeles, since their size and growth put them in a class by themselves. These areas will require, in addition to new satellite communities, either the construction of new cities on the scale of 500,000 to 1,000,000 each, or the enlargement of existing suburban centers in order to accommodate the inevitable growth in population. The only other comparable metropolitan area is Chicago, though I have classified it in the second category, since it shares more of the characteristics of metropolitan areas like Philadelphia, Detroit, Washington, St. Louis, Baltimore, Atlanta, San Francisco, Houston, Miami, and Dallas than it does those in the first category. All the metropolitan areas in the second category grew by more than 250,000 people between 1960 and 1970, and all have black populations in excess of 250,000. The first seven of these areas have central cities that are over thirty percent black, and each urgently requires a new satellite community program to desegregate, protect the environment, and accommodate part of the expected population growth.

The thirteen metropolitan areas in the first two categories contain nearly sixty percent of the total urban population living in cities of over 500,000 people and seventy percent of the black population. A minimum program consisting of satellite cities and communities for New York and Los Angeles and new satellite communities for the other eleven metropolitan areas would, nevertheless, accommodate a very large percentage of the people affected by the urban crisis.

If we add a third category, consisting of the other metropolitan areas with populations over 500,000 which grew by more than 100,000 people in the last decade and which have black populations exceeding 100,000, we will have accounted for 72% of the population living in metropolitan areas of over 500,000 people, 83% of the black population of such areas, and 72% of the growth (see Table 2). These metropolitan areas are Boston, Cleveland, Milwaukee, Cincinnati, Kansas City, Indianapolis, New Orleans, Tampa, Columbus, Louisville, and Norfolk.

Not all of these metropolitan areas require a new satellite community program of 30,000 to 100,000 in size. In fact, some of the smaller areas could best be served by satellite neighborhoods of 8 to 10,000 people. It is not our purpose here to plan all the metropolitan areas, but rather to emphasize the need for focusing on areas of urgent need with a concentration of people and problems.

A fourth category might consist of those metropolitan areas with substantial growth but relatively small black populations. Minneapolis, Seattle, San Diego, Phoenix, Denver, and Portland would make up this list. A new satellite communities program in these areas would relate to the goals of managing growth, rationalizing mass transportation, protecting the environment, and providing an alternative to ordinary subdivision development. Desegregation in these areas can readily be achieved by other measures.

A fifth category, very similar to the fourth, could include such smaller metropolitan areas as Rochester, San Antonio, Dayton, Sacramento, Fort Worth, Hartford, Honolulu, and Salt Lake City. They all had substantial growth, and none of them had a very large black population. Some could use a desegregation program, and —while a satellite neighborhood program would undoubtedly be beneficial—the urgency and the scale of the need for such a program in these areas is minimal.

Cities like Pittsburgh and Buffalo form a category of their own. Each had an insignificant record of growth between 1960 and 1970. There would be very little reason for a new communities program in

TABLE 2

Metropolitan Areas	Total Population 1970	Black Population 1970	Growth 1960-1970
Group 1			
New York	15,354,000	2,367,000	1,172,000
Los Angeles	9,588,000	824,000	2,036,000
Total	24,942,000	3,191,000	3,208,000
Group 2			
Chicago	7,608,000	1,340,000	814,000
Philadelphia	4,816,000	844,000	473,000
Detroit	4,196,000	756,000	434,000
San Francisco	4,181,000	348,000	890,000
Washington	2,861,000	704,000	797,000
St. Louis	2,364,000	379,000	259,000
Baltimore	2,071,000	490,000	267,000
Houston	1,983,000	383,000	565,000
Miami	1,888,000	267,000	619,000
Dallas	1,556,000	249,000	437,000
Atlanta	1,390,000	311,000	373,000
Total	34,914,000	6,071,000	5,928,000
Group 3			
Boston	2,754,000	127,000	159,000
Cleveland	2,064,000	333,000	155,000
Milwaukee	1,404,000	107,000	125,000
Cincinnati	1,385,000	152,000	127,000
Kansas City	1,254,000	151,000	161,000
Indianapolis	1,110,000	137,000	166,000
New Orleans	1,046,000	324,000	139,000
Tampa	1,013,000	109,000	241,000
Columbus	916,000	106,000	161,000
Louisville	827,000	101,000	102,000
Norfolk	681,000	168,000	102,000
Total	14,454,000	1,815,000	1,638,000
Group 4			
Minneapolis	1,814,000	32,000	332,000
Seattle	1,422,000	42,000	315,000
San Diego	1,358,000	62,000	325,000
Denver	1,228,000	50,000	299,000
Portland	1,009,000	23,000	187,000
Phoenix	968,000	33,000	304,000
Total	7,799,000	242,000	1,762,000
Group 5			
Rochester	883,000	58,000	150,000
San Antonio	864,000	60,000	148,000
Dayton	850,000	94,000	123,000
Sacramento	801,000	38,000	175,000
Forth Worth	762,000	83,000	189,000
Hartford	664,000	51,000	115,000
Honolulu	629,000	7,000	129,000
Salt Lake City	560,000	4,000	112,000
Total	6,013,000	395,000	1,141,000

these areas, unless possibly to provide for replacement of sub-standard units—which is, of course, a program quite different from the one I described here.

AMOUNT OF LAND NEEDED

How much land, then, do we need for new satellite communities? The population of the United States as a whole grew by some 23,500,000 people between 1960 and 1970, and about 15,000,000 people were added to the metropolitan areas of over 500,000 in that period. As was noted, Congress expects the population to increase by another 75,000,000 between 1970 and the year 2000, though some experts believe that a growth of 50 to 60,000,000 is a more realistic figure in view of the declining birth rate. For our purposes, however, it is reasonable to assume that each of the metropolitan areas will grow by approximately three times the growth it experienced between 1960 and 1970, but at a declining rate.

In order to calculate how many new communities we will need, let us assume that only half the population growth in each metropolitan area should be accommodated in new satellite communities. The balance of the growth can fill up vacant land in existing suburbs or vacant industrial land in the cities (perhaps as new towns-in-town). Moreover, to anticipate population growth in these new satellite communities, land should be made available so that they can reach their ultimate size over a thirty-year period. In that case, no additional land would be required to replace substandard housing, since the new units would originally be built with this purpose in mind. The substandard units thus vacated in the cities can themselves become the sites of new dwellings in the city.

A minimum new satellite community program would call for the acquisition of land in the metropolitan areas listed in Groups 1 through 3. One-half the estimated population growth during the next thirty years in those areas amounts to approximately 16,000,000 people. A maximum program, on the other hand, would involve acquiring land for satellite communities in all the metropolitan areas listed in Groups 1 through 5. One-half the estimated thirty-year growth in these areas would amount to almost 21,000,000 people.

At 21 people (or approximately six units) to the acre—a relatively low density—1,000,000 acres of land would be required to house 21,000,000 people. The land, in appropriate locations, may cost from $1 to $3,000 per acre. Therefore, a maximum of only $3 billion is required for the entire land acquisition program I have described. Even at lower densities, or with more land allocated for industrial parks and commercial use, $3 billion should be adequate. The minimum program, of course, would cost even less.

At densities of 33 persons per acre (including land for recreational, commercial, industrial, and other uses) the program would cost at most $2 billion. Densities, of course, will vary from one metropolitan area to another. In smaller metropolitan areas, available land for satellite neighborhoods and communities will be less expensive, and lower densities may be preferred, while land will be most expensive in areas like New York and Los Angeles, and therefore somewhat higher densities will be desirable.

Funds for land development are an initial expense rather than an annual expenditure, and the land ought to be resold for development at a profit. Moreover, $3 billion spent once is a surprisingly small figure compared to the $10 billion *annual* expense of the federal government for housing.

LOCATION OF
SATELLITE NEW COMMUNITIES

The metropolitan areas most in need of a new satellite communities program have already been identified. The next problem to be considered is where, within each metropolitan area, should they be located? While each metropolitan area requires a specific plan in order to adequately answer that question, there are certain definable principles to be considered in the acquisition of land.

The first step is to determine where *not* to build. In this regard, one can do no better than to paraphrase David Wallace's list of *don'ts:*

> *(1) Don't build on the ocean beach.* The shore and a substantial area next to it should be preserved in a natural state and be for public use and enjoyment.

(2) Don't build at the river's edge or violate the riverside or the river's setting with inappropriate development. The quality and quantity of clean water, a vital natural resource, flood control, and public enjoyment require application of controls in all areas of the river landscapes.

(3) Don't build on flood plains except under strict controls. Common sense (which does not usually prevail) would dictate limiting development where frequent floods occur.

(4) Don't build on steep slopes or denude areas of forests. Amenity and water management, both vital to future urbanization, argue for this caveat.

(5) Don't build in areas of great recreational value or unique visual personality. These areas can become the most valuable for people as open space close to urbanization.

(6) Don't allow seriously harmful development in violation of principles of natural conservation to remain. We must not be forever victims of mistakes of the past. Over time, uses that are clearly not in the public interest should be phased out, and nature reestablished in our cities.

These environmental safeguards actually remove very little land from circulation in comparison with the total amount available. Even in the Atlantic Region, with its present low-density consumption rates, enough land is left after securing these lands to accommodate all the development anticipated for the next fifty years.

If the removal of certain lands from development defines part of the form which the metropolitan area will take, the rest is defined by the design of the transportation system. This element of the metropolitan plan should be considered next—not only because the nature of the transportation system will determine the location of the new satellite communities, but also because the existence or absence of a mass transportation system has a major impact on the employment opportunities available to their residents.

Automobiles serve only low-density areas efficiently. If sufficient streets and parking spaces are provided to accommodate the automobiles of all the residents of medium-density areas, very little land is left over for the buildings. More urban land is presently used for transportation facilities than for any other function outside of residential. Most suburban zoning codes now require two parking spaces per apartment unit or town house, and over two-thirds of the land area in a typical suburban shopping center is earmarked for parking. A typical community college usually provides about 150 square feet of parking space per student, or one parking space for every two students. Though not all neighborhood obsolescence can be attributed to the automobile, it is certainly true that no urban neighborhood or industrial area developed prior to World War II has the space or design to accommodate a majority of its residents traveling by auto.

In the past, mass transit has been largely built to catch up with population growth, since only high-density population centers can support it. However, both Stockholm and Toronto have shown us that mass transportation systems (in these instances, subways) can determine future growth. Unless the origin and destination of the transit system are in clustered locations, it is not possible to have a mass transit system at all. The automobile, on the other hand, lends itself rather to low-density, single-family housing. The process, then, is circular—the more expressways that are constructed, the more single-family homes are encouraged. Conversely, if subways or high-speed buses are provided, high-density developments are generated at their nodes. Since this appears to be one of the facts of life, good planning suggests that such subway stations or mass transit stops become the foci of new satellite communities with shopping, commercial, and municipal facilities, together with apartments near the core and single-family homes and townhouses located at the periphery.

A good deal of technological study has gone into the development of new mass transportation systems; it is not necessary at this stage to choose among them. It is sufficient to state that what is needed in most of our metropolitan areas is a series of new satellite communities linked to the central city and to each other by some form of high-speed mass transportation—whether it be subway, bus, monorail, or Gravity Vacuum Tube. Planning for transportation must be done simultaneously with determining the location of satellite communities, since each is closely related to the other.

In our larger metropolitan areas, the time now involved in traveling from one place to another exceeds the time devoted to any other single activity, except for sleep and education. The very least that will be accomplished by a good mass transportation system will be to provide inner-city residents with an opportunity to find their way quickly and inexpensively to the sites of the industrial jobs in outlying areas, while suburbanites similarly are able to commute to jobs downtown. Hopefully, if people eventually are able, through the development of new satellite communities, to live near their work, the total burden on the mass transportation system will be considerably diminished. It is even possible that recapturing those millions of hours of commuting time will contribute toward an improvement in the quality of life for the residents of both our cities and the new satellite communities.

THE INNER CITY

This focus on metropolitan areas does not deny that rural problems such as poverty, hunger, disease, poor education, and substandard housing also require urgent attention. Our description of urban racism as a conflict of black and white is but a simplification of a larger problem involving Spanish-speaking people, Indians, Appalachian whites, and other minority groups in many of our larger cities. And, finally, our emphasis on new satellite communities does not imply that inner-city housing problems should be neglected. Industries able to provide significant employment for ghetto residents should be induced to locate there. The test for the industrial use of the land relates to the number of jobs created per acre. Only a few industries will qualify. The Brooklyn Navy Yard is an example of land of this type. In more instances, the land will be suitable for construction of new towns-in-town.

Until a sufficient number of new satellite communities are built to relieve crowding in the ghetto, emphasis there should not be on urban renewal programs that displace more persons than they accommodate, but on programs that seek to improve the quality of education, to reduce unemployment, and to give ghetto residents a voice in the public decisions that affect their lives. What should be built in the ghetto now are medical facilities, schools, libraries, neighborhood centers, and adult job-training facilities. After new satellite communities have relieved the pressure of overcrowding, substandard housing in the ghetto can be eliminated, and new towns-in-town can be built on the vacated land. With overcrowding ended, ghetto rents and the artificially high value of slum buildings will decrease.

Creation of new satellite communities can serve as the catalyst for economic and social solutions to the nation's urban problems. Current policies and legislation, however, do not provide for these communities in sufficient number in the most critical locations or in a manner that will assure desegregation. Consequently, I have suggested a program of land acquisition now by an agency of the federal government. Most of the land needed could be acquired within two years of the passage of the necessary legislation through the government's power of eminent domain.

I am aware of the inevitable resistance to the idea of removing one million acres of land from private control by this method. But historic precedents for federal land utilization in the public interest exist. Land grants to colleges expanded public education; additional grants to railroads opened the West. And contemporary administrations have continued to acquire land for other public purposes such as construction of highways, slum clearance, and urban renewal. I have tried to demonstrate that the creation of satellite communities by federal land acquisition would give us a new arena for the solution of our most critical urban problems. I believe that no other domestic public purpose deserves a higher priority.

19

Epilogue: Suburbia in the Seventies
. . . and Beyond

LOUIS H. MASOTTI

□ THE MAJOR FOCUS OF THIS BOOK has been on the process of suburban urbanization—i.e., how suburban growth and development in the postwar period has resulted in an "outer city" that is rapidly becoming more like the inner city in social structure, economy, problems, and life style. The "quiet crisis" of suburbanization, discovered by the President's Task Force on Suburban Problems in the late sixties, is becoming louder in the seventies, and a serious concern for the future is beginning to develop. It is a concern shared by suburbanites and city dwellers alike, for the future form and direction of suburban development has serious implications for the survival potential of the central city and the viability of the entire metropolitan area.

Forecasting the future is always risky. With a topic as complex as suburbia, the probability of being accurate is low. There are so many factors involved, with parameters so dynamic, that getting a fix on the future of suburbia is not only difficult, but potentially misleading. However, it should be possible to identify some alternative suburban futures about which there is some consensus, as well as the most significant variables which will ultimately determine the one which obtains.

The future of most complex social phenomena, including suburbia, can best be viewed as a series of if-then propositions. However, most forecasting about the future of the suburbs to date has consisted of extrapolations from the present, or wishful thinking about preferred end states. Conjecture has focused primarily on such physical aspects of growth as population distribution and density. Some concentrate on social structure and social problems—e.g., racial and class integration and service-level disparities. Other prognosticators, especially new town advocates, concern themselves with life style and quality of life questions.

A comprehensive approach to the future of the suburban community must consider all three—population growth, social and economic equity, and the quality of life—as well as the resources and strategies likely to influence the suburban future.

ALTERNATIVE URBAN FUTURES

There would seem to be four alternative scenarios for the future development of suburbia. The first assumes that the present pattern of suburban sprawl in ever-widening circles around central cities will continue. Another is predicated on the closing of the suburban frontier and increased densities in the existing ring. A third envisions the development of comprehensively planned new communities and satellite cities in the metropolitan perimeter. And the fourth offers a view of the future in which suburban growth is stabilized or even reversed as the central city is resuscitated.

Bear in mind that these are being suggested as dominant themes of the future based on presently available options; they are not to be considered mutually exclusive. It is probable that some elements of each of these themes will be present in any suburban future.

CONTINUED SPRAWL

It is clear that the dominant theme of suburbanization to date has been the uncoordinated acquisition and development of open land, first on the rim of the city, and, more recently, on the rim of suburbia—in a word, "sprawl." It started when the population's housing preferences, government policy (e.g., FHA mortgages) and private development practices collectively encouraged it, and the weakness or absence of comprehensive land-use planning permitted it. It has continued because landowners and land developers make money, migrating suburbanites still want to fulfill the "American Dream," even in condominiums or mobile homes, and governments do little to regulate it.

Barring a significant decrease in suburbanization, sprawl is a fair bet to dominate the future as it has the past, because both the public and private sectors support it. The private sector (especially investors, mortgage bankers, and developers) encourages the "ever-outward," return-to-nature frontier spirit and benefits economically from the migration, while government agencies are either unable or unwilling to stem the flow. Some federal agencies appear to support suburban growth with a hope—rather than a plan—that an increasing proportion of the urban population will be able to secure "a decent home and suitable living environment." Many suburban munici-palities and school districts encourage a policy of "selective sprawl," which promotes the migration of high tax-producing, low service-consuming populations and activities and excludes others.

The absence of effective regional land-use agencies, state inter-vention, or federal policy permits the continued development of the suburban frontier. Sprawl might be more tolerable if some noble social purpose, such as equality of opportunity, were served. But this does not appear to be the case.

INTENSIFICATION

If effective land-use control of open land on the periphery can be achieved, either by government regulation (possible) or private sector restraint (improbable), the existing suburban ring will experience a significant increase in density. The pressure of new waves of urban migrants on suburban community homogeneity will grow as the development of new residential alternatives slows or stops. Suburbia will be forced into a more intense use of space and will rapidly become more like the central city than an alternative to it.

PLANNED SUBURBAN GROWTH

A third scenario for the suburban future envisions planned growth and development. Government agencies at the national, state, and regional level would establish guidelines and provide assistance to municipalities and private developers who design and execute residential, commercial, and industrial projects which achieve the planned public purposes of the metropolitan area. New "satellite communities" on the metropolitan margins and "accelerated growth centers" built around existing small towns and cities in the hinterland would absorb most of the new suburban growth. The rest would be encouraged to resettle in nonmetropolitan new towns, or "new-towns-in-town" (i.e., within the central cities). Such a future assumes a comprehensive plan for functional growth, including green belts, and regulatory agencies willing and able to enforce it.

RESETTLEMENT OF THE CENTRAL CITY

Another view of the future holds that redevelopment of the center city will become more attractive as suburbia becomes more urbanized. If the quality of city life can be significantly improved, suburbia may stabilize, and the city may be rejuvenated. A sizeable vested interest in the existing city, the federally assisted new-town-in-town program of the 1968 Housing and Urban Redevelopment Act and the cultural vacuum of suburbia could stimulate this alternative. The major deterrents would be the continued high costs of inner-city land, declining levels of public service (especially security) and the exodus of employment opportunities to the suburbs.

As suggested above, these alternative futures are not mutually exclusive. In point of fact, all of them are going on simultaneously. Sprawl continues as the dominant theme, but density of existing suburbs is increasing, satellite communities are being developed, and some urban resettlement is taking place. The future may well be an extrapolation of the present with modifications. But the direction of future development is dependent upon a number of complex factors.

FACTORS IN THE
FUTURE SHAPE OF SUBURBIA

Not the least of the difficulties in forecasting the suburban future is determining the relative weight to place on the myriad of variables likely to influence it. Major shifts in any one of them, or minor changes in some combination of them, could significantly alter the outcome.

In the final report of the President's Task Force on Suburban Problems, four factors were considered to be fundamental influences on the future of the metropolitan suburb; "land, financing, qualified manpower, and research to develop deeper understanding and new solutions" (Haar, 1972: 262). Each is tied to a major Task Force recommendation concerning federal government policy and programs in the area of suburban development—grants and loans to assist in the development of State Land Development Corporations, creation of an Urban Development Bank (URBANK) to provide long-term loans and technical assistance for community facilities, and major increases in federal urban training grants and R & D investments.

These are all obviously important, but, with the exception of land development, one might question how "fundamental" they are. With the advantage of hindsight and hopefully some insight, we would offer a somewhat different list of fundamental influences on the suburban future organized into four groups. No attempt is made here to assess their relative weight. We are simply attempting to identify those variables which are most likely to have a significant impact on the future shape of suburbia. Proceeding on this assumption, we have identified the following groups of influences: personal decisions, land-use controls, fiscal policies, and government organization.

PERSONAL ATTITUDES AND DECISIONS

Ultimately, barring the forced resettlement of populations, the future of the metropolitan area will be shaped by personal decisions about such things as the timing of marriage in the life cycle, family size, and residential preferences, including housing type and location. These decisions are influenced by perceptions and assessments of the relative quality of life styles in urban neighborhoods and suburban communities, which are in turn affected by attitudes toward things like class and racial integration.

Projecting the future of suburban migration and relocation without some attention to personal values and preferences is as foolish as it is difficult. While all the data on this set of variables are not at hand, some reasonable assumptions can be made. If marriage age increases and the size of families moves toward ZPG (zero population growth), suburbanization may decrease. Further, residential preferences appear to be shifting. More suburbanites are living in multiunit structures, and there is a marked increase in the number of people who would prefer to live in a small town rather than a suburb if given a choice.

Despite these trends, there is no strong evidence that attitudes on race and class have altered appreciably in recent years to offset the strong preferences people have, whether in cities or suburbs, to live in communities "with people like themselves." Thus, as the cities becomes poorer and blacker and the inner ring of suburbs more integrated, it seems probable that whites and the middle class of both races will continue to choose a suburban alternative, and some suburbanites will attempt to relocate in the suburban fringe.

LAND-USE CONTROLS

Almost everyone agrees that land is suburbia's most valuable resource. But there is considerable controversy about who ought to control it and how. The President's Suburban Task Force states the problem in terms of "finding a reasonable balance through democratic and legal procedures between the claims and rights of private property on the one hand and the interests and need of the community on the other" (Haar, 1972: 271). With all due respect to the Task Force, I suggest that the problem is more complicated than that.

Historically, control over land and its use has been in the hands of private enterprise. Despite the assertion of the public interest in the twentieth century, primarily through local zoning codes, subdivision regulation, and eminent domain, the use of land in suburbia is still effectively determined by the preferences of the private sector, rather than the requirements of public purpose.

Governmental fragmentation in suburbia and a relatively inarticulate statement of the public interest make effective public land-use control difficult. This vacuum has been filled by private land developers, especially the giants of the industry, who offer specific projects to consume suburban space. The developers, in coalition with large-scale investors and mortgage bankers, have the freedom and flexibility to bargain effectively among municipalities and even across metropolitan areas to market their product—tax-producing residential and commercial construction. Their advantage vis-à-vis the individual municipality is enhanced, and their proposals influenced, by the location decisions of highway and utility planners. Revenue-hungry municipal officials are relatively easy prey to development proposals, although the more astute may bargain successfully for public amenities such as schools, parks, roads, and utilities.

The continuing confrontation between municipalities and private land developers is only one of two arenas within which the struggle over control of land use is being fought. The other is intragovern-mental and involves all three branches of government and public agencies throughout the federal system. While individual munici-palities may make land-use decisions they consider to be in their best interests, these decisions are not always in the best interest of the metropolitan area. As Rubinowitz has argued in Chapter 13, various regional, state, and federal agencies have attempted to influence municipal land-use policies in an effort to bring about reform of social policy, particularly in the area of racial and class equality of opportunity. The results are not all in on these efforts, but to date they are not impressive. Some progress is being made in the state courts as a result of constitutionally based legal challenges, but these have not yet been upheld in the Supreme Court. Legislative and executive interposition has not been very successful even with the carrot of federal financial support. That inducement has now been withdrawn as the federal government moves to abdicate its inter-ventionist role in favor of revenue-sharing and local control, which is bound to insure racial and class exclusion in vast sections of suburbia.

FISCAL INEQUITIES

Land-use policy in the suburbs has two major effects—private sector hegemony and racial and class exclusion as discussed in the preceding section, and fiscal inequities among municipalities.

In spite of a movement toward the municipal income tax, suburban governments, including school districts, throughout the country are dependent on residential and commercial property tax revenue to finance the public services which determine the quality of life in a community and influence the migration patterns within the metropolitan area. For some suburban migrants, the metropolitan area is viewed as a large supermarket; they shop for the best bargain in combined residential quality, high service level, and low taxes. Some of the suburbs can offer none of these, most can afford one or two, and the fortunate few have all three. The first group will become the suburban slums, the third group attracts growth and development, and the middle group, the majority, is in serious financial trouble as taxes fail to keep up with the costs of government.

The problem is one of maldistribution of suburban property tax resources. The costs of suburban growth—congestion, pollution, loss of open space, and the like—are widespread, but the benefits are confined to specific municipalities in the form of tax income. A good example is the large shopping center on the edge of one community but next to another; the former gets all of the tax revenue and some of the congestion; the latter gets none of the taxes and a great deal of the congestion.

Some of this inequity results from historical accident, some from astute negotiation, and a great deal from an inequitable tax distribution policy and its cause, fragmented governmental structure. The Minneapolis-St. Paul metropolitan area is currently experimenting with a tax-sharing plan in which the suburb of location receives sixty percent and the affected communities share forty percent of the tax revenue produced by a major project like a regional shopping center. If such a system becomes widespread, in addition to a more equitable distribution of tax dollars, there is likely to be a rather marked shift in the politics of development location which could have a significant influence on suburbia's future shape.

There are two other fiscal developments of some considerable import for the future of suburbs. One of these is the emergence of federal revenue-sharing as a replacement for categorical grant programs; the other concerns the implications of the Serrano and Rodriguez decisions for the quality of public services.

Revenue-sharing will have a differential effect in suburbia. Relatively affluent suburbs have never benefitted directly from federal categorical programs, although they have obviously gained from open space, highway, sewer, and other public works grants in metropolitan areas. Under revenue-sharing, these suburbs will receive a check which will represent "new money." The question of how it will be used is as yet unanswered; reduced property taxes and increased services are the two obvious choices. For the less-affluent suburbs which have been participating in categorical aid programs like manpower development and child care, revenue-sharing may represent a net loss in income. Thus, future fiscal disparities among suburbs may be greater under revenue-sharing than at present; the rich get richer and the poor get poorer. How this situation will affect migration and resettlement patterns throughout the metropolitan area remains to be seen.

The Serrano and Rodriguez cases (and the other similar decisions) have implications directly opposite to revenue-sharing. While these decisions are limited to the property tax and education, the principle involved—that the quality of local public services and, presumably, the level of spending in them should not be function of local wealth—could logically be extended to apply to all local public services. If this were to occur, there could be a drastic redistribution of tax revenue within the metropolitan area and across the state. Although all disparity would not disappear among communities, the local tax base would become a less-significant factor in the quality of local services. Who would be helped and hurt by such a change, and how much, is less clear. To the extent that service levels now influence the settlement decisions of suburbanites, some changes in the future might be expected.

THE CHANGING ROLES OF GOVERNMENTS

It has been inferred in what has been said above that the problems of suburban land-use regulation and fiscal inequity are in part at least a function of governmental structure and the roles played by governmental actors at the local, state, and federal levels and in the legislative, executive, and judicial branches. Thus, it would appear evident that significant changes in either the structure or role of government will affect the suburban future.

If the past is any indicator of the future, the changes are likely to be incremental and evolutionary rather than radical or revolutionary. Localism is a well-entrenched value in American government, as attested to by the general failure of the long-standing movement to metropolitanize urban area governments and the recent efforts to remove the "federal presence" from local affairs. Nonetheless, the vicissitudes of politics are such that the future role of government is uncertain and highly unpredictable. What is certain and predictable is that governmental activities will continue to influence the direction suburbia goes and its speed in getting there.

We argued in the preface to this book that rapid growth and development is transforming suburbia into a new form of urban civilization. The variables discussed in this section will play a major role in determining what shape this new urban form will take and its relationship to the historic city. Whether the new urbanism will prove to be better or worse than the one it replaces will depend on how successful it is in improving the quality of life and the equality of opportunity throughout the metropolitan area.

REFERENCE

HAAR, C. M. [ed.] (1972) The End of Innocence: A Suburban Reader. Glenview, Ill.: Scott, Foresman.

BIBLIOGRAPHY

BIBLIOGRAPHY

A

ABRAMS, C. (1951) Forbidden Neighbors. New York: Harper.

ADRIAN, C. R. and C. PRESS (1968) "Suburbia," pp. 41-69 in Governing Urban America. New York: McGraw-Hill.

AGGER, R. E., D. GOLDRICH, and B. E. SWANSON (1964) The Rulers and the Ruled. New York: John Wiley.

ALEXANDER, R. (1972) "Fifteen state housing agencies in review." J. of Housing (January): 9-17.

ALLEN, F. L. (1954) "The big change in suburbia, part I." Harper's 208 (June): 21-28.

ALLEN, F. L. (1954) "Part II, crisis in the suburbs." Harper's 209 (July): 47-53.

ALOI, F. and A. A. GOLDBERG (1971) "Racial and economic exclusionary zoning: the beginning of the end?" Urban Law Annual: 9-62.

ALTSHULER, A. A. (1970) Community Control: The Black Demand for Participation in Large American Cities. New York: Western.

ANTON, T. (1969) "Politics and planning in a Swedish suburb." J. of Amer. Institute of Planners 35 (July): 253-263.

ARCHER, R. W. (1969) "From new towns to metrotowns and regional cities, II." Amer. J. of Economics and Sociology 28 (October): 385-398.

ARIES, P. (1962) Centuries of Childhood. New York: Alfred A. Knopf.

ARMSTRONG, R. B. (1972) The Office Industry. New York: Regional Plan Association.

ARNOLD, J. L. (1971) The New Deal in the Suburbs: A History of the Greenbelt Town Program, 1935-54. Columbus: Ohio State Univ. Press.

ARNOVICI, C. (1914) "Suburban development." Annals of Amer. Academy of Pol. and Social Sci. 51: 234-238.

ARON, J. B. (1969) The Quest for Regional Cooperation: A Study of the New York Metropolitan Regional Council. Berkeley: Univ. of California Press.

AXELROD, M. (1956) "Urban structure and social participation." Amer. Soc. Rev. 21 (February): 13-18.

B

BABCLAY, H. B. (1963) "Muslim religious practice in a village suburb of Khartoum." Muslim World 53: 205-211.

BABCOCK, R. F. (1972) "Let's stop romancing regionalism." Planning (July).

BABCOCK, R. F. and F. P. BUSSELMAN (1967) "Citizen participation: a suburban suggestion for the central city." Law and Contemporary Problems 32 (Spring): 220-231.

BABCOCK, R. F. and F. P. BUSSELMAN (1963) "Suburban zoning and the apartment boom." University of Pennsylvania Law Review 3: 1040-1091.

BACHRACH, P. and M. S. BARATZ (1970) Power and Poverty: Theory and Practice. New York: Oxford Univ. Press.

BAERWALD, D. H. and S. CHIZUKO (1972) The Urban Way. Monticello, Ill.: Council of Planning Librarians.

BAHL, R. W. (1968) "The determinants of public expenditures: a review," pp. 198-220 in S. Muchkin et al. (eds.) Functional Federalism: Grants-in-aid and PPB Systems. Washington, D.C.: George Washington University.

BAHL, R. W. (1969) Metropolitan City Expenditures. A Comparative Analysis. Lexington: Univ. of Kentucky Press.

BAHL, R. W. (1970) "Public policy and the urban fiscal problem: piecemeal vs. aggregate solutions." Land Economics 46 (February): 41-50.

BAILEY, S. K. et al. (1962) Schoolmen and Politics. Syracuse: Syracuse Univ. Press.

BAILYN, B. (1960) Education in the Forming of American Society. Chapel Hill: Univ. of North Carolina Press.

BAKER, C. A. (1910) "Population and costs in relation to city management." J. of Royal Statistics Society 73 (December): 73-79.

BALAZS, E. (1964) Chinese Civilization and Bureaucracy. New Haven, Conn.: Yale Univ. Press.

BALDINGER, S. (1971) Planning and Governing the Metropolis. New York: Praeger.

BALTZELL, E. D. (1964) The Protestant Establishment: Aristocracy and Caste in America. New York: Alfred A. Knopf.

BALTZELL, E. D. (1958) Philadelphia Gentlemen: The Making of a National Upper Class. New York: Free Press.

BANFIELD, E. C. (1965) Big City Politics. New York: Random House.

BANFIELD, E. C. (1968) The Unheavenly City: The Nature and the Future of Our Urban Crisis. Boston: Little, Brown.

BANKS, J. A. (1954) Prosperity and Parenthood: A Study of Family Planning among the Victorian Middle Classes. London: Routledge & Kegan Paul.

BANOVETZ, J. M. (1965) "Metropolitan subsidies—an appraisal." Public Administration Rev. 25 (December): 297-301.

BARBER, J. D. (1966) Power in Committees: An Experiment in The Governmental Process. Chicago: Rand McNally.

BARCLAY, D. (1959) "Adolescents in suburbia." National Education Association J. 48 (January): 10-11, 76-79.

BARDIS, P. D. (1961) "Familism among Jews in suburbia." Social Sci. 36 (June): 190-196.

BARTELL, G. D. (1970) "Group sex among the mid-Americans." J. of Sex Research 6 (May): 113-130.

BEATTY, J. L. and O. A. JOHNSON [eds.] (1958) Heritage of Western Civilization. Englewood Cliffs, N.J.: Prentice-Hall.

BEEGLE, A. J. (1947) "Characteristics of Michigan's fringe population." Rural Sociology 12 (September): 254-263.

BELL, C. (1969) "A new suburban politics." Social Forces 47 (March): 280-287.

BELL, W. (1968) "The city, the suburb, and a theory of social choice," pp. 132-168 in Scott Greer et al., The New Urbanism. New York: St. Martin's Press.

BELL, W. (1956) "Familism and suburbanization: one test of the social choice hypothesis." Rural Sociology 21 (September-December): 276-283.

BELL, W. (1958) "Social choice, life styles, and suburban residence," pp. 225-247 in W. M. Dobriner (ed.) The Suburban Community. New York: G. P. Putnam's.

BELL, W. and M. D. BOAT (1957) "Urban neighborhoods and informal social relations." Amer. J. of Sociology 62 (January): 391-398.

BENDINER, R. (1969) The Politics of Schools. New York: Harper & Row.

BENSMAN, J. and B. ROSENBERG (1962) "The culture of the new suburbia." Dissent 9 (Summer): 267-270.

BENSON, C. S. (1968) The Economics of Public Education. Boston: Houghton Mifflin.

BENSON, P. H., A. BROWN, Jr., and L. M. SHEEHY (1956) "A survey of family difficulties in a metropolitan suburb." Marriage and Family Living 18 (August): 249-253.

BERGER, B. M. (1961) "The myth of suburbia." J. of Social Issues 17 (November): 38-49.

BERGER, B. M. (1966) "Suburbia and the American dream." Public Interest 2 (Winter): 80-92.

BERGER, B. M. (1971) "Suburbs, subcultures, and styles of life," pp. 165-187 in Looking for America. Englewood Cliffs, N.J.: Prentice-Hall.

BERGER, B. M. (1960) Working Class Suburb: A Study of Auto-Workers in Suburbia. Berkeley: Univ. of California Press.

BERGER, P. L. and T. LUCKMANN (1966) The Social Construction of Reality. Garden City, N.Y.: Doubleday.

BERK, R. A., R. W. MACK, and J. L. McKNIGHT (1971) Race and Class Differences in Per Pupil Staffing Expenditures in Chicago Elementary Schools, 1969-1970. Evanston, Ill: Northwestern University Center for Urban Affairs.

BERKE, J. S. and J. J. CALLAHAN (1972) "Serrano v. Priest: milestone or millstone for school finance." J. of Public Law: 23-71.

BERKE, J. S., A. K. CAMPBELL, and R. J. GOETTEL [eds.] (1972) Financing Equality of Educational Opportunity: Alternatives for State Finance. Berkeley: McCutchan.

BERKE, J. S. et al. (1971) Federal Aid to Public Education: Who Benefits? Syracuse: Syracuse University Research Corporation.

BERLIN, G. L. and J. R. LANCASTER (1971) "Industrial suburbanization." Exchange Bibliography 223. Monticello, Ill.: Council of Planning Librarians.

BERRY, B.J.L. (1970) "The geography of the United States in the year 2000." Transactions of Institute of British Geographers 51.

BERTSCH, D. and A. SHAFOR (1971) "A regional planning commission experience." Planners Notebook (April): 1-8.

BIDERMAN, A. D. (1966) "Social indicators and goals," pp. 68-153 in R. A. Bauer (ed.) Social Indicators. Cambridge, Mass.: MIT Press.

BINZEN, P. (1970) Whitetown USA. New York: Random House.

BIRCH, D. L. (1970) The Economic Future of City and Suburb. New York: Committee for Economic Development.

BIRKHEAD, G. S. (1962) Metropolitan Issues: Social, Governmental, Fiscal. New York: Maxwell Graduate School of Citizenship and Public Affairs.

BISH, R. L. (1971) Public Economy of Metropolitan Areas. Chicago: Markham.

BLAKE, P. (1963) "The suburbs are a mess." Saturday Evening Post 236 (October 5): 14-16.

BLIZZARD, S. E. and W. F. ANDERSON (1952) "Problems in rural-urban fringe research: conceptualization and delineation." Pennsylvania State College Agricultural Experiment Progress Report 89.

BLOOMBERG, W., Jr., M. SUNSHINE, and T. J. FANAVO (1963) Suburban Power Structures and Public Education, A Study of Values, Influences, and Tax Effort. Syracuse: Syracuse Univ. Press.

BLUMBERG, L. and M. LALLI (1966) "Little ghettos: a study of Negroes in the suburbs." Phylon 27 (Summer): 117-131.

BLUMENFELD, H. (1949) "On the growth of metropolitan areas." Social Forces 28 (October): 59-64.

BLUMROSEN, A. et al. (1970) Enforcing Fair Housing Laws: Apartments in White Suburbia. Springfield, Va.: National Technical Information Service.

BOELAERT, R. (1970) "Political fragmentation and inequality of fiscal capacity in the Milwaukee SMSA." National Tax J. 23 (March): 83-88.

BOGUE, D. J. (1954) A Few Facts About Chicago's Suburbs. Chicago: Chicago Community Inventory.

BOGUE, D. J. (1950) Metropolitan Decentralization: A Study of Differential Growth. Oxford, Ohio: Scripps Foundation for Research in Population Problems; Chicago: University of Chicago Population Research and Training Center.

BOGUE, D. J. [ed.] (1953) Needed Urban and Metropolitan Research. Oxford,

Ohio: Scripps Foundation for Research in Population Problems; Chicago: University of Chicago Population Research and Training Center.

BOGUE, D. J. (1953) Population Growth in Standard Metropolitan Areas, 1900-1950. Washington, D.C.: Government Printing Office.

BOGUE, D. J. (1949) The Structure of the Metropolitan Community. Ann Arbor: Horace H. Rackham School of Graduate Studies.

BOGUE, D. J. (1955) "Urbanism in the United States, 1950." Amer. J. of Sociology 60 (March): 471-486.

BOGUE, D. J. and E. SEIM (1956) "Components of population change in suburban and central city populations of Standard Metropolitan Areas: 1940-1950." Rural Sociology 21 (September-December): 267-275.

BOLLENS, J. C. (1961) Exploring the Metropolitan Community. Berkeley: Univ. of California Press.

BOLLENS, J. C. and H. J. SCHMANDT (1965) "Metropolitan models and types," pp. 58-72 in The Metropolis. New York: Harper & Row.

BOORSTIN, D. J. (1948) The Lost World of Thomas Jefferson. New York: Holt, Rinehart & Winston.

BOOTH, C. (1889) Life and Labour of the People, First Edition. London and Edinburgh: Williams and Norgate.

BOOTH, C. (1902-1903) Life and Labour of the People of London. London and New York: Macmillan.

BOOTH, D. A. (1963) Metropolitics: The Nashville Consolidation. East Lansing: Michigan State University Institute for Community Development and Services.

BORSODI, R. (1933) Flight from the City: The Story of a New Way to Family Security. New York: Harper.

BOSKOFF, A. (1966) "Social and cultural patterns in a suburban area: their significance for urban change." J. of Social Issues 22 (January): 85-100.

BOSKOFF, A. (1970) The Sociology of Urban Regions. New York: Appleton-Century-Crofts.

BOSS, S. T. (1957) "Family size and social mobility in a California suburb." Eugenics Q.: 208-213.

BOSSELMAN, V. and M. CALLIES (1971) The Quiet Revolution in Land Use Control. Washington, D.C.: Council on Environmental Quality.

BOSWELL, J. (1950) London Journal, 1762-1763. New York: McGraw-Hill.

BOULDING, K. E. (1956) The Image. Ann Arbor: Univ. of Michigan Press.

BOWEN, F. (1856) The Principles of Political Economy. Boston: Little, Brown.

BRADBURN, N. M., S. SUDMAN, and G. L. GOCKEL (1971) Side by Side: Integrated Neighborhoods in America. Chicago: Quandrangle.

BRADFORD, D. F. (1970) An Econometric Model of the Flight to the Suburbs. Springfield, Va.: National Technical Information Service.

BRANYAN, R. L. and L. H. LARSON [eds.] (1971) Urban Crisis in Modern America. Boston: D. C. Heath.

BRAZER, H. (1959) City Expenditures in the United States. New York: National Bureau of Economic Research.

BRAZER, M. C. (1967) "Economic and social disparities between central cities and their suburbs." Land Economics 43 (August): 294-302.

BRECKENFELD, G. (1972) "Downtown has fled to the suburbs." Fortune (October).

BREMNER, R. H. (1956) From the Depths: The Discovery of Poverty in the United States. New York: New York Univ. Press.

BRESSLER, M. (1968) "To suburbia with love." Public Interest 10 (Winter): 97-103.

BRIDENBAUGH, C. (1938) Cities in the Wilderness: Urban Life in America, 1625-1742. New York: Ronald Press.

BRIDENBAUGH, C. and J. BRIDENBAUGH (1942) Rebels and Gentlemen: Philadelphia in the Age of Franklin. New York: Reynal & Hitchcock.

BRODERICK, C. B. (1966) "Socio-sexual development in a suburban community." J. of Sex Research 2 (April): 1-24.

BRODY, D. (1960) Steelworkers in America: The Nonunion Era. Cambridge, Mass.: Harvard Univ. Press.

BROOKS, M. (1972) Lower Income Housing: the Planners' Response. Chicago: American Society of Planning Officials.

BROWDER, L. H., Jr. (1970) "A suburban school superintendent plays politics," pp. 191-212 in M. W. Kirst (ed.) The Politics of Education at the Local, State and Federal Levels. Berkeley: McCutchan.

BROWN, W. H., S. C. JACKSON, and J. H. POWELL, Jr. (1971) Open or Closed Suburb: Corporate Location and the Urban Crisis. White Plains, N.Y.: Suburban Action Institute.

BROWNE, W. P. and R. H. SALISBURY (1971) "Organized spokesmen for cities: urban interest groups," pp. 255-278 in Urban Affairs Annual Reviews 6. Beverly Hills: Sage Pubns.

BRUNNER, E. S. and W. C. HALLENBECK (1955) American Society: Urban and Rural Patterns. New York: Harper.

BRY, M. (1969) The Suburbs of San Francisco. San Francisco: Chronicle.

BUCK, D. G. et al. (1956) Friendship and Social Values in a Suburban Community: A Working Report. Eugene: Univ. of Oregon Press.

BUDER, S. (1967) Pullman: An Experiment in Industrial Order and Community Planning, 1880-1930. New York: Oxford Univ. Press.

BURCHARD, J. and A. BUSH-BROWN (1961) The Architecture of America: a Social and Cultural History. Boston: Little, Brown.

BURKE, J. F. (1967) The Suburbs of Pleasure. New York: Delacorte.

BURTON, H. (1955) "Trouble in the suburbs." Saturday Evening Post 228 (September 17): 19-21; 228 (September 24): 32-33; 228 (October 1): 30.

BUSHNELL, H. (1864) Work and Play; or Literary Varieties. New York: Harper.

C

CALHOUN, D. H. (1965) Professional Lives in America: Structure and Aspiration, 1750-1850. Cambridge, Mass.: Harvard Univ. Press.

CALIGURI, J. P. and D. V. LEVINE (1970) "A study of the use of inter-ethnic materials in suburban schools in a major metropolitan area." Phylon, 31 (Fall): 220-230.

CAMPBELL, A. K. [ed.] (1970) The States and the Urban Crisis. Englewood Cliffs, N.J.: Prentice-Hall.

CAMPBELL, A. K. and H. SCHUMAN (1972) "A comparison of black and white attitudes and experiences in the city," pp. 97-110 in C. H. Haar (ed.) The End of Innocence. A Suburban Reader. Glenview, Ill.: Scott, Foresman.

CAMPBELL, A. K. and S. SACKS (1967) Metropolitan America: Fiscal Patterns and Governmental Systems. New York: Free Press.

CAMPBELL, A. and H. C. COOPER (1956) Group Differences in Attitudes and Votes. Ann Arbor: Univ. of Michigan Survey Research Center.

CAPLOW, T. and R. FORMAN (1950) "Neighborhood interaction in a homogeneous community." Amer. Soc. Rev. 15 (June): 357-366.

CARCOPINO, J. (1940) Daily Life in Ancient Rome: The People and the City at the Height of the Empire. New Haven, Conn.: Yale Univ. Press.

CAREY, M. (1830) Miscellaneous Essays. Philadelphia: Carey & Hart.

CARLOS, S. (1970) "Religious participation and the urban-suburban continuum." Amer. J. of Sociology 75 (March): 742-759.

CARMICHAEL, S. and C. V. HAMILTON (1967) Black Power: The Politics of Liberation in America. New York: Random House.

CARRE, M. H. (1964) "Lyson's Greater London." History Today 14 (September): 634-641.

CARTER, P. A. (1971) The Spiritual Crisis of the Guilded Age. De Kalb: Northern Illinois Univ. Press.

CARTER, R. E., Jr., and P. CLARKE (1963) "Suburbanites, city residents and local news." Journalism Q. 40 (Autumn): 548-558.

CARTER, R. E., Jr., and P. CLARKE (1962) "Why suburban news attracts reader interest." Journalism Q. 39 (Autumn): 522-525.

CARTER, R. F. (1963) Voters and Their Schools. Stanford: Institute for Communication Research.

CARTER, R. F. and J. SUTTHOFF (1960) Communities and Their Schools. Stanford: Institute for Communication Research.

CARVER, H. (1965) Cities in the Suburbs. Toronto: Univ. of Toronto Press.

CASSIDY, R. (1971) "G.S.A. plays the suburban game on a grand scale." City (Fall): 12-14, 72.

CASSIDY, R. (1972) "Moving to the suburbs." New Republic (January 22).

CAWELTI, J. G. (1965) Apostles of the Self-Made Man. Chicago: Univ. of Chicago Press.

CHANDLER, A. D., Jr., and S. SALSBURY (1971) Pierre S. du Pont and the Making of the Modern Corporation. New York: Harper & Row.

Chicago Community Inventory (1954) Demographic and Socio-Economic Characteristics of the Population of the City of Chicago and of the Suburbs and Urban Fringe: 1950. Chicago.

Chicago Urban League (1971) Linking Black Residence to Suburban Employment Through Mass Transportation. Chicago.

CHILDE, V. G. (1950) "The urban revolution." Town Planning Rev. 21: 3-17.

CHILDE, V. G. (1954) What Happened in History. Baltimore: Penguin.

CHINITZ, B. [ed.] (1964) City and Suburb: The Economics of Metropolitan Growth. Englewood Cliffs, N.J.: Prentice-Hall.

Citizens' Commission on Maryland Government (1971) "A responsible plan for the financing, governance and evaluation of Maryland's Public Schools." Baltimore, November.

City (1971) "The suburbs: frontier of the 70's." Volume 5 (January/February).

CLARK, D. (1955) "The church in the suburbs." Social Order 5 (January): 26-29.

CLARK, F. P. (1954) "Office buildings in the suburbs." Urban Land 13 (July/August): 7.

CLARK, K. B. (1965) Dark Ghetto: Dilemmas of Social Power. New York: Harper & Row.

CLARK, R. (1970) Crime in America: Observation on Its Nature, Causes, Prevention, and Control. New York: Simon & Schuster.

CLARK, S. D. (1966) The Suburban Society. Toronto: Univ. of Toronto Press.

CLARK, T. N. (1972) "Community social indicators and models." Presented at the Conference on Social Indicator Models, Russell Sage Foundation, New York, July 12-15.

CLAWSON, M. (1960) "Suburban development districts." J. of Amer. Institute of Planners 26 (May): 69-83.

CLAWSON, M. (1971) Suburban Land Conversion in the United States. Baltimore: Johns Hopkins Press.

CLOWSER, M. (1962) "Urban sprawl and speculation in suburban land." Land Economics 38 (May): 99-112.

COHEN, Y. S. (1972) "Diffusion of an innovation in an urban system." University of Chicago Department of Geography Research Paper 140.

COLEMAN, J. S. (1957) Community Conflict. New York: Free Press.

COLEMAN, J. et al. (1966) Equality of Educational Opportunity, Washington, D.C.: Government Printing Office.

Commission on Population Growth and the American Future (1972) Population and the American Future. Washington, D.C.: Government Printing Office.

Committee for Economic Development (1972) Reducing Crime and Assuring Justice. New York: Committee for Economic Development.

CONANT, J. B. (1961) Slums and Suburbs: A Commentary on Schools in the Metropolitan Areas. New York: McGraw-Hill.

Conference on City and Regional Planning (1954) Moving People in Metropolitan Areas. Berkeley: University of California.

CONKIN, P. K. (1959) Tomorrow a New World: The New Deal Community Program. Ithaca: Cornell Univ. Press.

COONS, J. E., W. H. CLUNE, and S. D. SUGARMAN (1970) Private Wealth and Public Education. Cambridge, Mass.: Harvard Univ. Press.

COSSE, M. V. (1928) The Suburban Weekly. New York: Columbia Univ. Press.

COULTER, P. B. (1967) Politics of Metropolitan Areas. New York: Thomas Y. Crowell.

Council of State Governments (1956) The States and the Metropolitan Problem: A Report to the Governor's Conference. Chicago.

COUNTS, G. S. (1927) The Social Composition of Boards of Education. Chicago: Univ. of Chicago Press.

COX, F. and L. BRICKSUR (1967) Retail Decentralization. East Lansing: Michigan State Univ. Press.

COX, H. (1965) The Secular City. New York: Macmillan.

COX, H. (1971) "The ungodly city: a theological response to Jacques Ellul." Commonweal (July 9): 351-357.

COX, K. R. (1968) "Suburbia and voting behavior in the London metropolitan area." Annals of Association of Amer. Geographers 58 (March): 111-127.

CRANDELL, R. F. (1961) This Is Westchester: A Study of Suburban Living. New York: Sterling.

CREMIN, L. A. (1964) The Transformation of the School: Progressivism in American Education, 1876-1957. New York: Vintage.

CREVECOEUR, J. H. (1912) Letters from an American Farmer. London: J. M. Dent.

CROSBY, J. A. et al. (1965) The Urban Fringe Influx: A Study of Real Estate Marketability. Fresno, California: Fresno State College School of Business Bureau of Business Research.

CROUCH, W. W. and B. DINERMAN (1968) "Decentralization in megalopolis," pp. 194-200 in M. D. Speizman (ed.) Urban America in the Twentieth Century. New York: Thomas Y. Crowell.

CROUCH, W. and B. DINERMAN (1964) Southern California Metropolis: A Study of Development of Government for a Metropolitan Area. Los Angeles: Univ. of California Press.

CUNNINGHAM, C. and J. SCOTT (1970) Land Development and Racism in Fairfax County. Fairfax, Va.: Washington Suburban Institute.

CUZZORT, R. P. (1955) Suburbanization of Service Industries within Standard Metropolitan Areas. Oxford, Ohio: Scripps Foundation for Research in Population Problems; Chicago: University of Chicago Population Research and Training Center.

D

DAHL, R. (1960) "The analysis of influence in local communities," in C. R. Adrian, Social Science and Community Action. East Lansing: Michigan State Institute for Community Development and Services.

DALAND, R. T. (1954) "Extraterritorial jurisdiction as an approach to suburban problems of small cities." Southwestern Social Sci. Q. 35 (December): 235-243.

DALY, C. U. [ed.] (1968) The Quality of Inequality: Urban and Suburban Public Schools. Chicago: University of Chicago Center for Policy Study.

DANERELL, R. (1968) Triumph in White Suburbia. New York: William Morrow.

DAUMARD, A. (1970) Les bourgeois de Paris au XIXe siècle. Paris: Flammarion.

DAVIDOFF, L., P. GOLD, and N. GOLD (1971) "The suburbs have to open their gates." New York Times Magazine (November 7): 40-44, 46, 48, 50, 55, 58, 60.

DAVIDOFF, P. and N. GOLD (1970) "Exclusionary zoning." Yale Rev. of Law and Social Action 1 (Winter): 56-63.

DAVIDOFF, P., L. DAVIDOFF, and N. GOLD (1970) "Suburban action: advocate planning for an open society." J. of Amer. Institute of Planners 36 (January): 12-21.

DAVIE, M. R. (1951) "The pattern of urban growth," pp. 244-259 in P. K. Hatt and A. J. Reiss (eds.) Reader in Urban Sociology. New York: Free Press.

DAVIS, H. E. (1953) Urban Transportation—Service or Chaos. Berkeley: University of California Institute of Transportation and Traffic Engineering.

DE GRAZIA, S. (1948) The Political Community: A Study in Anomie. Chicago: Univ. of Chicago Press.

DEMOS, J. (1970) A Little Commonwealth: Family Life in Plymouth Colony. New York: Oxford Univ. Press.

DEUTSCHER, I. and E. THOMPSON (1968) Down Among the People: Encounters with the Poor. New York: Basic Books.

DEWEY, R. (1948) "Peripheral expansion in Milwaukee County." Amer. J. of Sociology 54 (September): 118-125.

DEWEY, R. (1960) "The rural-urban continuum: real but relatively unimportant." Amer. J. of Sociology 66 (July): 60-66.

DICKINSON, W. B. (1960) "Suburban migration." Educational Research Report (July 20): 525-541.

DIETRICH, T. S. (1960) "Nature and direction of suburbanization in the South." Social Forces 39 (December): 181-186.

DIXON, R. G., Jr. (1968) Democratic Representation: Reapportionment in Law and Politics. New York: Oxford Univ. Press.

DIXON, T., Jr. (1896) The Failure of Protestantism in New York and its Causes. New York: Strauss & Rehn.

DIXON, T., Jr. (1905) The Life Worth Living: A Personal Experience. New York: Doubleday, Page.

DOBRINER, W. M. (1963) Class in Suburbia. Englewood Cliffs, N.J.: Prentice-Hall.

DOBRINER, W. M. (1958) "Local and cosmopolitan as contemporary suburban character types," pp. 132-143 in W. M. Dobriner (ed.) The Suburban Community. New York: G. P. Putnam's.

DOBRINER, W. M. (1960) "The natural history of a reluctant suburb." Yale Rev. 49 (Spring): 398-412.

DOBRINER, W. M. [ed.] (1958) The Suburban Community. New York: G. P. Putnam's.

DODSON, D. W. (1958) "Suburbanism and education." J. of Educational Sociology 32 (September): 2-7.

DONALDSON, S. (1968) "City and country: marriage proposals." Amer. Q. 20 (Fall): 547-566.

DONALDSON, S. (1969) The Suburban Myth. New York: Columbia Univ. Press.

DONALDSON, S. (1969) "Suburbia: last frontiersville." Interplay 3 (June/ July): 36-38.

DORNBUSCH, S. M. (1952) A Typology of Suburban Communities: Chicago Metropolitan District, 1940. Chicago: Chicago Community Inventory.

DOTSON, F. and L. O. DOTSON (1956) "Urban centralization and decentralization in Mexico." Rural Sociology 21 (March): 41-49.

DOUGLASS, H. P. (1925) The Suburban Trend. New York and London: Century.

DOUGLASS, H. P. (1939) "The suburban trend," pp. 89-94 in W. M. Dobriner (ed.) The Suburban Community. New York: G. P. Putnam's.

DOUGLASS, H. P. (1934) "Suburbs." Encyclopedia of Social Sciences 14: 433-435.

DOWNES, B. T. [ed.] (1971) Cities and Suburbs. Belmont, Calif.: Wadsworth.

DOWNES, B. T. (1969) "Issue conflict, factionalism, and consensus in suburban city councils." Urban Affairs Q. 4 (June): 477-497.

DOWNES, B. T. (1969) "Suburban differentiation and municipal choices: a comparative analysis of suburban political systems," pp. 243-268 in T. N. Clark (ed.) Community Structure and Decision-making. San Francisco: Chandler.

DOWNES, B. T. and K. R. GREENE (1970) "Black demands for open housing: the response of local governments in three Michigan cities." Presented at the American Political Science Association Annual Meeting, Los Angeles, September.

DOWNING, A. J. (1844) A Treatise on the Theory and Practice of Landscape Gardening. New York: Wiley & Putnam.

DOWNING, A. J. (1850) The Architecture of Country Houses. New York: D. Appleton.

DOWNING, A. J. (1853) Rural Essays. New York: Leavitt & Allen.

DOWNING, P. B. (1969) "Extension of sewer service at the urban-rural fringe." Land Economics 45 (February): 103-111.

DOWNS, A. (1972) Opening Up the Suburbs. New Haven, Conn.: Yale Univ. Press.

DOWNS, A. (1971) "Residential segregation: its effects on education." Educational Digest 36 (April): 12-15.

DOWNS, A. (1969) Urban Problems and Prospects. Chicago: Markham.

DROETTBOOM, T., Jr., R. J. McALLISTER, E. J. KAISER, and E. W. BUTLER (1971) "Urban violence and residential mobility." J. of Amer. Institute of Planners 37 (September): 319-325.

DUHL, L. (1963) The Urban Condition. New York: Basic Books.

DUNCAN, B. (1954) Demographic and Socio-economic Characteristics of the Population of the City of Chicago and the Suburbs and Urban Fringe, 1950. Chicago: Chicago Community Inventory.

DUNCAN, B. (1956) "Factors in work-residence separation: wage and salary workers, Chicago, 1951." Amer. Soc. Rev. 21 (February): 48-56.

DUNCAN, O. D. (1959) "Human ecology and population studies," pp. 678-716 in P. Hauser and O. D. Duncan (eds.) The Study of Population. Chicago: Univ. of Chicago Press.

DUNCAN, O. D. (1951) "The optimum size of cities," pp. 632-645 in P. K. Hatt and A. H. Reiss (eds.) Reader in Urban Sociology. New York: Free Press.

DUNCAN, O. D. and A. J. REISS, Jr. (1956) Social Characteristics of Urban and Rural Communities, 1950. New York: John Wiley.

DUNCAN, O. D. and A. J. REISS, Jr. (1956) "Suburbs and urban fringe," pp. 117-133 in Social Characteristics of Urban Rural Communities. New York: John Wiley.

DWORKIN, R. J. (1968) "Segregation and suburbia," pp. 190-234 in R. W. Mack (ed.) Our Children's Burden. New York: Random House.

DYE, T. R. (1965) "City-suburban social distance and public policy." Social Forces 44 (September): 100-106.

DYE, T. R. (1963) "Local-cosmopolitan dimension and the study of urban polities." Social Forces 41 (March): 263-269.

DYE, T. R. (1962) "Popular images of decision-making in suburban communities." Sociology and Social Research 47 (October): 75-83.

DYE, T. R. (1972) Understanding Public Policy. Englewood Cliffs, N.J.: Prentice-Hall.

E

EBERHARD, J. P. (1966) "Technology for the city." International Science and Technology 57: 18-31.

EDIN, N. J. (1966) Residential Location and Mode of Transportation to Work: A Model of Choice. Chicago: Chicago Area Transportation Study.

ELAZAR, D. J. (1966) American Federalism. New York: Thomas Y. Crowell.

ELKIN, F. and W. A. WESTLEY (1955) "Myth of adolescent culture." Amer. Soc. Rev. 20 (December): 680-684.

ELKINS, S. (1959) Slavery: A Problem in American Institutional and Intellectual Life. Chicago: Univ. of Chicago Press.

ELLICKSON, R. (1967) "Government housing assistance to the poor." Yale Law J. (January): 508-544.

ELLUL, J. (1964) The Technological Society. New York: Alfred A. Knopf.

ELLUL, J. (1970) The Meaning of the City. Grand Rapids, Mich.: Eerdmans.

ELMAN, R. M. (1967) Ill-At-Ease in Compton. New York: Pantheon.

EMPSON, W. (1950) Some Versions of the Pastoral. London: Chatto & Windus.

ENGEL, J. F. and R. D. BLACKWELL (1969) "Attitudes of affluent suburbia toward the Negro neighbor." Michigan State University Business Topics 14 (Fall): 42-49.

ENNIS, P. H. (1967) "Criminal victimization in the United States: a report of a national survey." Presented to the President's Commission on Law Enforcement and Administration of Justice. Washington, D.C.: Government Printing Office.

ENNIS, P. (1958) "Leisure in the suburbs: research prolegomenon," pp. 248-270 in W. Dobriner (ed.) The Suburban Community. New York: G. P. Putnam's.

ERICKSEN, E. G. (1954) Urban Behavior. New York: Macmillan.

ERIKSON, E. H. (1968) Identity; Youth and Crisis. New York: W. W. Norton.

EULAU, H. and K. PREWITT [eds.] (1970) The Urban Governors. Indianapolis: Bobbs-Merrill.

EVERSLEY, D.E.C. (1965) "Population, economy and society," in D. V. Glass and D.E.C. Eversley (eds.) Population in History: Essays in Historical Demography. Chicago: Aldine.

EYESTONE, R. (1970) The Threads of Public Policy: A Study in Policy Leadership. Indianapolis: Bobbs-Merrill.

F

FAGIN, H. (1958) "Problems of planning in the suburbs," pp. 362-371 in W. M. Dobriner (ed.) The Suburban Community. New York: G. P. Putnam's.

FALTERMEYER, E. K. (1968) "Controlling the suburban explosion," in Redoing America: A Nationwide Report on How to Make Our Cities and Suburbs Livable. New York: Harper & Row.

FALTERMEYER, E. K. (1966) "We can cope with the coming suburban explosion." Fortune 74 (September): 147-151.

FANTINI, M., M. GITTELL, and R. MAGAT (1970) Community Control and the Urban School. New York: Praeger.

FARLEY, R. (1970) "The changing distribution of Negroes within metropolitan areas—the emergence of black suburbs." Amer. J. of Sociology 75 (January): 512-529.

FARLEY R. (1964) "Suburban persistence." Amer. Soc. Rev. 29 (February): 38-47.

FARNSWORTH, R.W.C. (1883) A Southern California Paradise, in the Suburbs of Los Angeles, Being a Historic and Descriptive Account of Pasadena, San Gabriel, Sierra Madre, and La Cañada; With Important Reference to Los Angeles and All Southern California. Pasadena: Pacific.

FARWELL, P. T. (1907) "The social work of a suburban church." Annals of Amer. Academy of Pol. and Social Sci. 30 (November): 68-74.

FAVA, S. F. (1958) "Contrasts in neighboring: New York City and a suburban county," pp. 122-131 in W. M. Dobriner (ed.) The Suburban Community. New York: G. P. Putnam's.

FAVA, S. F. (1956) "Suburbanism as a way of life." Amer. Soc. Rev. 21 (February): 34-37.

FELDMAN, S. and G. THIELBAR [eds.] (1972) Life Styles: Diversity in American Society. Boston: Little, Brown.

FINE, B. (1958) "Education problems in the suburbs," pp. 317-325 in W. M. Dobriner (ed.) The Suburban Community. New York: G. P. Putnam's.

FINE, J., N. D. GLENN, and J. K. MONTS (1971) "The residential segregation of occupational groups in central cities and suburbs." Demography 8 (February): 91-101.

FISHMAN, J. A. (1963) "Moving to the suburbs: its possible impact on the role of the Jewish minority in American community life." Phylon 24 (Summer): 146-153.

FISHMAN, J. A. (1961) "Some social and psychological determinants of intergroup relations in changing neighborhoods—an introduction to the Bridgeview study." Social Forces 40 (October): 42-51.

FISHMAN, R. U. and R. W. EVERSON (1971) "The implications of the differing perceptions of urban and suburban citizens for environmental comprehensive health planning." Amer. J. of Public Health 61 (June): 1126 ff.

FITCH, L. C. (1956) "Fiscal and political problems of increasing urbanization." Pol. Sci. Q. 71 (March): 71-90.

FLINN, T. A. (1964) "Party responsibility in the states: some causal factors." Amer. Pol. Sci. Rev. 58 (March): 60-71.

FOGELSON, R. M. (1967) The Fragmented Metropolis: Los Angeles 1850-1930. Cambridge, Mass.: Harvard Univ. Press.

FOLEY, D. L. (1957) The Suburbanization of Administrative Offices in the San Francisco Bay Area. Berkeley: University of California Bureau of Business and Economic Research.

FORCE, J. (1962) "The daily press in suburbia: trends in 15 metropolitan areas." Journalism Q. 39 (Autumn): 457-463.

FORM, W. H. (1945) "Status stratification in a planned community." Amer. Soc. Rev. 10 (October): 605-613.

FORM, W. H. et al. (1954) "The compatibility of alternative approaches to the delimitation of urban sub-areas." Amer. Soc. Rev. 19 (August): 434-440.

FOWLER, E. P. and R. L. LINEBERRY (1972) "The comparative analysis of urban policy: Canada and the United States," pp. 345-368 in H. Hahn (ed.) People and Politics in Urban Society. Beverly Hills: Sage Pubns.

FRANCIS, R. G. (1963) "Family strategy in middle class suburbia." Soc. Inquiry 33 (Spring): 157-164.

FRANKLIN, B. (1959) Papers. New Haven, Conn.: Yale Univ. Press.

FREDRICKSON, G. (1965) The Inner Civil War: Northern Intellectuals and the Crisis of the Union. New York: Harper & Row.

FRENCH, R. M. and J. K. HADDEN (1968) "Mobile homes: instant suburbia or transportable slums?" Social Problems 16 (Fall): 219-226.

FRIED, A. and R. M. ELMAN [eds.] (1968) Charles Booth's London. New York: Pantheon.

FRIESEMA, H. P. (1970) "Interjurisdictional agreement in metropolitan areas." Administrative Sci. Q. 15 (June): 242-252.

FRISCH, M. H. (1972) Town Into City: Springfield, Massachusetts, and the Meaning of Community, 1840-1880. Cambridge, Mass.: Harvard Univ. Press.

G

GABLER, L. R. (1971) "Population size as a determinant of city expenditures and employment—some further evidence." Land Economics 47 (May): 130-138.

GABLER, L. R. (1969) "Economies and diseconomies of scale in urban public sectors." Land Economics 45 (November): 425-434.

GAFFNEY, M. (1964) "Containment policies for urban sprawl," in Approaches to the Study of Urbanization. Lawrence: Univ. of Kansas Press.

Gallup, G. and Associates (1972) "The Gallup poll." Washington Post (December 12).

GALPIN, C. J. (1924) "Replanning the city as a place not to live," in Rural Social Problems. New York: Century.

GANS, H. J. (1967) "An anatomy of suburbia." New Society 10 (September 28): 423-431.

GANS, H. J. (1968) "The balanced community: homogeneity or heterogeneity in residential areas?" pp. 166-182 in People and Plans. New York: Basic Books.

GANS, H. J. (1968) "The disenchanted suburbanite," pp. 371-377 in People and Plans. New York: Basic Books.

GANS, H. J. (1963) "Effects of the move from city to suburb," pp. 184-198 in L. J. Duhl (ed.) The Urban Condition. New York: Basic Books.

GANS, H. J. (1959) The Levittowners. New York: Pantheon.

GANS, H. J. (1951) "Park Forest: birth of a Jewish community." Commentary 11 (April): 330-339.

GANS, H. J. (1968) People and Plans. New York: Basic Books.

GANS, H. J. (1968) "Planning for the everyday life and problems of suburban and new town residents," pp. 183-201 in People and Plans. New York: Basic Books.

GANS, H. J. (1961) "Planning and social life: an evaluation of friendship and neighborhood relations in suburban communities." J. of Amer. Institute of Planners 27 (May): 134-140.

GANS, H. J. (1957) "Progress of a suburban Jewish community." Commentary 23 (February): 113-122.

GANS, H. J. (1956) "Sociology of new towns: opportunities for research." Sociology and Social Research 40 (March): 231-239.

GANS, H. J. (1968) "Suburbia reclaimed," pp. 366-370 in People and Plans. New York: Basic Books.

GANS, H. J. (1968) "The suburban community and its way of life," pp. 132-140 in People and Plans. New York: Basic Books.

GANS, H. J. (1962) "Urbanism and suburbanism as ways of life: a re-evaluation of definitions," pp. 625-648 in A. Rose (ed.) Human Behavior and Social Processes. Boston: Houghton Mifflin.

GANS, H. J. (1962) The Urban Villagers. New York: Free Press.

GANS, H. J. (1968) "White exodus to suburbia steps up." New York Times Magazine (January 7): 24-25.

GARMIRE, B. L. (1972) "The police role in an urban society," pp. 1-11 in R. F. Steadman (ed.) The Police and the Community. Baltimore: Johns Hopkins Press.

Georgetown Law Journal (1970) "Symposium on restructuring metropolitan area government." Volume 58 (March-May): 663 ff.

GERDINE, M. W. and R. L. BRAGG (1970) "Referral patterns among mental health agents in three suburban communities." Amer. J. of Orthopsychiatry 40 (October): 841-849.

GERSH, H. (1954) "The new suburbanites of the 50's—Jewish division." Commentary 17 (March): 209-221.

GERSMAN, E. M. (1970) "Progressive reform of the St. Louis School Board, 1897." History of Education Q. 10: 8-15.

GHURYE, G. S. (1964-1965) "Bombay suburbanites: some aspects of the working life." Soc. Bull. 13 (September): 73-83; 14 (September): 1-8.

GILBERT, C. E. (1967) Governing the Suburbs. Bloomington: Indiana Univ. Press.

GILMORE, H. W. (1953) Transportation and the Growth of the Cities. New York: Free Press.

GIST, N. P. (1952) "Developing patterns of urban decentralization." Social Forces 30 (March): 257-267.

GIST, N. P. (1952) "Ecological decentralization and rural-urban relationships." Rural Sociology 17 (December): 328-335.

GIST, N. P. (1952) "New urban fringe." Sociology and Social Research 36 (May): 297-302.

GIST, N. P. and S. F. FAVA (1964) "Urban and suburban residential areas," pp. 177-208 in Urban Society. New York: Thomas Y. Crowell.

GITCHOFF, G. T. (1969) Kids, Cops and Kilos: A Study of Contemporary Suburban Youth. San Diego: Malter-Westerfield.

GLAAB, C. N. (1963) The American City: A Documentary History. Homewood, Ill.: Dorsey.

GLAAB, C. N. (1968) "Metropolis and suburb: the changing American city," pp. 399-437 in J. Braemen et al. (eds.) Change and Continuity in Twentieth Century America: The 1920's. Columbus: Ohio State Univ. Press.

GLAAB, C. N. and A. T. BROWN (1967) A History of Urban America. New York: Macmillan.

GLASSAR, T. (1970) "Jewburbia: a portable community." New Society 16 (October 8): 630-631.

GLAZER, N. and D. P. MOYNIHAN (1964) Beyond the Melting Pot. Cambridge, Mass.: MIT Press.

GLEAVES, S. (1964) "Levittown revisited." New York Sunday Herald-Tribune Magazine (August 16): 9-12.

GODKIN, E. L. (1896) Problems of Modern Democracy. New York: Charles Scribners.

GOETTEL, R. J. (1971) "The relationship between selected fiscal and economic factors and voting behavior in school budget elections in New York State." Presented to the American Educational Research Association Annual Conference.

GOLD, M. (1930) Jews Without Money. London: N. Douglas.

GOLDSMITH, H. F. and E. G. STOCKWELL (1969) "Interrelationship of occupational selectivity patterns among city, suburban and fringe areas of major metropolitan centers." Social Forces 45 (May): 194-205.

GOLDSTEIN, S. (1965) "Rural-suburban-urban political redistribution in Denmark." Rural Sociology 30 (September): 267-277.

GOLDSTEIN, S. (1963) "Some economic consequences of suburbanization in the Copenhagen metropolitan area." Amer. J. of Sociology 68 (March): 551-564.

GOLDSTEIN, S. and K. B. MAYER (1965) "Impact of migration on the socioeconomic structure of cities and suburbs." Sociology and Social Research 50 (October): 5-23.

GOLDSTON, R. C. (1970) Suburbia: Civic Denial: A Portrait in Urban Civilization. New York: Macmillan.

GOODING, J. (1972), "Roadblocks ahead for the great corporate move-out." Fortune (June).

GORDON, A. I. (1959) Jews in Suburbia. Boston: Beacon.

GORDON, L. (1969) "Suburban consensus formation and the race issue." J. of Conflict Resolution 13 (December): 550-556.

GORDON, R. E. et al. (1962) The Split-Level Trap. New York: Dell.

GORDON, R. E. and K. K. GORDON (1968) "Emotional disorders of children in a rapidly growing suburb." International J. of Social Psychiatry 4 (Autumn): 85-97.

GORDON, R. E. and K. K. GORDON (1960) "Social psychiatry of a mobile suburb." International J. of Social Psychiatry 6 (Summer): 89-100.

GOWANS, A. (1964) Images of American Living: Four Centuries of Architecture and Furniture as Cultural Expression. Philadelphia: J. B. Lippincott.

GRANT, D. R. (1968) "The metropolitan approach—should, can, and will it prevail?" Urban Affairs Q. 3 (March): 103-110.

GRANT, D. R. (1955) "Urban and suburban Nashville: a case study in metropolitanism." J. of Politics 17 (February): 82-99.

GRAS, N.S.B. (1922) "The development of metropolitan economy in Europe and America." Amer. Historical Rev. 27: 695-708.

GRAS, N.S.B. (1926) "The rise of the metropolitan community," in E. W. Burgess (ed.) The Urban Community. Chicago: Univ. of Chicago Press.

GREELEY, A. M. (1959) The Church and the Suburbs. New York: Sheed & Ward.

GREELEY, A. M. (1961) "Suburbia revisited." Social Order 1 (October): 371-337.

GREEN, C. M. (1967) The Secret City: A History of Race Relations in the Nation's Capital. Princeton: Princeton Univ. Press.

GREENFIELD, R. W. (1961) "Factors associated with attitudes toward desegregation in a Florida residential suburb." Social Forces 40 (October): 31-41.

GREENSTEIN, F. I. and R. WOLFINGER (1958-1959) "The suburbs and shifting party loyalties." Public Opinion Q. 22 (Winter): 473-482.

GREER, S. (1962) The Emerging City. New York: Free Press.

GREER, S. (1963) Metropolitics, A Study of Political Culture. New York: John Wiley.

GREER, S. (1960) "The social structure and political process of suburbia." Amer. Soc. Rev. 25 (August): 514-526.

GREER, S. (1962) "The social structure and political process of suburbia: an empirical test." Rural Sociology 27 (December): 438-459.

GREER, S. (1962) "The suburbs: republics in miniature," pp. 83-104 in Governing the Metropolis. New York: John Wiley.

GREER, S. (1972) The Urbane View. New York: Oxford Univ. Press.

GREER, S. (1956) "Urbanism reconsidered: a comparative study of local areas in a metropolis." Amer. Soc. Rev. 21 (February): 19-25.

GREER, S. and E. KUBE (1959) "Urbanism and social structure: a Los Angeles study," in M. Sussman (ed.) Community Structure and Analysis. New York: Thomas Y. Crowell.

GREER, S. and P. ORLEANS (1962) "The mass society and the parapolitical structure." Amer. Soc. Rev. 27 (October): 634-646.

GREVEN, P. J., Jr. (1970) Four Generations: Population, Land and Family in Colonial Andover, Massachusetts. Ithaca: Cornell Univ. Press.

GRIFFITH, T.L.C. (1965) "Evolution and duplication of a pattern of urban growth." Economic Geography 41 (April): 133-156.

GRISWOLD, A. W. (1952) Farming and Democracy. New Haven, Conn.: Yale Univ. Press.

GRISWOLD, E. N. (1925) "Pro and con of annexation of suburbs, illustrated by the situation in Cleveland." National Municipal Rev. 14 (February): 86-90.

GRODZENS, M. (1958) The Metropolitan Area as a Racial Problem. Pittsburgh: Univ. of Pittsburgh Press.

GRUENBERG, S. M. (1955) "The challenge of the new suburbs." Marriage and Family Living 17 (May): 133-137.

GUSFIELD, J. R. (1963) Symbolic Crusade: Status Politics and the American Temperance Movement. Urbana: Univ. of Illinois Press.

GUTMAN, R. (1963) "Population mobility in the American middle class," pp. 172-183 in L. J. Duhl (ed.) The Urban Condition. New York: Basic Books.

H

HAAR, C. M. [ed.] (1972) The End of Innocence: A Suburban Reader. Glenview, Ill.: Scott, Foresman.

HADDEN, J. K. (1969) "Use of ad hoc definitions," in E. F. Borgatta (ed.) Sociological Methodology. San Francisco: Jossey-Bass.

HALL, P. (1966) The World Cities. New York: McGraw-Hill.

HAMMOND, M. (1972) The City in the Ancient World. Cambridge, Mass.: Harvard Univ. Press.

HAMOVITCH, W. and A. LEVENSON (1969) "Projecting suburban employment." Urban Affairs Q. 4 (June): 459-476.

HANDLIN, O. (1962) "Motives for Negro migration to the suburbs," pp. 125-130 in The Newcomers. Garden City, N.Y.: Doubleday Anchor.

HANDLIN, O. and J. BURCHARD [eds.] (1963) The Historian and the City. Cambridge, Mass.: MIT Press.

HARE, E. H. and G. K. SHAW (1965) Mental Health on a New Housing Estate; A Comparative Study of Health in the Two Districts of Croydon. London and New York: Oxford Univ. Press.

Harper's Bazaar (1900) "Rapid transit and home life." Volume 33 (December): 2001-2005.

HARRIS, C. C., Jr. (1966) Suburban Development as a Stochastic Process. Berkeley: University of California Center for Real Estate and Urban Economics.

HARRIS, C. D. (1943) "Suburbs." Amer. J. of Sociology 49 (July): 1-13.

HARRIS, C. D. and E. L. ULLMAN (1945) "The nature of cities." Annals of Amer. Academy of Pol. and Social Sci. 242 (November): 7-17.

Harris, Louis et al., Inc. (1970) "The Virginia Slims American women opinion poll."

HARTNETT, H. D. (1971) A Locational Analysis of those Manufactoring Firms that Have Located and Relocated Within the City of Chicago, 1955-1968. Champaign-Urbana, Illinois.

HATRY, H. P. (1970) "Measuring the effectiveness of nondefense public programs." Operations Research 18 (September/October): 772-784.

HAVENS, M. C. (1964) "Metropolitan areas and Congress: foreign policy and national security." J. of Politics 26 (November): 758-774.

HAWKINS, B. W. (1968) "Fringe-city life-style distance and fringe support of political integration." Amer. J. of Sociology 74 (November): 248-255.

HAWKINS, B. W. (1967) "Life-style, demographic distance and voter support of city-county consolidation." Southwestern Social Sci. Q. 48 (December): 325-337.

HAWKINS, B. W. and T. R. DYE (1970) "Metropolitan 'fragmentation': a research note." Midwest Rev. of Public Administration 4 (February): 17-24.

HAWLEY, A. H. (1956) The Changing Shape of Metropolitan America: Deconcentration Since 1920. New York: Free Press.

HAWLEY, A. H. (1953) Interstate Migration in Michigan, 1935-1940. Ann Arbor: University of Michigan Institute of Public Administration.

HAWLEY, A. H. (1971) Urban Society. New York: Ronald Press.

HAWLEY, A. H. and B. G. ZIMMER (1970) The Metropolitan Community. Its People and Government. Beverly Hills: Sage Pubns.

HEIN, C. J. (1961) "Metropolitan government: residents outside the central urban area." Western Pol. Q. 14 (September): 764-769.

HENDERSON, H. (1953) "The mass-produced suburbs." Harper's 207 (November): 25-32; 207 (December): 80-86.

HIGBEE, E. (1960) "Suburbia," pp. 89-138 in The Squeeze: Cities Without Space. New York: William Morrow.

HIGHAM, J. (1958) "Another look at nativism." Catholic Historical Rev. 44: 147-158.

HIGHAM, J. (1963) Strangers in the Land: Patterns of American Nativism, 1860-1925. New York: Atheneum.

HIGHAM, J. (1966) "American anti-Semitism historically reconsidered," in C. H. Stember (ed.) Jews in the Mind of America. New York: Basic Books.

HIGHAM, J. (1968) "Immigration," in C. V. Woodward (ed.) The Comparative Approach to American History. New York: Basic Books.

HIGHAM, J. (1969) From Boundlessness to Consolidation: The Transformation of American Culture, 1848-1860. Ann Arbor: William L. Clements Library.

HILBERT, M. S. (1954) "Development of sanitary districts for water, sewage, drainage and refuse control." Amer. J. of Public Health 44 (April): 467-522.

HILLS, S. L. (1969) "The planned suburban community." Land Economics 45 (May): 277-282.

HIMES, N. E. (1938) Medical History of Contraception. Baltimore: Williams & Wilkins.

HINDEN, S. J. (1969) "Politics in the suburbs," in C. P. Cotter (ed.) Practical Politics in the United States. New York: Allyn & Bacon.

HINDERSMAN, C. H. (1960) "The evolving downtown—suburban retail patterns." J. of Marketing 25 (October): 59-62.

HINMAN, A. G. (1931) "An inventory of housing in a suburban city." J. of Land and Public Utility Economics 7 (May): 169-180.

HIRSCH, H. (1968) "Suburban voting and national trends: a research note." Western Pol. Q. 21 (September): 508-514.

HIRSCH, W. Z. (1959) "Expenditure implications of metropolitan growth and consolidation." Rev. of Economics and Statistics 41 (August): 232-241.

HODGE, P. L. and P. M. HAUSER (1968) The Challenge of America's Metropolitan Population Outlook, 1960-1985. New York: Frederick A. Praeger.

HODGE, R. W. and D. J. TREIMAN (1968) "Social participation and social status." Amer. Soc. Rev. 33 (October): 722-739.

HOFSTADTER, R. (1955) The Age of Reform. New York: Alfred A. Knopf.

HOLDEN, M. Jr. (1964) "The governance of the metropolis as a problem in diplomacy." J. of Politics 26 (August): 627-647.

HOLLI, M. G. (1969) Reform in Detroit: Hazen S. Pingree and Urban Politics. New York: Oxford Univ. Press.

HOLMES, W. A. (1968) Tomorrow's Church, A Cosmopolitan Community; A Radical Experiment in Church Renewal. Nashville: Abington.

HOLTMAN, A. G. (1968) "Migration to the suburbs, human capital and city income tax losses: a case study." National Tax J. 21 (September): 326-331.

HOOVER, E. M. and R. VERNON (1959) Anatomy of a Metropolis: New York Metropolitan Region Study I. Cambridge, Mass.: Harvard Univ. Press.

HORTON, J. W. and W. E. THOMPSON (1962) "Powerlessness and political negativism: a study of defeated local referendums." Amer. J. of Sociology 67: 485-493.

Housing Authority of Cook County and Real Estate Research Corporation (1949) Scoring Community Desirability. Chicago.

HOWARD, E. (1965) Garden Cities of Tomorrow. Cambridge, Mass.: MIT Press.

HOYES, H. (1960) Spiritual Suburbia: The Church in a New and Growing Community. New York: Vantage.

HOYT, H. (1950) "The influence of highways and transportation in the structure and growth of cities and urban land values," pp. 201-206 in J. Labatut and W. J. Lane (eds.) Highways in Our National Life. Princeton: Princeton Univ. Press.

HUCKSHORN, R. J. and C. E. YOUNG (1960) "Study of voting splits in city councils in Los Angeles County." Western Pol. Q. 13 (June): 479-497.

HUDSON, B. J. (1952) The Urban-Fringe Problem: A Bibliography. Berkeley: University of California Bureau of Public Administration.

HUGGINS, N. I. (1971) Protestants Against Poverty: Boston's Charities, 1870-1900. Westport, Conn.: Greenwood.

HUTHMACHER, J. J. (1962) "Urban liberalism and the age of reform." Mississippi Valley Historical Rev. 49: 231-241.

HUTHMACHER, J. J. (1968) Senator Robert F. Wagner and the Rise of Urban Liberalism. New York: Atheneum.

HUXTABLE, A. L. (1964) "Clusters instead of slurbs." New York Times Magazine (February 9): 36-37, 40, 42, 44.

I

IANNACCONE, L. and F. W. LUTZ (1970) Politics, Power and Policy: The Governing of Local School Districts. Columbus: Charles Merrill.

IPPOLITO, D. S. (1969) "Political perspectives of suburban party leaders." Social Sci. Q. 49 (March): 800-815.

IPPOLITO, D. S. and L. BOWMAN (1969) "Goals and activities of party officials in a suburban community." Western Pol. Q. 22 (September): 572-580.

ISHAK, S. T. (1972) "Consumers' perception of police performance. Consolidation vs. deconcentration: the case of Grand Rapids, Michigan metropolitan area." Ph.D. dissertation. Indiana University.

ISSEL, W. H. (1970) "Modernization in Philadelphia school reform, 1882-1905." Pennsylvania Magazine of History and Biography 94: 381-382.

J

JACKSON, K. T. (1972) "Metropolitan government versus political autonomy: politics on the crabgrass frontier," in K. T. Jackson and S. K. Schultz (eds.) Cities in American History. New York: Alfred A. Knopf.

JACKSON, K. T. (1970) "Urban deconcentration and suburbanization in the nineteenth century." Presented at the Conference on the New Urban History, University of Wisconsin.

JACKSON, S. (1972) "New communities." HUD Challenge (August): 1-4.

JACO, E. G. and I. BELKNAP (1953) "Is a new family form emerging in the urban fringe?" Amer. Soc. Rev. 18 (October): 551-557.

JACOB, P. E. and J. V. TOSCANO [eds.] (1964) The Integration of Political Communities. Philadelphia: J. B. Lippincott.

JAHER, F. C. (1968) "The Boston brahmins in the age of industrial capitalism," in F. C. Jaher (ed.) The Age of Industrialism in America. New York: Free Press.

JAMES, H. T. et al. (1966) Determinants of Educational Expenditures in Large Cities. Stanford: Stanford University School of Education.

JANOSIK, G. E. (1956) "The new suburbia: political significance." Current History 31 (August): 91-95.

JANOSIK, G. E. (1955) "Suburban balance of power: threat to political machine." Amer. Q. 7 (Summer): 123-141.

JEFFERSON, T. (1955) Notes on the State of Virginia. Chapel Hill: Univ. of North Carolina Press.

JENCKS, C. (1972) Inequality: A Reassessment of the Effect of Family and Schooling in America. New York: Basic Books.

JOHNSON, E. S. (1951) "The function of the central business district in the metropolitan community," pp. 480-491 in P. K. Hatt and A. J. Reiss, Jr. (eds.) Cities and Societies. New York: Free Press.

JONASSEN, C. T. (1955) The Shopping Center Versus Downtown. Columbus: Ohio State University Bureau of Business Research.

JONES, L. W. (1955) "The hinterland reconsidered." Amer. Soc. Rev. 20 (February): 40-44.

JUERGENS, G. (1966) Joseph Pulitzer and the New York World. Princeton: Princeton Univ. Press.

K

KAIN, J. F. (1970) "The distribution and movement of jobs and industry," pp. 1-43 in J. Wilson (ed.) The Metropolitan Enigma. Garden City, N.Y.: Doubleday Anchor.

KAIN, J. F. (1968) "Housing segregation, Negro unemployment and metropolitan decentralization." Q. J. of Economics 82 (May): 175-197.

KAPLAN, H. (1967) Urban Political Systems: A Functional Analysis of Metro Toronto. New York: Columbia Univ. Press.

KATZ, M. (1968) The Irony of Early School Reform: Educational Innovation in Mid-Nineteenth Century Massachusetts. Cambridge, Mass.: Harvard Univ. Press.

KEATS, J. (1956) The Crack in the Picture Window. Boston: Houghton Mifflin.

KEE, W. S. (1967) "Suburban population growth and its implications for core city finance." Land Economics 42 (May): 202-211.

KELLEY, A. (1958) "Suburbia, is it a child's utopia?" New York Times Magazine (February 2): 22, 35, 38.

KELLY, J. (1957) "Strange interlude on the 5:28." New York Times Magazine (October 6): 35-36, 41.

KENNEDY, A. J. (1962) The Zone of Emergence. Cambridge, Mass.: Harvard Univ. Press.

KENNEDY, D. M. (1971) Birth Control in America: The Career of Margaret Sanger. New Haven, Conn.: Yale Univ. Press.

KENWARD, J. (1955) The Suburban Child. Cambridge, Eng.: Cambridge Univ. Press.

KESSEL, J. (1962) "Governmental structures and political environment." Amer. Pol. Sci. Rev. 56 (September): 615-621.

KEY, W. H. (1952) A Social Profile of Detroit: 1952. Ann Arbor: Univ. of Michigan Press.

KHALDUN, I. (1969) The Maquaddimah. Princeton: Princeton Univ. Press.

KINZER, D. L. (1964) An Episode in Anti-Catholicism: The American Protective Association. Seattle: Univ. of Washington Press.

KIRKLAND, E. C. (1956) Dream and Thought in the Business Community, 1860-1900. Ithaca: Cornell Univ. Press.

KISH, L. (1954) "Differentiation in metropolitan areas." Amer. Soc. Rev. 19 (August): 388-398.

KITAGAWA, E. M. and D. J. BOGUE (1955) Suburbanization of Manufacturing Activity within Standard Metropolitan Areas. Oxford, Ohio: Scripps Foundation for Research in Population Problems; Chicago: University of Chicago Population Research and Training Center.

KLAIN, A. (1971) "Zoning in suburbia: keep it, reject it, or replace it?" Exchange Bibliography 180. Council of Planning Librarians.

KNEELAND, D. E. (1971) "From 'tin can on wheels' to the 'mobile home.' " New York Times Magazine (May 9): 18, 48, 51, 53, 56, 58, 64-66.

KNOOP, H. (1966) "Some demographic characteristics of a suburban squatting community of Leopoldville: a preliminary analysis." Cahiers Economiques et Sociaux 4 (June): 119-150.

KNOWLES, L. L. and K. PREWITT [eds.] (1969) Institutional Racism in America. Englewood Cliffs, N.J.: Prentice-Hall.

KOHN, C. F. (1957) "Spatial patterns of suburban commercial retail services: the problem and present shopping habits." Annals of Association of Amer. Geographers 47 (June): 166-167.

KOISTINEN, P.A.C. (1967) "The 'industrial-military complex' in historical perspective: World War I." Business History Rev. 41 (Winter): 819-839.

KOMAROVSKY, M. (1946) "The voluntary associations of urban dwellers." Amer. Soc. Rev. 11 (December): 686-698.

KRAMER, J. [ed.] (1972) North American Suburbs: Politics, Diversity, and Change. Berkeley: Glendessary.

KRISTENSEN, C. et al. (1971) The Suburban Lock-out Effect. White Plains, N.Y.. Suburban Action Institute.

KRISTOL, I. (1970) "Urban civilization and its discontents." Commentary (July): 29-35.

KRISTOL, I. (1970) "Is the urban crisis real?" Commentary (November): 44-47.

KRUTCH, J. W. (1965) "If you don't mind my saying so (the American sloburb)." Amer. Scholar 34 (Spring): 162-164.

KTSANES, T. and L. REISSMAN (1959-1960) "Suburbia—new homes for old values." Social Problems 7 (Winter): 187-195.

KURLAND, P. (1968) "Equal educational opportunity: the limits of Constitutional jurisprudence undefined." University of Chicago Law Rev. 35: 583.

KURTZ, R. A. and J. B. EICHER (1958) "Fringe and suburb: a confusion of concepts." Social Forces 37 (October): 32-37.

L

LAAS, W. (1950) "The suburbs are strangling the city." New York Times Magazine (June 18): 22-23, 52-53.

LADD, E. C., Jr. (1969) Ideology in America: Change and Response in a City, a Suburb, and a Small Town. Ithaca: Cornell Univ. Press.

LANE, R. (1967) Policing the City: Boston, 1821-1885. Cambridge, Mass.: Harvard Univ. Press.

LANGDON, G. (1952) "Delimiting the Main Line District of Philadelphia." Economic Geography 28 (January): 57-65.

LASLETT, P. (1965) The World We Have Lost. New York: Charles Scribner's.

LASSWELL, H. (1958) Politics: Who Gets What, When, How. New York: World.

LAZERWITZ, B. (1960) "Metropolitan community residential belts." Amer. Soc. Rev. 25 (April): 245-252.

LAZERWITZ, B. (1960) "Suburban voting trends: 1948-56." Social Forces 39 (October): 29-36.

LE CORBUSIER, (1967) "To live! (to breathe); suburbs must be eliminated and nature brought into the cities themselves," pp. 104-111 in The Radiant City. London: Faber & Faber.

LEE, D. (1963) "Suburbia reconsidered: diversity and the creative life," pp. 122-134 in E. Geen et al. (eds.) Man and the Modern City. Pittsburgh: Univ. of Pittsburgh Press.

LEONARD, W. N. (1958) "Economic aspects of suburbanization," pp. 181-194 in W. M. Dobriner (ed.) The Suburban Community. New York: G. P. Putnam's.

LEONE, R. A. (1972) "The role of data availability in intrametropolitan workplace location studies." Annals of Economic and Social Measurement: 171-182.

LERNER, D. (1958) "Comfort and fun: morality in a nice society." Amer. Scholar 27 (Spring): 153-165.

LERNER, M. (1957) "The suburban revolution," pp. 172-182 in America as a Civilization. New York: Simon & Schuster.

LEVIN, B. et al. (1972) Public School Finance: Present Disparities and Fiscal Alternatives. Washington, D.C.: Urban Institute.

LEVINE, E. M. (1966) The Irish and Irish Politicians. Notre Dame, Ind.: Univ. of Notre Dame Press.

LIEBERSON, S. (1962) "Suburbs and ethic residential patterns." Amer. J. of Sociology 67 (May): 673-681.

LIEBMAN, C. (1961) "Functional differentiation and political characteristics of suburbia." Amer. J. of Sociology 66 (March): 485-490.

LINEBERRY, R. L. and I. SHARKANSKY (1971) Urban Politics and Public Policy. New York: Harper & Row.

Little, Arthur D. Inc. (1969) East Cleveland: Response to Urban Change. Washington, D.C.: Communication Service Corporation.

LOCKRIDGE, K. A. (1970) A New England Town at First Hundred Years: Dedham, Massachusetts, 1636-1736. New York: W. W. Norton.

London County Council (1915) Comparative Municipal Statistics, 1912-1913. London: Local Government and Statistical Department.

LONG, N. (1972) The Unwalled City: Reconstituting the Urban Community. New York: Basic Books.

Look Magazine, Editors of (1968) Suburbia, The Good Life in Our Exploding Utopia. New York: Cowles Education Corporation.

LOTCHIN, R. (1972) "San Francisco: the patterns of chaos and growth," in K. T. Jackson and S. K. Schultz (eds.) Cities in American History. New York: Alfred A. Knopf.

LOTH, D. (1967) Crime in the Suburbs. New York: William Morrow.

LOVELOCK, C. H. [ed.] (1971) Balanced Transportation Planning for Suburban and Academic Communities: A Case Study of the Midpeninsula Region of the San Francisco Bay Area. Stanford: Stanford Workshops on Political and Social Issues.

LOWENSTEIN, L. K. (1963) "Impact of new industry on the fiscal revenues and expenditures of suburban communities: case study of three Philadelphia townships: Lower Merion, Upper Merion, and Radnor." National Tax J. 16 (June): 113-136.

LOWRY, R. (1955) Toward a Sociology of Suburbia. Berkeley: Berkeley Publications in Society and Institutions.

LUBOVE, R. (1963) Community Planning in the 1920's: The Contribution of the Regional Planning Association of America. Pittsburgh: Univ. of Pittsburgh Press.

LUNDBERG, G. A., M. KOMARSKY, and M. A. McINERY (1934) Leisure—A Suburban Study. New York: Columbia Univ. Press.

LYNES, R. (1954) The Tastemakers. New York: Harper.

LYONS, W. E. and R. L. ENGSTROM (1971) "Life-style and fringe attitudes toward the political integration of urban governments." Midwest J. of Pol. Sci. 15 (August): 475-494.

M

MacNAMARA, D.E.J. (1950) "American police administration at midcentury." Public Administration Rev. 10 (Summer): 181-189.

MAIR, L. (1964) Primitive Government. Baltimore: Penguin.

MALINOWSKI, B. (1929) The Sexual Life of Savages in North-Western Melanasia. New York: Horace Liveright.

MANDELKER, D. R. (1963) "The suburban community: change vs. stability," pp. 729-856 in Managing our Urban Environment. Indianapolis: Bobbs-Merrill.

MANGIN, W. (1957) "Latin American squatter settlements: a problem and a solution." Latin American Research Rev. 2 (Summer): 65-98.

MANHEIM, U. (1957) "Residential growth patterns in metropolitan areas." Urban Land 16 (March): 3-7.

MANIS, J. G. (1959) "Annexation: the process of reurbanization." Amer. J. of Economics and Sociology 18 (July): 353-360.

MANIS, J. G. (1968) "Urbanism and annexation attitudes in two similar suburban areas." Amer. J. of Economics and Sociology 27 (October): 347-363.

MANIS, J. G. and L. STINE (1958) "Suburban residence and political behavior." Public Opinion Q. 22 (Winter): 483-489.

MARCUS, N. and M. W. GROVES [eds.] (1970) The New Zoning: Legal, Administrative and Economic Concepts and Techniques. New York: Praeger.

MARTIN, R. C. (1962) Government and the Suburban School. Syracuse: Syracuse Univ. Press.

MARTIN, R. C. (1963) Metropolis in Transition. Washington, D.C.: Government Printing Office.

MARTIN, R. C. et al. (1961) Decisions in Syracuse. Bloomington: Indiana Univ. Press.

MARTIN, R. G. (1950) "Life in the new suburbia: Long Island development." New York Times Magazine (January 15): 16, 40-42.

MARTIN, W. (1952) "A consideration of differences in the extent and location of the formal associational activities of rural-urban fringe residents." Amer. Soc. Rev. 17 (December): 687-694.

MARTIN, W. T. (1957) "Ecological change in satellite rural areas." Amer. Soc. Rev. 22 (April): 173-183.

MARTIN, W. T. (1953) The Rural-Urban Fringe. Eugene: Univ. of Oregon Press.

MARTIN, W. T. (1953) "Some socio-psychological aspects of adjustment to residence location in the rural-urban fringe." Amer. Soc. Rev. 18 (April): 248-253.

MARTIN, W. T. (1956) "The structuring of social relationships engendered by suburban residence." Amer. Soc. Rev. 21 (August): 446-453.

MARX, L. (1964) The Machine in the Garden: Technology and the Pastoral Ideal in America. New York: Oxford Univ. Press.

MASLOW, A. H. (1959) "A theory of human motivation," in L. Gorlow and W. Katkovsky (eds.) Readings in the Psychology of Adjustment. New York: McGraw-Hill.

MASOTTI, L. H. (1967) Education and Politics in Suburbia. Cleveland: Western Reserve Press.

MASOTTI, L. H. (1968) "Political integration in suburban education communities," pp. 264-286 in S. Greer et al., The New Urbanism. New York: St. Martin's Press.

MASOTTI, L. H. and D. R. BOWEN (1971) "Communities and budgets: the sociology of municipal expenditures," pp. 314-326 in C. H. Bonjean et al. (eds.) Community Politics. A Behavioral Approach. New York: Free Press.

MASOTTI, L. H. and J. K. HADDEN [eds.] (forthcoming) Suburbia: The New America. Chicago: Quadrangle.

McARTHUR, R. E. (1971) Impact of City-County Consolidation of the Rural-Urban Fringe: Nashville-Davidson County, Tennessee. Washington, D.C.: Government Printing Office.

McCAUSLAND, J. L. (1972) "Crime in the suburbs," pp. 61-64 in C. H. Haar (ed.) The End of Innocence. A Suburban Reader. Glenview, Ill.: Scott, Foresman.

McINTOSH, J. R. (1970) Life Styles and Attitudes Just Beyond the Urban Fringe. Stroudsburg, Pa.: Tocks Island Regional Advisory Council.

McKAIN, W. C., Jr. (1963) "Rural suburbs and their people." J. of Cooperative Extension 1 (Summer): 76-84.

McKELVEY, B. (1963) The Urbanization of America, 1860-1915. New Brunswick, N.J.: Rutgers Univ. Press.

McKELVEY, B. (1968) The Emergence of Metropolitan America, 1915-1966. New Brunswick, N.J.: Rutgers Univ. Press.

McKENZIE, R. D. (1933) The Metropolitan Community. New York: McGraw-Hill.

McLELLAND, M. (1968) "Needs and services in an outer suburb." Australian J. of Social Issues 3 (April): 44-62.

McMAHON, J. R. (1917) Success in the Suburbs. New York and London: G. P. Putnam's.

MELTZER, J. (1972) "Loop betrays deeper ills of city." Chicago Sun-Times Viewpoint (October 22).

MERINO, J. A. (1968) "A great city and its suburbs: an attempt to integrate metropolitan Boston, 1865-1920." Ph.D. dissertation. University of Texas.

Metropolitan Council (1971) The Impact of Fiscal Disparity on Metropolitan Municipalities and School Districts. Minneapolis.

Metropolitan Housing and Planning Council (1957) Towards New Horizons; Chicago and Its Suburbs Plan Together. Chicago.

MEYER, A. E. (1967) An Educational History of the American People. New York: McGraw-Hill.

MEYER, H. M. (1953) "What we need to know about the internal structure of cities and metropolitan areas," in D. J. Bogue (ed.) Needed Urban and Metropolitan Research. Oxford, Ohio: Scripps Foundation for Research in Population Problems; Chicago: University of Chicago Population Research and Training Center.

MEYERS, M. (1957) The Jacksonian Persuasion: Politics and Belief. Stanford: Stanford Univ. Press.

MEYERSOHN, R. and R. JACKSON (1958) "Gardening in suburbia," pp. 271-286 in W. Dobriner (ed.) The Suburban Community. New York: G. P. Putnam's.

MIEL, A. (1967) The Short-changed Children of Suburbia. New York: Institute of Human Relations Press.

MILLER, P. (1967) Nature's Nation. Cambridge, Mass.: Harvard Univ. Press.

MILLER, Z. L. (1968) "Boss Cox's Cincinnati: a study in urbanization and politics, 1880-1914." J. of Amer. History 54 (March): 823-838.

MILLER, Z. L. (1968) Boss Cox's Cincinnati: Urban Politics in the Progressive Era. New York: Oxford Univ. Press.

MILLET, J. H. and D. PITTMAN (1958) "The new suburban voter: a case study in electoral behavior." Southwestern Social Sci. Q. 39 (June): 33-42.

MILLS, C. W. (1948) The New Men of Power. New York: Harcourt, Brace.

MILLS, E. S. (1970) "Urban density functions." Urban Studies 7 (February): 5-20.

MILLS, E. S. (1972) Urban Economics. Chicago: Scott, Foresman.

MINAR, D. (1966) "The community basis of conflict in school system politics." Amer. J. of Sociology 31 (December): 822-835.

MINAR, D. (1962) "Democracy in the suburbs." Northwestern Tri-Q. (Fall): 23-28.

MINAR, D. (1966) "Educational decision-making in suburban communities." Northwestern University Cooperative Research Project 2440.

MINAR, D. (1962) "School, community and politics in suburban areas," pp. 90-104 in B. J. Chandler et al. (eds.) Education in Urban Society. New York: Dodd, Mead.

MITGANG, H. (1955) "From the suburbs back to the city." New York Times Magazine (May 15): 17, 37.

MOGEY, J. M. (1955) "Changes in family life experienced by English workers moving from slums to housing estates." Marriage and Family Living 17 (May): 123-129.

MOTE, F. W. (1970) "The city in traditional Chinese civilization," in J.T.L. Liu and W. Tu (eds.) Traditional China. Englewood Cliffs, N.J.: Prentice-Hall.

MOWITZ, R. and D. S. WRIGHT (1962) Profile of a Metropolis. Detroit: Wayne State Univ. Press.

MOWRER, E. R. (1958) "The family in suburbia," pp. 147-164 in W. M. Dobriner (ed.) The Suburban Community. New York: G. P. Putnam's.

MOWRER, E. R. (1961) "Sequential and class variables of the family in the suburban area." Social Forces 40 (December): 107-112.

MUELLER, S. A. (1966) "Change in the social status of Lutheranism in ninety Chicago suburbs, 1950-1960." Soc. Analysis 27 (Fall): 138-145.

MUMFORD, L. (1961) The City in History: Its Origins, Its Transformations, and Its Prospects. New York: Harcourt, Brace & World.

MUMFORD, L. (1938) The Culture of Cities. New York: Harcourt, Brace.

MUMFORD, L. (1962-1963) "The future of the city." Architectural Record 132 (October): 121-128; 132 (November): 139-144; 132 (December): 101-108; 133 (January): 119-126; 133 (February): 119-126.

MUMFORD, L. (1961) "Suburbia and beyond," pp. 482-524 in The City in History. New York: Harcourt, Brace & World.

MUNSON, B. (1956) "Attitudes toward urban and suburban residence in Indianapolis." Social Forces 35 (October): 76-80.

N

NASH, D. and P. BERGER (1962) "The child, the family and the 'religious revival' in suburbia." J. for Scientific Study of Religion 2 (October): 85-93.

NASH, D. and P. BERGER (1963) "Church commitment in an American suburb: an analysis of the decision to join." Archives de Sociologie des Religions 7 (January-July): 105-120.

NASH, R. (1967) Wilderness and the American Mind. New Haven, Conn.: Yale Univ. Press.

National Center for Education Statistics, Office of Education (n.d.) "Bond sales for public school purposes." Annual reports.

National Committee Against Discrimination in Housing (1970) Jobs and Housing: A Study of Employment and Housing Opportunities for Racial Minorities in Suburban Areas of the New York Metropolitan Region. New York.

National Education Association (1970-1971) Estimates of School Statistics, 1969-70; Rankings of the States, 1970. Washington, D.C.

National Housing and Development Law Project (1970) "Handbook on Housing Law." Berkeley, California.

National Resources Committee (1939) Urban Government. Washington, D.C.: Government Printing Office.

NEENAN, W. B. (1970) "Suburban-central city exploitation thesis: one city's tale." National Tax J. 23 (June): 117-139.

NEUTZE, M. (1971) The Suburban Apartment Boom: Case Study of a Land Use Problem. Baltimore: Johns Hopkins Press.

New Jersey, State of (1970) "The housing crisis in New Jersey."

NEWMAN, D. K. (1967) "Decentralization of jobs." Monthly Labor Rev. 90: 7-13.

New Jersey, State of (1970) "The housing crisis in New Jersey."

New York Commission to Study the Quality, Costs, and Financing of Elementary and Secondary Education (1972) Report.

New York State Urban Development Corporation (1972) "Fair share."

NIEDERCORN, J. H. and J. F. KAIN (1963) Suburbanization of Employment and Population, 1948-1975. Santa Monica, Calif.: RAND Corporation.

Northeastern Illinois Metropolitan Area Planning Commission (1960) Suburban Factbook 1950-1960; A Socio-Economic Data Inventory for 100 Municipalities in Northeastern Illinois. Chicago.

NORTON, J. A. (1963) "Referenda voting in a metropolitan area." Western Pol. Q. 16 (March): 195-212.

NOYCE, G. B. (1970) The Responsible Suburban Church. Philadelphia: Westminster.

O

OGBURN, W. F. (1923) Social Change with Respect to Culture and Original Nature. New York: B. W. Heubsch.

OGBURN, W. F. (1937) Social Characteristics of Cities. Chicago: International City Managers' Association.

OKUDA, M. (1952) "Urbanization in suburbia." Japanese Soc. Rev. 13 (October): 21-31.

OLMSTEAD, F. L. and C. VAUX (1868) Preliminary Report Upon the Proposed Suburban Village at Riverside, Near Chicago. New York: Sutton Browne.

OPPENHEIM, A. D. (1955) Cost Factors in Suburban Development; An Analysis of the Effect of Physical Patterns of Community Development, Housing Types and Densities on the Cost of Selected Municipal Services and Required Streets and Frontage Utilities. Harrisburg, Pa.: Pennsylvania State Planning Board.

ORBELL, J. M. and K. S. SHERRILL (1969) "Racial attitudes and the metropolitan context, a structure analysis." Public Opinion Q. 32 (Spring): 46-54.

ORR, J. B. and F. P. NICHELSON (1970) The Radical Suburb. Philadelphia: Westminster.

OSBORN, F. S. and A. WHITTICK (1970) The New Towns. Cambridge, Mass.: MIT Press.

OSTROM, E. (1971) "Institutional arrangements and the measurements of policy consequences." Urban Affairs Q. 6 (June): 447-476.

OSTROM, E. (1972) "Metropolitan reform: propositions derived from two traditions." Social Sci. Q. (December).

OSTROM, E. and G. P. WHITAKER (1973) "Does local community control of police make a difference? Some preliminary findings." Midwest J. of Pol. Sci. (February).

OSTROM, E., W. BAUGH, R. GUARASCI, R. PARKS and G. P. WHITAKER (1973) Community Organization and the Provision of Police Services. Sage Professional Papers in Administrative and Policy Studies. Beverly Hills: Sage Pubns.

OSTROM, E., R. B. PARKS, and G. P. WHITAKER (1973) "Do we really want to consolidate urban police forces? a reappraisal of some old assumptions." Public Administration Rev.

OSTROM, V. (1973) The Intellectual Crisis in American Public Administration. University: Univ. of Alabama Press.

OSTROM, V. (1971) The Political Theory of a Compound Republic. Blacksburg: Virginia Polytechnic Institute Center for the Study of Public Choice.

OSTROM, V., C. M. TIEBOUT, and R. WARREN (1961) "The organization of government in metropolitan areas: a theoretical inquiry." Amer. Pol. Sci. Rev. 55 (December): 831-842.

Oxford University (1938) A Survey of the Social Services in the Oxford District. London: Oxford Univ. Press.

P

PACKARD, V. (1972) A Nation of Strangers. New York: David McKay.

PALEN, J. J. and L. F. SCHNORE (1965) "Color composition and city-suburban status differences: a replication and extension." Land Economics 41 (February): 87-91.

PALMER, C. B. (1955) "All suburbia is divided into three parts." New York Times Magazine (March 6): 14, 64-65.

PALMER, C. B. (1950) Slightly Cooler in the Suburbs. Garden City, N.Y.: Doubleday.

PALMER, D. (1963) All Things New; A Study of the Church's Approach to the New Communities. London: Mowbray.

PATTERSON, D. L. and S. J. SMITS (1972) "Reactions to inner-city and suburban adolescents to three minority groups." J. of Psychology 80 (January): 127-134.

PETERSEN, W. (1960) "The demographic transition in the Netherlands." Amer. Soc. Rev. 25: 334-347.

PETERSON, M. D. (1960) The Jefferson Image in the American Mind. New York: Oxford Univ. Press.

PHARES, D. (1971) "Racial change and housing values: transition in an inner suburb." Social Sci. Q. 52 (December).

PHILLIPS, H. S. (1942) "Municipal efficiency and town size." J. of Town Planning Institute 28 (May/June): 139-148.

PINKERTON, J. R. (1969) "City-suburban residential patterns by social class: a review of the literature." Urban Affairs Q. 4 (June): 499-519.

PIRENNE, H. (1956) Medieval Cities. Garden City, N.Y.: Doubleday.

POTTER, J. (1965) "The growth of population in America, 1700-1860," in D. V. Glass and D.E.C. Eversley (eds.) Population in History. London: Arnold.

POTVIN, D. J. (1969) "Subzoning ordinances and buildings codes: their effect on low and moderate income families." Notre Dame Lawyer 45 (Fall): 123-134.

POWERS, M. G. (1964) "Age and space aspects of city and suburban housing." Land Economics 40 (November): 381-387.

PRATT, S. (1957) "Metropolitan community development and change in sub-center economic functions." Amer. Soc. Rev. 22 (August): 434-440.

PRATT, S. and L. PRATT (1960) "The impact of some regional shopping centers." J. of Marketing 25 (October): 44-50.

PRATT, S. and L. PRATT (1958) Suburban Downtown in Transition; A Problem in Business Change in Bergen County, New Jersey. Rutherford, N.J.: Fairleigh Dickinson University Institute of Research.

PRATTER, J. (1969) "Legal implementation of a satellite city plan: the planned disposition of public land." Urban Law Annual 2: 1-21.

President's Commission on Law Enforcement and Administration of Justice (1967) The Challenge of Crime in a Free Society. New York: Avon.

President's Commission on Law Enforcement and Administration of Justice (1967) Task Force Report on the Police. Washington, D.C.: Government Printing Office.

President's Commission on School Finance (1972) Schools, People and Money: The Need for Educational Reform.

President's Task Force on Suburban Problems (1972) "Final report excerpt," pp. 13-17 in C. M. Haar, The End of Innocence: A Suburban Reader. Glenview, Ill.: Scott, Foresman.

PRESS, C. (1963) "Attitudes toward annexation in a small city area." Western Pol. Q. 16 (June): 271-278.

Public Administration Review (1970) "Symposium on governing megacentropolis." Volume 30 (September/October): 520, 531-543.

PUTNAM, R. D. (1966) "Political attitudes and the local community." Amer. Pol. Sci. Rev. 60 (September): 640-654.

Q

QUINN, J. A. (1955) Urban Sociology. New York: American.

R

RABINOVITZ, F. F. (1971) "The role of the Department of Housing and Urban Development in suburban and metropolitan development." Report of the U.S. House of Representatives Committee on Banking and Currency Subcommittee on Housing, Washington, D.C.

RAE, J. B. (1965) The American Automobile: A Brief History. Chicago: Univ. of Chicago Press.

RAKOVE, M. (1965) The Changing Patterns of Suburban Politics in Cook County. Chicago: Loyola University Illinois Center for Research in Urban Government.

RAMOS, C. P. (1961) "Manila, Quezon City and suburbs." EROPA Rev. 1 (October): 85-130.

RAPKIN, C. and W. G. GRIGSBY (1960) Residential Renewal at the Urban Core. Philadelphia: Univ. of Pennsylvania Press.

RAY, W. W. (1970) "The rural-urban fringe." Exchange Bibliography 133. Council of Planning Librarians.

Raymond and May Associates (1968) Zoning Controversies in the Suburbs: Three Case Studies. Washington, D.C.: Government Printing Office.

REEDER, L. G. (1952) "The central area of Chicago—a re-examination of the process of decentralization." Land Economics 28 (November): 369-373.

REEDER, L. G. (1955) "Industrial deconcentration as a factor in rural-urban fringe development." Land Economics 31 (August): 276-277.

REICHLEY, J. (1970) "As go the suburbs, so goes U.S. politics." Fortune 82 (September): 105-109, 155-156.

REICHLEY, J. (1959) The Art of Government. New York: Fund for the Republic.

REILLY, W. and S. SCHULMAN (1969) "The state urban development corporation: New York's innovation." Urban Lawyer: 129-146.

REISS, A. J., Jr. (1956) "Research problems in metropolitan population redistribution." Amer. Soc. Rev. 21 (October): 571-577.

REISS, A. J., Jr. (1959) "Rural-urban and status differences in interpersonal contacts." Amer. J. of Sociology 65 (September): 182-195.

REISSMAN, L. (1953) "Levels of aspiration and social class." Amer. Soc. Rev. 18 (June): 233-242.

REPS, J. W. (1965) The Making of Urban America: A History of City Planning in the United States. Princeton: Princeton Univ. Press.

Resettlement Administration (1936) Greenbelt Towns. Washington, D.C.: Government Printing Office.

RIESMAN, D. (1959) "Flight and search in the new suburb." International Rev. of Community Development 4: 123-126.

RIESMAN, D. (1956) "The found generation." Amer. Scholar 25 (Autumn): 421-436.

RIESMAN, D. (1957) "The suburban dislocation." Annals of Amer. Academy of Pol. and Social Sci. 312 (November): 123-147.

RIESMAN, D. (1958) "The suburban sadness," pp. 375-408 in W. M. Dobriner (ed.) The Suburban Community. New York: G. P. Putnam's.

RIEW, J. (1970) "State aids for public schools and metropolitan finance." Land Economics 46: 297-304.

ROBINSON, T. E. (1957) "Opportunity in suburbia." National Education Association J. 46 (April): 246-248.

ROBINSON, W. S. (1950) "Ecological correlation and the behavior of individuals." Amer. Soc. Rev. 15 (June): 351-357.

RODEHAVER, M. W. (1947) "Fringe settlement as a two-directional movement." Rural Sociology 12 (March): 49-57.

ROSE, A. N., F. J. ATELSEK, and L. R. McDONALD (1953) "Neighborhood reactions to isolated Negro residents: an alternative to invasion and succession." Amer. Soc. Rev. 18 (October): 497-507.

ROSENBERG, C. S. (1971) Religion and the Rise of the American City: The New York City Mission Movement, 1812-1870. Ithaca: Cornell Univ. Press.

ROSENBERG, H. (1959) The Tradition of the New. New York: Horizon.

ROSENBERG, W. (1970) "New York State's urban development corporation is an action agency." J. of Housing (December).

ROSSI, P. H. (1956) Why Families Move. New York: Free Press.

ROSSI, P. H. and R. A. DENTLER (1961) The Politics of Urban Renewal. New York: Free Press.

ROURKE, F. E. (1964) "Urbanism and American democracy." Ethics 74: 255-268.

Ruttenberg, Stanley H. & Associates (1972) Abandonment of the Cities. Washington, D.C.: National Urban Coalition.

S

SABAGH, G. et al. (1969) "Some determinants of intrametropolitan residential mobility: conceptual conceptual considerations." Social Forces 48 (September): 88-98.

SACKS, S. (1972) City Schools/Suburban Schools: A History of Fiscal Conflict. Syracuse: Syracuse Univ. Press.

SACKS, S. and D. C. RANNEY (1966) "Suburban education: a fiscal analysis." Urban Affairs Q. 2 (September): 103-119.

SACKS, S. and D. C. RANNEY (1967) "Suburban education: a fiscal analysis," pp. 60-76 in M. Gittell (ed.) Educating an Urban Population. Beverly Hills: Sage Pubns.

SADLER, M. (1832) The Sadler Report, Report from the Committee on the Bill to Regulate the Labour of the Children in the Mills and Factories of the United Kingdom. London: House of Commons.

SALISBURY, R. H. (1967) "Schools and politics in the big city." Harvard Education Rev. 37: 408-424.

SALISBURY, R. H. (1964) "Urban politics: the new convergence of power." J. of Politics 37: 775-797.

SANDEEN, E. R. (1970) The Roots of Fundamentalism: British and American Millenarianism, 1800-1930. Chicago: Univ. of Chicago Press.

SCAFF, A. (1952) "The effect of community on participation in community organizations." Amer. Soc. Rev. 17 (April): 215-220.

SCAMMON, R. M. [ed.] (1958) American Votes. New York: Macmillan.

SCAMMON, R. M. (1957) "Voting for President in the larger metropolitan areas, 1952-1956." Midwest J. of Pol. Sci. 1 (November): 330-333.

SCAMMON, R. M. and B. J. WATTENBERG (1970) The Real Majority. New York: Coward-McCann.

SCHAFFER, B. H. (1954) Growing Suburbs and Town Finance. Storrs: Univ. of Connecticut Press.

SCHILTZ, T. and W. MOFFITT (1971) "Inner-city/outer-city relationship in metropolitan areas: a bibliographic essay." Urban Affairs Q. 7 (September): 75-108.

SCHLESINGER, A. M. (1933) The Rise of the City, 1878-1898. New York: Macmillan.

SCHMID, A. A. (1970) "Suburban land appreciation and public policy." J. of Amer. Institute of Planners 36 (January): 38-43.

SCHMITT, P. J. (1969) Back to Nature: The Arcadian Myth in Urban America. New York: Oxford Univ. Press.

SCHNORE, L. F. (1962) "City-suburban income differentials in metropolitan areas." Amer. Soc. Rev. 27 (April): 252-255.

SCHNORE, L. F. (1958) "Components of population in large metropolitan suburbs." Amer. Soc. Rev. 23 (October): 570-573.

SCHNORE, L. F. (1956) "The functions of metropolitan suburbs." Amer. J. of Sociology 61 (March): 453-458.

SCHNORE, L. F. (1957) "The growth of metropolitan suburbs." Amer. Soc. Rev. 22 (April): 165-173.

SCHNORE, L. F. (1967) "Measuring city-suburban status differences." Urban Affairs Q. 3 (September): 95-108.

SCHNORE, L. F. (1957) "Metropolitan growth and decentralization." Amer. J. of Sociology 63 (September): 171-180.

SCHNORE, L. F. (1962) "Municipal annexation and the growth of metropolitan suburbs, 1950-60." Amer. J. of Sociology 67 (January): 406-417.

SCHNORE, L. F. (1957) "Satellites and suburbs." Social Forces 36 (December): 121-127.

SCHNORE, L. F. (1963) "The socio-economic status of cities and suburbs." Amer. Soc. Rev. 28 (February): 276-285.

SCHNORE, L. F. (1959) "The timing of metropolitan decentralization: a contribution to the debate." J. of Amer. Institute of Planners 25 (November): 200-206.

SCHNORE, L. F. (1965) The Urban Scene: Human Ecology and Demography. New York: Free Press.

SCHNORE, L. F. (1964) "Urban structure and suburban selectivity." Demography 1: 164-176.

SCHNORE, L. F. and R. R. ALFORD (1963) "Forms of government and socio-economic characteristics of suburbs." Administrative Sci. Q. 8 (June): 1-17.

SCHNORE, L. F. and J.K.O. JONES (1969) "The evolution of city-suburban types in the course of a decade." Urban Affairs Q. 4 (June): 421-442.

SCHNORE, L. F. and H. SHARP (1960) "Racial changes in metropolitan areas, 1950-1960." Social Forces 41 (March): 247-253.

SCHORR, A. L. (1970) "Housing and its effects," in R. Gutman and D. Popenoe (eds.) Neighborhood, City and Metropolis. New York: Random House.

SCHUBERT, G. (1965) Judicial Policy-Making. Glenview, Ill.: Scott, Foresman.

SCOTT, J. (1969) "The glassworkers of Carmaux, 1850-1900," in S. Thernstrom and R. Sennett (eds.) Nineteenth Century Cities: Essays in the New Urban History. New Haven, Conn.: Yale Univ. Press.

SCOTT, J. F. and L. H. SCOTT (1968) "They are not so much anti-Negro as pro-middle class." New York Times Magazine (March 24): 46-47, 107, 109-110, 117, 119-120.

SCOTT, M. (1969) American City Planning Since 1890. Berkeley: Univ. of California Press.

SCOTT, S. and E. L. FEDER (1957) Factors Associated with Variations in Municipal Expenditure Levels. Berkeley: University of California Bureau of Public Administration.

SCOTT, T. M. (1968) "The diffusion of urban governmental forms as a case of social learning." J. of Politics 30 (November): 1091-1108.

SCOTT, T. M. (1968) "Metropolitan government reorganization proposals." Western Pol. Q. 21 (June): 498-507.

SCULLY, V. J., Jr. (1953) "Romantic retionalism and the expression of structure in wood: Downing, Wheeler, Gardner, and the 'stick style.' " Art Bull. 35: 121-142.

SCULLY, V. J., Jr. (1955) The Shingle Style: Architectural Theory and Design from Richardson to the Origins of Wright. New Haven, Conn.: Yale Univ. Press.

SEELEY, J. R. et al. (1956) Crestwood Heights: The Culture of Suburban Life. New York: Basic Books.

SEIDMAN, D. and M. COUZENS (1972) "Crime, crime statistics, and the great American anti-crime crusade: police misreporting of crime and political pressures." Presented at American Political Science Association Annual Meetings, Washington, D.C., September.

SENN, M. (1962) "Race, religion and suburbia." J. of Intergroup Relations 3: 159-170.

SENNETT, R. (1970) Families Against the City: Middle Class Homes of Industrial Chicago, 1872-1890. Cambridge, Mass.: Harvard Univ. Press.

SENNETT, R. (1970) The Uses of Disorder: Personal Identity and City Life. New York: Alfred A. Knopf.

SHAPPELL, D. L., L. G. HALL, and R. B. TARRIER (1971) "Perceptions of the world of work: inner city versus suburbia." J. of Counseling Psychology 18 (January): 55-59.

SHARKANSKY, I. (1968) "Environment, policy, output, and impact." Presented at the American Political Science Association Annual Meetings, Washington, D.C., September.

SHARKANSKY, I. (1970) The Routines of Politics. New York: Van Nostrand Reinhold.

SHARKANSKY, I. (1966) "Voting behavior of metropolitan congressman: prospects for changes with reapportionment." J. of Politics 28 (November): 774-793.

SHARP, H. and L. F. SCHNORE (1962) "The changing color composition of metropolitan areas." Land Economics 38 (May): 168-185.

SHAW, L. A. (1954) "Impressions of family life in a London suburb." Social Rev. 2 (December): 179-194.

SHELLOW, R. et al. (1967) Suburban Runaways of the 1960's. Chicago: Univ. of Chicago Press.

SHIELDS, C. (1952) "The American tradition of empirical collectivism." Amer. Pol. Sci. Rev. 46: 104-120.

SHIPPEY, F. A. (1964) Protestantism in Suburban Life. New York: Abingdon.

SHRYOCK, H. S., Jr. (1956) "Population redistribution within metropolitan areas: evaluation of research." Social Forces 35 (December): 154-159.

SHUVAL, J. (1956) "Class and ethnic correlates of casual neighboring." Amer. Soc. Rev. 21 (August): 453-458.

SIKORSKY, I. (1972) "A-95: deterrent to discriminatory zoning." Civil Rights Digest (August): 17-19.

SIMON, E. (1962) "Suburbia—its effect on the American Jewish teenager." J. of Educational Sociology 36 (November): 124-133.

SINGLETON, G. H. (1973) " 'Mere middle-class institutions': urban Protestantism in nineteenth-century America." J. of Social History (June).

SJOBERG, G. (1960) The Preindustrial City, Past and Present. New York: Free Press.

SKLARE, M. and J. GREENBLAUM (1967) Jewish Identity on the Suburban Frontier; A Study of Group Survival in an Open Society. New York: Basic Books.

SKLARE, M. et al. (1969) Not Quite at Home: How an American Jewish Community Lives With Itself and Its Neighbors. New York: Institute of Human Relations Press.

SKOLER, D. L. and J. M. HETLER (1970) "Governmental restructuring and criminal administration: the challenge of consolidation," pp. 53-75 in Crisis in Urban Government. A Symposium: Restructuring Metropolitan Area Government. Silver Spring, Md.: Thos. Jefferson.

SKOLNICK, J. H. (1967) Justice Without Trial: Law Enforcement in Democratic Society. New York: John Wiley.

SMALLWOOD, F. (1965) Greater London: The Politics of Metropolitan Reform. Indianapolis: Bobbs-Merrill.

SMELSER, N. J. (1968) "Toward a general theory of social change," in Essays in Social Explanation. Englewood Cliffs, N.J.: Prentice-Hall.

SMITH, D. C. (1971) "The constitution of police legitimacy." Indiana University Department of Political Science Papers on Political Theory and Policy Analysis.

SMITH, H. A. (1960) Let the Crabgrass Grow: H. Allen Smith's Suburban Almanac. New York: Random House.

SMITH, H. L. (1943) "White Bluff: a community of commuters." Economic Geography 19 (April): 143-147.

SMITH, H. N. (1950) Virgin Land: The American West as Symbol and Myth. Cambridge, Mass.: Harvard Univ. Press.

SMITH, T. C. (1960) "The structure of power in a suburban community." Pacific Soc. Rev. 3 (Fall): 83-86.

SNOWISS, L. M. (1966) "Congressional recruitment and representation." Amer. Pol. Sci. Rev. 60 (September): 627-639.

SOFEN, E. (1966) The Miami Metropolitan Experiment. Garden City, N.Y.: Doubleday Anchor.

SOLOMON, B. M. (1956) Ancestors and Immigrants: A Changing New England Tradition. Cambridge, Mass.: Harvard Univ. Press.

SOWER, C. (1948) "Social stratification in suburban communities." Sociometry 11 (August): 235-243.

SPECTORSKY, A. G. (1955) The Exurbanites. New York: J. B. Lippincott.

SPENGLER, O. (1928) The Decline of the West. Vol. II: Perspectives of World-History. New York: Alfred A. Knopf.

STAUBER, R. L. (1965) "The swampy science of suburbia: a case for the sociology of knowledge." Kansas J. of Sociology 1 (Summer): 137-154.

STEIN, M. R. (1960) "Suburbia: dream or nightmare?" pp. 199-226 in The Eclipse of Community. Princeton: Princeton Univ. Press.

STEPHENS, G. R. (1969) "The suburban impact of earnings tax policy." National Tax J. 22 (September): 313-333.

STERNLIEB, G. S. (1972) "Death of the American dream house." Society 9 (February): 39-42.

STERNLIEB, G. S. et al. (1971) The Affluent Suburb: Princeton. New Brunswick, N.J.: Trans-action Books.

STERNLIEB, G. S. and W. P. BEATON (1972) The Zone of Emergence: A Case Study of Plainfield, New Jersey. New Brunswick, N.J.: Trans-action Books.

STIEBER, J. (1969) "Employee representation in municipal government," in Municipal Yearbook. Washington, D.C.: International City Management Association.

STIX, C. H. (1970) The Suburban Condition: Leadership and Lifestyle. Westchester, N.Y.: American Jewish Committee.

STONE, G. P. (1954) "City shoppers and urban identification: observations on the social psychology of city life." Amer. J. of Sociology 60 (July): 36-45.

STONIER, C. E. (1958) "Problems of suburban transport services," pp. 326-344 in W. M. Dobriner (ed.) The Suburban Community. New York: G. P. Putnam's.

STRAUSS, A. L. (1960) "The changing imagery of American city and suburbs." Soc. Q. 1 (January): 15-24.

STRAUSS, A. L. (1961) "Suburbia: the union of urbanity and rurality," pp. 230-245 in Images of the American City. New York: Free Press.

STUART, D. G. and R. B. TESKA (1971) "Who pays for what: a cost-revenues analysis of suburban land use alternatives." Urban Land 30 (March): 3-16.

SULIMSKI, J. (1964) "Population and social changes in the suburbs of Krakow." Polish Soc. Bull. 2: 130-133.

SULLIVAN, W. A. (1955) The Industrial Worker in Pennsylvania, 1800-1840. Harrisburg: Pennsylvania State Univ. Press.

SUSSNA, S. (1967) "Bulk control and zoning: the New York City experience." Land Economics 43 (May): 58-79.

SUTTON, S. B. [ed.] (1971) Civilizing American Cities: A Selection of Frederick Law Olmsted's Writings on City Landscapes. Cambridge, Mass.: MIT Press.

T

TAEUBER, K. E. (1968) "The effect of income redistribution on racial residential segregation." Urban Affairs Q. 4 (September): 5-14.

TAEUBER, K. E. and A. F. TAEUBER (1965) Negroes in Cities: Residential Segregation and Neighborhood Change. Chicago: Aldine.

TAEUBER, K. E. and A. F. TAEUBER (1964) "White migration and socio-economic differences between cities and suburbs." Amer. Soc. Rev. 29 (October): 718-729.

TAKAHASHI, Y. (1963) "On some theories of the suburb in America." Japanese Soc. Rev. 14 (August): 79-90.

TALLMAN, I. (1969) "Working-class wife in suburbia: fulfillment or crisis." J. of Marriage and Family 31 (February): 65-72.

TALLMAN, I. and R. MORGNER (1970) "Life style differences among urban and suburban blue-collar families." Social Forces 48 (May): 334-348.

TARVER, J. D. (1969) "Migration differentials in Southern cities and suburbs." Social Sci. Q. 50 (September): 298-324.

TARVER, J. D. (1957) "Suburbanization of retail trade in the Standard Metropolitan Areas of the U.S., 1948-54." Amer. Soc. Rev. 22 (August): 427-433.

TATE, H. C. (1954) Building a Better Home Town: A Program of Community Self-Analysis and Self-Help. New York: Harper.

TAYLOR, G. R. (1966) "The beginnings of mass transportation in urban America." Smithsonian J. of History 1 (Summer): 35-50.

TAYLOR, G. R. (1951) The Transportation Revolution, 1815-1860. New York: Holt, Rinehart & Winston.

TAYLOR, G. R. (1915) Satellite Cities. New York and London: D. Appleton.

Technical Committee for Fiscal Facts (1968) Central City Benefits to Milwaukee Suburbs. Milwaukee.

THERNSTROM, S. (1964) Poverty and Progress: Social Mobility in a Nineteenth-Century City. Cambridge, Mass.: Harvard Univ. Press.

THOMPSON, W. D. (1947) The Growth of Metropolitan Districts in the United States: 1900-1940. Washington, D.C.: Government Printing Office.

THOMPSON, W. R. (1965) A Preface to Urban Economics. Baltimore: Johns Hopkins Press.

THOMPSON, W. S. (1951) Migration Within Ohio, 1935-1940. Oxford, Ohio: Scripps Foundation for Research in Population Problems; Chicago: University of Chicago Population Research and Training Center.

TOMEII, A. (1964) "Informal group participation and residential patterns." Amer. J. of Sociology 70 (July): 28-35.

TRESSLER, I. D. (1952) "So you live in a suburb?" pp. 208-213 in P. G. Wodehouse and S. Meredith (eds.) Best of Modern Humor. New York: Metcalf Associates.

TROLDAHL, V. C. (1964) "Communicating to the suburbs." J. of Cooperative Extension 2 (Summer): 82-88.

TRYON, R. C. (1955) Identification of Social Areas by Cluster Analysis. Berkeley and Los Angeles: Univ. of California Press.

TUNNARD, C. (1947) "The romantic suburb in America." Magazine of Art 40 (May): 184-187.

TUNNARD, C. and B. PUSHKAREV (1963) Man Made America, Chaos or Control. New Haven, Conn.: Yale Univ. Press.

TURNER, D. L. (1924) Suburban Transit Problem. New York: M. B. Brown.

TYLER, P. [ed.] (1957) "Cities vs. suburbs," pp. 126-180 in City and Suburban Housing. New York: H. W. Wilson.

U

United States
Advisory Commission on Intergovernmental Relations (1967) Fiscal Balance in the American Federal System: Metropolitan Fiscal Disparities. Washington, D.C.: Government Printing Office.
Advisory Commission on Intergovernmental Relations (1965) Metropolitan Social and Economic Disparities: Implications for Intergovernmental Relations in Central Cities and Suburbs. Washington, D.C.: Government Printing Office.
Bureau of the Census (1968) Census of Governments: Government Organization. Washington, D.C.: Government Printing Office.
Bureau of the Census (1971) Social and Economic Characteristics of the Population in Metropolitan and Nonmetropolitan Areas: 1970 and 1960. Washington, D.C.: Government Printing Office.
Bureau of the Census (1965) Statistical Abstract of the United States: Colonial Times to the Present. Washington, D.C.: Government Printing Office.
Commission on Civil Rights (1970) "Federal installations and equal housing opportunity." Washington, D.C.
Commission on Civil Rights (1971) "Hearings before the United States Commission on Civil Rights held in Washington, D.C., June 14-17, 1971." Washington, D.C.
Department of Housing and Urban Development (1971) Equal Opportunity in Housing. Englewood Cliffs, N.J.: Prentice-Hall.
House of Representatives, Committee of the Judiciary (1972) "Hearings before the Civil Rights Oversight Subcommittee of the Committee of the Judiciary, House of Representatives, on federal government's role in the achievement of equal opportunity in housing." Washington, D.C.
Savings and Loan League (1971) Yearbook. Chicago.
University of Pennsylvania Institute of Urban Studies (1954) Accelerated Urban Growth in a Metropolitan Area, A Study of Urbanization, Suburbanization, and the Impact of the Fairless Works Steel Plant in Lower Bucks County, Pennsylvania. Philadelphia.
UYEKI, E. S. (1966) "Patterns of voting in a metropolitan area." Urban Affairs Q. 2 (June): 65-77.

V

VAN DOREN, C. (1938) Benjamin Franklin. New York: Viking.
VERBA, S. (1961) Small Groups and Political Behavior: A Study of Leadership. Princeton: Princeton Univ. Press.

VERNON, R. (1965) "Production and distribution in the large metropolis," pp. 85-88 in C. E. Elias, Jr. et al. (eds.) Metropolis: Values in Conflict. Belmont, Calif.: Wadsworth.

VICKREY, W. S. (1963) "Pricing in urban and suburban transport." Amer. Economic Rev. Papers and Proceedings 53 (May): 452-465.

VOGEL, E. F. (1967) Japan's New Middle Class: The Salary Man and His Family in a Tokyo Suburb. Berkeley: Univ. of California Press.

W

WADE, R. (1959) "Urbanization," in C. V. Woodward (ed.) The Comparative Approach to American History. New York: Basic Books.

WALLACE, D. (1964) "Suburbia: predestined republicanism," in First Tuesday: A Study of Rationality in Voting. Doubleday.

WALLIS, C. P. and R. MALIPHANT (1967) "Delinquent areas in the county of London: ecological factors." British J. of Criminology 7 (July): 250-284.

WALTER, B. and F. M. WIRT (1972) "Social and political dimensions of American suburbs," pp. 109-111 in B.J.L. Berry (ed.) City Classification Handbook. New York: Wiley-Interscience.

WARD, D. (1971) Cities and Immigrants: A Geography of Change in Nineteenth Century America. New York: Oxford Univ. Press.

WARD, D. (1964) "Comparative historical geography of streetcar suburbs in Boston, Massachusetts and Leeds, England." Annals of Association of Amer. Geographers 54 (December): 477-489.

WARD, J. W. (1955) Andrew Jackson: Symbol for an Age. New York: Oxford Univ. Press.

WARD, J. W. (1969) Red, White and Blue: Men, Books and Ideas in American Culture. New York: Oxford Univ. Press.

WARDEN, G. B. (1970) Boston, 1689-1776. Boston: Little, Brown.

WARE, N. (1924) The Industrial Worker, 1840-1860. Boston: Houghton Mifflin.

WARNER, S. B., Jr. (1962) Streetcar Suburbs: The Process of Growth in Boston, 1870-1900. Cambridge, Mass.: Harvard Univ. Press.

WARNER, S. B., Jr. (1968) The Private City: Philadelphia in Three Periods of its Growth. Philadelphia: Univ. of Pennsylvania Press.

WARREN, D. J. (1970) "Suburban isolation and race tension: the Detroit case." Social Problems 17 (Winter): 324-339.

WATKINS, M. (1967) "White skins, dark skins, thin skins." New York Times Magazine (December 3): 127, 129, 137, 139.

WATSON, R. A. and J. H. ROMANI (1961) "Metropolitan government for metropolitan Cleveland: an analysis of the voting record." Midwest J. of Pol. Sci. 5 (November): 365-390.

WATSON, W. B., E.A.T. BARTH, and D. P. HAYES (1965) "Metropolitan decentralization through incorporation." Western Pol. Q. 18 (March): 198-206.

WATTEL, H. L. (1958) "Levittown: a suburban community," pp. 287-313 in W. M. Dobriner (ed.) The Suburban Community. New York: G. P. Putnam's.

WEBER, A. F. (1899) The Growth of Cities in the Nineteenth Century. New York: Columbia Univ. Press.

WEBER, A. F. (1898) "Suburban annexations." North Amer. Rev. 166 (May): 612-617.

WECHTER, D. (1963) The Hero in America. Ann Arbor: Univ. of Michigan Press.

WEHRWEIN, G. S. (1942) "Rural-urban fringe." Economic Geography 18 (July): 217-228.

WEICHER, J. C. (1970) "Determinants of central city expenditures: some overlooked factors and problems." National Tax J. 23 (December): 379-396.

WESTLEY, W. A. and F. ELKIN (1957) "Protective environment and adolescent socialization." Social Forces 35 (March): 243-249.

WHEATLEY, P. (1971) The Pivot of the Four Quarters: A Preliminary Inquiry into the Origins and Character of the Ancient Chinese City. Chicago: Aldine.

WHEELER, H. (1971) "The phenomenon of God." Center Magazine (March/April): 7-12.

WHETTEN, N. L. (1971) "Suburbanization as a field of sociological research." Rural Sociology 16 (December): 319-330.

WHETTEN, N. L. (1939) "Wilton: a rural town near metropolitan New York." Studies of Suburbanization in Connecticut 3, Connecticut State College Agricultural Experiment Station Bulletin.

WHETTEN, N. L. and E. C. DEVEREUX, Jr. (1936) "Windsor: a highly developed agricultural area." Studies of Suburbanization in Connecticut 1.

WHETTEN, N. L. and R. F. FIELD (1938) "Norwich: an industrial part-time farming area." Studies of Suburbanization in Connecticut 2.

WHITE, M. and L. WHITE (1962) The Intellectual versus the City. Cambridge, Mass.: Harvard Univ. Press and MIT Press.

WHITLEY, O. R. (1964) "Suburbia: demi-paradise or Babylonian captivity for the Church?" pp. 83-107 in Religious Behavior: Where Sociology and Religion Meet. Englewood Cliffs, N.J.: Prentice-Hall.

WHYTE, W. H., Jr. (1964) Cluster Development. New York: American Conservation Association.

WHYTE, W. H., Jr. (1954) "The consumer in the new suburbia," pp. 1-14 in L. H. Clark (ed.), Consumer Behavior. New York: New York Univ. Press.

WHYTE, W. H., Jr. (1968) The Last Landscape. Garden City, N.Y.: Doubleday.

WHYTE, W. H., Jr. (1957) The Organization Man. New York: Doubleday.

WHYTE, W. H., Jr. (1953) "The transients." Fortune 47 (May): 112-117, 221-226; 47 (June): 126-131, 186-196; 47 (July): 84-89, 160; 48 (August): 120-122, 186-190.

WHYTE, W. H., Jr. (1957) "Urban Sprawl," pp. 115-139 in Fortune Magazine (eds.) The Exploding Metropolis. Garden City, N.Y.: Doubleday Anchor.

WIEBE, R. H. (1967) The Search for Order, 1877-1920. New York: Hill & Wang.

WILENSKY, H. (1961) "Review of working class suburb by Bennett Berger." Amer. Soc. Rev. 26 (April): 310-312.

WILLIAMS, O. P. (1967) "Life-style values and political decentralization in metropolitan areas." Southwestern Social Sci. Q. 48 (December): 299-310.

WILLIAMS, O. P. (1971) Metropolitan Political Analysis. New York: Free Press.

WILLIAMS, O. P. and C. R. ADRIAN (1963) Four Cities: A Study in Public Policy Making. Philadelphia: Univ. of Pennsylvania Press.

WILLIAMS, O. P. et al. (1965) Suburban Differences and Metropolitan Policies. Philadelphia: Univ. of Pennsylvania Press.

WILLIAMS, R. (1958) Culture and Society, 1750-1950. New York: Columbia Univ. Press.

WILLIAMS, R. M., Jr. (1959) "Friendship and social values in a suburban community: an exploratory study." Pacific Soc. Rev. 2 (Spring): 3-10.

WILLIS, C. L. (1967-1968) "Analysis of voter response to school financial proposals," Public Opinion Q. 31: 648-651.

WILLMOTT, P. (1963) The Evolution of a Community. London: Routledge & Kegan Paul.

WILMOTT, P. and M. YOUNG (1960) Family and Class in a London Suburb. New York: Penguin.

WILNER, D., R. WALKLEY, T. PINKERTON, and M. TAYBACK (1962) Housing Environment and Family Life. Baltimore: Johns Hopkins Press.

WILSON, J. Q. (1966) "The war on cities." Public Interest 3 (Spring).

WILSON, J. Q. [ed.] (1968) The Metropolitan Enigma: Inquiries into the Nature and Dimensions of America's 'Urban Crisis.' Cambridge, Mass.: Harvard Univ. Press.

WILSON, J. Q. (1969) "Policing the suburbs," in Varieties of Police Behavior. Cambridge, Mass.: Harvard Univ. Press.

WILSON, J. Q. and E. C. BANFIELD (1965) "Voting behavior on municipal expenditures: a study in rationality and self-interest," pp. 74-91 in J. Margolis (ed.) The Public Economy of Urban Communities. Washington, D.C.: Resources for the Future.

WINCH, R. F. (1971) The Modern Family. New York: Holt, Rinehart & Winston.

WINCH, R. F., S. GREER, and R. L. BLUMBERG (1967) "Ethnicity and extended familism in an upper-middle class suburb." Amer. Soc. Rev. 32 (April). 265-272.

WINGO, L. (1972) "Introduction," pp. 1-8 in L. Wingo (ed.) Minority Perspectives. Washington, D.C.: Resources for the Future.

WINSBOROUGH, H. H. (1963) "An ecological approach to the theory of suburbanization." Amer. J. of Sociology 68 (March): 565-570.

WINSLOW, O. E. (1968) John Eliot, "Apostle to the Indians." Boston: Houghton Mifflin.

WINTER, G. (1961) The Suburban Captivity of the Churches. Garden City, N.Y.: Doubleday.

WIRT, F. M. (1972) "The allocational politics of New York State aid to education," in J. Berke and M. W. Kirst (eds.) Federal Aid to Education: Who Benefits? Who Governs? Lexington, Mass.: D. C. Heath.

WIRT, F. M. (1965) "The political sociology of American suburbia." J. of Politics 27 (August): 647-666.

WIRT, F. M. and M. W. KIRST (1972) The Political Web of American Schools. Boston: Little, Brown.

WIRT, F. M. and B. WALTER (1971) "The political consequences of suburban variety." Social Sci. Q. 52 (December).

WIRT, F. M. and B. WALTER (1972) "Social and political dimensions of American suburbs," in B.J.L. Berry (ed.) Classification of Cities: New Methods and Alternative Uses. New York: John Wiley.

WIRT, F. M., B. WALTER, F. F. RABINOVITZ and D. R. HENSLER (1972) On the City's Rim: Politics and Policy in Suburbia. Lexington, Mass.: D.C. Heath.

WISSINK, G. A. (1962) American Cities in Perspective; With Special Reference to the Development of Their Fringe Areas. Assen, Eng.: Van Gorcum.

WITHINGTON, W. A. (1956) "Suburban migration: a case study of Winchester, Massachusetts, 1930 to 1950." Annals of Association of Amer. Geographers 47 (June): 281.

WOHL, R. R. (1953) "The rags to riches story: an episode of secular idealism," in R. Bendix and S. M. Lipset (eds.) Class, Status and Power. New York: Free Press.

WOHL, R. R. (1969) "The 'country boy' myth and its place in American urban culture: the nineteenth-century contribution." Perspectives in Amer. History 3: 77-156.

WOLF, E. and M. RAVITZ (1964) "Lafayette Park: new residents in the core city." J. of Amer. Institute of Planners 30 (August): 234-239.

WOLFINGER, R. E. and J. O. FIELD (1966) "Political ethos and the structure of city government." Amer. Pol. Sci. Rev. 60 (June): 306-326.

WOOD, R. C. (1963) "The American suburb: boy's town in a man's world," pp. 112-121 in E. Green et al. (eds.) Man and the Modern City. Pittsburgh: Univ. of Pittsburgh Press.

WOOD, R. C. (1961) 1400 Governments. Cambridge, Mass.: Harvard Univ. Press.

WOOD, R. C. (1969) "The ghettos and metropolitan politics," in R. L. Warren (ed.) Politics and the Ghettos. New York: Atherton.

WOOD, R. C. (1958) "The governing of suburbia," pp. 165-180 in W. M. Dobriner (ed.) The Suburban Community. New York: G. P. Putnam's.

WOOD, R. C. (1960) "The impotent suburban vote." Nation 190 (March 26): 271-274.

WOOD, R. C. (1958) "Metropolitan government, 1975: an extrapolation of trends: the new metropolis: green belts, grass roots, or Gargantua?" Amer. Pol. Sci. Rev. 52 (March): 108-122.

WOOD, R. C. (1958) Suburbia: Its People and Politics. Boston: Houghton Mifflin.

WOODBURY, C. (1955) "Suburbanization and suburbia." Amer. J. of Public Health 45 (January): 2-9.

WOODS, R. A. and A. J. KENNEDY (1962) The Zone of Emergence: Observations of the Lower Middle and Upper Working Class Communities of Boston, 1905-1914. Cambridge, Mass.: MIT Press.

WOODWARD, C. V. (1938) Tom Watson: Agrarian Rebel. New York: Macmillan.

WRIGHT, A. F. (1967) "Changan," in A. Toynbee (ed.) Cities of Destiny. New York: McGraw-Hill.

WRIGHT, C. D. (1883) "The factory system of the United States," in U.S. Census Office, Report on the Manufactures of the United States at the Tenth Census, June 1, 1880. Washington, D.D.: Government Printing Office.

WRONG, D. H. (1967) "Suburbs and the myth of suburbia," pp. 358-364 in D. H. Wrong and H. L. Gracey (eds.) Readings in Introductory Sociology. New York: Macmillan.

WUERSCHING, T.K.H. (1968) "Indicators of suburban sprawl in a metropolitan area." Rocky Mountain Social Sci. J. 5 (October): 55-57.

WURMAN, R. S. (1971) Making the City Observable. Cambridge, Mass.: MIT Press.

WYDEN, P. (1962) Suburbia's Coddled Kids. New York: Avon.

WYLLIE, I. G. (1954) The Self-Made Man in America: The Myth of Rags to Riches. New Brunswick, N.J.: Rutgers Univ. Press.

Y

YOUNG, D. (1972) "Industry flees decaying city." Chicago Tribune (May 7).

YUAN, D. Y. (1966) "Chinatown and beyond: the Chinese population in metropolitan New York." Phylon 27 (Winter): 321-332.

Z

ZDEP, S. M. (1971) "Educating disadvantaged urban children in suburban schools: an evaluation." J. of Applied Social Psychology 1: 173-186.

ZELAN, J. (1968) "Does suburbia make a difference: an exercise in secondary analysis," pp. 401-408 in S. Fava (ed.) Urbanism in World Perspective. New York: Thomas Y. Crowell.

ZELOMEK, A. W. (1959) "Suburbia: urbia—sub, ex and inter," pp. 132-160 in A Changing America: At Work and Play. New York: John Wiley.

ZEUL, C. R. and C. R. HUMPHREY (1971) "The integration of blacks in suburban neighborhoods: a re-examination of the contact hypothesis." Social Problems 18 (Spring): 462-474.

ZIBBELL, C. (1961) "Suburbia and Jewish community organization." J. of Jewish Communal Service 38: 69-79.

ZIKMUND, J. II (1967) "A comparison of political attitude and activity patterns in central cities and suburbs." Public Opinion Q. 31 (Spring): 71-75.

ZIKMUND, J. II (1971) "Do suburbanites use the central city?" J. of Amer. Institute of Planners 37 (May): 192-195.

ZIKMUND, J. II (1968) "How suburbia votes." Nation 206 (June 24): 826-828.

ZIKMUND, J. II (1968) "Suburban voting in presidential elections: 1948-64." Midwest J. of Pol. Sci. 12 (May): 238-256.

ZIKMUND, J. II and R. SMITH (1969) "Political participation in an upper-middle class suburb." Urban Affairs Q. 4 (June): 443-458.

ZIMMER, B. G. and A. H. HAWLEY (1956) "Approaches to the solution of fringe problems: preferences in the Flint metropolitan area." Public Administration Rev. 16 (Autumn): 258-268.

ZIMMER, B. G. and A. H. HAWLEY (1968) "Evalution of schools—city and suburb," pp. 85-101 in Metropolitan Area Schools: Resistance to District Reorganization. Beverly Hills: Sage Pubns.

ZIMMER, B. G. and A. H. HAWLEY (1959) "The significance of membership in Associations." Amer. J. of Sociology 65 (September): 196-201.

ZIMMER, B. G. and A. H. HAWLEY (1959) "Suburbanization and church participation." Social Forces 37 (May): 348-354.

ZIMMER, B. G. and A. H. HAWLEY (1961) "Suburbanization and some of its consequences." Land Economics 37 (February): 88-93.

ZSCHOCK, D. K. [ed.] (1969) Economic Aspects of Suburban Growth: Studies of the Nassau-Suffolk Planning Region. Stony Brook: State University of New York at Stony Brook Economic Research Bureau.

ZSCHOCK, D. K. (1971) "Black youth in suburbia." Urban Affairs Q. 7 (September): 61-74.

ZUCKERMAN, M. (1970) Peaceable Kingdoms: New England Towns in the Eighteenth Century. New York: Alfred A. Knopf.

THE AUTHORS

THE AUTHORS

RICHARD F. BABCOCK is an attorney with the firm of Ross, Hardies, O'Keefe, Babcock and Parsons in Chicago. He has served as president of the American Society of Planning Officials, board member of the Northeastern Illinois Planning Commission, and chairman of the Advisory Committee for the American Law Institute project to draft a model land development code. He authored THE ZONING GAME (1966).

JOSEF J. BARTON is Assistant Professor of History at the University of Virginia. His research and teaching interests are in the areas of American urban and social history and comparative approaches to social history. From 1968 to 1970, he held a grant from the Fund for the Advancement of Education to study four generations of Italian peasant families in villages and cities, 1850 to 1950, and immigration and American urban culture, 1920 to 1950. He is the author of A SAFE LODGING: THREE IMMIGRANT COMMUNITIES IN AN AMERICAN CITY, 1890-1950 (1973).

BRIAN J.L. BERRY is Irving B. Harris Professor of Urban Geography and Director of Training Programs, Center for Urban

Studies, at the University of Chicago. He has published several books, including SPATIAL ANALYSIS (1968), GEOGRAPHIC PERSPECTIVES ON URBAN SYSTEMS (1970), and CITY CLASSIFICATION HANDBOOK (1972). Forthcoming works include THE HUMAN CONSEQUENCES OF URBANIZATION (1973), THE GEOGRAPHY OF ECONOMIC SYSTEMS (1974), and MANAGING THE URBAN ENVIRONMENT (1974).

YEHOSHUA S. COHEN is Visiting Assistant Professor of Geography at Simon Fraser University. He received his Ph.D. from the University of Chicago in 1971 and taught at the University of Illinois, Chicago Circle. He has published two monographs, coauthored DECENTRALIZATION OF INDUSTRY IN THE UNITED STATES, 1950-1962, with Brian Berry and is preparing a volume on diffusion of innovation for Oxford University Press.

BRYAN T. DOWNES is Associate Professor of Political Science at the University of Missouri, St. Louis. He has written two books on suburban politics–CITIES AND SUBURBS: SELECTED READINGS IN LOCAL POLITICS AND PUBLIC POLICY (1971) and THE POLITICS OF CHANGE IN CENTRAL CITIES AND SUBURBS: AN INQUIRY INTO AMERICA'S URBAN CRISIS (forthcoming in 1974). His other areas of interest include American urban politics, policy-making and implementation, policy analysis and decision-making, political change, and black and minority group politics.

H. PAUL FRIESEMA is Associate Professor of Political Science and Urban Affairs at Northwestern University. A former lawyer, he has been a consultant to Iowa water pollution agencies, urban governments and educational television. He is currently directing the Public Lands Policy Project, which deals with the urban-generated political pressures on the public lands and natural resources administering agencies of the United States. He is the author of METROPOLITAN POLITICAL STRUCTURES (1972).

NORVAL D. GLENN is Professor of Sociology at University of Texas, Austin. He has a grant from the National Science Foundation to study aging and conservatism and published some of his findings in "Further Evidence on Aging and Party Identification" in the PUBLIC OPINION QUARTERLY. His current research and teaching interests are in the areas of sociology of aging and the life cycle, urban sociology, political sociology, and social stratification.

SCOTT A. GREER is Professor of Sociology, Political Science and Urban Affairs at Northwestern University. His research and teaching interests are in the areas of kinship, ethnicity and voluntary associations, and the sociology of the family. His most recent book is THE URBANE VIEW (1972). He has also written several volumes of poetry.

JEFFREY K. HADDEN is Professor of Sociology at the University of Virginia. His teaching and research interests are in the areas of urban sociology and the sociology of religion. He is the coeditor of SUBURBIA IN TRANSITION (1973), RELIGION IN RADICAL TRANSITION (1971), and METROPOLIS IN CRISIS (second edition, 1971), and the author of THE GATHERING STORM IN THE CHURCHES (1969).

HARLAN HAHN is Visiting Professor of Political Science at the University of Southern California. His research and teaching interests are American politics, political behavior, and social psychology. Recent publications include POLICE IN URBAN SOCIETY (1972) and GHETTO REVOLTS (with Joe R. Feagin, 1973). He also edited Volume VI of this series, PEOPLE AND POLITICS IN URBAN SOCIETY (1972).

HARVEY H. MARSHALL, Jr., is Assistant Professor of Sociology at Purdue University. He is researching the relationship between migration and professional success of engineers and studying the effects of decentralization of various types of jobs on the labor force participation and employment patterns of central-city residents, especially blacks. His recently published articles include "Hidden

Deviance and the Labeling Approach: The Case for Drinking and Driving," with Ross Purdy, in SOCIAL PROBLEMS.

LOUIS H. MASOTTI is Director of the Center for Urban Affairs and Professor of Political Science and Urban Affairs at Northwestern University. He has published extensively in the fields of urban violence and suburban development; recent publications include SUBURBIA IN TRANSITION (1973). He has a two-year grant from NIMH for a comparative historical study of urban public order in the cities of London, Stockholm, Sydney, and Calcutta.

DAVID W. MINAR is Professor of Political Science and Urban Affairs and Chairman of the Department of Political Science at Northwestern University. He is continuing his research into the politics of educational decision-making in urban and suburban school systems. He is the coeditor of THE CONCEPT OF COMMUNITY (1969) and THE NEW URBANIZATION (1967).

ELINOR OSTROM is Associate Professor of Political Science at Indiana University. She worked as an employee relations manager and personnel analyst before joining the faculty. She is the principal investigator for a three-year NIMH grant from the Center for Studies of Metropolitan Problems to study community organization and urban institutions, especially the police.

ROGER B. PARKS is a Research Associate in Political Science at Indiana University, working on a study of the St. Louis police. He will receive his Ph.D. from Indiana in 1974. He has previously researched citizen attitudes toward police in Indianapolis and directed a similar study in the St. Louis metropolitan area. He has worked as a senior systems analyst with RCA Corporation and as an operations analyst with McDonnell-Douglas Corporation.

WILLIAM W. PENDLETON is presently Associate Director of the Center for Research in Social Change at Emory University, where he has been since 1967. His major areas of interest are in population, statistics, social organization, and social change. Before going to Emory, he was associated with Human Sciences Research, Inc., and with the Center for Medical Research and Training in New Orleans, Louisiana. He has written and spoken extensively in his fields.

LEONARD S. RUBINOWITZ is a Research Associate with the Center for Urban Affairs at Northwestern University. He is directing a project to develop locational alternatives for low- and moderate-income housing. He was formerly special assistant to the Midwest Regional Administrator of the U.S. Department of Housing and Urban Development.

THOMAS M. SCOTT is Associate Professor and Chairman of the Department of Political Science at the University of Minnesota. He is completing a project on developments in metropolitan governance for the Brookings Institution and is serving his eighth year on the Board of Trustees of the Minnesota League of Municipalities.

GREGORY H. SINGLETON is Assistant Professor of History at Northeastern Illinois University. He received his Ph.D. from the University of California, Los Angeles, in 1972. Recent publications include "Mere Middle-Class Institutions: Urban Protestantism in Nineteenth-Century America," in the JOURNAL OF SOCIAL HISTORY, and "The Dynamics of WASP Culture: From Ethnic Cohesion to the Organization Man," forthcoming in RIDE UPON THE WHIRLWIND, edited by Gilbert Osofsky.

WILBUR R. THOMPSON is Professor of Economics at Wayne State University. He is currently living in Phoenix, Arizona, and working as a freelance consultant, lecturer, and writer on urban affairs. He has lectured in the Urban Policy Conferences of the Brookings Institution and from 1968 to 1970 served as Chairman of the Board of the Southeastern Michigan Transportation Authority. He is the author of A PREFACE TO URBAN ECONOMICS (1965).

BERNARD WEISSBOURD is President of Metropolitan Structures, Inc., a residential development corporation based in Chicago, and an Adjunct Professor at the Center for Urban Studies of the University of Illinois, Chicago Circle. A former lawyer and chemist, he formed Metropolitan Structures in 1959 and is currently engaged in the creation of two pioneering "new-towns-in town"—one in Montreal and the other in downtown Chicago.

FREDERICK M. WIRT is Professor of Political Science at the University of Maryland, Baltimore County. He coauthored two books in 1972: ON THE CITY'S RIM: POLITICS AND POLICY IN SUBURBIA and THE POLITICAL WEB OF AMERICAN SCHOOLS, and he is nearing completion of a volume exploring decision-making in San Francisco. He plans research on local government officials' perceptions of community power structures and its consequences for their work and a fifty-state study of the locus of power in state-local educational policy-making.

JOSEPH ZIKMUND II is Associate Professor and Chairman of the Department of Political Science at Albion College. He spent two recent summers in Yugoslavia studying urbanization and city planning. His research and teaching interests are in the fields of American politics, urban studies, and the political geography of American suburbs. Two forthcoming books are POLITICAL CULTURE IN AMERICA and BLACK MAN IN PHILADELPHIA POLITICS.